Springer Proceedings in Materials

Volume 9

Springer Proceedings in Materials publishes the latest research in Materials Science and Engineering presented at high standard academic conferences and scientific meetings. It provides a platform for researchers, professionals and students to present their scientific findings and stay up-to-date with the development in Materials Science and Engineering. The scope is multidisciplinary and ranges from fundamental to applied research, including, but not limited to:

- Structural Materials
- Metallic Materials
- Magnetic, Optical and Electronic Materials
- Ceramics, Glass, Composites, Natural Materials
- Biomaterials
- Nanotechnology
- Characterization and Evaluation of Materials
- Energy Materials
- Materials Processing

To submit a proposal or request further information, please contact one of our Springer Publishing Editors according to your affiliation:

European countries: **Mayra Castro** (mayra.castro@springer.com)
India, South Asia and Middle East: **Priya Vyas** (priya.vyas@springer.com)
South Korea: **Smith Chae** (smith.chae@springer.com)
Southeast Asia, Australia and New Zealand: **Ramesh Nath Premnath** (ramesh.premnath@springernature.com)
The Americas: **Michael Luby** (michael.luby@springer.com)
China and all the other countries or regions: **Mengchu Huang** (mengchu.huang@springer.com)

More information about this series at http://www.springer.com/series/16157

Swarup Bag · Christ Prakash Paul ·
Mayuri Baruah
Editors

Next Generation Materials and Processing Technologies

Select Proceedings of RDMPMC 2020

Springer

Editors
Swarup Bag
Department of Mechanical Engineering
Indian Institute of Technology Guwahati
Guwahati, Assam, India

Mayuri Baruah
National Institute of Technology
Jamshedpur
Jamshedpur, Jharkhand, India

Christ Prakash Paul
Laser Material Processing Division
Raja Ramanna Centre for Advanced
Technology
Indore, Madhya Pradesh, India

ISSN 2662-3161 ISSN 2662-317X (electronic)
Springer Proceedings in Materials
ISBN 978-981-16-0181-1 ISBN 978-981-16-0182-8 (eBook)
https://doi.org/10.1007/978-981-16-0182-8

This Springer imprint is published by the registered company Springer Nature Singapore Pte Ltd.
The registered company address is: 152 Beach Road, #21-01/04 Gateway East, Singapore 189721, Singapore

Preface

This book presents the papers submitted to the Online National Conference on Research and Developments in Material Processing, Modelling and Characterization 2020 (RDMPMC-2020) held on August 26 and 27, 2020, organized by the Department of Metallurgical and Materials Science in association with Department of Production and Industrial Engineering, National Institute of Technology, Jamshedpur, Jharkhand, India. The book can be broadly segregated as Materials, Modelling and Simulation, Characterization, and Material Processing. Out of 206 papers accepted for presentation in the conference, 44 papers were accepted as chapters for this book. Featuring the various papers accepted from the conference, this book will be brief of interest to experts in various fields whose work involves in different domains of materials, metallurgical, mechanical, and manufacturing engineering. The book highlights some of the emerging technologies and the current capability gaps of the materials and its fabrication techniques. Innovative new technology may lead breakthroughs in materials design and fabrication which may lead to the improvement in energy productivity of manufacturing. Thus, innovative revolution in materials with defined interrelations between its properties and microstructure may result in promising technologies with energy-efficient processes, enhanced quality, and high economic benefits.

Guwahati, India — Swarup Bag
Indore, India — Christ Prakash Paul
Jamshedpur, India — Mayuri Baruah

Contents

About the Editors

Dr. Swarup Bag completed his Bachelors of Engineering in Mechanical Engineering from the Jalpaiguri Government Engineering College, Jalpaiguri, West Bengal, India, in the year 2000, M.E. in Mechanical Engineering in 2002 from Bengal Engineering College, Shibpur, India (now Indian Institute of Engineering Science and Technology), in the year 2002, and Ph.D. on "Development of Bi-directional Heat Transfer and Fluid Flow Model for Reliable Design of GTA and Laser Welding Processes" in 2009 from IIT Bombay. Later he has worked at the Center for Material Forming (CEMEF), MINES Paris Tech, France, for 20 months in Metallurgy, Structure and Rheology (MSR) group. In the year 2011, he joined the Department of Mechanical Engineering, Indian Institute of Technology Guwahati, as a faculty member. He worked as a visiting scholar at Center for Advanced Materials Joining, University of Waterloo, Canada, from May 2018 to July 2018. His primary area of research is fundamental process modeling of welding and joining technologies, optimization of manufacturing processes and recrystallization in metal forming processes. Dr. Bag has published about 50 journal papers, 50 conference papers and 10 book chapters related to welding and joining processes. He is the author of the book "Computational models for GTA and laser welding processes" and the recipient of "Royal Arc Award 2009" from Indian Institute of Welding for the best Ph.D. thesis in welding. Till now, five students have received doctorate degree and seven research scholars are working under his guidance. Thirteen M.Tech. projects and 11 B.Tech. projects have completed so far. Two research projects have completed so far in his credit.

Dr. Christ Prakash Paul is leading a research group in the area of Laser Additive Manufacturing (LAM) as Head, Laser Additive Manufacturing Laboratory at Raja Ramanna Centre for Advanced Technology (RRCAT). He has over 15 years of experience in research and development of LAM, from system development to its applications in engineering and prosthesis. His research programs focus on next generation additive manufacturing system and processes. To this end, his group explores novel techniques to develop advanced materials, innovative products, modeling and simulation tools, and holistic in-situ and ex-situ characterization techniques. He is instrumental in promoting LAM technology in India through collaborations with other national labs and reputed academic institutes by extending the indigenous developed facility for experiments. He has contributed chapters to twelve books and published over 100 research papers. His extensive experience and leadership have allowed him to address LAM challenges holistically and promoting this strategic research area. His efforts have been recognized with many awards by Indian and Canadian Governments: The indigenous development of first Metal LAM machines in India (for both schemes—Direct Energy Deposition and Powder Bed Fusion) is at his credit and is awarded with DAE Group Achievement Award for the year 2015. As recognition to exemplary works in the development and deployment of laser-baser metal additive manufacturing, he has been conferred with highest DAE Excellence in Science, Technology and Engineering award "Homi Bhabha Science and Technology Award" for the year 2017 NSERC-Discovery Research Grant (Individual) award for repair of Tungsten carbide tools in the year 2007 by Canadian Government during his stay at the University of Waterloo; Ontario Centre of Excellence-Conference Travel Award during the year 2006 by Ontario Provential Government for the laser additive manufacturing process development.

Dr. Mayuri Baruah is currently Assistant Professor at the Department of Production and Industrial Engineering, NIT Jamshedpur, Jharkhand, since 2018. She obtained her B.E. (Civil) from Dibrugarh University, Assam, and Integrated Ph.D. from the Indian Institute of Technology Guwahati, Assam. Her major areas of research interests include welding, additive manufacturing, numerical modeling, characterization and optimization of manufacturing processes. She has published 11 papers in respected international journals and 10 conferences. Till now, 2 research scholars are working under her guidance. 4 M.Tech. projects and 12 B.Tech. projects have completed so far. One minor seed grand research project is running under her.

Materials

Development of High Strength Low Carbon Lean Micro-alloyed Steel with Optimized Toughness

Md. Serfraj Alam, Ratnesh Gupta, Ghanshyam Das, Vinod Kumar, and Bimal Kumar Jha

1 Introduction

Flat rolled micro-alloyed steel plates have wide range of applications requiring high strength, toughness and weldability. Normally, micro-alloying additions of Nb, V and Ti are made singly or in combination to achieve the specified mechanical properties. Cr and Mo are added to improve hardenability of steel and thereby support formation of bainite microstructure. Traditionally, hot rolled high strength plates (YS $\leq$ 500 MPa) relied on fine grained ferrite–pearlite microstructure through use of micro-alloying and TMCP rolling. Increasing strength levels more than 550 MPa with adequate toughness (CIE: >120 J at 0 °C, %EL: 20 minimum) required a shift towards a ferrite/acicular ferrite microstructure through a logical alloy composition and optimizing the reduction per pass and cooling rate [1–6].

Some researchers have focused primarily on the strengthening by grain refinement using the thermo-mechanical control process (TMCP). This could not be supported for getting yield strength >500 MPa. One researcher found that thermo-mechanical processing with online accelerated cooling (TMCP) exhibited an admirable combination of strength and toughness, i.e. yield strength of 555 MPa and tensile strength of 652 MPa but provided high yield ratio of 0.85, representing poor formability. This combination of properties was achieved by acicular ferrite, lesser bainite, fine precipitates and some amount of martensite–austenite (M–A) constituents [7].

Md.Serfraj Alam (✉) · V. Kumar
Research & Development Centre for Iron and Steel (RDCIS),
Steel Authority of India Limited (SAIL), Ranchi 834002, India
e-mail: serfraj.alam21@sail.in

R. Gupta · G. Das · B. K. Jha
Department of Materials and Metallurgical Engineering, National Institute of Foundry and Forge Technology, Ranchi 834003, India

S. Bag et al. (eds.), *Next Generation Materials and Processing Technologies*,
Springer Proceedings in Materials 9, https://doi.org/10.1007/978-981-16-0182-8_1

In this present work, two micro-alloyed steel grades with low carbon, one having Nb–V–Mo–Ti combination and other having Nb–Ti only, were commercially produced in a modern high power plate mill. Nb–Ti steel plates were processed with high reduction per pass (up to 34%) and high cooling rate of 16 °C/s. The TMCP so employed resulted in final microstructure consisting of predominantly acicular ferrite with bainite and very fine precipitates of Nb, Ti carbides. This microstructure leads to an attractive property in terms of YS: 584–592 MPa, UTS: 763–803 MPa, %EL: 29–31, YS/UTS: 0.72–0.77 and CIE: 210–222 J. Acicular ferrite microstructures within the steel have a greater potential to enhance strength, toughness, H_2S resistance and fatigue behaviour compared to ferrite–pearlite microstructure by grain refinement, increased dislocation density and simply controlled precipitates [8, 9].

The paper discusses the influence of TMCP parameters on the final microstructures and thereby the properties of two micro-alloyed steel grades processed in a high power plate mill.

2 Experimental Procedure

The development of high strength low carbon micro-alloyed steel was carried out by producing Nb–V–Mo–Ti added steel. The steel was made through basic oxygen furnace (BOF) followed by Ladle furnace (LF) and RH treatment. The steel compositions are shown in Table 1. Steel was cast into slabs of 250 × 1500 mm size. After soaking inside the reheating furnace, slabs were discharged, and further, steel rolling was done as per designed process parameters in plate of thickness 15.5 mm.

Salient features of processing consisted of (a) alloy design with Nb, V, Mo and Ti content, (b) reheating temperature, (c) reduction per pass, (d) finishing temperature, (e) temperature after accelerated cooling (ACC), and (f) cooling rate. Control parameters of the trials were (a) soaking zone temperature: 1200 °C, (b) reduction per pass: up to 18%, (c) finishing temperature: 800 °C, (d) temperature after ACC: 585 °C and (e) cooling rate: 7 °C/s for plates of thickness 15.5 mm.

Further, lean micro-alloyed steel with Nb and Ti was designed to reduce the cost of production. The steel was made through basic oxygen furnace (BOF) followed by LF and RH treatment. The chemical compositions of the steel are shown in Table 2. Steel was cast into slabs of 250 × 1500 mm size. After soaking in reheating furnace, slabs were discharged, and rolling was done as per designed process parameters in plate of thickness 17.5 mm.

Table 1 Steel chemistry of Nb, V, Ti and Mo micro-alloyed steel (wt%)

C	Mn	P	S	Si	Mo	Nb	V	Ti	Al
0.05	1.45	0.012	0.004	0.326	0.2	0.05	0.047	0.015	0.020

Table 2 Steel chemistry of Nb and Ti micro-alloyed steel (wt%)

C	Mn	P	S	Si	Nb	Ti	Al
0.08	1.57	0.02	0.007	0.27	0.05	0.021	0.040

Table 3 Critical temperature

	(Nb–V–Mo–Ti)	(Nb–Ti)
Nb dissolution temperature (°C)	1167	1208
Tnr (°C)	977	973

Modified processing parameters include (a) alloy design with Nb and Ti content, (b) reheating temperature, (c) reduction per pass, (d) finishing temperature, (e) temperature after ACC and (f) cooling rate. Control parameters of the Nb–Ti micro-alloyed steel were (a) soaking zone temperature: 1220 °C, (b) reduction per pass: up to 34%, (c) finishing temperature: 805 °C, (d) temperature after ACC: 540 °C and (f) cooling rate: 16 °C/s for plates of thickness 17.5 mm.

Critical temperatures related to Nb–V–Ti–Mo and Nb–Ti had been calculated based on empirical formulas and are shown in Table 3.

The property diagram of Nb–Ti micro-alloyed steel is drawn through ThermoCalc software (TCFE7) to find out the Ar3 and Ar1 temperature and is shown in Fig. 1.

The critical temperatures based on empirical formulas and ThermoCalc software were supportive to design the rolling processing parameters (slab heating to final plate finishing temperature).

3 Results and Discussion

The test results of tensile and Charpy toughness values related to first experimental steel of Nb–V–Mo–Ti micro-alloyed are shown in Table 4. It may be noted that in plates of thickness 15.5 mm, yield strength varied between 487 and 490 MPa, and tensile strength varied between 566 and 576 MPa. %EL and YS/UTS values were in the range of 32–34 and 0.84–0.86, respectively. Rolled plates exhibited excellent Charpy impact toughness values in the range of 328–344 J at 0 °C.

While in case of lean Nb–Ti micro-alloyed steel, the mechanical properties achieved were very attractive. It is worth mentioning that yield strength varied between 584 and 592 MPa, while tensile strength varied between 763 and 803 MPa with % EL and YS/UTS values in the range of 29–31 and 0.72–0.77. Rolled plates exhibited good Charpy impact toughness values 210–222 J at 0 °C [9]. Table 5 summarizes the mechanical properties achieved in the experimental Nb–Ti micro-alloyed steel. Samples were prepared in transverse direction for tensile and Charpy properties evaluation and tested as per ASTM E8 and ASTM E23.

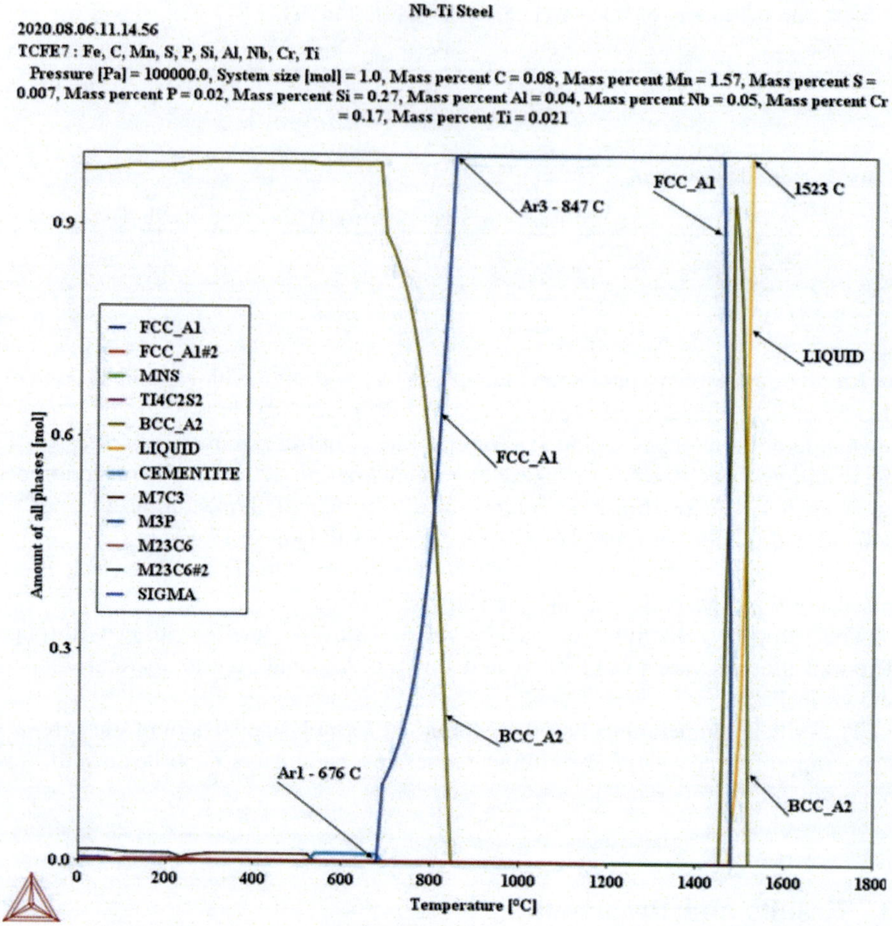

Fig. 1 Property diagram of Nb–Ti micro-alloyed steel

Table 4 Mechanical properties of Nb–V–Mo–Ti micro-alloyed steel

YS, MPa	UTS, MPa	%EL	CIE, J	YS/UTS
487–490	566–576	32–34	328–344	0.84–0.86

Table 5 Mechanical properties of Nb–Ti micro-alloyed steel

YS, MPa	UTS, MPa	%EL	CIE, J	YS/UTS
584–592	763–803	29–31	210–222	0.72–0.77

3.1 Microstructural Evolution

The samples were observed using optical, SEM and TEM, and silent microstructural features are outlined below.

The typical microstructures of Nb–V–Mo–Ti and Nb–Ti micro-alloyed steels of thickness 15.5 mm and 17.5 mm are shown in Figs. 2 and 3, respectively. Microstructure of Nb–V–Mo–Ti micro-alloyed steel is polygonal ferrite–bainite type with average ferrite grain size of around 8 μm.

In the case of Nb–Ti micro-alloyed steel, the microstructure is fine acicular ferrite and bainite type. The microstructure is obtained due to the modifications in process parameters viz. furnace heating temperature, reduction schedule, finishing temperature, temperature after ACC and cooling rate [9–12].

Figures 4 and 5 show the scanning electron micrographs of Nb–V–Mo–Ti and Nb–Ti micro-alloyed steel, respectively. It may be noted that Nb–V–Mo–Ti steel shows predominantly polygonal ferrite structure with less percentage of bainite. This in fact led to lower yield strength values of 487–490 MPa. As far as Nb–Ti micro-alloyed steel is concerned, it is worth mentioning that in this steel no pearlite and hard phase like cementite could be observed, it mainly consisted of fine acicular ferrite and bainite phase. This microstructure has given rise to high strength (YS: 584–592 MPa) and high toughness values (210–222 J). Acicular ferrite–bainite microstructure leading to improved combination of strength and ductility has been reported by other researchers also [9–11].

Figure 6 shows a transmission electron micrograph and an EDS spectrum of carbides in Nb–Ti micro-alloyed steel. The precipitate size of less than 80 nm was found. Titanium and niobium were detected within the fine carbides. High dislocation density is associated with ferrite lath. Nb–Ti micro-alloyed steel is found to be strengthened because of such nanometre-sized carbides containing niobium and titanium and dislocations present within the steel [7, 12–14].

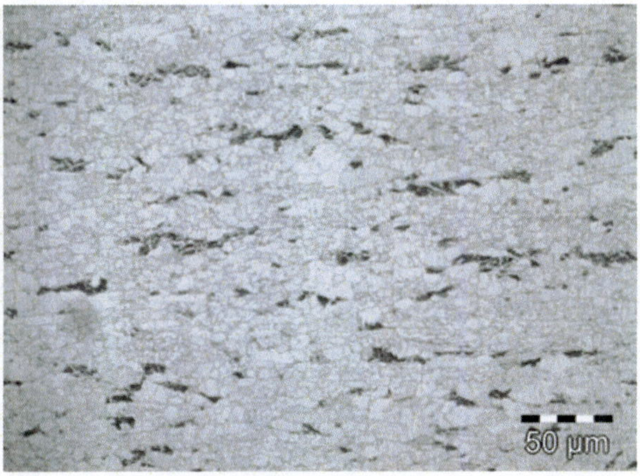

Fig. 2 Typical micrograph of Nb–V–Mo–Ti micro-alloyed Steel

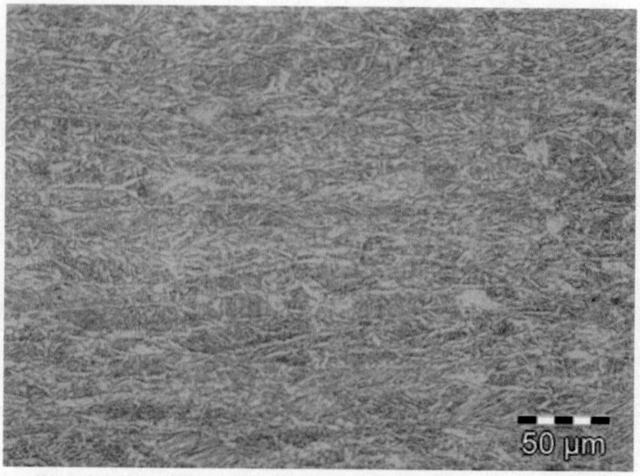

Fig. 3 Typical micrograph of Nb–Ti micro-alloyed steel

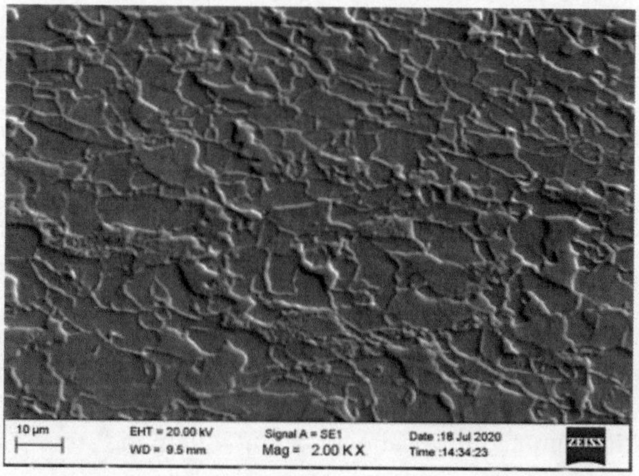

Fig. 4 SEM micrograph of Nb–V–Mo–Ti micro-alloyed steel

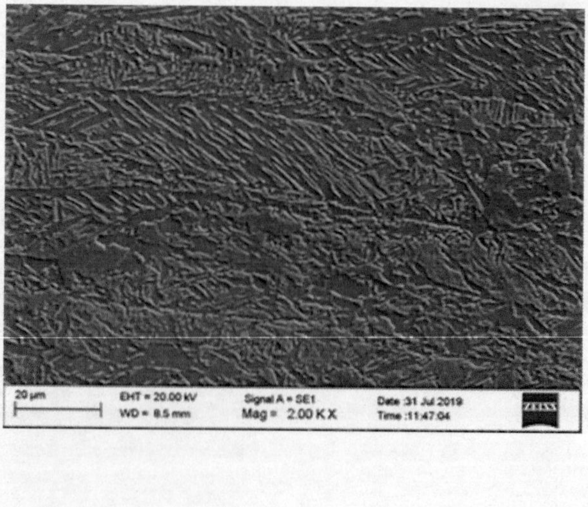

Fig. 5 SEM micrograph of Nb–Ti micro-alloyed steel

Fig. 6 Transmission electron micrograph showing carbides in steel **a** bright field image of fine carbides **b** EDS spectrum of fine carbides and **c** bright field image showing very fine carbides embedded within the matrix

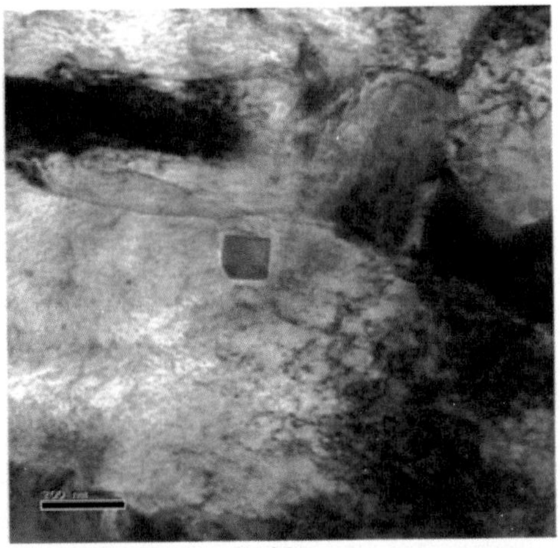

(a)

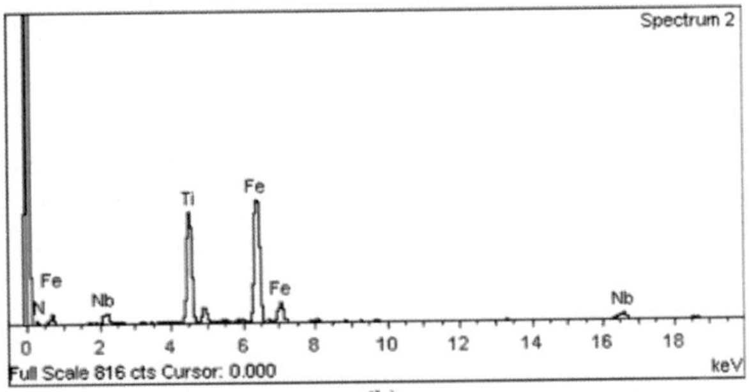

(b)

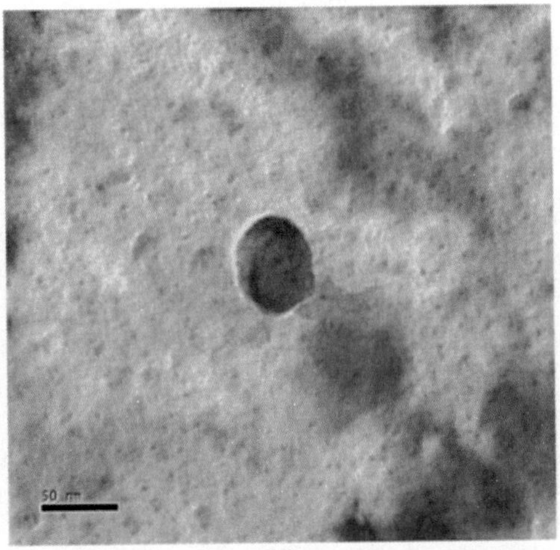

(c)

With a significant amount of acicular ferrite microstructure within the steel as reportedly has an optimized combination of mechanical properties in micro-alloyed steel when compared to alloy with bainitic structure [7]. Microstructure of acicular ferrite has the capacity of combining high strength and high toughness within the steel [13]. This feature happens because the plates of acicular ferrite nucleate at non-metallic inclusions intra-granularly inside large austenite grains and then they diffuse in different directions from those inclusions though maintaining an orientation relationship with the austenite grains.

Figure 7a–e shows the effect of processing parameters on the properties of two studied steel grades, and an all round improvement in the strength–toughness properties of Nb–Ti micro-alloyed steel can be noticed. It is interesting to report that YS and UTS values for Nb–Ti steel have increased by 17 and 27% compared to that of Nb–V–Mo–Ti. The sharp increase in UTS has led to lower YS/UTS ratio (13%) also. Lower YS/UTS ratios in fact favour improved formability, and Nb–Ti steel on the present study not only shows an improved strength–toughness combination but also depicts improved formability [14].

Properties so achieved have been related to microstructure in the succeeding section. The microstructural observations clearly depict that high strength was achieved in the Nb–Ti micro-alloyed steel because of fine acicular ferrite, bainite and fine carbides [9–12, 15–18]. In the case of Nb–V–Mo–Ti micro-alloyed steel, the strength was lesser than Nb–Ti micro-alloyed steel because of polygonal ferrite and fewer bainite phases within the steel.

The soaking temperature of Nb–Ti micro-alloyed steel was maintained around 1220 °C, i.e. 20 °C more when compared to Nb–V–Mo–Ti micro-alloyed steel, and soaking time of 5 h adopted within the current research basically controls the composition and austenite grain size to confirm the possible dissolution of micro-alloying carbides, nitrides and carbonitrides in austenite phase. During this high soaking temperature, the austenite grain growth would be restricted by fine precipitates of Nb and/or Ti. An optimized process of precipitation of carbides, nitrides and carbonitrides shaped during the TMCP is liable for the formation of fine grained ferritic microstructure. The finishing temperature of plate rolling is maintained under the nonrecrystallization temperature (Tnr) of austenite in order to stop the static recrystallization of austenite. Therefore, because of rolling at nonrecrystallized temperature, the grains become pancaked. Hence, in the absence of static recrystallization, the dynamic recrystallization is probably going to happen as strain accumulation occurs by the number of roll passes during rolling [17]. The rolling in nonrecrystallization region accumulates more strain, i.e. austenite grains with more dislocations, and this type of strain supports for the acicular ferrite grain refinement by promoting nucleation site for transformation of γ-α [8]. The typical acicular ferrite grain size is because of pancaked austenite grains, alloying elements that lower austenite to ferrite transformation and fast cooling rate (16 °C/s) from austenite region. The grain refinement effect is satisfactorily achieved by the

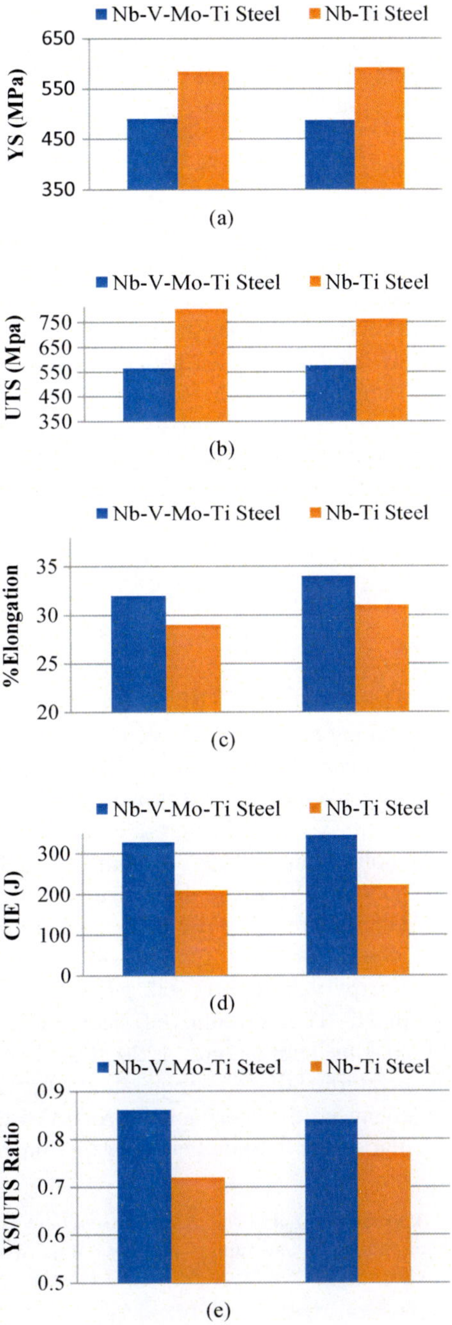

Fig. 7 Effect of processing parameters on **a** YS, **b** UTS, **c** %EL, **d** CIE and **e** YS/UTS values on Nb–V–Mo–Ti and Nb–Ti micro-alloyed steel

thermo-mechanical controlled rolling process (TMCP) at high temperatures, and low transformation temperature microstructures are formed by the accelerated cooling process followed after rolling. Fast cooling process is encouraging to the formation of ferrite nuclei and leads to the fine acicular ferrite grains with bainite (Figs. 3 and 5). In general, by increased cooling rate, the quality and morphology of ferrite alter from polygonal ferrite, i.e. high cooling rate promotes the formation of acicular ferrite (Figs. 3 and 5). In line with number of researchers, a combination of low temperature after ACC and high cooling rate in micro-alloyed steels supports to provide the fine acicular ferrite + bainitic microstructure, making higher strength steels. A steel having acicular ferrite type microstructure has the capacity of getting combination of high strength and high toughness, because a crack will have to follow a more complex path through the acicular ferrite microstructure [10]. Hence, the improved mechanical properties are associated with the microstructures obtained, i.e. fine acicular ferrite + bainite combination, solid solution strengthening by the addition of micro-alloying elements in low carbon steels, precipitation strengthening and dislocation strengthening [9, 12, 19–21].

4 Conclusions

- TMCP consisting of soaking temperature, reduction per pass and accelerated cooling plays critical role in evolution of microstructure and resultant properties of micro-alloyed steel. Acicular ferritic structure has been found to yield attractive mechanical properties in comparison with polygonal ferritic structure.
- In Nb–V–Mo–Ti micro-alloyed steel, higher temperature after ACC and lower cooling rate (7 °C/s) in comparison with Nb–Ti micro-alloyed steel led to the polygonal ferrite–bainite structure and hence lower strength value.
- Lean micro-alloyed Nb–Ti steel resulted in an attractive plate property of YS: 584–592 MPa, UTS: 763–803 MPa, %EL: 29–31, YS/UTS: 0.72–0.77 and Charpy values: 210–222 J at 0 °C.
- In Nb–Ti micro-alloyed steel, the modified processing parameters viz. steel soaking temperature: 1220 °C, reduction per pass: up to 34%, finishing temperature: 805 °C, temperature after ACC: 540 °C and cooling rate: 16 °C/s enhanced the mechanical properties through microstructural evolution of acicular ferrite–bainite and very fine precipitation of Nb–Ti carbide.

Acknowledgements The authors are grateful to the management of RDCIS, SAIL, for extending kind permission and support to carry out the work and also for giving permission to publish this work.

References

1. Tiana Y, Wanga H-T, Lia Y, Wanga Z-D, Wanga G-D (2017) The analysis of the microstructure and mechanical properties of low carbon microalloyed steels after ultra fast cooling. Mater Res 20(3):853–859
2. Li XL, Lei CS, Deng XT, Wang ZD, Yu YG, Wang GD, Misra RDK (2016) Precipitation strengthening in titanium microalloyed high strength steel plates with new generation thermomechanical controlled processing. J Alloys Compd 689:542–553
3. Chen Y, Zhang DT, Liu YC, Li HJ, Xu DK (2013) Effect of dissolution and precipitation of Nb on the formation of acicular ferrite—bainite ferrite in low carbon HSLA steels. Mater Charact 84:232–239
4. Challa VSA, Zhou WH, Misra RDK, O'Malley R, Jansto SG (2014) The effect of coiling temperatures on the microstructure and mechanical properties of Niobium–Titanium micro-alloyed steel processed via thin slab casting. Mater Sci Eng A 595:143–153
5. Seich GR (1981) In: Kot RA, Bramfitt BL (eds) Fundamentals of dual-phase steels. AIME, Chicago, 3
6. Belladi H (2004) Ultrafine ferrite formation in steels through thermomechanical processing. Ph.D. Thesis, Deakin University, Geelong, Victoria, Australia
7. Zuo X, Zhou Z (2015) Study of pipeline steels with acicular ferrite microstructure and ferrite-bainite dual-phase microstructure. Mater Res 18(1):36–41
8. Kim YW, Song SW, Seo SJ, Hong SG, Lee CS (2013) Development of Ti and Mo micro-alloyed hot-rolled high strength sheet steel by controlling thermomechanical controlled processing schedule. Mater Sci Eng A 565:430–438
9. Zhao H, Wynne BP, Palmiere EJ (2018) Conditions for the occurrence of acicular ferrite transformation in HSLA steels. J Mater Sci 53:3785–3804
10. Pedrosa IRV, Castro RS, Yadava YP, Ferreira RAS (2013) Study of phase transformations in API 5L X80 steel in order to increase its fracture toughness. Mater Res 16(2):489–496
11. Kim YM, Lee H, Kim NJ (2008) Mater Sci Eng A 478:361–370
12. Zou H, Kirkaldy JS (July 1991) Carbonitride precipitate growth in titanium/niobium microalloyed steels. Metall Trans A 22A:1511–1524
13. Lee CH, Bhadeshia HKDH, Lee HC (2003) Effect of plastic deformation on the formation of acicular ferrite. Mater Sci Eng A 360:249–257
14. Funakawa Y, Shiozaki T, Tomita K, Yamamoto T, Maeda E (2004) ISIJ Int 44:1945–1951
15. Umemoto M, Todaka Y, Tschuiya K (2003) Nanocrystallization of steels by various severe plastic deformation processing and properties of structural nanomaterials. MS&T 126–132
16. Lowe TC, Valiev RZ (Oct 2004) The use of severe plastic deformation techniques in grain refinement. JOM 64–68
17. Shukla R, Das RK, Ravi Kumar B, Ghosh SK, Kundu S, Chatterjee S (2012) An ultra low C, thermomechanically controlled processed microalloyed steel: microstructure and mechanical properties. Metall Mater Trans A 43(12):4835–4845
18. Kejian H, Baker TN (1997) In: Proceedings of conference on 'Titanium technology in microalloyed steels', the Institute of Materials, London, p 115
19. Kneissl AC, Garcia CI, DeArdo AJ (1991) In: Proceedings of international conference on the 'Processing, microstructure and properties of microalloyed and other modern high strength low alloy steels', ISS, Warrendale, PA, p 145
20. DeArdo AJ (2001) In: Conference proceedings on 'Niobium science and technology', Niobium 2001 Limited, Orlando, Florida, p 427
21. Kostryzhev AG, Marenych OO, Killmore CR, Pereloma EV (2015) Strengthening mechanisms in thermomechanically processed NbTi-microalloyed steel. Metall Mater Trans A 46(8):3470–3480

The Pros and Cons of an Energy-Efficient Q&P Approach to Develop Advanced Steels

G. K. Bansal, A. K. Chandan, Chiradeep Ghosh, V. Rajinikanth, V. C. Srivastava, Monojit Dutta, and S. Ghosh Chowdhury

1 Introduction

Steel is defined as an alloy of iron with up to 2.1 wt% carbon. Some other alloying elements, such as Mn, Ni, Cr, Al and Si, are also added in this alloy to achieve the desired microstructure and properties. Some of the critical factors such as cost-effectiveness, ease of recycling and mass manufacturing, make steel a predominant engineering material for use in manmade structures and machine components. Keeping this in view, the Indian National Steel Policy 2017 [1] stipulated an increase in Indian steel production from ~125 million tons (MT) at present to 300 MT by 2030. Recently, emphasis has also been given on the development of high-end value-added steel grades with improved properties and performance compared to the existing grades. Therefore, substantial efforts are required to fully comprehend the composition–structure–property correlation in existing grades of steel. This can help in designing novel steel composition and processing routes for improving the properties to the desired levels.

The suitability of a steel grade for a specific application is dictated by its ability to withstand the given load, resistance to fracture, performance against corrosion and wear, ease of bulk manufacturing, cost, etc. In this regard, numerous grades of steel are being manufactured industrially and used by consumers worldwide. The premise on which various grades of structural steel are classified into conventional

G. K. Bansal (✉) · A. K. Chandan · V. C. Srivastava · S. Ghosh Chowdhury
Academy of Scientific and Innovative Research (AcSIR), Ghaziabad 201002, India

G. K. Bansal · A. K. Chandan · V. Rajinikanth · V. C. Srivastava · S. Ghosh Chowdhury
Materials Engineering Division, CSIR-National Metallurgical Laboratory, Jamshedpur 831007, India

C. Ghosh · M. Dutta
Research and Development Division, Tata Steel Limited, Jamshedpur 831001, India

S. Bag et al. (eds.), *Next Generation Materials and Processing Technologies*, Springer Proceedings in Materials 9, https://doi.org/10.1007/978-981-16-0182-8_2

and different generations (first, second and third generation) is the strength–ductility combination, i.e. the product of ultimate tensile strength (UTS) and total elongation (TE). This is also identified as the energy absorption capability of the steel during the course of tensile deformation. Therefore, an increase in either strength or ductility or both leads to improved performance of the component during service. However, for most of the steel grades, an increase in strength is generally accompanied by a concurrent decrease in the ductility; and vice versa is also true. As a result, the development of steel grades with simultaneous enhancement of both strength and ductility, by applying novel processing routes, has fascinated the researchers in recent times.

The strength–ductility combinations achieved in conventional and different generations of steels are shown in Fig. 1. The data for an approximate range of strength and ductility has been taken from advanced high-strength steels application guidelines version 6.0 of world auto steel [2]. The steel grades, which were developed at an early stage, are known as conventional steels, e.g. IF, Mild, BH, HSLA, FB, etc. Although these grades achieve higher ductility (up to ~60%) and formability, their tensile strength is quite low due to the ferrite-based microstructure. These steels are in full-fledge commercial production at present and have almost reached saturation in terms of further research possibility for a significant improvement. Subsequently, new grades of steel (mostly martensite-based) were developed to improve the strength levels, e.g. TRIP, DP, CP, Martensitic, HF, etc. These steel grades come under the category of first generation steels, which can achieve tensile strength as high as 2000 MPa. However, their total elongation is limited to about 10% at such high-strength levels. Therefore, the energy absorption capability (UTS × TE) for various steel grades within first generation is restricted to about 15 ± 10 GPa% [3]. Consequently, the need for steels with improved

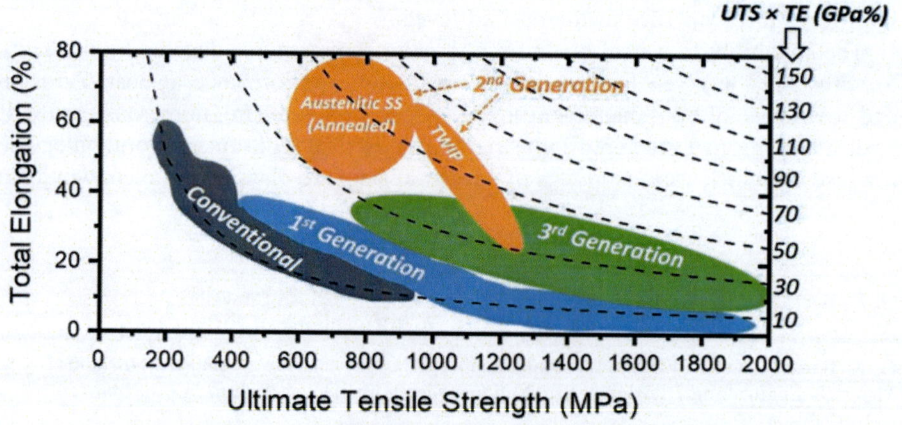

Fig. 1 Strength and ductility achieved in different grades of steels, adapted from [2] (each dotted isopleth represents a constant value of energy absorption capability (theoretical), as indicated in the plot)

strength–ductility combinations led to the evolution of second generation steels, i.e. TWIP and austenitic stainless steels. These steels possess a complete austenitic structure at room temperature due to the addition of austenite stabilizers in a significant amount. The austenite improves the ductility, and the desired strength is achieved by the higher alloying content. In addition, the TWIP mechanism of work hardening, i.e. the twinning-induced plasticity, also contributes to the enhancement of both strength and ductility. As a result, the energy absorption capability of these steels has been elevated to about 60 ± 10 GPa% [3], which is a fourfold increase in comparison with the first generation steels. However, significant alloying additions in these steels lead to certain issues such as higher cost and the issues in steelmaking, processing, fabrication and welding.

In view of the above, the development of steel grades with reduced alloying additions is of paramount importance to achieve the strength–ductility combinations superior to the first generation and close to the second generation steels, using novel processing routes. Such scope of strength–ductility combinations has been marked as third generation steels in Fig. 1. The steel with higher strength will allow the use of a thin section and, therefore, a reduction in the weight of the component and fuel consumption without compromising on safety. On the other hand, an improvement in the ductility will prevent early failure of the component. Hence, the development of steels with increasing strength levels while retaining a sufficient ductility has been the major focus area of research in recent times. A possible strategy that has been commonly applied till date to achieve the same is the attainment of a multiphase microstructure mainly containing retained austenite, martensite and bainite. Some of the examples in this category include advanced/nano-bainitic steels [4–6], medium-Mn steels [3, 7–10], quench and partitioned (Q&P) steels [3, 11–33], etc. Therefore, the present paper brings out an overview of Q&P and related concepts, which have been applied in hot-rolled steels. The different sections have been devoted to elaborate the different aspects of these steels, such as prospects and limitations in terms of industrial implementation, microstructure evolution and mechanical properties.

2 Q&P and Related Approaches in Hot-Rolled Steels

In an industrial hot rolling mill, the cast steel slabs are first austenitized and hot-rolled to the desired thickness with the finish rolling temperature generally above Ae_3. This is followed by quenching on the run-out table, by laminar cooling to the desired temperature and coiling. The final microstructure depends highly on the time-temperature transformation (TTT) and continuous cooling transformation (CCT) diagrams of the given alloy composition. Therefore, the freedom to select different cooling rates during run-out table quenching and the targeted quench/coiling temperature provides enormous scope to develop a variety of microstructures (ferrite-, bainite- or martensite-based) in the single alloy composition. It is well-known that ferrite-based microstructures exhibit excellent ductility and formability; however, their strength is quite low. The improvement of strength via bainite and/or martensite-based microstructures leads to reduced ductility, in comparison with the ferrite-based microstructure. Therefore, a majority of the steel research in recent times has been concentrated on the development of steel grades with high strength and improved ductility and formability characteristics.

The different processes (conventional as well as recently developed) to achieve a martensite and/or bainite-dominated microstructures in hot-rolled steel are schematically shown in Fig. 2. To achieve a complete martensitic structure, the steel after finish rolling is directly quenched to room temperature. The alloy composition and the run-out table quenching rates are selected such that the formation of ferrite and bainite during cooling is circumvented. Although a complete martensitic structure can give rise to significantly high strength, their ductility is quite low due to the supersaturated carbon trapped in the interstitial sites, particularly for steels with high carbon content. This brittleness of martensite is generally relieved by an additional tempering treatment, during which the steel is held isothermally at a higher temperature (generally below Ae_1) for a suitable time, followed by cooling to room temperature. During the isothermal holding, the supersaturated carbon trapped in the interstitial sites is precipitated in the form of carbides, thereby providing a low-carbon martensitic matrix with a uniform distribution of carbide precipitate. Although the tempering process has been shown to improve the toughness and ductility, the strength and hardness of the steel are compromised.

It can be inferred from the above that the new processing/heat treatment methods have to be identified to achieve a good combination of strength and ductility. In this

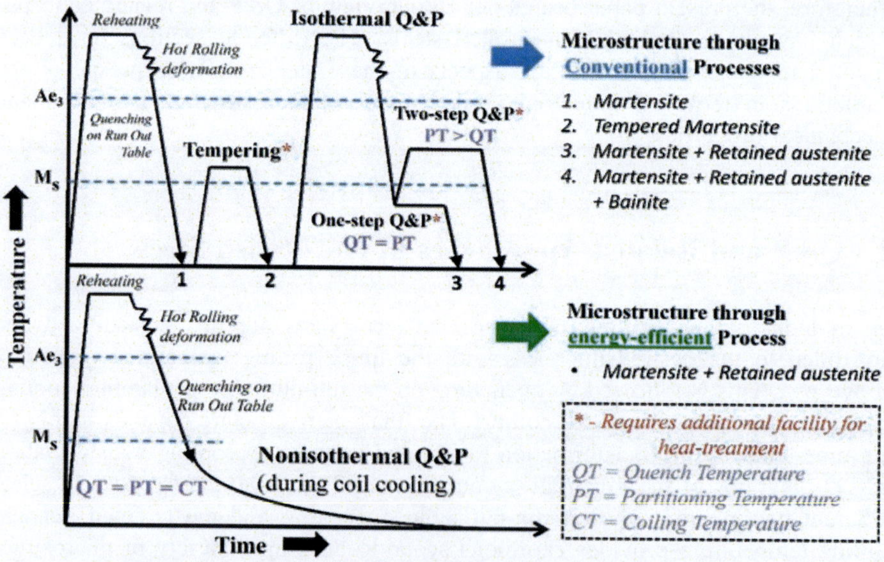

Fig. 2 Schematic illustration of the various heat treatment schedules applied on hot-rolled steel

regard, a recently developed Q&P process could be a potential approach to develop a multiphase microstructure mainly containing martensite, bainite and retained austenite. The retained austenite in these steels plays a crucial role in the simultaneous enhancement of both strength and ductility through transformation-induced plasticity (TRIP) mechanism of work hardening. As shown in Fig. 2, based on the processing steps, the Q&P process can be broadly categorized into two types, i.e. isothermal and nonisothermal Q&P. The isothermal Q&P, which is the conventional approach, was originally proposed by Speer et al. [11] in the year 2003. Most of the process steps in the isothermal Q&P process are similar to that followed for achieving a complete martensitic structure in a hot rolling mill, except the quench temperature. In the isothermal Q&P process, the steel is quenched to a temperature within the M_s–M_f range, so as to form some amount of martensite with remaining untransformed austenite, instead of room temperature quenching that leads to a complete martensitic structure. Immediately after quenching in the M_s–M_f range, the steel is kept isothermally at the quench temperature for the transport of supersaturated carbon in the martensite to the remaining untransformed austenite, followed by cooling to room temperature. The diffusion of carbon from martensite to the untransformed austenite happens due to a significantly higher solubility of the carbon in the austenite ($\sim$2.1 wt%), in comparison with the martensite (max. 0.02 wt%). This increased carbon content in the untransformed austenite lowers its M_s temperature and thus avoids its decomposition to martensite and/or bainite during cooling to room temperature. As a result, the microstructure containing low-carbon martensite and retained austenite is generally observed at room temperature. In the above process, the quench temperature is the same as the partitioning temperature, and hence, it is called a one-step Q&P process. Another variant of the isothermal Q&P process is two-step Q&P process (Fig. 2), in which the carbon partitioning is performed at a temperature higher than the quench temperature. Although the purpose of partitioning at higher temperatures is to speedup the process of carbon diffusion, a partial transformation of remaining austenite to bainite and carbide precipitation is also observed in the two-step Q&P steels, therefore, providing a multiphase microstructure containing martensite, bainite, retained austenite and carbides.

As an alternative to the isothermal holding in one- and two-step Q&P processes, the slow cooling of a hot-rolled coil may be utilized for the diffusion of carbon atoms to the austenite and its subsequent retention at room temperature. The cooling profile of a hot-rolled coil was measured by B. Nelson from Arcelor Mittal, Dofasco, and has been mentioned by G. Thomas in [12]. The results highlight that it takes about 45 h for the hot-rolled coil to reach room temperature after coiling at 700 °C. Furthermore, even if the coiling temperature is in the range of 200–300 °C, which is generally employed in the Q&P process, the achievement of room temperature by the hot-rolled coil still takes more than 30 h. Therefore, this extremely slow cooling might serve the purpose of carbon diffusion, similar to that achieved during the isothermal Q&P process. The lead-off work in this direction was performed by G. Thomas in the year 2009 [12]. The result of their study revealed comparable austenite retention by both the isothermal and nonisothermal Q&P

processes. Thereafter, some other researchers, including the present authors, have shown the austenite retention through the nonisothermal Q&P process [13–28]. The various aspects of these steels, in terms of industrial implementation, microstructure evolution and mechanical properties, have been discussed in succeeding sections.

3 Pros and Cons of Various Approaches Applied to Hot-Rolled Steels

The foregoing sections highlighted the merits and limitations of different grades of steel. In addition, some of the processing schedules that have been applied to the hot-rolled steels were also discussed. The present section, therefore, will focus on the advantages and constraints of these approaches, with a special emphasis on the energy-efficient nonisothermal Q&P approach. The pros and cons of these steel grades, in terms of industrial implementation, microstructure evolution and mechanical properties, have been systematically listed in Table 1 and discussed hereafter. It is evident from Fig. 2, that the steel with a complete martensitic structure can be produced through conventional hot rolling mills, without any need of the external heat treatment facility. The high strength in a martensitic structure requires high carbon content, which subsequently leads to increase tetragonality (c/a ratio) due to the supersaturation of carbon in the interstitial sites. Therefore, the tempering treatment is performed to bring out this carbon in the form of carbides, which relieves the internal stresses and brittleness. As a result, the tempered steels show better ductility and toughness with a concurrent decrease in the strength, in comparison with the steels with a martensitic structure. Moreover, the tempering treatment can be performed within a wide temperature range (generally below Ae_1) and holding durations. Therefore, it provides a possibility to engineer carbide precipitation (the type of carbides and their amount, size, distribution, morphology, composition, etc.), and hence, multiple combinations of mechanical properties can be achieved. However, tempering treatment requires an additional furnace for heat treatment. Although this additional process does not hamper the continuous hot rolling mill operations as it is performed as a separate post-treatment, the requirement of heat treatment furnace and energy adds to the production cost, and thereby, increasing the cost of steel.

Similar to the tempering treatment, recently developed isothermal Q&P processes (both one-step and two-step) also necessitate supplementary furnaces for isothermal holding. Moreover, this isothermal holding has to be performed during the continuous operation in a hot rolling mill and not as a separate treatment. For example, immediately after the run-out table quenching in the M_s–M_f range, the steel has to be kept isothermally at the same temperature (one-step Q&P) or at a relatively higher temperature (two-step Q&P), which leads to the interruption of the continuous operations in a hot rolling mill. However, these steels after isothermal Q&P treatment have been shown to exhibit improved mechanical properties, in

Table 1 Pros and cons of various processes applied to the hot-rolled steels

S. No.	Process	Pros	Cons
1	Direct quenching	• No additional treatment • Low production cost	• Presence of internal stresses and brittleness • Reduces ductility and toughness
2	Tempering	• Relieves internal stresses and brittleness • Improves toughness and ductility • Possibility to employ a wide range of temperatures and holding times, leading to a variety of microstructures	• Additional heat treatment facility and energy, leading to increased production cost • Reduces strength and hardness
3	Isothermal Q&P	• Freedom to select different temperatures for quenching and partitioning in the two-step process • Isothermal holding at a constant temperature and hence, a uniform and higher driving force for carbon diffusion • Improves strength–ductility–toughness combinations, in comparison with the direct quenched and/or tempered steels	• Additional heat treatment facility and energy, leading to increased production cost • Interruption of continuous hot rolling operation and therefore reduced productivity
4	Nonisothermal Q&P	• Utilizes the remaining heat of the hot-rolled coil and thereby, eliminates the need of additional furnace, making it an energy-efficient approach • Increases productivity and reduced cost of steel • Austenite retention comparable to the isothermal Q&P • Strength–ductility–toughness combinations better than the direct quench and/or tempered steels and comparable to the isothermal Q&P steels	• The continuous drop in temperature during coil cooling, which reduces the driving force for carbon diffusion • Possible variation in the microstructure and properties due to non-uniform cooling across the coil • Limited scope for microstructure variation due to the restriction of only one temperature for quenching, partitioning and coiling

comparison with the direct quenched and/or tempered steels [16, 29–32]. The isothermal Q&P treatment also has the opportunity to employ a wide range of temperatures and holding time for quenching and partitioning steps. For example, in the one-step Q&P process, different quench/partitioning temperatures (below M_s) and holding durations can be employed to vary the fraction of martensite and subsequently, the tendency towards carbon partitioning. Quenching to a temperature near M_s leads to the formation of low amount of martensite with higher levels of untransformed austenite. As a result, this untransformed austenite will be enriched with less amount of carbon, despite having a higher driving force for carbon diffusion at higher quench/partitioning temperatures. The vice versa is true for lower quench/partitioning temperature. The two-step Q&P process provides an additional advantage of performing partitioning step at temperatures, which are higher than the quench temperature. Therefore, lower temperatures can be selected for quenching to increase the martensite fraction, providing enough supply of carbon for the untransformed austenite. Subsequently, the partitioning can be performed at higher temperatures, and thereby, eliminating the issue with poor diffusivity. Therefore, the isothermal Q&P process offers huge scope for microstructure engineering, which leads to the development of steels with diverse mechanical properties.

As shown in Fig. 2, in the nonisothermal Q&P approach, the remaining heat of the hot-rolled coil is used for the carbon diffusion from martensite to the remaining austenite. Therefore, the elimination of additional heat treatment facilities, which were required in the tempering and isothermal Q&P processes, makes it an energy-efficient and cost-effective process and also easy to implement in a conventional hot rolling mill. However, in contrast to the isothermal Q&P process, this approach suffers from the restriction of selecting only one temperature for quenching and partitioning steps, which is actually equal to the coiling temperature. This limits the scope of extensive microstructure engineering in steel treated by the nonisothermal Q&P approach. Additionally, the steel is cooled slowly during coil cooling, and therefore, a continuous drop in the driving force for the carbon diffusion may limit the austenite retention at room temperature. Despite these process limitations, the steels treated by this energy-efficient nonisothermal Q&P process have shown the microstructure evolution and/or tensile properties close to that achieved after the isothermal Q&P process [12–16, 33].

It is also important to note that in the isothermal Q&P process, the isothermal holding at a constant temperature provides a constant driving force for carbon partitioning, and therefore, a consistent microstructure and mechanical properties can be expected at the different section of the hot-rolled coil. However, during the process of coil cooling in the nonisothermal Q&P steels, the outer surface of the hot-rolled coil is exposed to the atmosphere and hence will cool at a faster rate than the interior part of the coil. Therefore, different sections of the coil will experience a significant variation in the cooling rate during coil cooling. This non-uniformity in the cooling rate may alter the extent of carbon partitioning and subsequent austenite retention at room temperature. As a result, there may a possible variation in the mechanical properties at different sections of the coil. This heterogeneity will have serious issues while the steel is in use and may lead to premature failure of the component from the weak links, i.e. the locations of poor mechanical properties. In this regard, to the best of the author's knowledge, only Li et al. [17] have recently reported the result of the first industrial coil, which was produced through the nonisothermal Q&P process at Qian'an Iron and Steel Company, China. Although the authors have performed detailed dilatometry experiments to investigate the effect of quench temperature and cooling rate (during slow cooling) using small size samples, only the variation in the tensile properties was reported at different sections (across the length) of the industrial coil. The results of their dilatometry study show that the variation in the coil cooling rate from 0.05 to 10 °C/s lead to the austenite retention and its carbon content in the range of 10–13.5 vol.% and 1.05–1.26 wt%, respectively [17]. Moreover, the variation in the quench temperature from 240 to 400 °C led to the change in austenite retention and its carbon content in the range of 11–15 vol.% and 1.1–1.4 wt%, respectively [17]. These results of dilatometry investigation indicated only a small variation in the amount of austenite and its carbon content. The tensile properties, which were measured at a distance of 25 m across the 500 m long hot-rolled coil, showed a maximum difference of 61 MPa and 2.3% in the tensile strength and total elongation, respectively [17]. Based on these findings, it is expected that the non-uniform cooling of the hot-rolled coil should not be a concern in terms of heterogeneity in microstructure evolution and mechanical properties at different sections. However, further efforts are required to demonstrate its repeatability and acceptability by investigating multiple hot-rolled coils, which are produced industrially or at least at a pilot-scale, which can mimic the actual industrial conditions.

4 Future Prospects and Outlook

As mentioned in the preceding sections, the quench and nonisothermal partitioning process has shown its potential to develop steel with improved mechanical properties. The possible applications, that have been explored till date, are automotive and wear-resistant application. However, further efforts are required or ongoing to explore the potential of these steels for other structural applications. In terms of processing and subsequent microstructure evolution, there is enormous scope to further enhance the mechanical properties of these steel by microstructure engineering. This is due to the fact that the mechanical properties are a great function of amount and the characteristics (size, distribution, morphology, composition, etc.) of microstructural constituents. Among all the phases in the microstructure (i.e. martensite, retained austenite and carbides), the retained austenite amount and its composition has significant effect on the mechanical properties. This is due to the initiation and the progress of transformation-induced plasticity effect during tensile straining or in the actual application. For example, a very low stability of retained austenite will trigger the transformation-induced plasticity effect at very low strain or stress levels, and therefore, its beneficial effects on mechanical properties would not be utilized, and vice versa is also true. Therefore, it is important to consider not only the amount but also the stability of retained austenite during the alloy design and heat treatment.

5 Conclusions

The paper provided a broad overview of the benefits and limitations of different grades of steel, which have been considered in the recent past for several applications. In addition, various processes that have been applied to hot-rolled steel have been discussed in terms of process outline, industrial feasibility, microstructure evolution and mechanical properties. Some major conclusions drawn from the present review are as follows:

1. The limitation of lower strength in ferrite-based conventional grades of steel could be overcome by martensite-dominated first generation steels. However, a concurrent decrease in the ductility in these steels led to the evolution of the second generation steels with a complete austenitic structure, which showed excellent strength–ductility combinations. However, these steels suffer from the issues in steelmaking, processing, welding and fabrication due to higher alloying additions and therefore paved the way for future research to develop third generation steels with reduced alloying addition and search for novel processing routes.
2. The quenching and partitioning (Q&P) process on low alloy steels is one such approach for the stabilization of retained austenite in a martensitic matrix, leading to improved strength–ductility combinations. The isothermal Q&P process has shown to improve the strength–ductility–toughness combination, in comparison with the direct quenched and/or tempered steels. However, the need for additional furnace for isothermal holding interrupts the continuous operation of the hot rolling mill and requires additional energy, which increases the cost of steel.
3. A recently developed energy-efficient nonisothermal Q&P approach, which utilizes the remaining heat of the hot-rolled coil for carbon partitioning, has shown to achieve strength–ductility–toughness combinations comparable to the isothermal Q&P and superior to the direct quenched and/or tempered steels. The elimination of the additional heat treatment facility in this process increases the production capability, and therefore, provides the steel at a lower cost.

References

1. National Steel Policy (NSP), 2017. https://steel.gov.in/national-steel-policy-nsp-2017
2. AHSS Application Guidelines 6.0. https://www.worldautosteel.org/downloads/599700/
3. Cao W, Shi J, Wang C, Wang C, Xu L, Wang M, Weng Y, Dong H (2011) The 3rd generation automobile sheet steels presenting with ultrahigh strength and high ductility. In: Advanced steels. Springer, Berlin, Heidelberg, pp 209–227. https://doi.org/10.1007/978-3-642-17665-4_22
4. Caballero FG, Bhadeshia HKDH (2004) Very strong bainite. Curr Opin Solid State Mater Sci 8:251–257. https://doi.org/10.1016/j.cossms.2004.09.005

5. Bhadeshia HKDH (2010) Nanostructured bainite. Proc R Soc A 466:3–18. https://doi.org/10.1098/rspa.2009.0407
6. Bansal GK, Srivastava VC, Chowdhury SG (2019) Role of solute Nb in altering phase transformations during continuous cooling of a low-carbon steel. Mater Sci Eng A 767 (138416). https://doi.org/10.1016/j.msea.2019.138416
7. Bansal GK, Madhukar DA, Chandan AK, Ashok K, Mandal GK, Srivastava VC (2018) On the intercritical annealing parameters and ensuing mechanical properties of low-carbon medium-Mn steel. Mater Sci Eng A 733:246–256. https://doi.org/10.1016/j.msea.2018.07.055
8. Chandan AK, Bansal GK, Kundu J, Chakraborty J, Chowdhury SG (2019) Effect of prior austenite grain size on the evolution of microstructure and mechanical properties of an intercritically annealed medium manganese steel. Mater Sci Eng A 768(138458). https://doi.org/10.1016/j.msea.2019.138458
9. Mishra G, Chandan AK, Kundu S (2017) Hot rolled and cold rolled medium manganese steel: mechanical properties and microstructure. Mater Sci Eng A 701:319–327. https://doi.org/10.1016/j.msea.2017.06.088
10. Chandan AK, Mishra G, Mahato B, Chowdhury SG, Kundu S, Chakraborty J (2019) Stacking fault energy of austenite phase in medium manganese steel. Metall Mater Trans A 50:4851–4866. https://doi.org/10.1007/s11661-019-05367-x
11. Speer JG, Streicher AM, Matlock DK, Rizzo F, Krauss G (2003) Quenching and partitioning: a fundamentally new process to create high strength trip sheet microstructures. In: Austenite formation and decomposition, pp 505–522. ISBN: 087339559X
12. Thomas GA (2009) Simulation of hot-rolled advanced high strength sheet steel production using a Gleeble system. MS Thesis, Colorado School of Mines, Golden, Co
13. Thomas GA, Speer JG, Matlock DK (2011) Quenched and partitioned microstructures produced via gleeble simulations of hot-strip mill cooling practices. Metall Mater Trans A 42:3652–3659. https://doi.org/10.1007/s11661-011-0648-5
14. Tan X, Xu Y, Yang X, Liu Z, Wu D (2014) Effect of partitioning procedure on microstructure and mechanical properties of a hot-rolled directly quenched and partitioned steel. Mater Sci Eng A 594:149–160. https://doi.org/10.1016/j.msea.2013.11.064
15. Tan XD, Xu YB, Yang XL, Hu ZP, Peng F, Ju XW, Wu D (2015) Austenite stabilization and high strength-elongation product of a low silicon aluminum-free hot-rolled directly quenched and dynamically partitioned steel. Mater Charact 104:23–30. https://doi.org/10.1016/j.matchar.2015.03.022
16. Bansal GK, Rajinikanth V, Ghosh C, Srivastava VC, Kundu S, Chowdhury SG (2018) Microstructure–property correlation in low-Si steel processed through quenching and nonisothermal partitioning. Metall Mater Trans A 49:3501–3514. https://doi.org/10.1007/s11661-018-4677-1
17. Li YJ, Chen D, Kang J, Li XL, Yuan G, Misra RDK, Wang GD (2019) Kinetics of carbon partitioning during non-isothermal treatment and mechanical properties of the first industrial coils of hot rolled quenched and partitioned steel. Steel Res Int 90:1–11. https://doi.org/10.1002/srin.201900122
18. Li Y, Li X, Yuan G, Kang J, Chen D, Wang G (2016) Microstructure and partitioning behavior characteristics in low carbon steels treated by hot-rolling direct quenching and dynamical partitioning processes. Mater Charact 121:157–165. https://doi.org/10.1016/j.matchar.2016.10.005
19. Li Y, Chen D, Li X, Kang J, Yuan G, Misra RDK, Wang G (2018) Microstructural evolution and dynamic partitioning behavior in quenched and partitioned steels. Steel Res Int 89:1–11. https://doi.org/10.1002/srin.201700326
20. Li YJ, Kang J, Zhang WN, Liu D, Wang XH, Yuan G, Misra RDK, Wang GD (2018) A novel phase transition behavior during dynamic partitioning and analysis of retained austenite in quenched and partitioned steels. Mater Sci Eng A 710:181–191. https://doi.org/10.1016/j.msea.2017.10.104

21. Jian K, Chao W, Yunjie L, Guo Y, Guodong W (2016) Effect of direct quenching and partitioning treatment on mechanical properties of a hot rolled strip steel. J Wuhan Univ Technol Mat Sci Edit 31:178–185. https://doi.org/10.1007/s11595-016-1349-0
22. Li YJ, Chen D, Liu D, Kang J, Yuan G, Mao QJ, Misra RDK, Wang GD (2018) Combined thermo-mechanical controlled processing and dynamic carbon partitioning of low carbon Si/Al–Mn steels. Mater Sci Eng A 732:298–310. https://doi.org/10.1016/j.msea.2018.07.021
23. Kang J, Li YJ, Wang XH, Wang HS, Yuan G, Misra RDK, Wang GD (2019) Design of a low density Fe–Mn–Al–C steel with high strength-high ductility combination involving TRIP effect and dynamic carbon partitioning. Mater Sci Eng A 742:464–477. https://doi.org/10.1016/j.msea.2018.11.044
24. Li YJ, Liu D, Chen D, Kang J, Wang XH, Yuan G, Misra RDK, Wang GD (2019) Response of retained austenite to quenching temperature in a novel low density Fe–Mn–Al–C steel processed by hot rolling-air cooling followed by non-isothermal partitioning. Mater Sci Eng A 753:197–207. https://doi.org/10.1016/j.msea.2019.03.043
25. Sun T, Li J, Song H (2018) Microstructure and mechanical properties of a hot pressing-dynamic partition treated steel. Mater Sci Technol 34:1070–1078. https://doi.org/10.1080/02670836.2017.1421021
26. Kantanen P, Somani M, Kaijalainen A, Haiko O, Porter D, Kömi J (2019) Microstructural characterization and mechanical properties of direct quenched and partitioned high-aluminum and high-silicon steels. Metals 9(256). https://doi.org/10.3390/met9020256
27. Bansal GK, Junior LP, Ghosh C, Rajinikanth V, Tripathy S, Srivastava VC, Bhagat AN, Chowdhury SG (2020) Quench temperature-dependent phase transformations during nonisothermal partitioning. Metall Mater Trans A 51:3410–3424. https://doi.org/10.1007/s11661-020-05779-0
28. Bansal GK, Rajinikanth V, Ghosh C, Srivastava VC, Dutta M, Chowdhury SG (2020) Effect of cooling rate on the evolution of microstructure and mechanical properties of nonisothermally partitioned steels. Mater Sci Eng A 788(139614). https://doi.org/10.1016/j.msea.2020.139614
29. Wolfram PC (2013) The microstructural dependence of wear resistance in austenite containing plate steels. MS Thesis, Colorado School of Mines, Golden, Co
30. Abian MK, Hanzaki AZ, Abedi HR, Manesh SH (2016) Micro and macro-mechanical behavior of a transformation-induced plasticity steel developed by thermomechanical processing followed by quenching and partitioning. Mater Sci Eng A 651:233–240. https://doi.org/10.1016/j.msea.2015.10.116
31. Bagliani EP, Santofimia MJ, Zhao L, Sietsma J, Anelli E (2013) Microstructure, tensile and toughness properties after quenching and partitioning treatments of a medium-carbon steel. Mater Sci Eng A 559:486–495. https://doi.org/10.1016/j.msea.2012.08.130
32. Yan S, Liu X, Liu WJ, Lan H, Wu H (2015) Comparison on mechanical properties and microstructure of a C–Mn–Si steel treated by quenching and partitioning (Q&P) and quenching and tempering (Q&T) processes. Mater Sci Eng A 620:58–66. https://doi.org/10.1016/j.msea.2014.09.047
33. Bansal GK, Pradeep M, Ghosh C, Rajinikanth V, Srivastava VC, Bhagat AN, Kundu S (2019) Evolution of microstructure in a low-Si micro-alloyed steel processed through one-step quenching and partitioning. Metall Mater Trans A 50:547–555. https://doi.org/10.1007/s11661-018-5039-8

Ductility Improvement in Commercially Pure Aluminium by Friction Stir Processing

Abhishek Kumar Jha, Md. Mofeed Alam, Shitanshu S. Chakraborty, Kashif Hasan Kazmi, Prakash Kumar, and Sumanta Mukherjee

1 Introduction

Friction stir processing (FSP) is a versatile technique which can be employed for altering the microstructural and metallurgical properties of native material. It is a solid-state processing technique, where the heating of the workpiece material takes place as a result of the sliding friction between two surfaces. Although the generated heat softens the material and enables severe plastic deformation, the temperature remains restricted below the melting point of the workpiece. This technique was mainly developed for processing of aluminium and its alloys, but the advancement in technology allowed for further expansion of the material palette to process magnesium (Mg), titanium (Ti), copper (Cu) and even different types of steels to improve their mechanical as well as metallurgical properties [1].

Friction stir processing technique uses a non-erodible cylindrical tool consisting of a small probe. When the rotating tool plunges down into the native material, the probe enters the native material and heat is generated due to the sliding friction between the probe and workpiece. Very large amount of localized heat is produced at the rotating tool and workpiece interface, which softens the native material. Thus, plastic deformation takes place under severe shear strain accompanied by heat accumulation. A traversing motion is also provided to the tool. The side, in which direction of traverse speed is along the direction of tool rotation, is known as advancing side (AS), while the other side, in which the direction of tool rotation is opposite to the direction of traverse speed, is known as retreating side (RS). As apart from providing frictional heat, the tool shoulder also provides an additional downward plunge force which helps in proper forging of material. Sufficient

A. K. Jha · Md. Mofeed Alam · K. H. Kazmi · P. Kumar · S. Mukherjee (✉)
Production Engineering Department, BIT Sindri, Dhanbad, India

S. S. Chakraborty
Materials Processing and Microsystems Laboratory, CSIR-Central Mechanical Engineering Research Institute, Durgapur, India

S. Bag et al. (eds.), *Next Generation Materials and Processing Technologies*, Springer Proceedings in Materials 9, https://doi.org/10.1007/978-981-16-0182-8_3

amount of heat is required to obtain a defectless processed zone [2]. The temperature produced at the interface zone largely depends upon the ratio of rotational speed to traverse speed [3]. Total heat generation in the processed zone depends upon rotational speed and traverse speed of tool [4]. The schematic diagram of friction stir processing technique is shown in Fig. 1.

Four different microstructural zones are produced in the processed cross section that are stir zone (SZ), thermo-mechanically affected zone (TMAZ), heat affected zone (HAZ) and base material (BM) as shown in Fig. 1. The flow of material takes place from the retreating side to the advancing side along the processed zone [4–7].

The microstructural changes induced by FSP is not uniform across the different regions as mentioned above. The HAZ region undergoes no plastic deformation of the grains, and the thermal exposure may cause grain coarsening and softening of the material. On the other hand, the SZ consists of highly refined and almost uniform grains resulting from dynamic recrystallization due to severe plastic deformation at elevated temperature. However, the thermo-mechanical energy input at the TMAZ is not sufficient to induce complete dynamic recrystallization, and as a result, both recrystallized refined grains and grains elongated following the stirring direction are observed. Further, the unequal distribution of strain energy in the AS and RS regions leads to heterogeneous distribution of grain size in those zones [8]. Therefore, it is crucial to control the FSP parameters very specifically so that the resulting microstructures can effectively bring about the desired changes in the mechanical properties of the material.

Commercially pure aluminium (AA 1100) has specific strength of about 4.26×10^4 $(m/s)^2$, which is significantly higher than that of stainless steel AISI 304 $(\sim 3.8 \times 10^4$ $(m/s)^2)$. But still, it has less commercial application because of lower

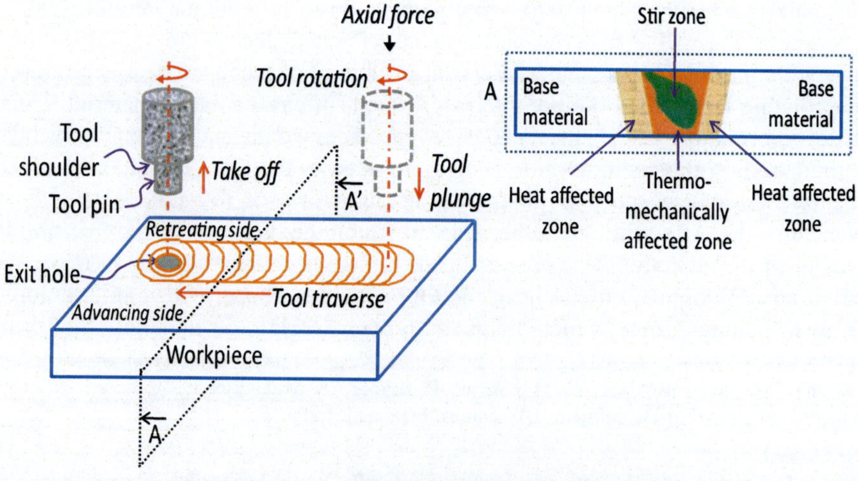

Fig. 1 Schematic diagram of friction stir processing (FSP) technique and four microstructural zones in transverse cross section of the material

ductility (~10%) and wear resistance. In this work, we have performed single pass friction stir processing (FSP) technique on commercially pure aluminium to improve its % elongation (indicating Ductility) and other mechanical properties and also study the effects of various process parameters on the process at room temperature.

2 Materials and Methods

Experiments were performed using a rigid HMT make milling machine transformed for FSW. Commercially, pure aluminium was used as workpiece whose composition and mechanical properties as provided by the manufacturer are shown in Table 1.

A H13 steel tool having 18 mm shoulder diameter, a tapered cylindrical probe with 5.5 mm diameter at the larger end and 4.5 mm at the smaller end, and 1.7 mm probe length was used for the processing. Depth of shoulder penetration into the native material was 0.2 mm, and the tool tilt angle was kept 0°. The schematic diagram of tool is presented in Fig. 2a.

Single-pass FSP was performed on commercially pure aluminium sheet of dimension 250 mm × 100 mm × 2 mm at nine different parameter combinations after conducting pilot experiments to obtain the suitable processing window. In the first step, the range of the FSP parameters can produce surfaces with minimum burr and other visible defects, and the highest and lowest values of those parameters were considered. Then, an intermediate value in the range was selected for both the parameters. The rotational speed (*N*) of the tool was varied from 560 to 1400 RPM, and the traverse speed (*V*) was varied from 100 to 200 mm/min (Table 2). The maximum and minimum ratios of rotational speed to traverse speed were 14 and 2.8, respectively. The experiments were performed in triplicate. Representative images of the processed samples are shown in Fig. 2b–d.

Tensile test specimens were prepared as per ASTM E8 standard, and the tests were carried out using a universal tensile testing machine (model H50KS, make: Tiniu Olsen). Microhardness values at 10 points across the friction stir processed tracks were measured using a Vickers hardness tester (model: FALCON 500, make: INNOVA TEST). Carl Zeiss Axio Imager M2M optical microscope was used to observe the microstructure of the sample after polishing and etching.

Table 1 Mechanical properties of the AA 1100 aluminium sheet used in the experiment

Workpiece material (Dimensions in mm)	Ultimate tensile strength (MPa)	Elongation at break (%)	Microhardness (HV 1/10)
Aluminium AA 1100 (250 × 100 × 2)	115.3 ± 1	11.8 ± 0.4	41 ± 2

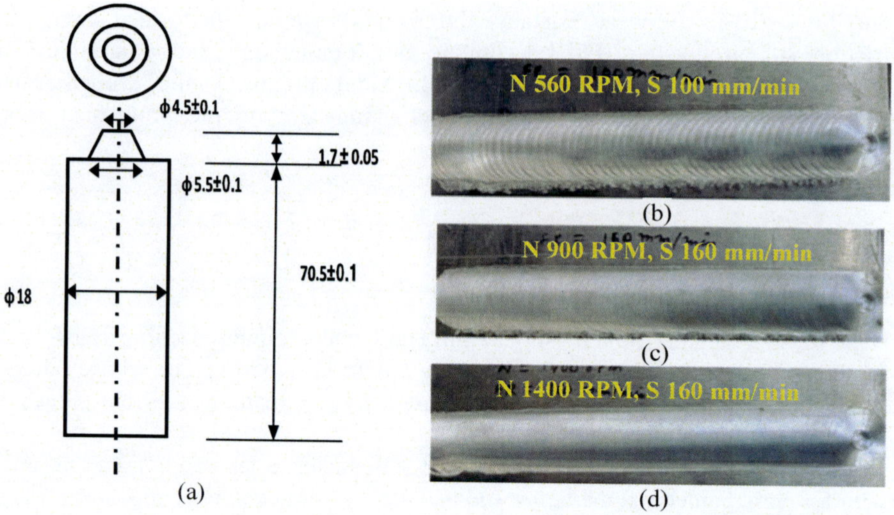

Fig. 2 **a** Schematic diagram of the tool (dimensions are in mm), and **b–d** FSP-ed samples

Table 2 Different tool rotational speed (*N*) and tool traverse speed (*V*) combinations explored in this study

Parameter combination no.	1	2	3	4	5	6	7	8	9
N (RPM)	560	560	560	900	900	900	1400	1400	1400
V (mm/min)	100	160	200	100	160	200	100	160	200

3 Results and Discussion

3.1 Mechanical Properties

Tensile properties. The performance of the samples under tensile condition was analysed from the ultimate tensile strength (UTS), 0.2% proof stress, elongation at break and work hardening parameters (Table 3). A power law relationship was assumed between the stress and the strain values as per the Ramberg–Osgood equation (Eq. 1)—

$$\varepsilon = \frac{\sigma}{E} + K\left(\frac{\sigma}{E}\right)^n \tag{1}$$

where ε, σ, E, K and n refer to the strain, stress, Young's Modulus, strength coefficient and strain hardening exponent, respectively [6–9]. In the equation, the $\frac{\sigma}{E}$ part corresponds to the linear elastic region of the stress-strain curve, and the $K\left(\frac{\sigma}{E}\right)^n$

Table 3 Variation of tensile properties of the FSP-ed samples for different process parameter combinations

Parameter combination no.	UTS (MPa)	Elongation at break (%)	Strain hardening exponent	Strain at peak stress (%)	0.2% proof stress (MPa)
1	78.9 ± 1.2	17.3 ± 0.3	0.105 ± 6e−3	9.38 ± 0.8	68.6 ± 1.1
2	137.5 ± 7.4	21.7 ± 1.2	0.110 ± 8e−3	11.01 ± 1.3	103.7 ± 6.1
3	145.5 ± 5	18.3 ± 6.4	0.116 ± 13e−3	11.76 ± 1.3	109.9 ± 4.6
4	132 ± 10.4	15.7 ± 0.3	0.097 ± 7e−3	7.76 ± 0.9	78.3 ± 9.3
5	137.9 ± 5.7	16.2 ± 4.7	0.103 ± 12e−3	8.24 ± 0.9	95.7 ± 5.2
6	138.2 ± 1.4	17.4 ± 0.4	0.110 ± 8e−3	8.81 ± 1.0	96.7 ± 1.7
7	132.6 ± 7.3	17.3 ± 1.8	0.122 ± 7e−3	9.34 ± 0.9	80.2 ± 6.5
8	129.8 ± 1.3	21.4 ± 4.6	0.131 ± 10e−3	9.49 ± 1.1	83.3 ± 1.3
9	130.6 ± 1.3	18.7 ± 4.1	0.138 ± 7e−3	9.76 ± 1.0	84.9 ± 1.1

represent the plastic deformation behaviour. The fitted values of the strain hardening exponents were considered as the indicator of the formability of the samples [9–11]. Under uniaxial loading, the ultimate stress is reached when $\varepsilon = n$ in the idealized form. Beyond this strain at maximum stress necking initiates. Therefore, higher value of this parameter refers to material properties allowing a higher degree of formability. The variations of the UTS and the elongation at break of the processed samples under different processing conditions are given in Table 3 and are graphically presented in Fig. 3.

It can be observed from the table and the figures (Fig. 3) that the ductility of the processed samples was generally improved from the base material, and increasing tool traverse speed generally augmented those properties. However, no concrete trends of variation of properties were noticeable with varying rotational speed.

The extent of plastic deformation before initiation of necking was relatively higher for increased tool traverse speeds, as noted from the strain at maximum strain (Fig. 4a) as well as the strain hardening exponent (Fig. 4b).

From the workability point of view, the 0.2% proof stress of the processed samples gradually reduced for higher tool rotational speeds, denoting higher workability. However, the workability deteriorated for increasing tool traverse speed (Fig. 4c).

Higher amount heat is accumulated at the FSP-ed zone when the tool traverse speed is low, and therefore, the temperature at the FSP-ed zone reduces with increasing tool traverse speed. The strain hardened nature of the processed material remains unaltered at lower temperature, but the elevated temperature at lower tool traversed speed may adversely affect the strain hardened properties. Therefore, the higher strain hardening at higher tool traverse speed may induce some brittleness to the processed material, reducing the elongation at break.

Microhardness of the processed zone. No discernible softening of the material was noticed post-FSP, and nominal increase in microhardness was observed with

Fig. 3 Effect of processing parameters on **a** UTS and **b** % elongation

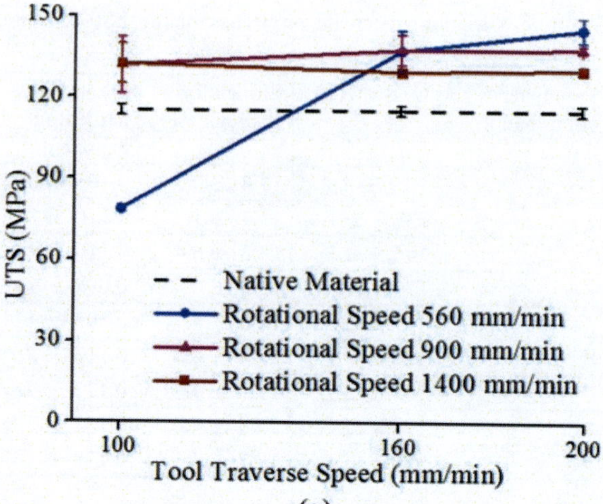

(a)

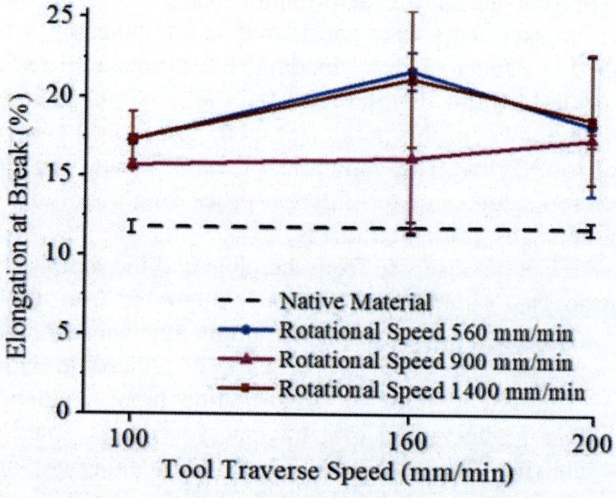

(b)

increasing tool traverse speed (Table 4; Fig. 5). Again, the tool rotation speed did not influence the microhardness of the processed samples with a specific trend.

Although the average of microhardness measurements taken over the entire width of the processed zone did not indicate any statistically significant variation with varying traverse speed, the distribution of the microhardness values across the width of the processed track did show that increased traverse speed resulted in higher microhardness at the central zone of the FSP tracks (Fig. 6). It can be assumed that the higher heat accumulation at lower tool traverse speed negatively affects the strain hardened behaviour of the processed material. Therefore, the low

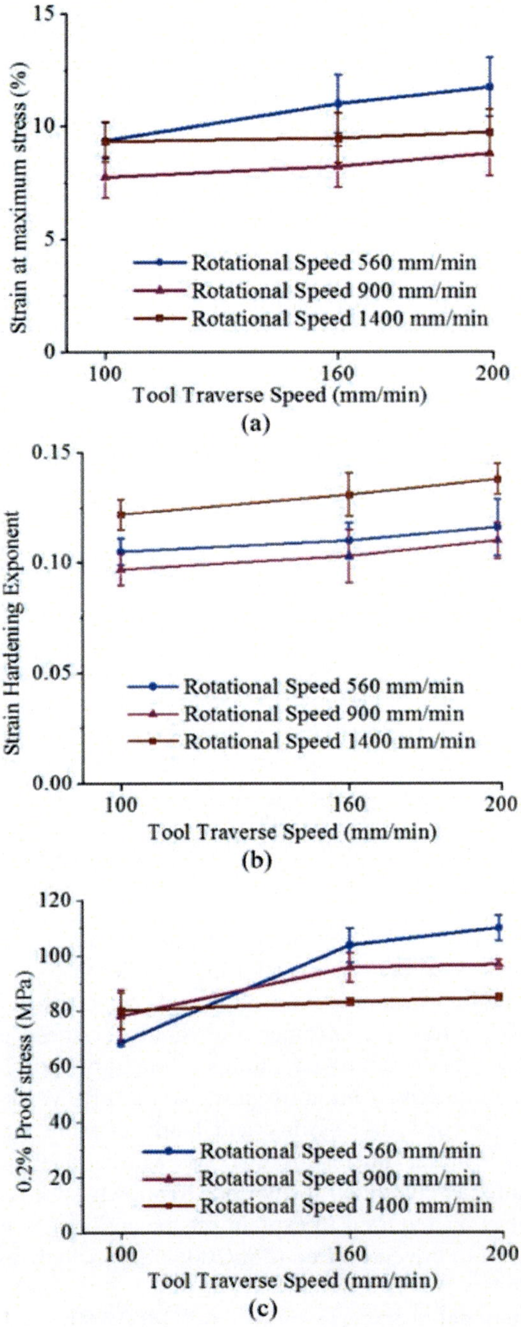

Fig. 4 Effect of processing parameters on **a** strain at maximum stress, **b** strain hardening exponent, and **c** 0.2% proof stress of the processed samples

Table 4 Variation of microhardness of the FSP-ed samples for different process parameter combinations

Parameter combination no.	Microhardness (HV)
1	39.44 ± 0.84
2	39.62 ± 0.83
3	41.92 ± 1.13
4	43.36 ± 0.47
5	41.45 ± 0.91
6	40.02 ± 0.97
7	39.82 ± 1.13
8	43.98 ± 1.3
9	43.19 ± 0.42

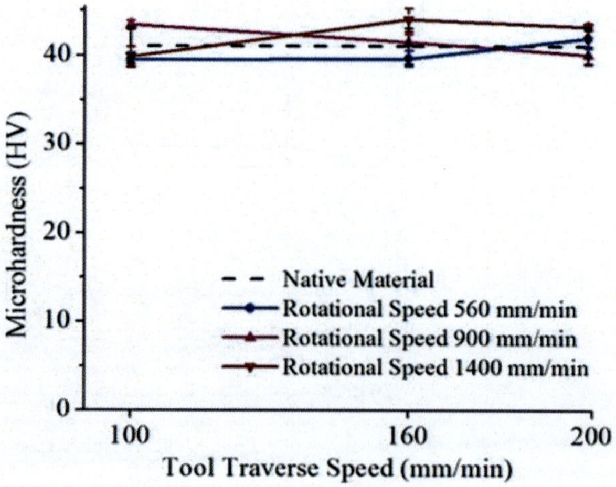

Fig. 5 Effect of processing parameters on microhardness of the processed samples

temperature rise at higher tool traverse speed helps to maintain the strain hardened properties of the processed materials, and a higher microhardness is observed.

The aim of having improved ductility should not cost sacrifice of ultimate tensile strength and microhardness. At all other tool rotational speed except 900 RPM, higher tool traverse speed improved strength, ductility and hardness. This is beneficial from the productivity point of view. At 1400 RPM rotational speed, the quality responses viz. strength, ductility and hardness were mostly found to be better than the base material. Interestingly 900 RPM rotational speed was found to deteriorate the results as compared to the tool rotational speeds of 560 and 1400 RPM. Further analysis is required to explain this behaviour. At lower value of the ratio of tool rotational to traverse speed (560/200 = 2.8), which resulted in less heat input, higher strength was noticed. Better ductility and microhardness were found at higher value of rotational to traverse speed ratio (1400/160 = 8.75). This indicates simultaneous possibility of grain growth and more oxide formation at the surface at 1400 RPM.

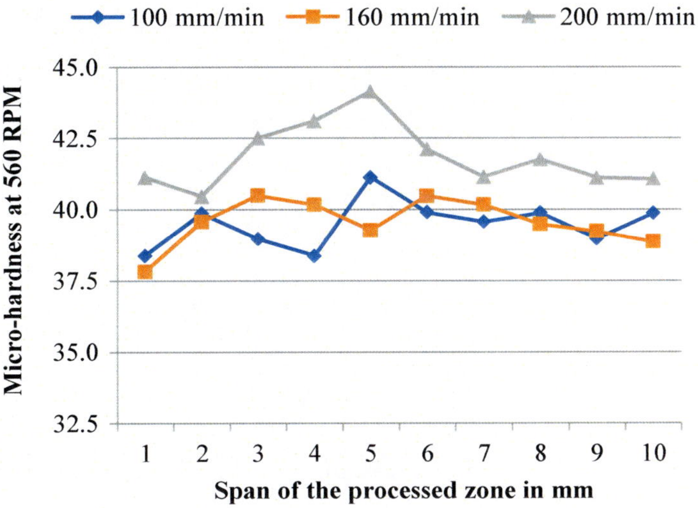

Fig. 6 Variation of microhardness across the width of the FSP-ed track

3.2 *Microstructure*

A significant number of pores were present in the native Al sample (Fig. 7a), and the friction stir processing helped to reduce the porosity of the material (Fig. 7b), as observed from the micrographs. The grains were also smaller for the processed sample, as the dynamic recrystallization under intense plastic deformation leads to grain refinement during FSP [12, 13], and as a result, the ductility improves.

This microstructural study ascertains the capability of FSP to simultaneously improve the conflicting quality responses: hardness and ductility. This is possible due to elimination of porosity.

3.3 *Grey Relational Analysis and Regression Analysis*

To help choosing the parameter combination giving better strength, ductility and microhardness, simultaneously, grey relation analysis was performed as per the standard procedure given in [14] and other literature [15, 16]. As all three responses were higher the better type, they were normalized using Eq. (2).

$$x_i^*(k) = \frac{x_i(k) - \min x_i(k)}{\max x_i(k) - \min x_i(k)} \tag{2}$$

Fig. 7 Porosity and grain distribution of **a** native AA 1100 and **b** FSP-ed sample (N = 560 RPM, V = 100 mm/min)

(a)

(b)

where i = 1, 2, …, m and k = 1, 2, …, n; m = Number of experimental data (here, 9) and n = number of responses (here, 3). $x_i(k)$ denotes the original data, $x_i^*(k)$ is the normalized data, max $x_i(k)$ is the maximum value, min $x_i(k)$ is the minimum value. Grey relational coefficient $\xi_i(k)$ is calculated using Eq. (3), from the above-calculated normalized value.

$$\xi_i(k) = \frac{\Delta_{\min} + \psi\Delta_{\max}}{\Delta_{oi}(k) + \psi\Delta_{\max}} \tag{3}$$

where Δ_{oi} is deviation sequence of reference sequence and compatibility sequence and is given by $x_0(k) - x_i(k)$. The value of ψ is taken as 0.5. Finally, giving equal weight of 1, grey relational grade (GRG) is calculated using Eq. (4).

$$\gamma_i = \frac{1}{n}\sum_{k=1}^{n} \xi_i(k) \tag{4}$$

Table 5 elaborates the grey relational analysis procedure enlisting the values obtained using Eqs. (2)–(4) for each of the 9 different parameter combinations in sequence as listed in Table 1.

As evident from Table 5, parameter combination no. 8, i.e., 1400 RPM tool rotational speed and 160 mm/min traverse speed gives the best result. Most significant achievement with this parameter combination is the 83.9% improvement in ductility over the base material. It is noticeable that ductility is the only response variable which improved for all the parameter combinations tried. The trend of variation of this response is nonlinear as evident from Fig. 3b. Hence, a regression model was developed fitting a quadratic equation relating % elongation at break, En, to the tool rotational speed (*N*) and traverse speed (*V*) as given by Eq. (5) [9]. Table 6 presents the values of the coefficients along with the R^2 value which indicates the degree of fitting of the model. The response surface obtained from the model is shown in Fig. 8.

$$\mathrm{En} = A_0 + A_1N^2 + A_2NV + A_3V^2 + A_4N + A_5V \tag{5}$$

Table 5 Normalized values, grey relational coefficient and rank based on grey relational grade

Parameter combination no.	Normalized values			Grey relational coefficients			Rank
	Microhardness	% Elongation	UTS	Microhardness	% Elongation	UTS	
1	0.000	0.000	0.267	0.333	0.333	0.405	9
2	0.040	0.881	1.000	0.342	0.808	1.000	2
3	0.546	1.000	0.433	0.524	1.000	0.469	3
4	0.863	0.799	0.000	0.785	0.714	0.333	5
5	0.443	0.887	0.083	0.473	0.816	0.353	6
6	0.128	0.892	0.283	0.364	0.822	0.411	7
7	0.084	0.808	0.267	0.353	0.723	0.405	8
8	1.000	0.767	0.950	1.000	0.682	0.909	1
9	0.826	0.779	0.500	0.742	0.693	0.500	4

Table 6 Values of the regression coefficients and R^2

A_0	A_1	A_2	A_3	A_4	A_5	R^2
11.022	1.576e-05	3.918e-06	−9.083e-4	−3.146e-2	0.282	0.793

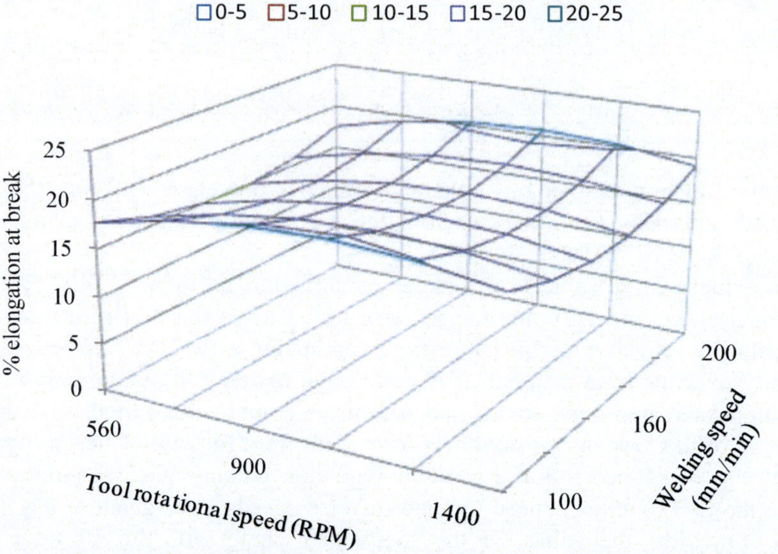

Fig. 8 Response surface obtained using the regression model

4 Conclusion

Friction stir processing of commercially pure aluminium alloy (AA 1100) has been explored varying easily controllable parameters, tool rotational speed and tool traverse speed. Using grey relational analysis, 1400 RPM and 160 mm/min were found to be the optimum parameter giving a blend of high ductility (as indicated by % elongation at break) and good strength and microhardness. Compared to the base material, ductility improved for all the combinations of parameters tried. At 1400 RPM tool rotational speed and 160 mm/min tool traverse speed, % elongation of the processed sample was found to be 83.9% higher than the base material. A quadratic regression model relating % elongation to the tool rotational speed and traverse speed could be developed with R^2 value of 0.793. The workability and formability of the processed materials were also improved for the friction stir processed samples, as observed from the 0.2% proof stress and strain hardening behaviour of the samples. Microstructural analysis revealed the capability of FSP to simultaneously improve hardness and ductility by reducing porosity. Further investigation into this may prove beneficial in widening the application domain of AA 1100 aluminium.

Acknowledgements Funding from a project (File no. SP/YO/2019/1123) sanctioned by the Department of Science and Technology (DST), Government of India, under Scheme for Young Scientists and Technologists (SYST), is gratefully acknowledged by the authors.

References

1. Ma ZY (2008) Friction stir processing technology: a review. Metall Mater Trans A 39:642–658. https://doi.org/10.1007/s11661-007-9459-0
2. Mishra RS, Ma ZY (2005) Friction stir welding and processing. Mater Sci Eng R Rep 50(1–2):1–78. https://doi.org/10.1016/j.mser.2005.07.001
3. Yadav D, Bauri R (2012) Effect of friction stir processing on microstructure and mechanical properties of aluminium. Mater Sci Eng A 539:85–92. https://doi.org/10.1016/j.msea.2012.01.055
4. Khayyamin D, Mostafapour A, Keshmiri R (2013) The effect of process parameters on microstructural characteristics of AZ91/SiO_2 composite fabricated by FSP. Mater Sci Eng A 559:217–221. https://doi.org/10.1016/j.msea.2012.08.084
5. Węglowski MS (2018) Friction stir processing—state of the art. Arch Civil Mech Eng 18:114–129. https://doi.org/10.1016/j.acme.2017.06.002
6. Ramberg W, Osgood W (1943) Description of stress–strain curves by three parameters. Technical notes, National Advisory Committee for Aeronautics: NACA. https://ntrs.nasa.gov/citations/19930081614
7. Nejad RM, Tohidi M, Darbandi AJ, Saber A, Shariati M (2020) Experimental and numerical investigation of fatigue crack growth behavior and optimizing fatigue life of riveted joints in Al-alloy 2024 plates. Theor Appl Fract Mech 108:102669. https://doi.org/10.1016/j.tafmec.2020.102669
8. Ali A, Brown MW, Rodopoulos CA (2008) Modelling of crack coalescence in 2024-T351 Al alloy friction stir welded joints. Int J Fatigue 30(10–11):2030–2043. https://doi.org/10.1016/j.ijfatigue.2008.02.014
9. Kumar P, Singh A (2017) Investigation of mechanical properties and fracture simulation of solution-treated AA 5754. J Mater Eng Perform 26:4689–4706. https://doi.org/10.1007/s11665-017-2802-8
10. Bressan JD, Wang Q, Simonetto E, Ghiotti A, Bruschi S (2020) Formability prediction of Ti6Al4V titanium alloy sheet deformed at room temperature and 600 °C. Int J Mater Form. https://doi.org/10.1007/s12289-020-01546-z
11. Ma JL, Chan TM, Young B (2015) Material properties and residual stresses of cold-formed high strength steel hollow sections. J Constr Steel Res 109:152–165. https://doi.org/10.1016/j.jcsr.2015.02.006
12. Li Y, Qin F, Liu C, Wu Z (2017) A review: effect of friction stir welding on microstructure and mechanical properties of magnesium alloys. Metals 7(12):524. https://doi.org/10.3390/met7120524
13. Sakai T, Belyakov A, Kaibyshev R, Miuraa H, Jonas JJ (2014) Dynamic and post-dynamic recrystallization under hot, cold and severe plastic deformation conditions. Prog Mater Sci 60:130–207. https://doi.org/10.1016/j.pmatsci.2013.09.002
14. Julong D (1989) Introduction to grey system theory. J Grey Syst 1(1):1–24
15. Panda A, Sahoo A, Rout R (2016) Multi-attribute decision making parametric optimization and modeling in hard turning using ceramic insert through grey relational analysis: a case study. Decis Sci Lett 5(4):581–592. https://doi.org/10.5267/j.dsl.2016.3.001
16. Montgomery DC (2001) Design and analysis of experiments, 5th edn. Wiley, New York, pp 427–500

Development of Composites Using Bamboo Waste and Polymers for Industrial Application

Atosh Kumar Sinha, Rahul Kanti Nath, John Deb Barma, and Mitali Saha

1 Introduction

In recent days, the development of composite materials is becoming popular owing to their properties like easy to fabrication [1], chemical and corrosion resistance, high strength and stiffness, economically friendly nature [2], and cost-effective properties [3, 4]. Wood reinforced polymer composite (WPC) is one of them because of their lot of application in different sectors due to their significant property such as less tool wear, low material cost, low density, and high specific strength [3] compared with conventional and mineral filler reinforced composite [5]. WPC is employed in building structural interior decoration such as designed roof, flooring, decking, beams and columns, railings, fences, and in marine, aerospace, and automotive industry owing to it enables to reduces the mass of the component [6–11].

Environment-friendly polymer composites are trying to develop by different industries and researchers using natural lignocellulose as a reinforcing material. Thermoplastic and thermoset, both types of polymers, have used as matrix material with lignocellulose reinforcement material [2, 12–15]. The epoxy polymer is widely used as a matrix material to develop wood polymer composites. Epoxy, thermoset polymer, has a high cross-linked structure on curing, provides phenomenal growth in certain properties like shear strength, toughness, high modulus, low creep, economically cheap, reducing friction coefficient and stretching the service life of

A. K. Sinha · R. K. Nath · J. D. Barma
Department of Mechanical Engineering, NIT Agartala, Agartala, Tripura 799046, India

M. Saha
Department of Chemistry, NIT Agartala, Agartala, Tripura 799046, India

A. K. Sinha (✉)
Department of Metallurgical and Materials Engineering, NIT Durgapur, Durgapur, West Bengal 713209, India
e-mail: aks.19mm1103@phd.nitdgp.ac.in

S. Bag et al. (eds.), *Next Generation Materials and Processing Technologies*, Springer Proceedings in Materials 9, https://doi.org/10.1007/978-981-16-0182-8_4

the component. Bagasse, rice husk, jute, kenaf, hemp, sisal, flax, sugarcane, banana, ramie, etc., are various natural biomass substances employed as reinforcement in different research work to develop polymer composites with the help of epoxy resin [16–25]. The size of filler particles, i.e., reinforcement plays a decisive role in boosting up the physical properties and mechanical performance of the polymer composites. More than 100-micron filler particles have been carried out in most of the research work to prepare the composite [26–28]. 20–75-micron particle size has been employed a short time ago with polylactic acid to fabricate polymer composites and better mechanical and dispersive properties have been found [29]. Wood-based epoxy polymer matrix has been reported by different kinds of literature. Wood polymer composite prepared by pine wood particles and thermoset epoxy polymer has reported that dust particle maintains good filler characteristics and the wear resistance of the samples also improved [26]. Tensile and flexural strength shown an increasing trend with the addition of filler material of wood apple shell particle epoxy composite [30]. Samples contain 0, 2.5, 5, 7.5, 10, and 12.5% (wt%) filler material of wood-based flour epoxy composite has fabricated and found that the highest tensile strength obtains samples contain 7.5% filler material [31].

The main purpose of this investigation is to draw up a wood matrix composite with waste material bamboo dust. Wood polymer composite is prepared by employing three different bamboo dust particle sizes, i.e., 60 microns, 110 microns, and 220 microns are mixed in three different weight ratios with epoxy, i.e., 60:40, 50:50, and 40:60, respectively. The composite is fabricated under controlled temperature and pressure by applying compressed molding technology. After making the composite, the microstructural evaluation of the composite is executed by using SEM and polarization optical microscope. A tensile test is implemented to evaluate the diversity of strength of the prepared composite with different sizes and different mixing ratios. The hardness test is done to calculate the variation of hardness with different particle sizes of bamboo dust and the mixing ratio of epoxy resin.

2 Experimental Methodology

Micro-sized bamboo particle and epoxy resin are used to produce the composite samples. R-31 epoxy is used with an H-31 hardener in this research work. Bamboo dust is collected from a local industry bamboo industry. To screen three different sized bamboo dust, five distinct sizes of the sieve are used, i.e., 53, 63, 125, 180, and 250 μm. The clouds of bamboo dust are then put in sieves which are placed in descending order (from 250 to 53 μm) to separate different sizes of bamboo particles. Three different sizes of 60, 110, and 220 μm of bamboo dust are detected. A magnet is placed in between the sieves during handling the screening process of bamboo particles to spelled out any ferrous particles if it comes in the industry through the grinding process. Then, the bamboo dust is cleaned through distilled water several times, and finally, it is washed with acetone to remove impurities.

After that, the wet dust particles are dried for 4 h at 60 °C. The set of temperature and time of drying the bamboo dust are finalized after several trial experiments.

For preparing the composite, epoxy resin and bamboo dust size particle are used. 60 μm, 110 μm, and 220 μm are the three different sizes, and the bamboo dust has taken 40%, 50%, and 60% (wt%) as reinforcement material to mix with 60%, 50%, and 40% (wt%) epoxy (matrix), respectively. According to the instruction of the epoxy manufacturer, initially, the epoxy is mixed with a hardener in a 3:1 ratio, and it is stirred well for some moments to remove the bubble which has formed during the mixing process of epoxy and hardener. Then, the dried bamboo dust is added with the epoxy mixture, and it is blended at high speed for 10–12 min. Later, the blended material is kept on a die shown in Fig. 1 which is kept on a hydraulic press at 80 °C, and it is covered with the lid of the die which is shown in Fig. 2. 2 MPa pressure is applied on the blended material by the hydraulic press for 1 min, and then, the pressure is released. After that to solidify and cool down the compressed blended material, it is left 2 h to solidify and then the prepared composite is withdrawn from the die. By this way, samples are prepared in nine different states: 60 μm particle size is reinforced with epoxy in 60:40, 50:50, and 40:60 condition. Samples of 110 and 220 μm bamboo dust particles are also fabricated in a similar fashion. The parameters are listed in Table 1.

After fabricating the wood polymer composites, the specimens are tested at room temperature for their water absorption property. The test is accomplished by weighing the all type 220 μm composite specimens (3 cm length, 2 cm width, and 1 cm thickness) prior and later soaking in water for 24 h. They are allowed to stay in for 24 h, after which they were removed from the water and reweighed. Water absorption is expressed as the ratio of the difference in weight (after and before water absorption) to the original weight of the sample. The thickness swell test is executed by calibrating the thickness of the specimens before and after 24 h in water at room temperature. It is expressed as the ratio of the difference in thickness (after and before water soaking) to the original thickness of the sample.

Microstructure evaluation is done on the cross section of the specimen which is taken from each fabricated composite sample. Optical microscope (Leica polarized) and SEM (Carl Zeiss EVO 50 WSEM) have been used to analyze the surface of the

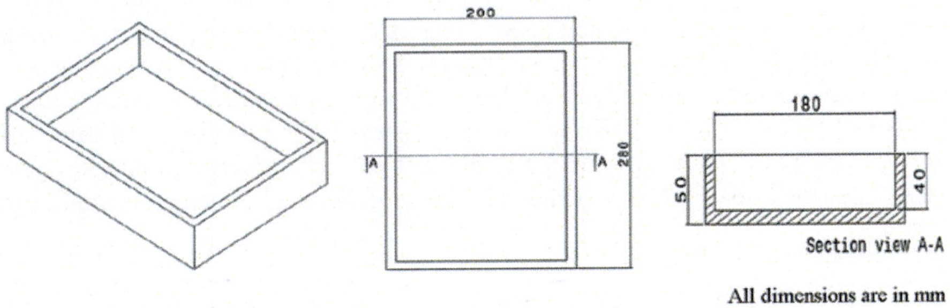

Fig. 1 Schematic diagram of the die

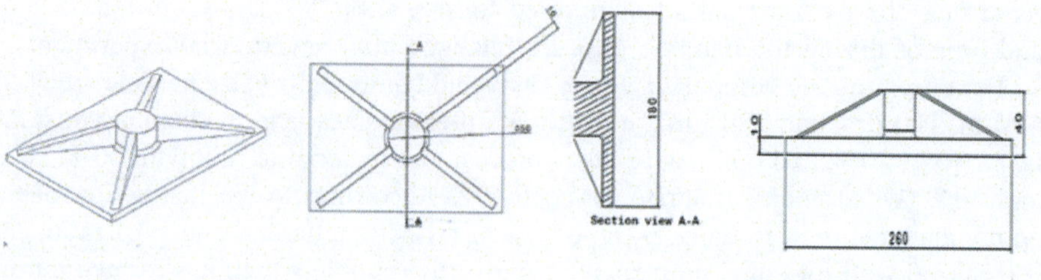

Fig. 2 Schematic diagram of the die's lid

Table 1 Parameters for the fabrication of the composite

Reinforcement	Bamboo dust (60, 110, 220 µm)
Matrix	R-31 epoxy
Hardener	H-31
Epoxy: Hardener	3:1
Reinforcement: Matrix mixture (wt. ratio)	60:40, 50:50, 40:60
The temperature during the fabrication process	80 °C
Pressure applied for compression	2 MPa
Solidification time after applied pressure	120 min

prepared composite. As the polymer is a non-conductive material, the gold coating is employed on the cross section of the specimens (1 m × 1 m) to convert the samples conductive so that electron can pass through the samples during SEM analysis. Additionally, EDS has also been performed to evaluate the composition of the prepared composite.

Ultimate tensile strength is executed on INSTRON universal testing machine (UTM) with a load cell capacity of 50 kN, and the entire specimens are tested at 1 mm/min crosshead. A tensile testing machine, a load cell, a power supply, and an *x–y* recorder have consisted of the tensile system. Tensile test specimens are prepared according to ASTM D-638 TYPE V (63.5 mm × 9.53 mm × 3 mm with gauge length 7.62 mm and distance between two grip 25.4 mm) standard which is shown in Fig. 3. The specimen is approximately uniform over a gauge length. The specimen is gripped at both ends by a fixture shown in Fig. 4 which is designed to elongate the specimen along with the lengthwise unit it fractures. Vickers hardness test is carried out in MATSUZAWA microhardness testing machine. To determine the Vickers hardness (HV), indenter is pressed into a fabricated composite specimen with a 50 g load for 10 s. The hardness is measured at the cross section of the samples.

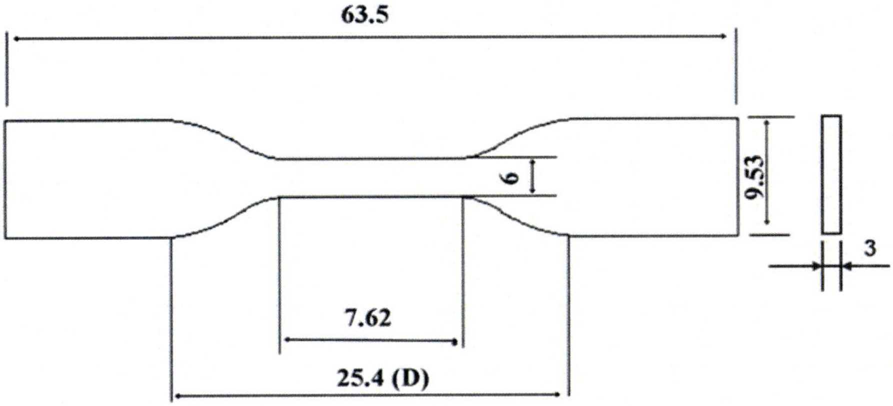

Fig. 3 Schematic diagram of tensile samples (All dimensions are in mm)

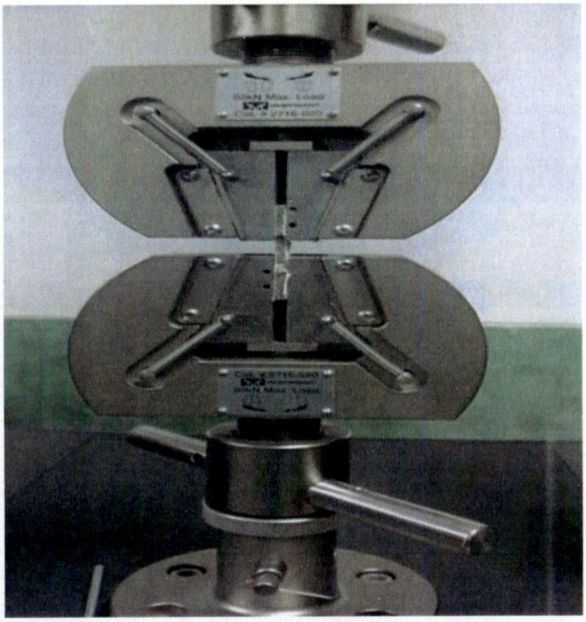

Fig. 4 Fixtures used to hold samples during the tensile test

3 Results and Discussion

3.1 Water Absorption and Thickness Swelling

Water absorptivity is the function of the degree of particleboard to humidity which can be termed as the ability to resist moisture. From the experiment of water absorption for 40%, 50%, and 60% epoxy composite sample, it is found that the specimens illustrate almost 58%, 45%, and 20% increase in water absorption, respectively. It reveals that with increasing the percentage of the epoxy resin in the composite specimens, cohesive, and binding forces of the wood polymer composite becoming stronger and more suitable to water.

Thickness swelling refers to the dimensional stability of a component in a moisture environment, a function of the composition, and the prevailing temperature. The specimens show a thickness surge of 21, 18, and 10 only with 40, 50, and 60% epoxy resin composite samples. It may be due to raising the epoxy percentage, and composites show stronger and relatively higher stable cohesive and adhesive forces characteristics compared to low percentage epoxy components.

3.2 *Microstructure*

From the optical microscope, the homogeneity of the prepared composite samples has been studied. Images obtained from an optical microscope are shown from Fig. 5a–d. From the results, it reveals that 60 μm bamboo size particle composites illustrate better homogeneity distribution compared to 110 and 220 μm composite specimens. Additionally, composites, which contain 60% epoxy as a matrix material, have shown more homogeneous characteristics with respect to composite prepared by 40 and 50% epoxy as a matrix material.

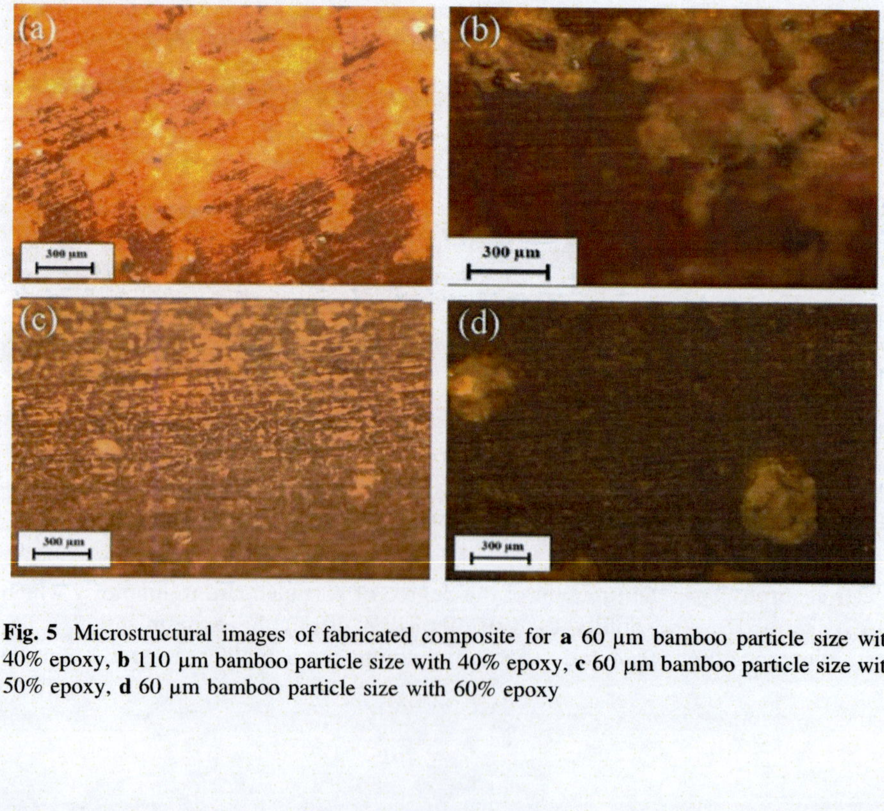

Fig. 5 Microstructural images of fabricated composite for **a** 60 μm bamboo particle size with 40% epoxy, **b** 110 μm bamboo particle size with 40% epoxy, **c** 60 μm bamboo particle size with 50% epoxy, **d** 60 μm bamboo particle size with 60% epoxy

The image obtained from the SEM analysis for the mixing ratio of 60:40 composite is presented in Fig. 6a–e. The image reveals the presence of minor cracks, lumps, and pores in the homogenous mixture of the bamboo dust within the polymer matrix. With increasing the epoxy percentage in the composite's cracks, lump and pores have reduced.

From the EDS analysis, it is found that carbon (approximate 56%) and oxygen (approximate 43%) dominated on the fabricated composite. A little amount of silicon (Si), sodium (Na), calcium (Ca), and hydrogen (H) have also been observed.

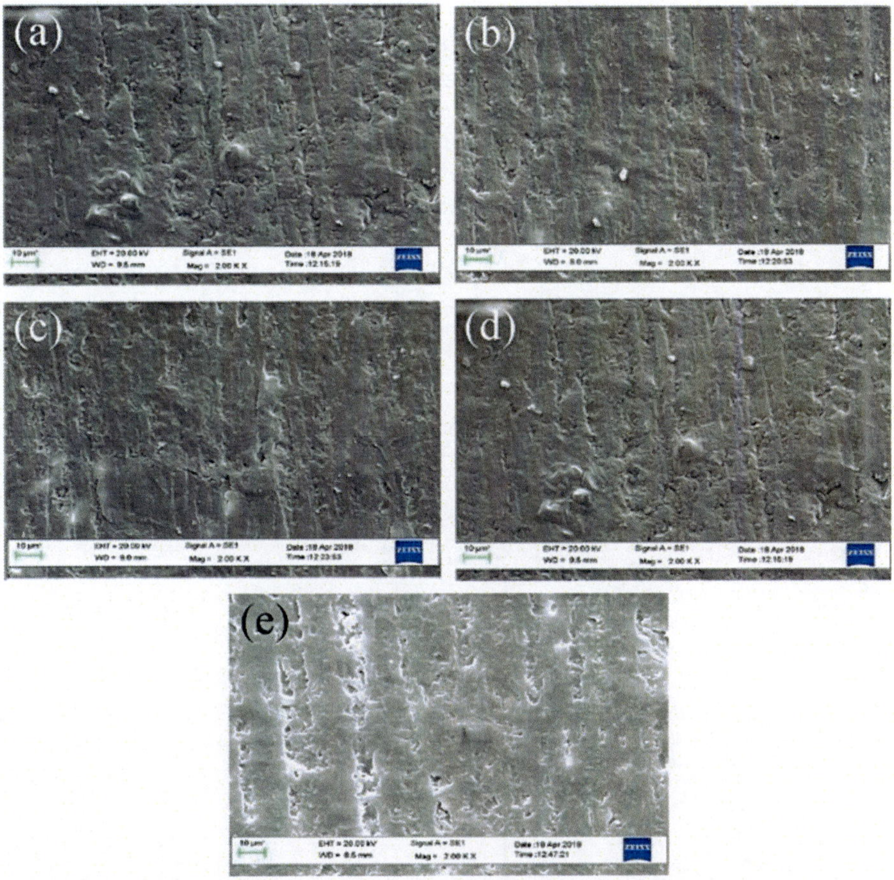

Fig. 6 SEM images of fabricated composite for **a** 60 μm bamboo particle size with 40% epoxy, **b** 110 μm bamboo particle size with 50% epoxy, **c** 220 μm bamboo particle size with 50% epoxy, **d** 60 μm bamboo particle size with 60% epoxy, **e** 220 μm bamboo particle size with 40% epoxy

3.3 Mechanical Performance Results

Ultimate tensile strength is referred to as the maximum strength obtained of the composite under the tensile load. The ultimate tensile strength is the representation of the maximum permissible load that the fabricated composite material can bear. The tensile test is conducted at an ambient temperature of 18 °C and relative humidity of 50%. Microhardness is measured on the cross-section surface of the specimens. The ultimate tensile strength and average microhardness of each specimen average tensile strength obtained from the tests are shown in Figs. 7 and 8, respectively. Ultimate strength ranges from 10.74 to 17.86 MPa and hardness ranges from $14.84HV_{0.05}$ to $27.2HV_{0.05}$.

From the record of tensile strength and hardness, it is observed that for the same mixing ratio, the mechanical performance of the specimens increases with reducing the size of the bamboo particles. It is known that the mechanical properties of a material increase when the grain size of the material decreases, and this may be the reason behind increasing tensile strength, while the size of the bamboo particles reduces in the composite samples. Additionally, a specific size of the composite specimen shows an increasing trend on mechanical performance by raising the epoxy content on the fabricated composite. In wood polymer matrix composite, mechanical performance of the samples mainly depends on the matrix material; so, in this case, when the content of the epoxy increases in the composite material, mechanical properties of the composites also boost up.

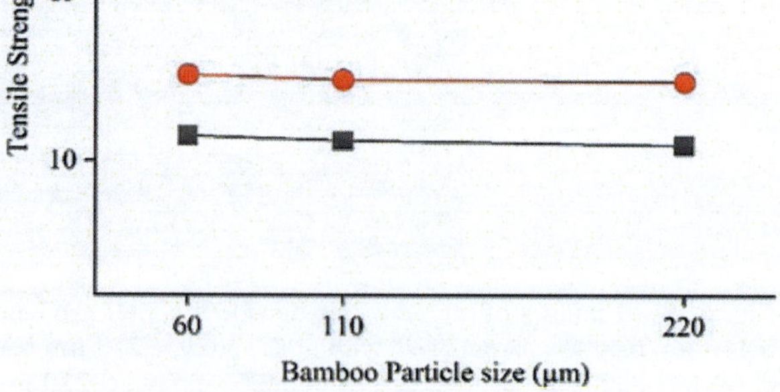

Fig. 7 Recorded ultimate tensile strength of the fabricated composites

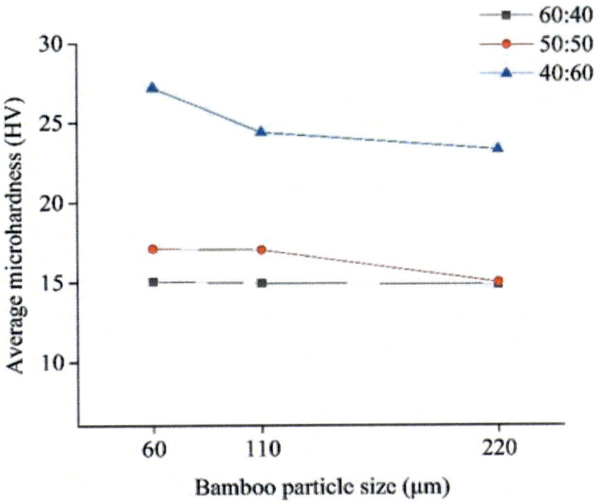

Fig. 8 The recorded average hardness of the fabricated composites

4 Conclusions

By using different sized bamboo dust particles with the different mixing ratio of thermosetting epoxy polymer, the wood polymer matrix specimen is successfully generated under controlled temperature and compressed pressure. The following major conclusion obtained based on the research work:

I 58%, 45%, and 20% water absorptivity has attributed for 40%, 50%, and 60% epoxy content wood polymer matrix, respectively. It admits that with increasing the epoxy percentage, then the prepared composite becomes stronger and more sustainable to water.

II The thickness of the 40%, 50%, and 60% epoxy fabricated composite swelled by 21%, 18%, and 10%, respectively. As the swelling thickness of the composites reduces with raising the epoxy content which indicates composites become stronger and demonstrate relatively higher stable cohesive and adhesive forces compared to low percentage epoxy components.

III Composite samples fabricated by finer bamboo dust particles show more homogeneity characteristics compared to composite samples which contain coarse bamboo dust. Additionally, the homogeneity of the samples increases when the percentage of epoxy raised on the fabricated composites. Moreover, cracks, lumps, and pores of the samples decrease when epoxy is increased on the composite samples.

IV The size of the bamboo particulates has affected the strength of the fabricated composites. The ultimate tensile strength shows an increasing trend with reducing the size of bamboo particles on the fabricated composites.

V Strengthening mainly depends on the epoxy percentage of the fabricated composites. It has been recorded that the ultimate tensile strength of the epoxy is enhanced by raising the epoxy content on the prepared composites.

References

1. Lebaron PC, Wang Z, Pinnavaia TJ (1999) Polymer-layered silicate nanocomposites: an overview. Appl Clay Sci 15(1–2):11–29
2. Christy A, Purohit R, Rana RS, Singh SK, Rana S (2017) Development and analysis of epoxy/nano SiO_2 polymer matrix composite fabricated by ultrasonic vibration assisted processing. Mater Today Proc 4(2):2748–2754
3. Saba N, Tahir PM, Jawaid M (2014) A review on potentiality of nano filler/natural fiber filled polymer hybrid composites. Polymers 6(8):2247–2273
4. Rostamiyan Y, Mashhadzadeh AH, Salmankhani A (2014) Optimization of mechanical properties of epoxy-based hybrid nanocomposite: effect of using nano silica and high-impact polystyrene by mixture design approach. J Mater 56:1068–1077
5. Kim JK, Pal K (2010) Recent advances in the processing of wood–plastic composites
6. Karakus K, Birbilen Y, Mengeloğlu F (2016) Assessment of selected properties of LDPE composites reinforced with sugar beet pulp. Meas J Int Meas Confed 88:137–146
7. Fan M, Fu F (2016) Introduction: a perspective—natural fibre composites in construction. In: Advanced high strength natural fibre composites in construction, pp 1–20
8. Klyosov AA (2007) Wood-plastic composites, pp 1–698
9. Siti S, Abdul HPS, Wan WO, Jawai M (2013) Bamboo based biocomposites material, design and applications. Mater Sci Adv Top
10. Koronis G, Silva A, Fontul M (2013) Green composites: a review of adequate materials for automotive applications. Compos Part B Eng 44(1):120–127
11. Ikada Y, Tsuji H (2000) Biodegradable polyesters for medical and ecological applications. Macromol Rapid Commun 21(3):117–132
12. Dunne R, Desai D, Sadiku R, Jayaramudu J (2016) A review of natural fibres, their sustainability and automotive applications. J Reinf Plast Compos 35(13):1041–1050
13. Thakur VK, Thakur MK (2014) Processing and characterization of natural cellulose fibers/ thermoset polymer composites. Carbohydr Polym 109:102–117
14. Mohanty AK, Misra M, Dreal LT (2001) Surface modifications of natural fibres and peformance of the resulting biocomposite. Compos Interfaces 8(5):313–343
15. Milanese AC, Cioffi MOH, Voorwald HJC (2012) Thermal and mechanical behaviour of sisal/phenolic composites. Compos Part B Eng 43(7):2843–2850
16. Mylsamy K, Rajendran I (2011) The mechanical properties, deformation and thermomechanical properties of alkali treated and untreated agave continuous fibre reinforced epoxy composites. Mater Des 32(5):3076–3084
17. Kumar R, Kumar K, Sahoo P, Bhowmik S (2014) Study of mechanical properties of wood dust reinforced epoxy composite. Procedia Mater Sci 6(May 2015):551–556
18. Lascano D, Quiles-Carrillo L, Torres-Giner S, Boronat T, Montanes N (2019) Optimization of the curing and post-curing conditions for the manufacturing of partially bio-based epoxy resins with improved toughness. Polymers 11(8)
19. De Vasconcellos DS, Touchard F, Chocinski-arnault L (2014) Tension—Tension fatigue behaviour of woven hemp fibre reinforced epoxy composite: a multi-instrumented damage analysis. Int J Fatigue 59:159–169
20. Doan T, Brodowsky H, Mäder E (2012) Jute fibre/epoxy composites: surface properties and interfacial adhesion. Compos Sci Technol 72(10):1160–1166
21. Fiore V, Bella G Di, Valenza A (2015) The effect of alkaline treatment on mechanical properties of Kenaf fibers and their epoxy composites. Compos Part B 68:14–21
22. Rong MZ, Zhang MQ, Liu Y, Yang GC, Zeng HM (2001) The effect of fiber treatment on the mechanical properties of unidirectional sisal-reinforced epoxy composites. Compos Sci Technol 61(10):1437–1447
23. Ochi S (2006) Development of high strength biodegradable composites using manila hemp fiber and starch-based biodegradable resin. Compos Part A Appl Sci Manuf 37(11):1879–1883

24. Surya Nagendra P, Prasad VVS, Ramji K (2017) A study on dynamic mechanical analysis of natural nano banana particle filled polymer matrix composites. Mater Today Proc 4(8):9081–9086
25. Hemath Kumar G, Babuvishwanath H, Purohit R, Sahu P, Rana RS (2017) Investigations on mechanical properties of glass and sugarcane fiber polymer matrix composites. Mater Today Proc 4(4):5408–5420
26. Kranthi G, Satapathy A (2010) Evaluation and prediction of wear response of pine wood dust filled epoxy composites using neural computation. Comput Mater Sci 49(3):609–614
27. Nayak R, Tarkes DP, Satapathy A (2010) A computational and experimental investigation on thermal conductivity of particle reinforced epoxy composites. Comput Mater Sci 48(3):576–581
28. Cho J, Joshi MS, Sun CT (2006) Effect of inclusion size on mechanical properties of polymeric composites with micro and nano particles. Compos Sci Technol 66(13):1941–1952
29. Nagarajan V, Mohanty AK, Misra M (2016) Biocomposites with size-fractionated biocarbon: influence of the microstructure on macroscopic properties. ACS Omega 1(4):636–647
30. Shakuntala O, Raghavendra G, Samir Kumar A (2014) Effect of filler loading on mechanical and tribological properties of wood apple shell reinforced epoxy composite. Adv Mater Sci Eng 2014(July)
31. Kumar R, Kumar K, Bhowmik S (2018) Mechanical characterization and quantification of tensile, fracture and viscoelastic characteristics of wood filler reinforced epoxy composite. Wood Sci Technol 52(3):677–699

Technologies

Design and Analysis of Membrane-Based Parabolic Space Reflector

Mehraj Shivendu

1 Introduction

A membrane-based inflatable parabolic reflector is a reflector is a reflective surface used to collect or project energy such as light, sound or radio waves. It has high directivity.

It transforms an incoming plane wave travelling along the axis into a spherical wave converging towards the focus [1–3].

There are two types of reflectors dominate for satellite:

(1) Rigid single piece reflector (RF cannot pass)
(2) Deployable mesh reflector or Canopy (RF can pass of 3.2–3.4 GHz frequency).

The inflatable reflector is used for telescopes, microstrips antennas, reflector antennas, arrays antennas, space-based solar power collection, solar cells, etc. We make a reflector which is a sculpture constructed of a lightweight material. It is supported by the inflatable torus and strut. The torus opens and releases the sculpture membrane reflector which self-inflates like a balloon. Sunlight reflects on to the sculpture and concentrates the sunlight on to the solar panel.

Inflatable antennas have the potential to provide high gains with a highly scalable, low cost, low weight and reliable deployment mechanism. Inflatable provide a means of 'Physical Amplification' that is scalable and known to be less complex [4, 5].

The studies were done keeping nearly the same reflective surface areas for each of the respective technologies. As can be seen, inflatable show maximum potential towards a promising technology.

M. Shivendu (✉)
Department of Production Engineering, BIT Sindri, Sindri, India

M. Shivendu
Vinoba Bhave University, Hazaribhagh, Jharkhand, India

S. Bag et al. (eds.), *Next Generation Materials and Processing Technologies*, Springer Proceedings in Materials 9, https://doi.org/10.1007/978-981-16-0182-8_5

The uniqueness of inflatable lies in the following factors [6–9]:

(1) High stowage efficiency: This parameter denotes the ratio of inflatable antenna diameter to the stowed volume. It occupies in stowed form. Inflatable has the potential natural vapour pressure at that temperature. This excludes the requirement of external pressure control measures making the process simple and reliable of extreme compression to be accommodated to low volume. The stowage efficiency of inflatable lied in the vicinity of 20:1.
(2) High Scalability: In terms of scalability, inflatable provides a pathway to scale up their sizes at relatively no additional mechanical complexities.

1.1 Problem Statement

The design process is to be done focusing on the principle challenges facing current inflatable technology. These are:

(1) Reliable deployment mechanism
(2) Maintaining shape accuracy upon deployment
(3) Structural reliability in external environment.

1.2 Scope of the Project

The scope of the present work extends to designing and building inflatable membranes, demonstration of effective shapes retention strategies on a laboratory scale and study possible paths towards lowering micrometeoroid impact damage probability.

Future space missions require high precision large space-borne antenna reflectors working up to KU band RF for spacecraft applications.

1.3 Objective of the Project

(1) To design, build and test an inflatable parabolic reflector concept.
(2) Demonstrate an efficient inflation process that deploys the reflector.
(3) Demonstrate methods of retaining desired shapes up on inflation in the laboratory.
(4) To reduce development time complexities, cost and increase reliability, critical care especially high curing temperature manufactured methods.

2 Membrane Theory

The plate in which the ratio $a/h > 80\ldots100$ is maintained over the plate where 'a' is a dimension of the plate in a plane and 'h' is a plate thickness or edge, such plate is referred as membrane and it remains free from flexural rigidity. Membranes carry the lateral load by axial tensile force and shear force acting in the plate middle surface. These forces are called membrane forces; they produce projection on a vertical axis and thus balance a lateral load applied to the plate membrane [10]. The fundamental assumptions used in the finite element analysis for the flat thin membrane material are as elastic, homogeneous and isotropic. The deflection of the mid-plate is to be taken as very small compared to that of membrane thickness. Middle surface remains unstrained even after bending, since the deflection is too small.

3 Methodology

To meet the objective of our proposed system, a thorough methodology was adopted our approach and was as follows [11–14]:

(1) Design and fabrication of an inflatable prototype using metallic foil laminates. Apart from being low cost and easy to handle. They have shown to be structurally robust for inflatable applications.
(2) Demonstration of a chemical sublimate-based inflation mechanism. This would allow for a simple mechanism that is self-pressure adjusting in its nature is low cost and nontoxic.
(3) Development and demonstration of a UV rigidzed polymer support structure. Rigidization process is to proceed up on deployment. Following is a detailed description of material selection and adopted methodology.

3.1 Membrane Material Selection

There are several polymeric materials available because of their lightweight (low density), ability to be formed into complex shapes, and the potential for compact storage leading to low-volume payloads.

The polyimide film possesses a unique combination of properties that make it ideal for a variety of applications in spacecraft. Kapton is synthesized by polymerizing an aromatic dianhydride and an aromatic Damien. It has excellent chemical resistance with known organic solvents for the film. Kapton does not melt or burn as it has the highest UL-94 flammability rating: V-0. The outstanding properties of Kapton permit it to be used at both high- and low-temperature extremes where other organic polymeric materials would not be functional (from DuPont). In this analysis, the membrane thickness is taken, one-third thick as a sheet of paper for the model in the dynamic analysis. The material parameters are taken identical for both the upper and lower surfaces of the present model. The mechanical properties of the Kapton material along with reflector dimensions used for the analysis are given as:

Kapton Membrane

$$\text{Density}\,(\rho) = 1418\,\text{kg/m}^3$$
$$\text{Young's Modulus}\,(E) = 2.59 \times 10^9\,\text{N/m}^2$$
$$\text{Poisson's Ratio}\,(\mu) = 0.34$$
$$\text{Thickness}\,(h) = 3.3867 \times 10^{-5}\,\text{m}$$
$$\text{Uniform pressure}\,(p) = 1729.69\,\text{N/m}^2$$

Parabolic Reflector

$$\text{Focal Length}\,(F) = 1.2192\,\text{m}$$
$$\text{Aperture Diameter}\,(D) = 4.8768\,\text{m}$$

3.2 Inflatable Design and Construction

Large-scale inflatable with a required degree of shape accuracy is built out of building blocks called gores. Gores stitched or bonded in patterns dictate the fundamental structural loads acting on the membrane during the process of inflation and deflation. Out gore design was based on the following consideration:

(1) Circumferential and meridional stresses induced into inflatable upon deployment.
(2) Gore sealing line/load tape configuration which affects the equilibrium stiffness attained by the membrane at characteristic pressures.
(3) Gore edge bonding techniques which define the boundary conditions applied on each gore.

Gore design involves transforming a desired three-dimensional unit to a projected two-dimensional space to allow for seamless integration. As shown in Fig. 1, a three-dimensional parabola was projected onto a two-dimensional surface to obtain a gore pattern. The following design formulas are useful for designing a parabolic reflector.

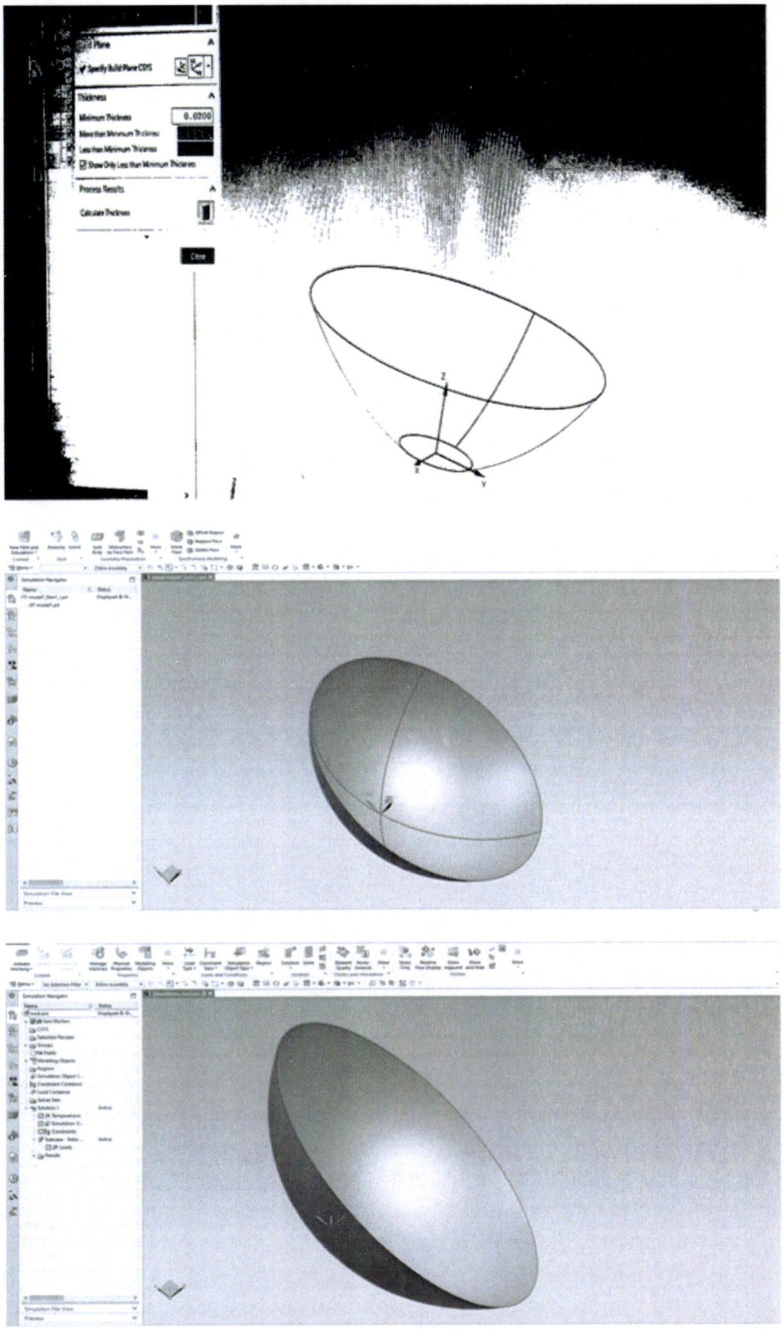

Fig. 1 FEM model of parabolic reflector

3.3 *Equation of a Parabola*

The equation of a parabola in terms of focal length f is $Y = ax^2$
Depth of a parabolic reflector
In designing a parabolic reflector, it is frequently convenient to use its depth d instead of its focal length. The formula for getting the depth is $d = D^2/16f$
Conversely, given a parabolic dish and its measurements for the diameter D and the depth d, then its focal length f is obtained with $f = D^2/16d$.

It is difficult to illuminate the uniformly with the feed inside the aperture plane. This is because waves arriving from opposite directions tend to cancel through superposition. So our eye peers in one direction only. And next, placing the focal point well outside the aperture plane modifies the chance of receiving unnecessary signals and noise. The feed point is not well protected, and this configuration increases the chance of loss. Signals from the feed horn may lose the edge of the dish. The ratio of the focal distance to the dish diameter denoted f/D is a standard component parameter used by systems installers.

$$\text{Efficiency}(\eta) = f/D = \text{focal/length}$$

3.4 *Software Versions*

The reason for this section is that when using computer-aided engineering (CAE) software the version has a great impact on the results produced. The software versions used for the computations in the thesis are displayed in Table.

Software	Version
MSC.PATRAN	V2007r2
MSC.NASTRAN	V2008rla
MATLAB	R2010a

When dealing with dynamic problems, it is essential to be aware that different versions of the same software can give the studied model specific behaviour. This is the case for NASTRAN when using the MFLUID card to include the added air mass effect. When MFLUID is utilized the peak in structural responses (acceleration, displacement, etc.) of a frequency response analysis is supposed to coincide with the eigen-frequency for the reflector with added air mass. NASTRAN v2005r2 was initially used in this thesis and even when using MFLUID NASTRAN v2005r2 produces peaks in the structural response at the eigen-frequencies without added air mass. This is of course a big problem since it gives non-physical behaviour when adding the air mass, which is an important part of the analysis, performed in this thesis. The solution to this problem was simply to upgrade to NASTRAN v2008r1a which handles MFLUID exactly the way it is supposed to.

4 Inflatable Membrane-Based Parabolic Reflector Model

The NASTRAN FE-model of the reflector is provided by RUAG Space and summarized as interesting nodes.

The nodes of interest for the analysis are taken. The nodes used in this thesis will be the same as the nodes used in the BEM-analysis, i.e. the sensor positions used in the physical test. Thereby, it is possible to correlate the model with available test and analysis data. The nodes are presented in Figs. 2 and 3.

In order to design an inflatable parabolic membrane-based reflector, a number of possible configurations were studied. Though this process it became evident that the inflatable would require surface reflective to radio waves that are shape controlled in addition to radio transparent surface to support the reflective surface.

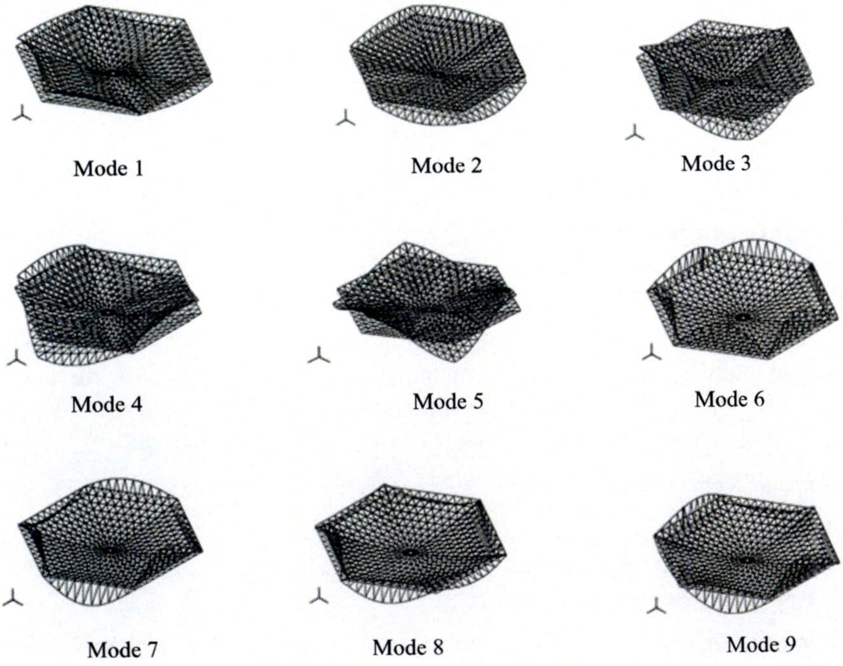

Fig. 2 Mode shapes of the parabolic reflector

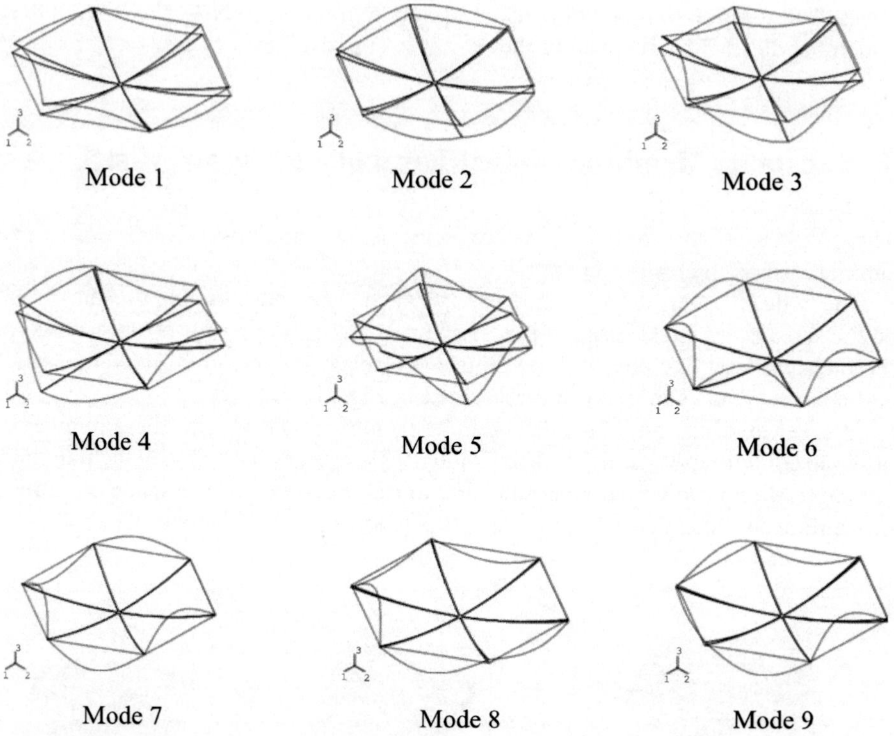

Fig. 3 Outline of mode shapes of reflector

5 Parametric Study

Membrane Thickness

As can be seen in the figure, membrane failure is highly sensitive to its thickness. Failure probability for inflatable membranes to converge to 65% for membrane thickness above 2 mil and between ultrathin foils (0.1–0.25 mils) and thick foils (2 mil and greater) is particularly usable to micrometeoroid damage.

Material Heat of Vaporization

Heat of vaporization of a material is a measure of a material's efficiency in transferring and dissipating incident energy. Shows chances of failure to decrease with heat of vaporization.

Material Density

A materials density bears the same correlation to impact damage probability as heat of vaporization. However, its sensitivity with regard to impact damage is much lower than membrane thickness.

Inflation Pressure

Similar to the material thickness, the internal inflation pressure is also one of the important factors in the structure design. The inflation pressure is changed from 500 to 3500 Pa, and the corresponding natural frequencies are obtained and the relation between the change of the inflation pressure and the first 10 natural frequencies are illustrated. The inflation pressure increase leads to strengthen the structure and increasing the natural frequencies, which means that the material stiffness is highly affected by the inflation pressure inside the structure.

6 Results and Conclusion

In this section, the result obtained during the analysis will be presented and explained. Main interest is in the acceleration power spectral densities of the nodes mentioned in the above section. In this section, the result of the analysis, and the model analysis is presented. The results are presented and compared to the result from the physical test.

Mode	Frequency (Hz)		
	6 ribs	12 ribs	24 ribs
1	3.14	2.97	2.85
2	3.14	2.97	2.85
3	5.09	4.55	4.01
4	5.09	4.55	4.01
5	6.74	6.60	5.96
6	6.89	6.60	5.96
7	7.12	8.61	8.04
8	7.47	8.61	8.04
9	7.47	9.41	10.18

Similarly, the different diameters and ribs are considered for the model analysis of the parabolic reflector and their respective frequencies are shown in the above table.

7 Conclusion

(1) Inflatable antennas provide packing efficiency of 20:1 and weight up to 10 times lower than conventional mechanically deployed parabolic antennas.
(2) Chemical sublimates have been demonstrated for our application on a laboratory scale.

(3) UV polymers have been demonstrated to rigidize vacuum A sample 30 mm 100 mm × 2 mm took about 2.5 h to completely rigidize.
(4) Probability of membrane failure due to micro-meteoroid impact was found to be most sensitive to membrane thickness and material density.

7.1 Future Work

(1) Development of simulation tools to understand what initial membrane shape would lead to desired inflated shapes.
(2) Studies into origami methods for optimal stowage and deployment of the membranes. The membrane needs to be packed at as low a volume as possible using the least number of creases and folds.
(3) Incorporation of demonstration of resin filled envelopes into the membrane to provide a rigid skeletal framework.
(4) Development of deployment system for the reflector and antenna.
(5) Optimal placement of patch antenna to enhance performance.

References

1. Seefeld P (2013) The preliminary design of the GOSSAMER-1 solar sail membrane and manufacturing strategies. In: The third international solar sailing symposium, Glasgow, June
2. Banks HT, Smith RC, Wang Y (1996) Smart material structures: modelling, estimation, and control. Wiley, New York
3. Callahan J, Baruh H (1999) Model sensing of circular cylinder shell using segment piezoelctric elements. Smart Mater Struct 8:125–135
4. Freeland RE (1998) Inflatable deployable space structures technology summary
5. Freeland RE, Bilyeu G (1993) In-step inflatable antenna experiment. Acta Astronautica 30 (1993):29–40
6. Jaffe L (1961) Project echo result. Astronautics, May
7. Wong YW, Pellegrino S (2006) Wrinkled membranes Part III: numerical simulations. J Mech Mater Struct 1(1):63–95. https://doi.org/10.2140/jomms.2006.1.63
8. Wang C, Tan H, Du X, Wan Z (2007) Wrinkling prediction of rectangular shell-membrane under transverse in-plane displacement. Int J Solids Struct 44:6507–6516. https://doi.org/10.1016/j.ijsolstr.2007.02.036
9. Lecieux Y, Bouzidi R (2010) Experimental analysis on membrane wrinkling under biaxial load comparison with bifurcation analysis. Int J Solids Struct 47:2459–2475. https://doi.org/10.1016/j.ijsolstr.2010.05.005
10. Bouzidi R, Lecieux Y (2012) A numerical method to optimize the design of a space inflatable membrane reflector. J Acta Astronautica 74:69–78. https://doi.org/10.1016/j.actaastro.2011.12.009
11. Srivastava A, Mishra BK, Jain SC (2008) Effect of enclosed fluid on the dynamic response of inflated torus. J Sound Vib 309:320–329. https://doi.org/10.1016/j.jsv.2007.01.019

12. Freeland RE, Bilyeu GD, Veal GR (1996) Development of flight hardware for a large, inflatable-deployable antenna experiment. Acta Astronaut 38(4–8):251–260. https://doi.org/10.1016/0094-5765(96)00030-6
13. Greschik G, Cassapakis C, Veal G, Mikulas MM (2001) Sensitivity study of precision pressurized membrane reflector deformations. AIAA J 39:308–314. https://doi.org/10.2514/2.1306
14. Gajbhiye SC, Upadhyay SH, Harsha SP (2012) Wrinkling dynamics of membrane based on user defined wrinkle pattern. Int J Comput Mater Sci Eng 1(4):1250034(11p). https://doi.org/10.1142/s2047684112500340

Development of Integrated Aerator combining Paddlewheel and Propeller Aspirator Aerators for Shrimp Farming

R. U. Roshan, R. Harini, and T. Anand

1 Introduction

Aquaculture is considered as one of the fastest developing sectors worldwide [1]. India is a country with tremendous resources and potential in the field of aquaculture, and it is the second-largest aquaculture producer in the world [7]. Global aquaculture production contributes around 110.1 million tones to world total fish production in 2016 [7]. Among the export commodities, the shrimp has a higher global market value over fish. Marine shrimp production by aquaculture is about 7 million tons per annum [2]. Countries like India, Thailand, Vietnam, China, which has adequate resources, started focusing on shrimp cultivation, which results in a massive raise in the aquaculture sector.

Later in the 1990s, the Black Tiger shrimp *Penaeus monodon* undergoes a severe outbreak to white spot diseases caused by white spot syndrome virus (WSSV) [6]. During that time, the farmers struggled a lot to culture this species and were not ready to invest money. That was the period with a more significant fall back for India in shrimp aquaculture. Then in 2010, CAA permitted to rear SPF vannamei brooders in Indian hatcheries and to grow this exotic species *Litopenaeus vannamei* in Indian farms after obtaining a license from it [5]. This species has many advantages over *P. monodon*. So the farmers started doing vannamei culture. *L. vannamei* later was changed to *Penaeus vannamei* also has fast growth and disease resistance compared to *P. monodon* [2]. The penaeid shrimp created dominancy among the crustaceans in both production and export. So the demand for shrimp started to increase. To meet out this demand, concept of intensive cultivation has developed to increase production.

The concept of intensive culture requires high capital cost and intensive management such as feeding, water quality, bio-security, aeration. A thriving culture

R. U. Roshan (✉) · R. Harini · T. Anand
College of Fisheries Engineering, Tamil Nadu Dr. J. Jayalalithaa Fisheries University, Nagapattinam, India

S. Bag et al. (eds.), *Next Generation Materials and Processing Technologies*, Springer Proceedings in Materials 9, https://doi.org/10.1007/978-981-16-0182-8_6

depends on proper culture practices and proper maintenance in the aspects mentioned above. Mainly the culture organisms require dissolved oxygen (DO) for their growth, to increase its metabolic activity and body maintenance, etc. The minimum dissolved oxygen level to be maintained in the aquaculture pond is 4–5 ppm [3]. The dissolved oxygen under steady partial pressure and temperature is not sufficient for the species in the aquaculture system. So oxygen should be added into the water through mechanical means, and this process is called aeration.

The mechanical aeration systems are broadly classified as surface, and subsurface aeration systems, and they are used in the ponds based on the requirement of the culture species. The widely used aquaculture aerators are paddlewheel aerators, diffused-air aerators, cascade aerators, propeller aspirator-pump aerators. These aerators are utilized for large-scale operation. Kumar et al. [8] suggested that, circular stepped cascade or pooled circular stepped cascade aerator is the most economical for the ponds with less than 1000 m^3 capacity. And for larger ponds with 5000 m^3 capacity and more, paddle wheel aerators (1 HP or 2 HP) are well suitable, and it is more efficient. These aeration systems are used not only for dissolved oxygen but also for maintaining the optimal temperature, for proper mixing, to circulate water, and to prevent stratification [4]. In most of the aquaculture farms, paddlewheel aerators with varied horsepower are available, which is a surface aeration system. Unlike *P. monodon*, the vital character of *P. vannamei* is that it occupies the entire water column. So, the whole water column needs aeration. In this project, the integration of two types of aeration systems was carried out (i.e., surface and subsurface aeration system). This system has some essential benefits like less investment, and increased DO levels throughout the water column. Also, it is easy to maintain and consumes a low level of power compared to other aerators.

2 Research Approach

2.1 *Problem Identification*

Aeration is an essential part in determining the success of the culture. In super-intensive culture techniques, among all the water quality parameters, dissolved oxygen is the most crucial parameter to be maintained, so that mortality can be reduced.

Dissolve Oxygen Stratification: In aquaculture ponds, commonly they use paddle wheel aerators for supplementary DO, which is a surface aeration system. These surface aeration systems supply dissolved oxygen efficiently up to half a meter depth and further DO penetration, and for proper mixing, it takes longer duration. But for a species that occupies the full depth water column like *P. vannamei* it is not possible to supply DO to the animal at the bottom by the surface aerator.

Thermal Stratification: It is a condition that occurs in the still water body like lakes and ponds. Especially in aquaculture ponds, thermal stratification causes stress to the animal as the animal gets exposed to varying temperature. It reduces the growth rate of the animal.

Water Current: In super-intensive culture techniques, continuous aeration is required. But in common culture practice, all the aerators will be put off during feeding operation. This is because the feed movement gets disturbed due to water current developed during aerator operation. During that time, all the animals suffer from oxygen insufficiency and come to the top to respire atmosphere diffused oxygen. Also, the minimum flow velocity to prevent the DO stratification is 5 cm/s [9].

3 Materials and Methodology

3.1 Development of Integrated Aerator (IA)

The regular 2 HP Paddle Wheel Aerator (PWA) consists of floats, four paddles, and a gearbox operated by 2 HP induction motor at 2880 rpm. The material specifications are listed below (Table 1). The gearbox comprises a worm-wheel and a worm-shaft, which distributes the motor power to the paddles to facilitate oxygenation. This gearbox is modified, and the dimension of the worm-shaft is altered to have the provisions for integrating another motor operated aeration system. Unlike the PWA gearbox, it has one extra output at the bottom. In this output, a propeller aspirator system is being setup in a way to draw the atmospheric air and inject it to the water column. This gearbox will also be operated by 2 HP motor but the number of paddles is reduced by 2 (Fig. 1).

Table 1 Material selection for the developed integrated aerator

S. No.	System parts (integrated aerator)	Material	Dimension		
			Length (cm)	Width (cm)	Diameter (cm)
1	Float	Synthetic polypropylene	172	23	
2	Paddlewheel	HDPE			60
3	Aerator frame	Stainless steel	104	36	
4	Worm-shaft (modified gearbox)	Stainless steel	28		2.3
5	Connecting shaft	Stainless steel	42		1.9
6	Propeller	Cast iron	8.5		4
7	Casing	Mild steel	40		6

Fig. 1 Regular gearbox (left) and modified gearbox (right) with vertical output

3.2 Experimental Setup

The aeration experiments were conducted in the small ponds of dimension 10 m 5 m × 1.5 m. The integrated aerators developed for testing consist of a paddle wheel aerator setup with two paddlewheels, one on each side and a propeller at the bottom connected to the gearbox which is modified and developed to integrate two different types of aerators (surface and subsurface). This integrated aerator is operated by an induction motor (2 HP, 3 Phase) at 2880 rpm and Nan Rong Hai Co. Ltd, Taiwan manufactures it. The propeller at the bottom weighs 700 g and a connecting shaft of length 420 mm connects it with the third output of the gearbox. A deflector is welded on the shaft casing whose length is up to 100 mm from the propeller. The integrated system is mounted at one end of the longer side of the pond. It is positioned and supported by the G.I. pipes of 45 mm diameter and 3 m in length.

In a similar way, a regular paddlewheel aerator with four wheels (two on each side) was mounted at the same position in another pond and supported by the G.I. pipes of the same dimension. 2 HP and three-phase induction motor operate this system at the speed of 2880 rpm. This entire system is manufactured by Nan Rong Hai Co. Ltd, Taiwan (Figs. 2 and 3).

3.3 Aeration Test

Field tests were conducted in two ponds, a Control Pond (CP) with Paddle Wheel Aerator (PWA) and an Experimental Pond (EP) with Integrated Aerator (IA). Before deploying the system, some essential water quality parameters were checked (Table 2), because the preexisting water quality parameters also affect the solubility of oxygen. Then the two aerators were installed in control and experimental ponds.

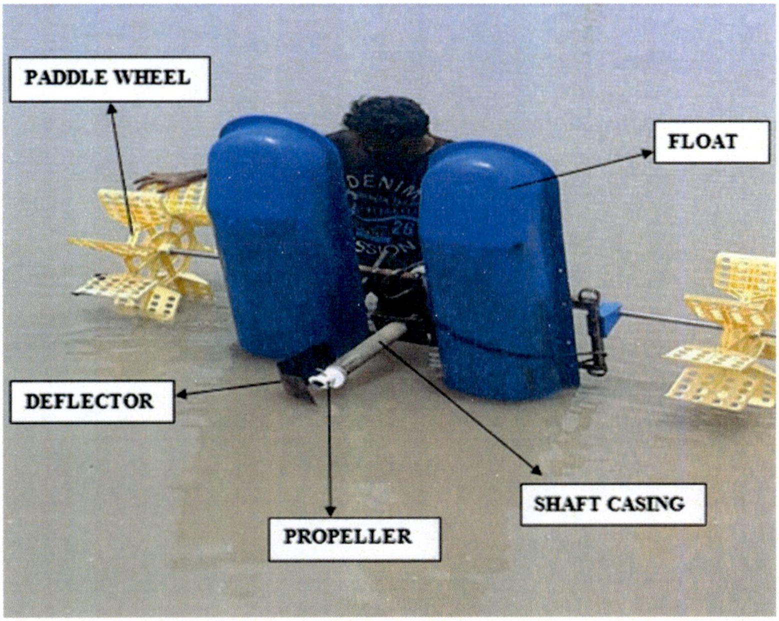

Fig. 2 Developed integrated aerator

Fig. 3 Regular paddle wheel aerator

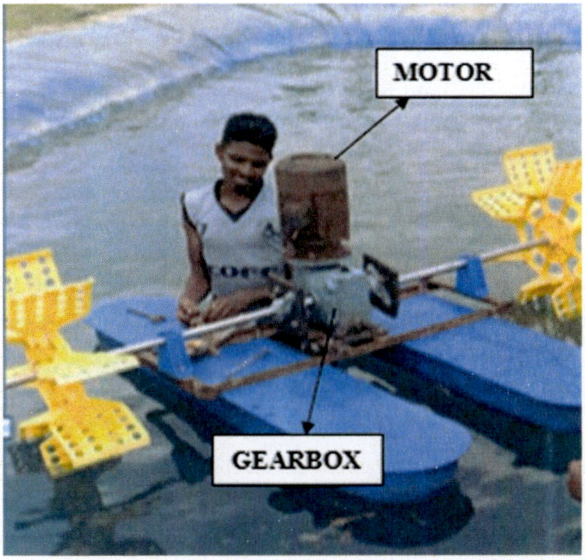

Table 2 Water quality parameters prior to operation

S. No.	Parameters	Experimental pond 1 (Regular aerator)	Experimental pond 2 (Integrated aerator)
1	pH	7.4	7.4
2	Salinity	45	45
3	CO_3	14	15
4	HCO_3	119	121
5	Total alkalinity	135	133
6	Total hardness	8600	8500
7	Ca/L	2800	3000
8	Mg/L	1409.4	1345.8

Two trials (48 hours each) were conducted, and reading such as dissolved oxygen, temperature, pH, water current was taken at an equal time interval of 1 hour.

Dissolved oxygen, temperature and pH readings were taken at two different levels of water column;

(a) At the top water column, the readings are taken 0.20 m below the water surface;
(b) At the bottom, water is collected from a closed plastic container and readings are obtained. Dissolved oxygen measurements were taken at a distance of 5 m from the aerator using Lutron DO meter. Temperature and pH were measured using Erma pH and temperature meter. Water current velocity was computed in a traditional method using a floating particle.

The obtained readings were plotted in the graph and studied for the effectiveness of the developed aerator in facilitating oxygenation. Standard deviations of DO and temperature values between the two trials of control pond (CP) and two trials of experimental pond (EP) were computed and analyzed for its performance. Also, the difference in the DO concentration and temperature between the water levels (surface and bottom) in the ponds was recorded and average deviation was computed for the data sets of both the trials in CP and EP, and the effectiveness of the developed aerator in minimizing the stratification was studied and compared with the regular paddle wheel aerator (PWA).

4 Result and Discussion

4.1 Dissolved Oxygen

Comparison of Surface Level DO: The dissolved oxygen concentration at the pond surface water level was measured for 48 hours. Two trials were carried out, and the comparison between two aerators is discussed below.

Trial 1: At the initial stage, the DO level is higher in the control pond, but after few hours of operations, two lines in the graph tend to meet at a point (Fig. 4). From that point, surface water DO levels in both the pond were nearly the same till the end of the experiment. So we can identify that the integrated aerator is competing with the regular paddle wheel aerator and impart the same DO levels in the ponds.

Trial 2: From the initial stage, the surface water DO concentrations were roughly the same (Fig. 5). At some instances, particularly during nights, DO level in the experimental pond is relatively higher.

Comparison of Bottom Level DO: The dissolved oxygen concentration at the pond bottom water was measured for 48 hours. Two trials were carried out, and the comparisons between two aerators are discussed below.

Trial 1: At the initial stage, both the ponds record similar dissolve oxygen concentration. After 5–6 hours of operation, the DO concentration in the experimental pond gradually increases (Fig. 6). From the point of its rise, till the end of experimentation, the lines in the graph did not meet.

Trial 2: This trail is also very similar to the first. After a few hours of operation, the DO concentration in the experimental pond starts to rise and maintains a range ahead of the control pond (Fig. 7).

Comparison of Average Deviation of DO between the Water Levels: Dissolved oxygen stratification is a condition occurs in the pond water body. Minimizing the stratification is one of the functions of the aerator. Two experimental trials were carried out for 48 hours each, and the average deviation between the surface and bottom water levels were compared and discussed below.

Trial 1: A higher deviation was recorded at the initial stage of operation in both the ponds. After a few hours, the deviation drops in the experimental pond and

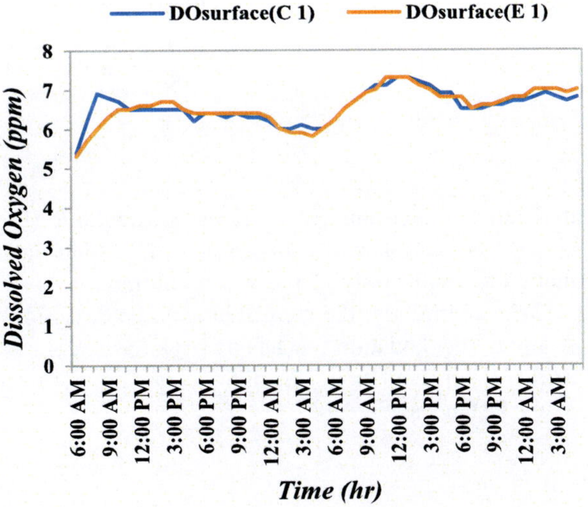

Fig. 4 Surface DO versus time—trial 1

Fig. 5 Surface DO versus time—trial 2

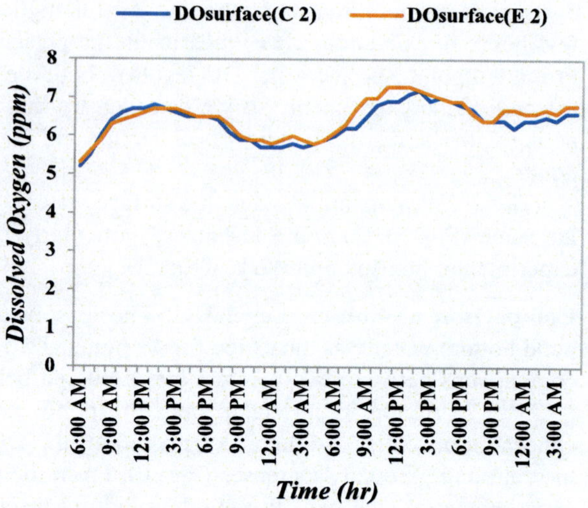

Fig. 6 Bottom DO versus time—trial 1

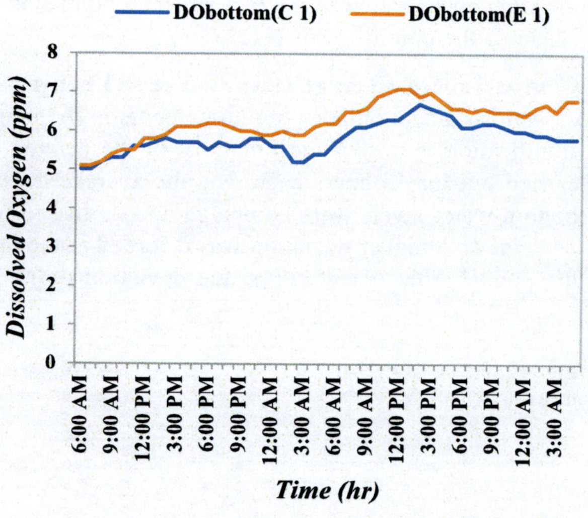

maintains a constant lower range throughout (Fig. 8). At some instances, the average deviation was recorded as 'nil,' which indicated that DO concentration is maintained uniformly in the water column.

Trial 2: Initially, the measured average deviations were equal in both the ponds. At noon, the deviation reaches its peak in the control pond, while it gradually drops in the experiment pond. Similar to the previous trial, there is zero deviations at many instances (Fig. 9).

Fig. 7 Bottom DO versus time—trial 2

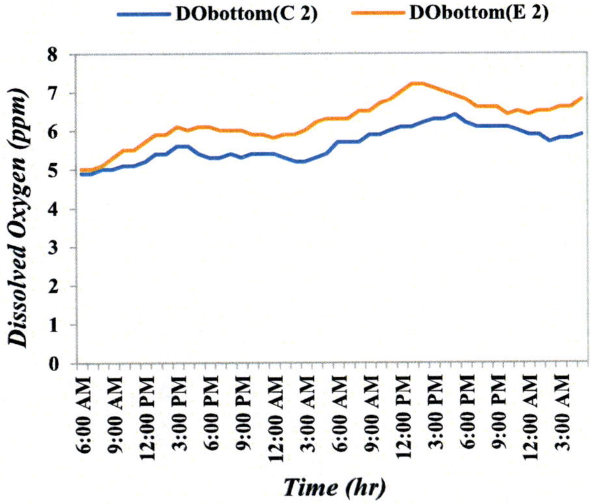

Fig. 8 Average deviation (DO) versus time—trail 1

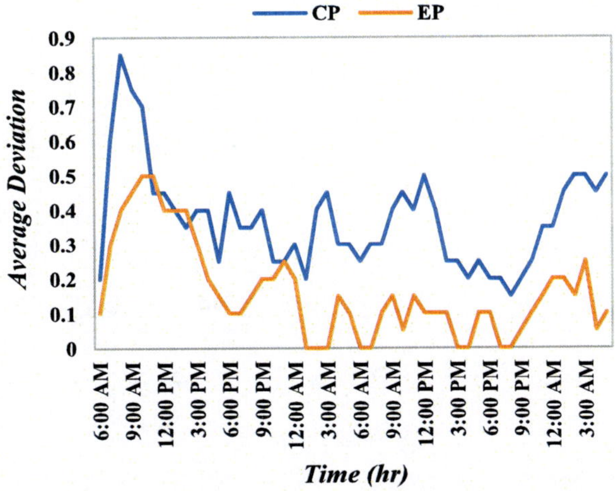

4.2 Temperature

Comparison of Average Deviation of Temperature between the Water Levels: Thermal stratification in the aquaculture pond is an unfavorable condition. Mainly during intensive culture practices, it makes the animal stressful as it exposes to the varying temperature. Experiments were carried to study the effect of paddle wheel aerator and integrated aerator in minimizing the stratification.

Trial 1: At the initial stage of the experiment, the deviation was rising in both the pond. In the control pond, the decrease in average deviation is recorded after 10 hours of operation (Fig. 10). Whereas in the experiment pond, it drops within

Fig. 9 Average deviation (DO) versus time—trail 2

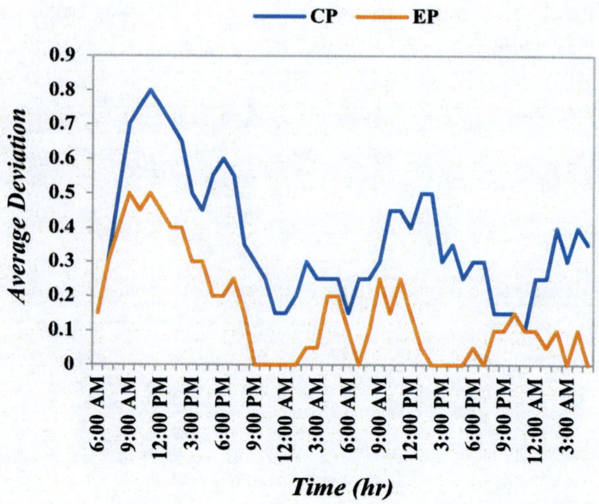

Fig. 10 Average deviation (*T*) versus time—trail 1

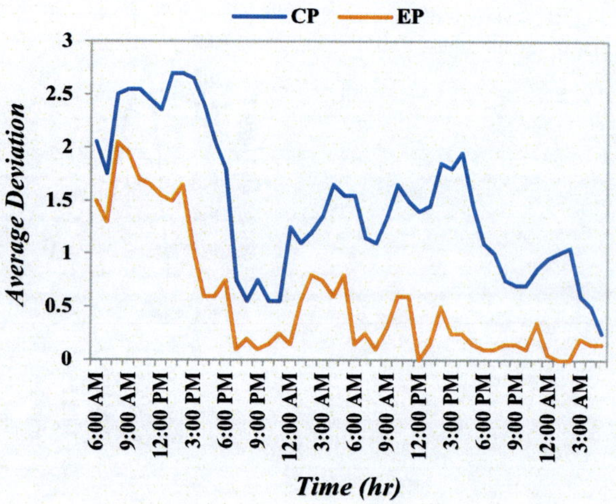

3 hours of operation, and the deviation is maintained comparatively less throughout the experiment.

Trial 2: The results of this trial are very similar to the first, where within 4 hours of operation, the average deviation drops and maintained low in the experimental pond (Fig. 11).

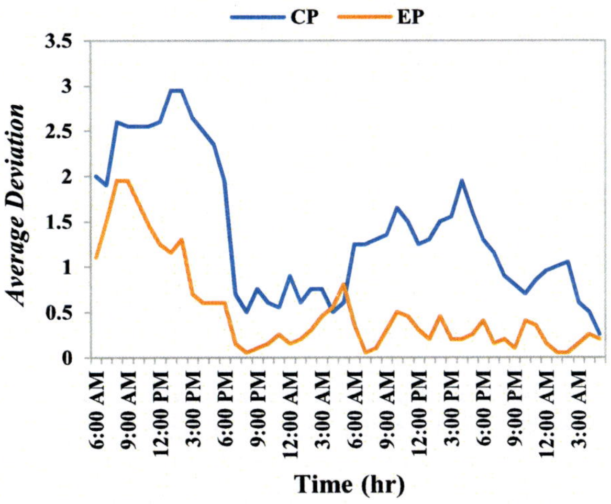

Fig. 11 Average deviation (*T*) versus time—trail 2

4.3 Standard Deviation

Standard Deviation for Dissolved Oxygen: In control pond with PWA, the standard deviation for the DO values in surface and bottom water level for two trials is found to be 0.1326, and similarly for experimental pond with IA, it is found to be 0.0309 (Table 3).

From the obtained values, it is very evident that, integrated aerator causes minimum DO stratification, which is suitable for the animal to get sufficient DO at any level of the pond. While in CP, the deviation is higher as the PWA agitates only the surface. So the DO concentration is low at the bottom, which is unsuitable for species like *P. vannamei*.

Standard Deviation for Temperature: In PWA installed control pond, the standard deviation for the temperature values for two trials is found to be 0.0545, while for IA installed pond, it is obtained as 0.0457 (Table 4).

It gives us a clear understanding that, temperature stratification is less in IA installed pond, which makes the pond environment suitable for the animal stock to survive without stress. The deviation value is high in the case of paddle wheel aerators, which indicates the improper mixing of pond water.

Table 3 Standard deviation for dissolved oxygen readings

Variables	Control pond		Experimental pond	
	Trial —1	Trial —2	Trial —1	Trial —2
Mean	6.189	6.001	6.405	6.361
Standard deviation	0.1326		0.0309	

Table 4 Standard deviation for temperature readings

Variables	Control pond		Experimental pond	
	Trial—1	Trial—2	Trial—1	Trial—2
Mean	29.96	30.04	30.95	31.01
Standard deviation	0.0545		0.0457	

4.4 Water Current

Comparison of Water Current In PWA and IA installed Ponds: The inference from this graph is, the velocity of water in regular paddle wheel aerator pond was always higher than the integrated aerator pond. The regular paddle aerator pushes the water very fast, which creates a significant wave action in the pond. Whereas in the integrated aerator, the water is slow and stable because the single power input is divided into two outputs (Fig. 12).

5 Conclusion

In the present scenario of shrimp aquaculture, *P. vannamei* is predominantly cultured all around. So the water quality parameter specifically dissolved oxygen and temperature play a vital role in efficient aquaculture production. A specific character between *P. monodon* and *P. vannamei* is, the first one lies only at the bottom of the

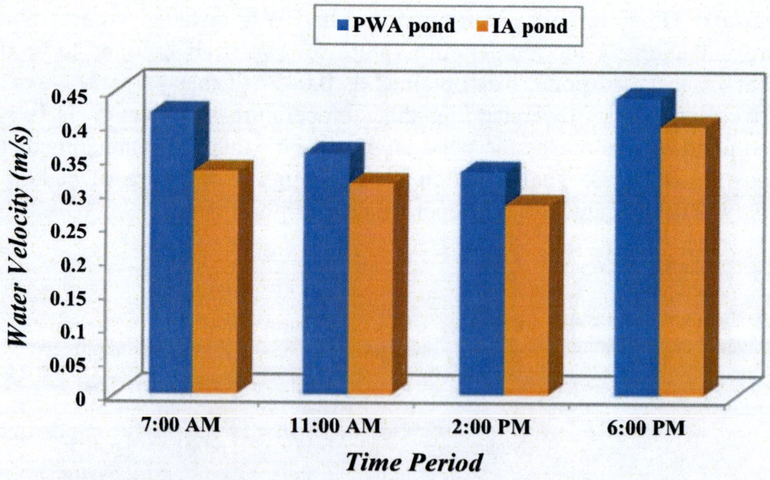

Fig. 12 Water current speed comparison

pond water column and the latter lies in entire water column. So, supplementary aeration should be supplied in a way that sufficient DO concentration is maintained throughout the water column. The three important significance of integrated aerator identified from the above experimentation are as follows;

The DO stratification is considerably high in PWA installed pond, and it is very minimum in integrated aerator pond. Because, unlike paddle wheel aerator, integrated aerator acts as a subsurface aerator with a propeller at the bottom. So, the stratification is decreased and it facilitates oxygen throughout the water column. If the deviation increases, there are chances for mortality, and if deviation is minimum, it creates a conducive environment for culture species.

The result shows that the thermal stratification is minimum in integrated aerator installed pond and maximum in PWA installed pond. Because the integrated aerator exhibits a thorough mixing of water so that the pond temperature is uniformly maintained in the water column. Higher thermal stratification causes thermal shock to the animal and makes the animal metabolically weaker, thereby growth rate is retarded. IA creates low thermal stratification suitable for aquaculture.

In super-intensive culture practices, continuous aeration is necessary even during feeding operation. But operating the regular PWA creates high water current due to its vigorous surface agitation. This disturbs the feed movement and will make the animal difficult for catching it. As a result, all the feed will get wasted as sludge. Integrated aerator operation creates an optimal water current, which will not affect the feed movement, and at the same time, culture species are supplied with adequate DO.

References

1. Aguiar E, Fuentes-Santos I, Labarta U, Álvarez-Salgado XA, Fernández-Reiriz MJ (2015) Empirical modelling of seston quality based on environmental factors in a mussel culture area (NW Iberian upwelling system). Mar Ecol Prog Ser 536:89–105
2. Bardera G, Usman N, Owen M, Pountney D, Sloman KA, Alexander ME (2019) The importance of behaviour in improving the production of shrimp in aquaculture. Rev Aquac 11 (4):1104–1132
3. Baylar A, Hanbay D, Ozpolat E (2007) Modeling aeration efficiency of stepped cascades by using ANFIS. Clean–Soil, Air, Water 35(2):186–192
4. Boyd CE, Martinson DJ (1984) Evaluation of propeller-aspirator-pump aerators. Aquaculture 36(3):283–292
5. CAA (2009) Guidelines for regulating coastal aquaculture. http://www.caa.gov.in/uploaded/doc/compendium_act.pdf
6. Chou H, Huang C, Wang C, Chiang H, Lo C-F (1995) Pathogenicity of a baculovirus infection causing white spot syndrome in cultured penaeid shrimp in Taiwan. Dis Aquat Org 23(3):165–173
7. FAO of United States (2020, Oct 20) The state of fisheries and aquaculture 2020. http://www.fao.org/state-of-fisheries-aquaculture
8. Kumar A, Moulick S, Mal BC (2013) Selection of aerators for intensive aquacultural pond. Aquacult Eng 56:71–78
9. Oswald WJ (1978) The engineering aspects of microalgae, vol 2. CRC Press, Boca Raton, pp 533–534

Experimental Investigation of Al 2024 Aluminum Alloy Joints by Underwater Friction Stir Welding for Different Tool Pin Profile

Surendra Kumar Lader, Mayuri Baruah, and Raj Ballav

1 Introduction

Welding of aluminum has been an important concern for several years. Friction stir welding is employed for the welding of aluminum alloys since decades and is still continued to join the alloys of aluminum with new dimensions and in different working environment. FSW is a solid-state welding process, also called a plastic state welding because the melting temperature is not reached in this process. The temperature generated in this process is 60–70% of base metal melting temperature [1–3]. Due to solid-state welding nature, FSW is one of the most eminent techniques for welding of similar and dissimilar aluminum alloys [4–7]. This technique is privileged to weld all series of aluminum alloys especially the AA2024 has highest strength-to-weight ratio which makes it more relevant for aviation industry such as fuselage and wing skin panels [8–10]. FSW process is shown in Fig. 1.

Recently, the advantages of friction stir welding are explored in different working environment in order to expand the horizon for its use. It is utilized for joining in a very cool environment such as liquid nitrogen in order to enhance the weld joint quality [11, 12]. FSW can also be carried out in underwater in order to enhance the weld zone mechanical and microstructural properties [13, 14]. Previous investigations confirmed the feasibility of underwater FSW [15]. Nowadays, underwater FSW is well-established process and considered as the advancement of FSW process. During underwater FSW, the peak temperature is controlled by cooling water and the grains at various heat-affected zones will not get enough time to grow. Hence, grain coarsening at weld region is comparatively lower and the

S. K. Lader (✉) · M. Baruah · R. Ballav
Production and Industrial Engineering Department, National Institute of Technology, Jamshedpur, India
e-mail: mayuri.prod@nitjsr.ac.in

R. Ballav
e-mail: rballav.prod@nitjsr.ac.in

S. Bag et al. (eds.), *Next Generation Materials and Processing Technologies*, Springer Proceedings in Materials 9, https://doi.org/10.1007/978-981-16-0182-8_7

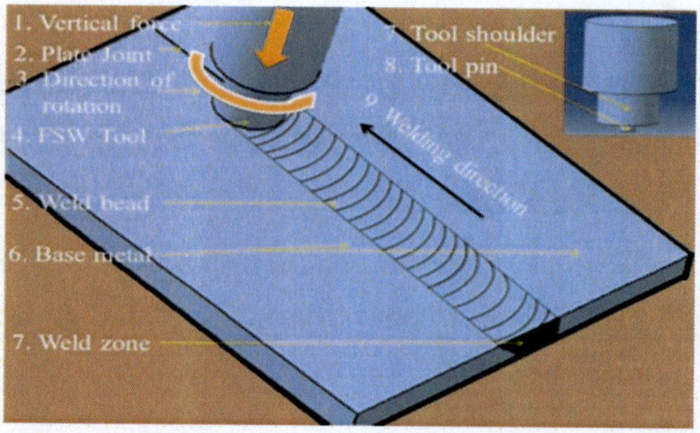

Fig. 1 FSW process

width of heat-affected zone (HAZ) and thermo-mechanically affected zone (TMAZ) is also reduced due to restriction in heat input [16]. The joints obtained from underwater FSW minimize defects such as porosity and solidification cracking which ultimately enhances the mechanical properties of the joint [17, 18]. Underwater FSW is also considered one of the best available methods for joining of heat-sensitive materials such as aluminum alloys where overheating of the weld joints deteriorates the mechanical properties [19]. Underwater FSW has wide range of applications in shipbuilding such as submarines fuel and oil tanks. Underwater FSW is considered one of the most appropriate methods due to its capability of producing good quality welds with respect to FSW and fusion welding [12, 15, 20].

The heat absorption capacity of water is more which reduces the coarsening of grains in weld region [17, 21, 22]. The weld surface texture as well as surface hardening is also found to enhance because water environment prevents oxidization and absorption of other elements from air at elevated temperature [18, 20, 23]. Thus, it can be effectively utilized for welding of aluminum alloys as they are prone to oxidation even at room temperature [24–26]. In order to enhance the joint strength, FSW is employed underwater. Moreover, it is considered as the great alternative to riveting which expands its range of application and applied extensively in shipbuilding industry, navy, and cryogenic fuel tanks [21, 27–29]. The advantage of FSW over riveting includes weight reduction, greater structural strength, and increased fatigue strength. The design of the tool plays vital role for successful FSW or underwater FSW. Flow of material, heat generation, and power requirement are the significant parameters during tool designing to obtain sound weld joint.

Tool pin geometry plays a significant role to enhance the material flow, stirring effect, heat generation; moreover, the role of tool shoulder is to produce large amount of frictional heat as well as prevents the weld zone material from escaping out. Ilangovan et al. [30] reported that the threaded type of pin profile in addition to the homogeneous flow of material enhanced mechanical and microstructural properties of weld joints are obtained. The significance of tool pin profile to enhance the weld joint quality is similar in conventional FSW as well as FSW in water environment. The flow of material in case of underwater FSW is complex; hence, the selection of correct tool geometry is necessary. Inappropriate selection of tool pin profile leads to the defect formation [31]. Lokesh et al. [32] evaluated the percentage contribution of cooling water in FSW and found that the contribution of optimized tool pin profile is 40% in enhancing the weld joint quality. Palanivel et al. [33] studied the effect of tool pin profile and found that the tapered profile produces tunnel defect and straight profile produces defect less weld. Moreover, the straight square profile gives highest joint strength. Akinlabi [34] studied the effect of shoulder size (diameter) on weld bead properties and found that 18 mm shoulder diameter shows optimum results compare to 15 and 25 mm shoulder diameter.

In addition to the tool design and material, selection of proper process parameters is an utmost necessity. The most important parameters in FSW are tool rotational speed and transverse speed. The effect of process parameters has been addressed by various researchers. Lee et al. [35] in their investigation found that with decrease in tool rotational speed the grains become more refined in SZ over and above smaller than the base metal grains size. Moreover, the average hardness value of SZ is 33% more as compared to base metals at tool rotational speed of 500 rpm. Liang et al. [4] examined during FSW process a soft region was formed adjacent to weld zone, fine grains are obtained at the stir zone and elongated grains are produced in TMAZ which reveals that the weld region is soft as compared to the hardness of base metals. Rodriguez et al. [36] concluded that by increasing the tool rotational speed the strength and material mixing of weld joint improves. However, a few literatures have come up to investigate the effect of weld joints formed by underwater friction stir welding, yet it is still in the premature stage and needs to be addressed properly. Kishta et al. [17] in their experimental investigation found that higher tool rotational and welding speed is required for producing sound weld in underwater FSW as compared to air FSW due to high thermal capacity and cooling rate of water. Furthermore, the frictional heat generation is more at higher tool rotational speeds and enhances the water circulation and cooling rate. Heirani et al. [37] demonstrated that heat-affected zone (HAZ) disappeared in the water-cooled specimen with optimized rotational and welding speed. Moreover, at optimized process parameters the hardness at the SZ of air-cooled samples was found 25% lower than the water-cooled samples. Liu et al. [28] divided the weld zone of water-cooled specimens in three different layers upper, middle, and lower and its homogeneity of mechanical properties were investigated and found that the homogeneity of middle and lower layer was superior than the upper layer. Hence, this improvement in homogeneity enhances the weld joint quality. According to the above literature survey, the pin profiles are selected square tapered, conical, and cylindrical for this study with a tool shoulder diameter of 18 mm.

The aim of this work is to study the influence of tool pin profiles at optimized process parameters. The optimization of process parameters are done in order to achieve stable welding zone for different environments in air and underwater FSW. Furthermore, the mechanical and macro-structural properties of friction stir welded 2024 aluminum alloys are compared for different pin profiles in air and underwater FSW.

2 Experimental Procedure

2.1 Workpiece Material

The material of choice in the present investigation is AA2024-T851 of dimension 150 × 80 × 5 mm is used for FSW process. The driving force behind using AA2024-T851 is the fact that T851 solution heat-treated aluminum alloy is resistant to stress corrosion cracking, high strength-to-weight ratio, good machining properties, stiffness, and resistant to fatigue stress. It has the property to be welded through FSW.

The chemical composition of AA2024 and the physical properties of the joint are given in Tables 1 and 2.

2.2 Tool Design

FSW tool composed of tool pin and shoulder as shown in Fig. 2, where H_1 and H_2 are the height of tool pin and shoulder and D_1 and D_2 are the diameter of pin and shoulder. Vertical force is applied on the rotating tool and slowly plunged on the plate which is to be joined. The combined action of tool pin and shoulder generates large amount of frictional heat which softens the material around pin produces stirring effect results in the mechanical mixing and bonding of the material [39].

The important factors of tool design are pin profile, diameter of shoulder including shape and size of tool [30, 40, 41]. Akinlabi [34] revealed the importance of shoulder size, with increase in diameter of shoulder the frictional surface area

Table 1 Chemical composition (wt%) of base metal (AA2024)

Elements	Cu	Mg	Mn	Si	Fe	Ti	Cr	Zr	Al
Base metal	4.3	1.6	0.4	0.3	0.2	0.08	0.05	0.10	Bal

Table 2 Physical properties of AA2024 [38]

S. No.	Properties	Values
1	Density	2780 kg/m^3
2	Thermal conductivity	121 W/mK
3	Melting point	775–911 K
4	Ultimate tensile strength	487 MPa
5	Yield strength	425.62 MPa
6	Young's modulus	73 GPa
7	Ductility	10.14%
8	Vickers hardness	146

Fig. 2 Tool geometry

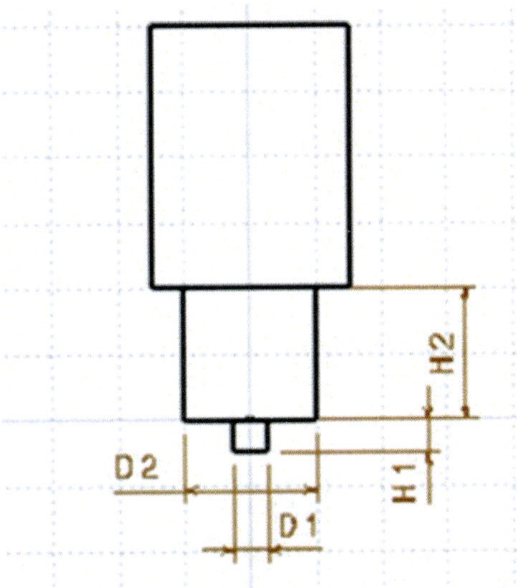

become more which leads to increase in temperature. The tool used for the present investigation is D3 tool or HCHCR. D3 tool steel is a high carbon high chromium tool steel with outstanding resistance to wear and high compressive strength. This is a heat treatable alloy and offers hardness value in the range of 58–64 HRC after heat treatment. In this study, tools of three different profiles are used for welding of AA2024-T851 plates and its effect on weld joint is investigated. The profiles are conical, straight cylindrical, and trapezoidal-type pin profile is designed depicted in Fig. 3. The tool geometry and pin profiles made are shown in Table 3, where D_S, H_S, D_P, H_P denote diameter of shoulder, height of shoulder, diameter of pin, height of pin.

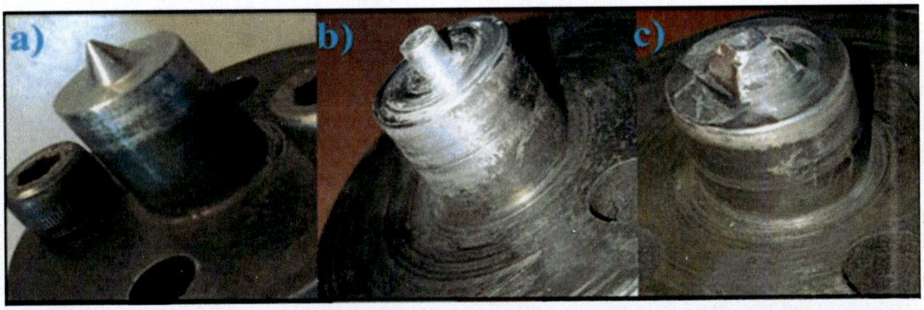

Fig. 3 Tool pin profiles: **a** conical; **b** straight cylindrical; **c** trapezoidal/square tapered type pin profile

2.3 Experimental Setup

The experiments would be carried out using a vertical milling machine with the necessary arrangements made for the welding process. Aluminum plate was clamped against the worktable with steel backup plate. Cast iron fixture is used for this process of friction stir welding. Fixture with a backing steel plate can be designed to prevent slipping of aluminum plate during welding. Nuts and bolts are used to clamp the plates on fixture to prevent shifting of plates. The plates are fixture together with side edge meet each other. The fixture will be fixed on the bed of vertical milling machine, and after setting of fixture on vertical milling machine, the further operation will be carried out the fixture used for the welding process for air and underwater is shown in Fig. 4. Underwater friction stir welding process is shown in Fig. 5.

2.4 Process Parameter

Optimization of process parameters is done for a successful weld joints. In this study, the optimization of tool rotational speed is carried out keeping the welding speed constant. The tool rotational speed is a key parameter for heat generation during welding process [17, 36]. Thus, its optimization becomes necessary, especially when the welding is done at water environment. For the optimization process, the selected rotational speed of tool for the welding process is 600, 800, 1000, 1200 rpm at constant tool transverse speed of 16 mm/min shown in Table 4.

The green zone of optimization table depicts the formation of stable weld zone. The red zone depicts no proper joining takes place due to the water environment large amount of frictional heat is carried away by water. At lower speeds, no proper stirring effect as well as groove defect is formed along the length of the weld. Such defects are formed due to lower frictional heat generation at lower tool rotational speeds; moreover, the generated heat was taken away by water results in the formation of groove defect shown in Fig. 6. By increasing the rotational speed of tool to 1200 rpm, higher frictional heat was generated and the turbulent flow of material as well as large weld area was observed which decreases the weld joint quality. At this speed due to higher frictional heat, overheating takes place along the advancing side of the tool which deteriorates the weld joint quality depicted in Fig. 7.

Table 3 Dimensions of tool pin profile

Tool	Pin profile	D_S (mm)	H_S (mm)	D_P (mm)	H_P (mm)
Tool 1	Conical pin (CP)	18	15	4	4.6
Tool 2	Straight cylindrical (SC)	18	15	4	4.6
Tool 3	Trapezoidal/Square tapered (ST)	18	15	5–3	4.6

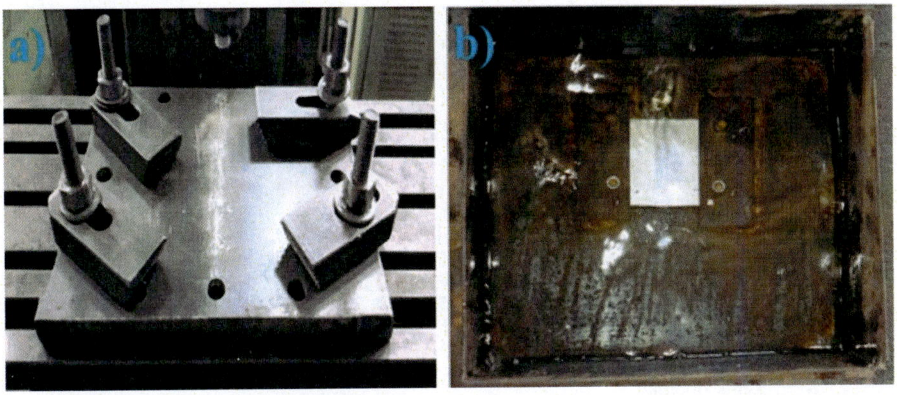

Fig. 4 FSW fixture arrangement: **a** air; **b** underwater

Fig. 5 Underwater FSW process

Table 4 Optimization of process parameters

N (Rotation speed in rpm), V (feed in mm per min.)	Underwater	Air
600,16	NO	YES
800,16	NO	YES
1000,16	YES	YES
1200,16	YES	YES

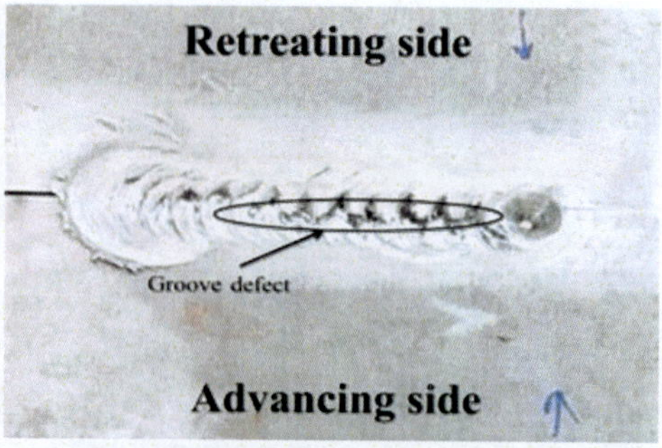

Fig. 6 Weld joints obtained at $N = 600$ rpm, $V = 16$ mm/min (underwater)

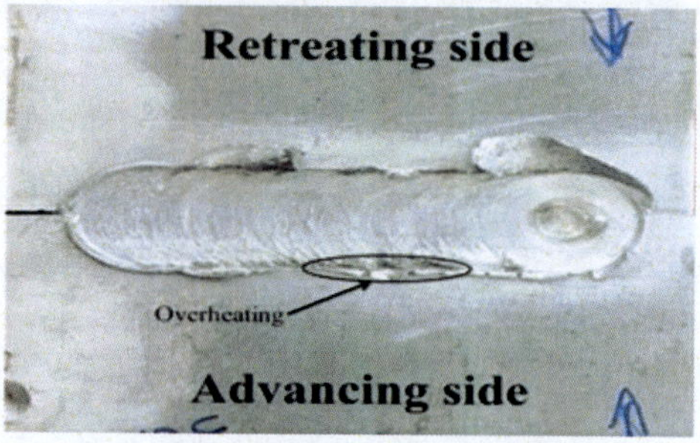

Fig. 7 Weld joints obtained at $N = 1200$ rpm, $V = 16$ mm/min (underwater)

So 1000 rpm was chosen as the rotation speed below it is not sufficient to cause the stirring effect. Proper welding is found at tool rotational speed of 1000 rpm and transverse speed of 16 mm/min, and this parameter is finally selected for further welding process and for testing the effect of tool pin profile. At this constant process parameter, 6 experiments are done from all the three types of pin profile in air and underwater friction stir welding. The weldment obtained by FSW in air and underwater exhibited stable welding with flash defects in surface and no other visible surface defects as shown in Figs. 8 and 9. The samples are considered for further analysis to check the mechanical and metallurgical properties of the weld joints.

3 Results and Discussion

In the current study, experiments were carried out on 5-mm-thick aluminum alloy by friction stir welding in air and underwater. The primary goal of the current study is to investigate the effect of design of tool pin profile on weld joints. The weld joints prepared were analyzed by microhardness test, micrographic analysis, and by tensile tests. A comparison is brought out to study characteristics of the joints in air and underwater.

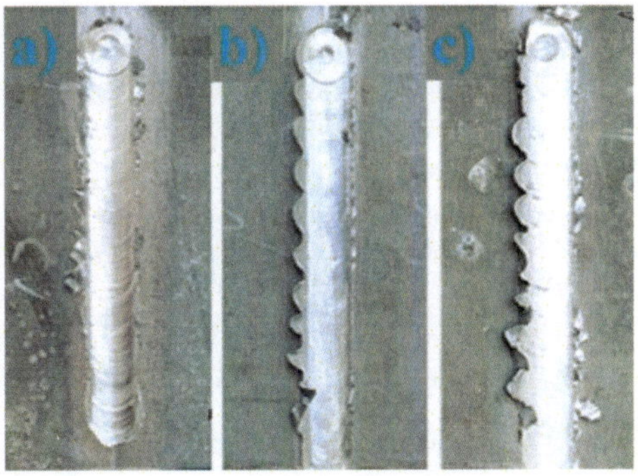

Fig. 8 Joints obtained at 1000 rpm and 16 mm/s in Air FSW by **a** conical pin; **b** cylindrical pin; **c** square tapered pin

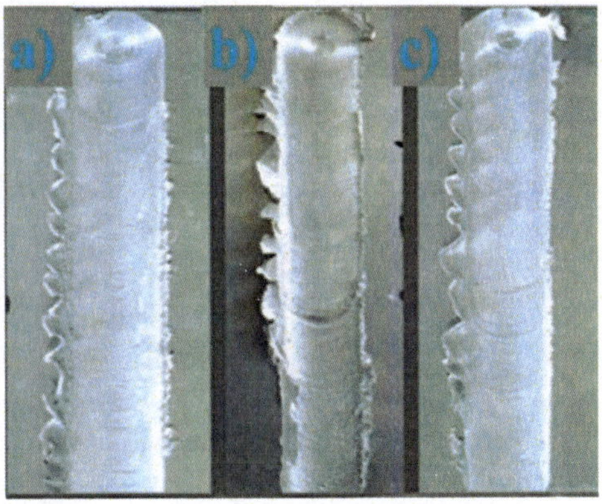

Fig. 9 Joints obtained at 1000 rpm and 16 mm/s by under water FSW where water is at room temperature around 25–30 °C by **a** conical pin; **b** cylindrical pin; **c** square tapered pin

3.1 Tensile Testing of the Joints

Transverse tensile test is carried out to evaluate the strength of the weldment. The ultimate tensile strength and elongation which is a measure of ductility are evaluated. The tensile property of the sample is evaluated according to ASTM E-08 standard. Two tensile specimens are prepared for each welding condition shown in Fig. 10, and the average value is considered here. The test was carried at a constant rate of 0.2 mm/min. The specimens are made by cutting in wire EDM. The tensile test is done in universal tensile testing machine to measure the yield strength, reduction in area, the elongation and ultimate tensile strength of the specimen. Sometimes the weld metal is stronger than parent metal and by applying the tensile force the failure occurs in the parent metal mainly at the heat-affected zone because the heat-affected zone is the weakest zone from all of the zone due to the fact the grain size is maximum here due to large amount of heat generation during welding there is growth in grains size.

In TMAZ thermo-mechanically affected zone, the grains size is bigger than the stirred zone and smaller than the heat-affected zone. The stirred zone is the strongest

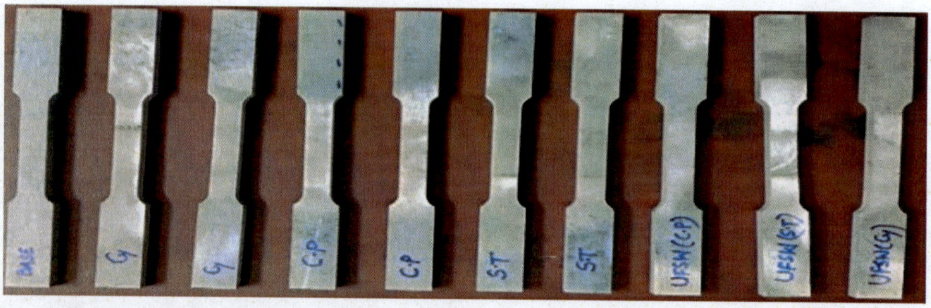

Fig. 10 Tensile test specimens for air and underwater

portion of the weld joint because the grains size is minimum here, and ultimately we can say that the strength of weld joint is inversely proportional to the grains size. It is found that most of the fracture occurs at HAZ zone in air FSW while case of water FSW the fracture occurred at weld zone. The stress–strain curve for the base metal and for different pin profiles is shown in Fig. 11. The tensile strengths as well as joint efficiency obtained from the several types of pin profiles are given in Tables 5 and 6.

Square tapered or trapezoidal-type pin profile shows maximum strength of joint efficiency of 65 and 83% for both air FSW and underwater FSW. While the conical pin profile shows the least joint efficiency due to the fact of lower stirring surface area of pin generates lower heat in the weld zone due to which the grain refinement is less as compared to tapered square pin profile.

Results shows that in underwater FSW there is 22.4% increase in tensile strength compared to the air FSW in tapered square pin profile. Comparison of tensile test graphs of weld joints obtained in square tapered pin profile for AFSW and UFSW are shown in Fig. 11h, respectively. Square tapered pin profiles show maximum result in both the cases for underwater and air FSW because the surface area of pin is higher than other profiles which generates large amount of heat required for proper fusion of parent material. Therefore, the joints obtained from square tapered pin profile are considered for further investigation and optical macrograph and hardness testing is carried out for this particular pin profile.

3.2 *Optical Macrograph of Weld Joints*

The optical macrograph of underwater friction stir welding UFSW and air friction stir welding are taken. Keller etchant was applied to the specimens for 15 s to expose the macrostructure of the joints after standard polishing down the specimens to 1 μm diamond paste. Figures 12 and 13 demonstrate the optical macrograph for FSW weld joints obtained in air and underwater by square tapered pin profile.

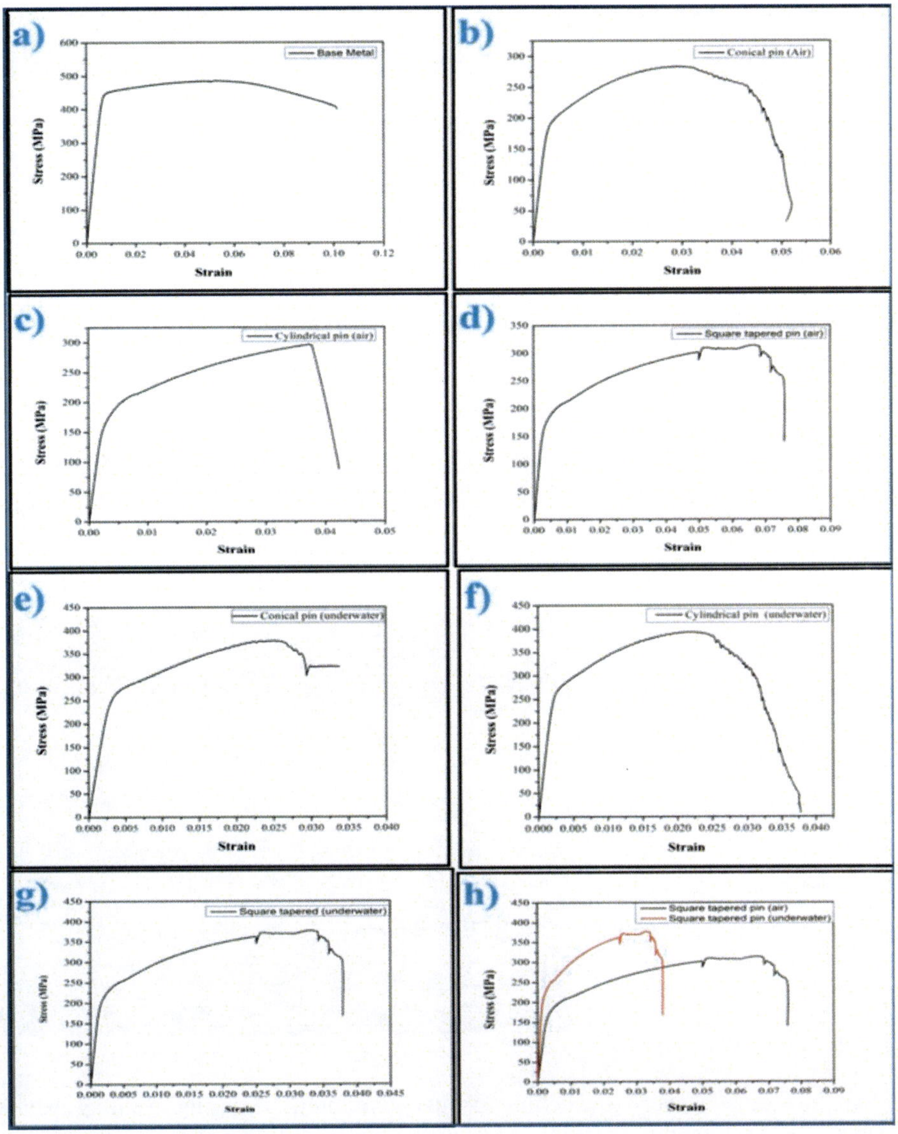

Fig. 11 Stress–strain curve for the base metal and for different pin profiles

The underwater welding shows improved mechanical properties as compared to the air FSW because the heat-affected zone in underwater FSW is absent. Moreover, the heat generated is absorbed by water and increases the cooling rate due to which grains gets very less time for its growth. Thus, adverse effect of grain growth is minimized in all region of underwater FSW which results in increase in mechanical properties of weld zone. However, the hardness of weld zone in case of underwater

Table 5 Tensile strength of the joints

	Conical	Square tapered	Cylindrical
Air FSW			
UTS—ultimate tensile strength (MPa)	284	314	297
YS—yield strength (MPa)	220	210	180
% Elongation	5	7.7	4.3
Underwater FSW			
UTS—ultimate tensile strength (MPa)	377	405	392
YS—yield strength (MPa)	280	201	175
% Elongation	2.3	2.6	2.3

Table 6 Joint efficiency of AFSW and UFSW

Pin profile	Conical	Square tapered	Cylindrical
Joint efficiency AFSW (%)	58	65	61
Joint efficiency UFSW (%)	77	83	80

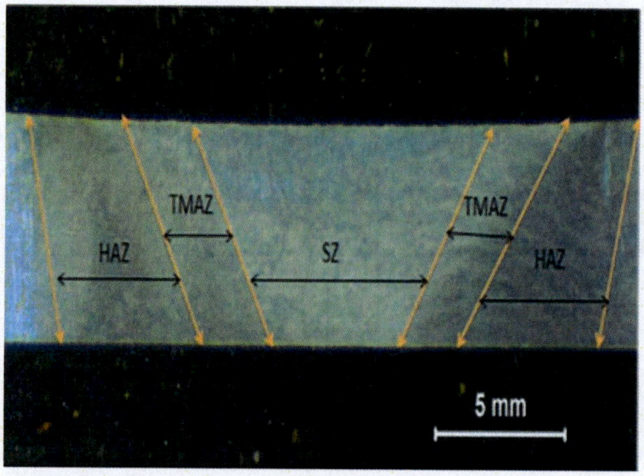

Fig. 12 Macrograph of air FSW

FSW increases because quenching occurs continuously during welding which makes the weld zone quite brittle and harder.

3.3 *Microhardness Test of Weld Joint*

Microhardness test was carried out, and 100 g of force is applied on the specimens for 15 s. The applied load and time may increase or decrease as per material hardness and mechanical properties every readings are taken 1.5 mm distance from

Fig. 13 Macrograph of underwater FSW

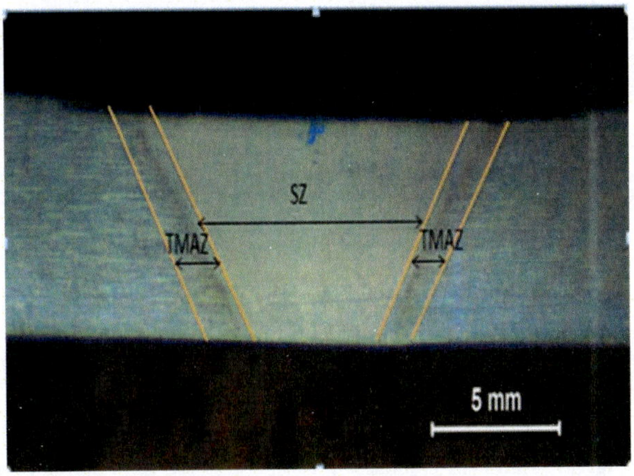

each other and total 13 indentation are done in the weld zone covering thermo-mechanically affected zone and heat-affected zone. The indentation image during testing is shown in Fig. 14. The maximum and minimum hardness comparisons of air FSW and UFSW are shown in Fig. 15. The hardness of joints obtained in AFSW and UFSW is found less than the base material hardness. The hardness obtained in underwater is 34% more than the air FSW. The elongation in case of underwater welding is less as compared to air friction stir welding because quenching occurs when the rotating tool moves forward the high-temperature zone comes to the direct contact of water which makes it brittle to some extent which ultimately increases its hardness. Formula for calculating Vicker's hardness (HV).

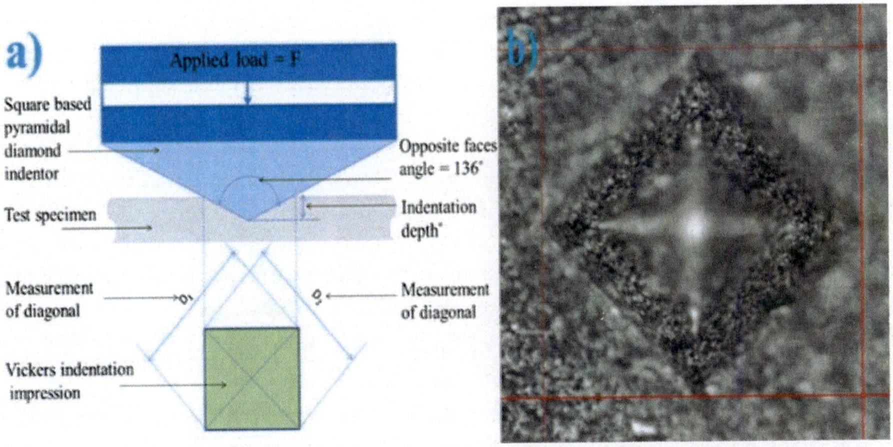

Fig. 14 **a** Vickers hardness test **b** indentation image during testing

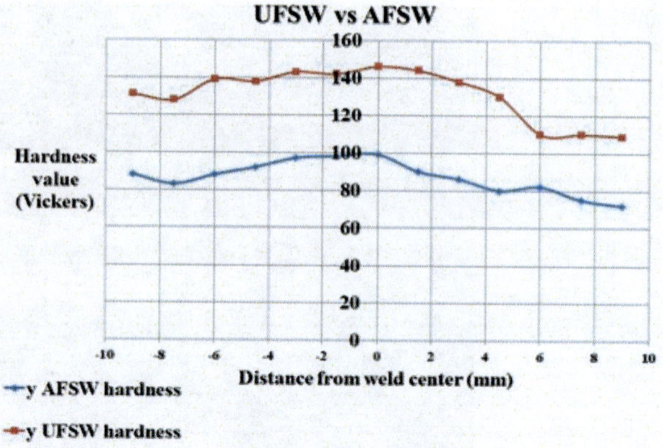

Fig. 15 Hardness comparison of air FSW (AFSW) and underwater FSW (UFSW)

Table 7 Average hardness value and percentage decrease of hardness in weld zone

	Average hardness value of weld zone (HRV)	% Decrease in hardness as compared to base metal (%)
Air FSW	86	40
Underwater FSW	131	11

$$(\text{HV}) = \frac{2F \times \text{Sin}(\theta/2)}{D_{\text{M}}}$$

where F = load applied, D_{M} = mean of the two diagonal imprints D_1 and D_2, θ = angle between two faces of indenter.

Table 7 shows the average hardness value and percentage decrease of hardness in weld zone of square pin profile in air and underwater friction stir welding.

4 Conclusions

The following conclusions are arrived at in the course of reviewing literatures, and conducting the experiments on friction stir welding process.

I Tool pin profile has a definite effect on weld joint characteristics. Tool geometry is important for producing sound welds.

II Based on the experiments undertaken, the straight cylindrical pin profile gives good surface finish but the other pin profiles produces flash defects at surface level.

III Optimum values of process parameters like rotational speeds and transverse speed are a must for good joints.
IV The square tapered pin profile shows higher tensile strength as compared to the other pin profiles.
V The microhardness test reveals that the hardness of underwater FSW is more than the air FSW.
VI The weldments showed no considerable distortion suggesting the robustness of the process and fixture adopted for the present investigation.
VII Optical macrograph reveals the absence of HAZ in UFSW which increases the strength.

References

1. Schneider JA, Nunes AC, Chen PS, Steele G (2005) TEM study of the FSW nugget in AA2195-T81. J Mater Sci 40:4341–4345. https://doi.org/10.1007/s10853-005-2808-8
2. Maeda M, Liu H, Fujii H, Shibayanagi T (2005) Temperature field in the vicinity of FSW-tool during friction stir welding of aluminium alloys. Weld World 49:69–75. https://doi.org/10.1007/BF03266478
3. Silva ACF, De Backer J, Bolmsjö G (2017) Temperature measurements during friction stir welding. Int J Adv Manuf Technol 88:2899–2908. https://doi.org/10.1007/s00170-016-9007-4
4. Liang XP, Li HZ, Li Z et al (2012) Study on the microstructure in a friction stir welded 2519-T87 Al alloy. Mater Des 35:603–608. https://doi.org/10.1016/j.matdes.2011.10.009
5. Xu W, Liu J, Luan G, Dong C (2009) Temperature evolution, microstructure and mechanical properties of friction stir welded thick 2219-O aluminum alloy joints. Mater Des 30:1886–1893. https://doi.org/10.1016/j.matdes.2008.09.021
6. Priya R, Subramanya Sarma V, Prasad Rao K (2009) Effect of post weld heat treatment on the microstructure and tensile properties of dissimilar friction stir welded AA 2219 and AA 6061 alloys. Trans Indian Inst Met 62:11–19. https://doi.org/10.1007/s12666-009-0002-4
7. Leitao C, Leal RM, Rodrigues DM et al (2009) Mechanical behaviour of similar and dissimilar AA5182-H111 and AA6016-T4 thin friction stir welds. Mater Des 30:101–108. https://doi.org/10.1016/j.matdes.2008.04.045
8. Miles MP, Pew J, Nelson TW, Li M (2005) Formability of friction stir welded dual phase steel sheets. Frict Stir Weld Process III—Proc a Symp Spons by Shap Form Comm Miner Met Mater Soc TMS 18:91–96
9. Bitondo C, Prisco U, Squillace A et al (2010) Friction stir welding of AA2198-T3 butt joints for aeronautical applications. Int J Mater Form 3:1079–1082. https://doi.org/10.1007/s12289-010-0958-y
10. De Backer J, Verheyden B (2009) Robotic friction stir welding for automotive and aviation applications. Mech Eng 57
11. Devaraju A, Kishan V (2018) Influence of cryogenic cooling (liquid nitrogen) on microstructure and mechanical properties of friction stir welded 2014-T6 aluminum alloy. Mater Today Proc 5:1585–1590. https://doi.org/10.1016/j.matpr.2017.11.250
12. Mofid MA, Abdollah-Zadeh A, Ghaini FM, Gür CH (2012) Submerged friction-stir welding (SFSW) underwater and under liquid nitrogen: An improved method to join Al alloys to Mg alloys. Metall Mater Trans A Phys Metall Mater Sci 43:5106–5114. https://doi.org/10.1007/s11661-012-1314-2
13. Fratini L, Buffa G, Shivpuri R (2009) In-process heat treatments to improve FS-welded butt joints. Int J Adv Manuf Technol 43:664–670. https://doi.org/10.1007/s00170-008-1750-8

14. Clark TD (2005) An analysis of microstructure and corrosion resistance in underwater friction stir welded 304L stainless steel. Brigham Young University
15. Hui-jie LIU, Hui-jie Z, Yong-xian H, Lei YU (2009) Mechanical properties of underwater friction stir welded 2219 aluminum alloy. Trans Nonferrous Met Soc China 20:1387–1391. https://doi.org/10.1016/S1003-6326(09)60309-5
16. Zhang HJ, Liu HJ, Yu L (2011) Microstructure and mechanical properties as a function of rotation speed in underwater friction stir welded aluminum alloy joints. 32:4402–4407. https://doi.org/10.1016/j.matdes.2011.03.073
17. Kishta EE, Darras B (2014) Experimental investigation of underwater friction-stir welding of 5083 marine-grade aluminum alloy. 1–8. https://doi.org/10.1177/0954405414555560
18. Hosseini M, Danesh Manesh H (2010) Immersed friction stir welding of ultrafine grained accumulative roll-bonded Al alloy. Mater Des 31:4786–4791. https://doi.org/10.1016/j.matdes.2010.05.007
19. Xu WF, Liu JH, Chen DL et al (2012) Improvements of strength and ductility in aluminum alloy joints via rapid cooling during friction stir welding. Mater Sci Eng A 548:89–98. https://doi.org/10.1016/j.msea.2012.03.094
20. Sree Sabari S, Malarvizhi S, Balasubramanian V, Madusudhan Reddy G (2016) Experimental and numerical investigation on under-water friction stir welding of armour grade AA2519-T87 aluminium alloy. Def Technol 12:324–333. https://doi.org/10.1016/j.dt.2016.02.003
21. Liu HJ, Zhang HJ, Yu L (2011) Effect of welding speed on microstructures and mechanical properties of underwater friction stir welded 2219 aluminum alloy. Mater Des 32:1548–1553. https://doi.org/10.1016/j.matdes.2010.09.032
22. Wang Q, Zhao Z, Zhao Y et al (2016) The strengthening mechanism of spray forming Al–Zn–Mg–Cu alloy by underwater friction stir welding. Mater Des 102:91–99. https://doi.org/10.1016/j.matdes.2016.04.036
23. Zhang HJ, Liu HJ, Yu L (2013) Thermal modeling of underwater friction stir welding of high strength aluminum alloy. Trans Nonferrous Met Soc China (English Ed) 23:1114–1122. https://doi.org/10.1016/S1003-6326(13)62573-X
24. Zhang Z, Xiao BL, Ma ZY (2015) Enhancing mechanical properties of friction stir welded 2219Al-T6 joints at high welding speed through water cooling and post-welding artificial ageing. Mater Charact 106:255–265. https://doi.org/10.1016/j.matchar.2015.06.003
25. Sinhmar S, Dwivedi DK (2017) Enhancement of mechanical properties and corrosion resistance of friction stir welded joint of AA2014 using water cooling. Mater Sci Eng A 684:413–422. https://doi.org/10.1016/j.msea.2016.12.087
26. Venkateswarulu D, Cheepu M, Krishnaja D, Muthukumaran S (2018) Influence of water cooling and post-weld ageing on mechanical and microstructural properties of the friction-stir welded 6061 aluminium alloy joints. Appl Mech Mater 877:163–176. https://doi.org/10.4028/www.scientific.net/amm.877.163
27. Jariyaboon M, Davenport AJ, Ambat R et al (2009) The effect of cryogenic CO_2 cooling on corrosion behaviour of friction stir welded AA2024-T351. Corros Eng Sci Technol 44:425–432. https://doi.org/10.1179/147842208X373173
28. Liu HJ, Zhang HJ, Yu L, et al (2011) Homogeneity of mechanical properties of underwater friction stir welded 2219-T6 aluminum alloy. 20:1419–1422. https://doi.org/10.1007/s11665-010-9787-x
29. Garg T, Mathur P, Singhal V et al (2014) Underwater friction stir welding: an overview. Int Rev Appl Eng Res 4:2248–9967
30. Ilangovan M, Rajendra Boopathy S, Balasubramanian V (2015) Effect of tool pin profile on microstructure and tensile properties of friction stir welded dissimilar AA6061–AA5086 aluminium alloy joints. Def Technol 11:174–184. https://doi.org/10.1016/j.dt.2015.01.004
31. Zhang H, Liu H (2012) Characteristics and formation mechanisms of welding defects in underwater friction stir welded aluminum alloy. 269–281. https://doi.org/10.1007/s13632-012-0038-4

32. Lokesh R, Senthil Kumar VS, Rathinasuriyan C, Sankar R (2015) Optimization of process parameters: tool pin profile, rotational speed and welding speed for submerged friction stir welding of AA6063 alloy. Int J Tech Res Appl 1:35–38
33. Palanivel R, Koshy Mathews P, Murugan N, Dinaharan I (2012) Effect of tool rotational speed and pin profile on microstructure and tensile strength of dissimilar friction stir welded AA5083-H111 and AA6351-T6 aluminum alloys. Mater Des 40:7–16. https://doi.org/10.1016/j.matdes.2012.03.027
34. Akinlabi ET (2012) Effect of shoulder size on weld properties of dissimilar metal friction stir welds. J Mater Eng Perform 21:1514–1519. https://doi.org/10.1007/s11665-011-0046-6
35. Lee WB, Yeon YM, Jung SB (2003) Evaluation of the microstructure and mechanical properties of friction stir welded 6005 aluminum alloy. Mater Sci Technol 19:1513–1518. https://doi.org/10.1179/026708303225008068
36. Rodriguez RI, Jordon JB, Allison PG et al (2015) Microstructure and mechanical properties of dissimilar friction stir welding of 6061-to-7050 aluminum alloys. Mater Des 83:60–65. https://doi.org/10.1016/j.matdes.2015.05.074
37. Heirani F, Abbasi A, Ardestani M (2017) Effects of processing parameters on microstructure and mechanical behaviors of underwater friction stir welding of Al5083 alloy. J Manuf Process 25:77–84. https://doi.org/10.1016/j.jmapro.2016.11.002
38. ASM Material Data Sheet. http://asm.matweb.com/search/SpecificMaterial.asp?bassnum=MA2024T851. Accessed 24 Feb 2019
39. Thomas WMW, Norris I, Nicholas ED et al (1991) Friction stir welding process developments and variant techniques. Google patents 1–21
40. Jamshidi Aval H (2015) Influences of pin profile on the mechanical and microstructural behaviors in dissimilar friction stir welded AA6082-AA7075 butt joint. Mater Des 67:413–421. https://doi.org/10.1016/j.matdes.2014.11.055
41. Jamshidi Aval H, Serajzadeh S, Kokabi AH, Loureiro A (2011) Effect of tool geometry on mechanical and microstructural behaviours in dissimilar friction stir welding of AA 5086–AA 6061. Sci Technol Weld Join 16:597–604. https://doi.org/10.1179/1362171811Y.0000000044

Performance Evaluation of Spiral Separator for Coal Cleaning Using Mixture Model

Vandana Kumari Jha and Soubhik Kumar Bhaumik

1 Introduction

The spiral classifier, recently, has emerged as the most effective gravity separator since its introduction. The spiral separator also known as spiral concentrator, continues to be the most effective gravity separator, useful in processing a wide variety of minerals such as coal, iron, chromite, gold, zircon and other heavy minerals ores [1, 2]. The popularity of the spiral separator is ascribed to the associated low capital investment, high efficiency, low energy demands, environment-friendly features and compact design [3–5]. With the increase in the demands of coal fines recovery, spiral separators are widely deployed in the concentration process of 0.1–3 mm coal. A recent survey shows spiral separators to be responsible for treating 6% of the coal in processing plants [6]. Apart from their use in coal and mineral processing plants, they are also effective in removing soil from other environmentally unwanted constituents [7].

A spiral separator consists of an open trough/sluice twisted downward in a helical arrangement about a central axis (Fig. 1a) [8]. Separation is caused by the centrifugal forces that combine with gravity to drive smaller or lighter particles towards the outer area of the channel, whereas the larger or denser particles remain in the inner region. A secondary circulation sets in across the trough that helps in the movement of particles from one zone to the other as shown in Fig. 1b, from inward at the base of trough to outward along the free surface. Gravity drives the slurry along the trough in the form of a shallow layer, entraining the mineral particles during flow. Other forces such as the drag force and Bagnold force also influence depending on particle size rather than weight [9].

Previous research works, focus on the development of spiral separators for cleaning coal and several other minerals. Both theoretical and experimental works

---

V. K. Jha · S. K. Bhaumik (✉)
Indian Institute of Technology (Indian School of Mines), Dhanbad, India

S. Bag et al. (eds.), *Next Generation Materials and Processing Technologies*,
Springer Proceedings in Materials 9, https://doi.org/10.1007/978-981-16-0182-8_8

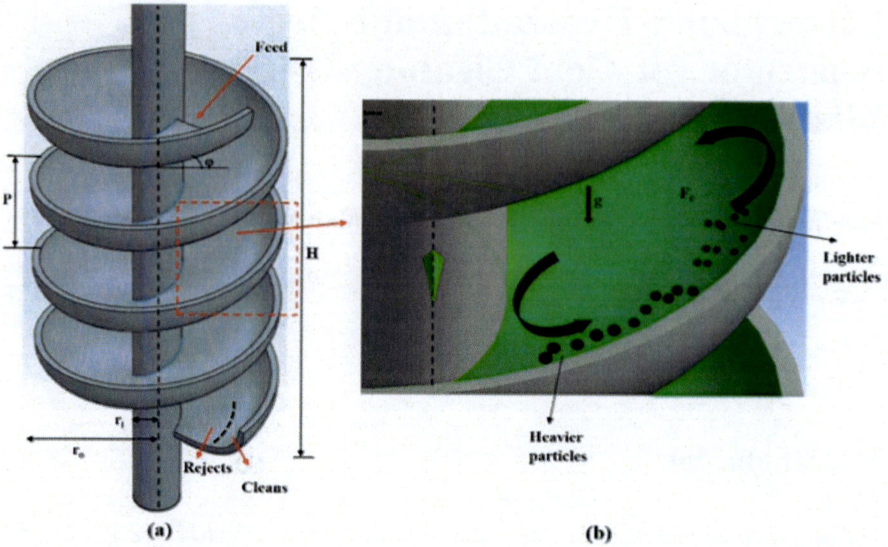

Fig. 1 Schematic showing **a** spiral separator; **b** particle distribution during separation on spiral trough

have been carried out to study the slurry behavior flowing over spiral troughs having the ability to achieve density-based separations [10]. Spiral performance for cleaning coal has been optimized using various empirical models. Mathematical correlations for the prediction of performance exist that are based on size of minerals, but are specific to the spiral and type of mineral being treated [11–14] and thus cannot be applied generally. Furthermore, the empirical approach fails to address the fundamental physics lying behind the particle separation of a spiral, thus putting more emphasis on numerical flow modeling. Numerous numerical works are reported in literature that can be analyzed to predict the spiral performance. The effect of the feed rate, pulp density, favorable feed rate for concentrate grade, recovery, etc., are performed using different techniques [15, 16]. Mechanistic models, codes, CFD modeling, dilute particle flow model and development of two-phase VOF model are carried out under certain assumptions for establishing a more reliable model for the spiral operation [17–20]. Despite these reported works in the literature on two-phase flow fields, there is still scope to investigate the fundamentals of particle separation that can help in the analysis of spiral performance.

In the above context, a predictive tool for flow characteristics and the separation characteristics is indispensable both for design optimization and determination of the operating range, for the existing spiral designs. CFD technology in modeling the slurry flow has been increasingly employed for visualizing the flow behavior and establishing trends that can help reduce the number of experimental trial runs, and thus the operational cost. As yet, works have been reported in literature based on shallow depths and main stream velocities using different modeling approaches. However, for coal slurry flow and separation behavior, mixture model has never been employed earlier. Use of CFD for design enhancements have to be further explored for parametric studies to check the influence of operating parameters on the flow that affect the separation efficiency. A dedicated work on this will be useful for optimizing the operating range.

In the present work, the potential of using mixture model for coal processing is examined computationally. CFD simulations are conducted to analyze the flow and separation characteristics of falling coal slurry along the spiral trough by employing a two-phase mixture model and using standard k-ε turbulence model. The performance study of spiral separator is carried out with the help of computational results by analyzing the two-phase flow field and influence of process parameter, i.e., feed rate on the distribution of coal slurry. The size of the coal is kept −0.5 mm and the feed rate is varied as 4, 6 and 8 m^3/hr. The performance study of the spiral separator in a five-turn assembly is carried out to obtain the desired quality product. Mixture model can be established as a useful numerical tool in order to understand the separation behavior of coal particles in a spiral separator.

2 Theory

2.1 Geometry

The spiral separator has an open channel twisted around a central column acting as support as shown in Fig. 1. The spiral geometry is specified by diameter, pitch, spiral length or number of turns, and shape or width of the trough. The geometrical parameters of the spiral separator considered in the present work are provided below:

(a) Inner radius = 70 mm; (b) outer radius = 350 mm; (c) trough width = 280 mm; (d) spiral pitch = 273 mm; (e) curvature = 0.75; and (f) number of turns = 5.

The feed consists of water and coal particle mixture that enters at the top of the spiral under gravitational effects. The main operating variables being the feed rate, the pulp density, the wash water addition and the position of the cutters or splitters that are used to separate the tailings, middlings and concentrate streams.

3 CFD Modeling

The CFD analysis is carried out using a commercial CFD package of Ansys Fluent 15.0. The dynamics of slurry flow is modeled with the help of the mixture model and turbulent flow interface.

3.1 Geometry and Mesh

The computational domain comprises of a five-turn spiral separator having a shallow width (0.001–0.020 m) along the spiral trough as shown in Fig. 2 along with the boundary conditions. The domain is meshed with structured hexahedral mesh elements with size varying in mainstream (*x*), cross-stream (*y*) and depth-wise (*z*) directions. Throughout the domain, mesh density is kept high to resolve the sharp interface, and at the wall to capture boundary layer effects. Along the axial direction, mesh size is kept reasonable in order to capture the curvature. A grid independent test is carried out for different grid sizes by comparing the primary velocity profiles at the outlet as shown in Fig. 3. The optimal mesh size is fixed at 67,500 number of cells through grid independent study, conducted on separate meshes of sizes leading to 45,000, 52,500, 67,500 and 75,000 number of cells, respectively. Refinement below this size showed no further change in the primary velocity profiles.

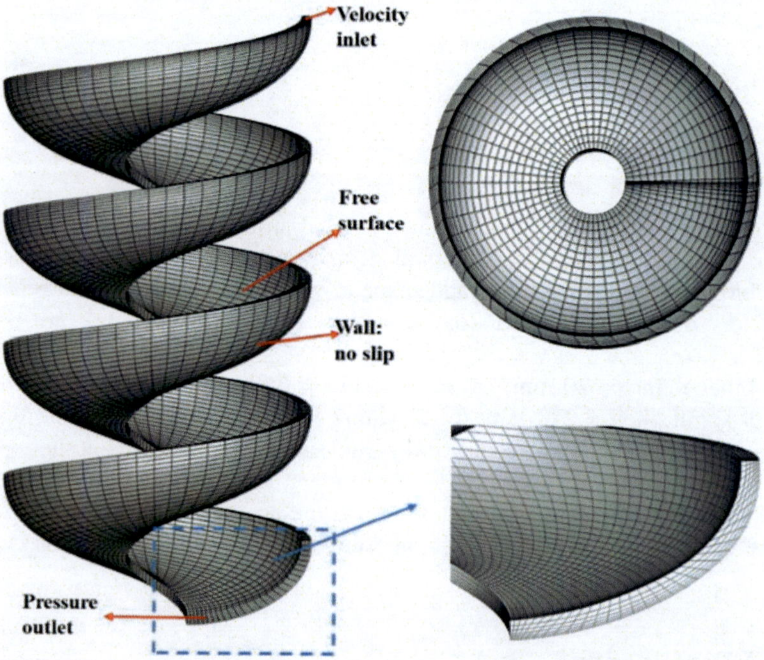

Fig. 2 Computational domain showing the meshed geometry of the spiral along with the applied boundary condition

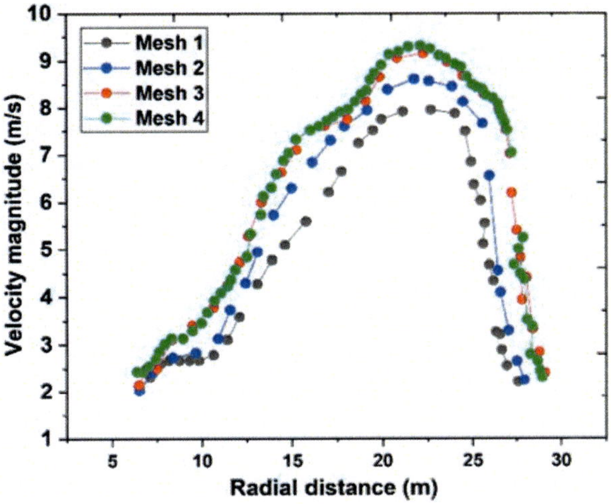

Fig. 3 Plot showing the primary velocity profiles at the outlet for different grid sizes

3.2 Governing Equations

The flow dynamics can be modeled using a momentum transport equation, continuity equation and a transport equation for the solid phase volume fraction considering the flow Newtonian and turbulent. The mixture model facilitates the movement of two phases at different velocities incorporating the concept of slip velocities. The governing equations can be written as follows:

Momentum equation:

$$\rho\frac{\partial j}{\partial t} + \rho(j.\nabla)j + \rho_c\varepsilon(j_{\text{slip}}.\nabla)j = -\nabla p - \nabla\mu\left[\nabla j + \nabla j^T\right] + \rho g \\ -\nabla.\left[\rho_c(1-\phi_c\varepsilon)u_{\text{slip}}j^T_{\text{slip}}\right] - \rho_c\varepsilon(j.\nabla)j_{\text{slip}} \quad (1)$$

Continuity equation:

$$\nabla.j = 0 \quad (2)$$

Transport equation for solid phase volume fraction:

$$\frac{\partial\phi_s}{\partial t} + \nabla.(\phi_s u_s) = 0 \quad (3)$$

3.3 Boundary Conditions

The spiral domain is surrounded by four boundaries, namely: inlet, outlet, solid trough wall and top wall. At inlet, velocity components and volume fractions of coal particles are specified. At outlet, velocity gradients are set to zero. At the trough wall, no-slip and no-penetration conditions are imposed. Equations (1)–(3) are solved subject to the following boundary conditions:

$$\text{Inlet}: \quad v = v_i \quad \text{(constant velocity)} \tag{4a}$$

$$\text{Outlet}: \quad P = P_{\text{atm}} \quad \text{(pressure outlet)} \tag{4b}$$

$$\text{Trough wall}: \quad v = 0 \quad \text{(no slip)} \tag{4c}$$

$$\text{Top wall}: \quad \tau = 0 \quad \text{(no shear)} \tag{4d}$$

4 Results and Discussion

4.1 Primary Velocity

Figure 4 shows the contour plots of primary velocity profiles of the primary velocity at a vertical plane cut across the middle of the spiral separator colored in total velocity magnitude. The predicted primary velocity profiles along the trough width for three different flow rates, i.e., 4, 6 and 8 m^3/hr show that the magnitude of primary velocity increases smoothly across the width of the trough in the radial direction away from the central supporting column or axis, qualitatively. Inset is provided to make the visibility clearer. High velocity zone (red) can be seen in the outer region of the trough. With the increase in flow rate, there is a slight effect on the flow velocity profile in the inner region than compared to the velocity profiles at the outer regions of the spiral trough. The results established are found to be in line with the results of the work carried out by Doheim et al. [21] that was based on VOF and turbulent modeling. Two graphs are plotted in order to support the velocity contour plots quantitatively that follow the same trend for a flow rate of 8 m^3/hr (Fig. 5).

4.2 Water Depth

It is observed that the water depth increases in the outward direction of the spiral trough following the trends of primary velocity profiles as shown in Fig. 5. As the slurry flows down the spiral, centrifugal and other forces mentioned previously act

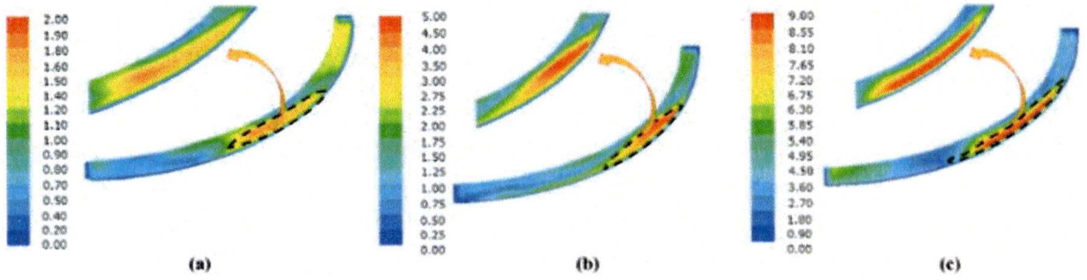

Fig. 4 Contours showing the primary velocity profiles for three different flow rates, i.e., **a** 4 m^3/hr; **b** 6 m^3/hr; **c** 8 m^3/hr

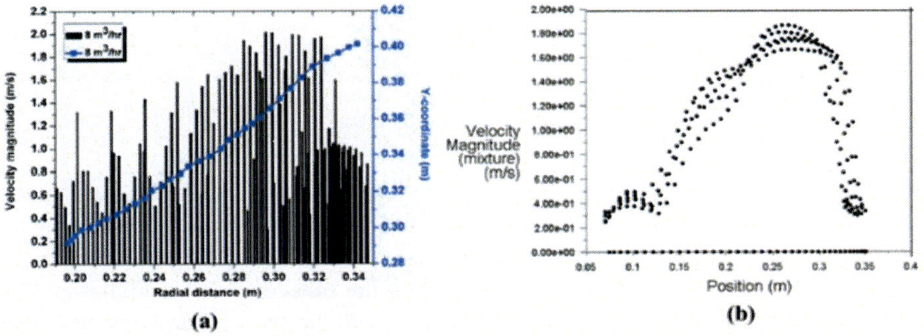

Fig. 5 Variation of primary velocity profiles along the width of the spiral trough for 4 m^3/hr shown through **a** column graph; **b** point graph

on it, resulting in the build-up of more water in the outer region of the trough. This causes the increase in water depth along the radial distance from the central axis. A quantitative comparison is drawn between the numerical simulations of water depth for the three different flow rates of 4, 6 and 8 m^3/hr, respectively (Fig. 6). It can be seen that the span of water depth increases with the flow rate on the trough width, and the water depth also increases with the increase in flow rate. However, the effect of flow rate is lesser on the water depth in the inner region than compared to the outer region of the trough.

5 Conclusions

Two-phase flow field simulations of coal-water slurry are conducted using mixture model in which the flow–particle interactions are taken into account by employing a standard k-ε turbulence model. The following conclusions can be drawn based on the above analysis:

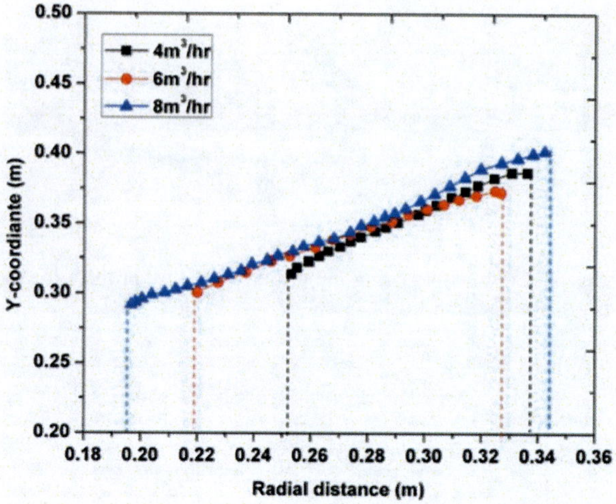

Fig. 6 Plot showing the span variation of water depth profiles along the width of the spiral trough for three different flow rates

- Water velocity profiles on the spiral trough can be predicted using the present mixture model. Primary velocity across the spiral trough increases along the radial distance from the vertical central axis. The effect of the flow rate on the primary velocity can be observed mainly in the outer zone of trough width.
- The water depth on spiral trough increases along the radial distance of the trough, i.e., from the central axis of the spiral separator. Also, as the slurry flow rate increases, the depth of water increases. Variation of water depth is more pronounced in the outer region of the trough compared to the inner region.
- The suggested computational model is applicable for any type of spiral separator just by modifying its geometrical parameters. Furthermore, it is not a very costly turbulence model as compared to other models that require high run-time and storage memory.

The above results can be validated and thus can prove helpful in the design optimization of the spiral separator by carrying out experiments more elaborately, as a future work.

Acknowledgements The authors acknowledge the assistance provided by the Coal Preparation Division of CIMFR, Digwadih campus, for the useful discussions and comments provided.

References

1. Wills BA (1996) Mineral processing technology, Fifth edn. Pergamon Press
2. Atasoy Y, Spottiswood DJ (1995) A study of particle separation in a spiral concentrator. Mineral Eng 8(10):1197–1208
3. Bazin C, Proulx M (2000) Distribution of reagents down a flotation bank to improve the recovery of coarse particles. Int J Miner Process 61(1):1–12

4. Kempnich RJ (2003) Coal preparation—a world review. In: Proceedings of the 20th international coal preparation conference, Lexington, Kentucky, pp 15–40
5. Sadeghi M, Bazin C, Devin PO, Lavoie F, Hodouin D, Renaud M (2016) Control of a spiral classifiers for the concentration of iron ore. Submitted to XXVIII international mineral processing congress (IMPC 2016), Quebec City, QC, Canada
6. Holland-Batt AB, Holtham PN (1991) Particle and fluid motion on spiral separators. J Minerals Eng 4(3–4):457–482
7. Holtham PN (1992) Primary and secondary fluid velocities on spiral separators. J Minerals Eng 5(1):79–91
8. Tucker P, Lewis KA, Hobba WJ, Wells D (1985) A mathematical model of spiral concentration, as part of a generalized gravity-process simulation model, and its application at two Cornish tin operations. In: 15th International mineral processing congress, Cannes, France, pp 3–15
9. Kelly EG, Gomer JS, Pillai KJ, Bull WR, Spottiswood DJ (1989) The development and application of mathematical models of concentrating spirals
10. Holland-Batt AB (1989) Spiral separation: theory and simulation. Trans Instn Min Metall (Sect C Min Process Extr Metall) 98:C46–C60
11. Holland-Batt AB, Holtham PN (1991) Particle and fluid motion on spiral separators. Mineral Eng 4:457–482
12. Jancar ML, Holtham PN, Davis JJ, Fletcher CAJ (1995) Approaches to the development of coal spiral models. In: An international symposium, Society for Mining, Metal and Exploration Littliton, USA, pp 335–345
13. Jancar T, Fletcher CAJ, Holtham PN, Reizes JA (1995) Computational and experimental investigation of spiral separator hydrodynamics. In: Proceedings XIX international mineral processing congress, San Francisco, USA
14. MacNamara L, Addison F, Miles NJ, Bethell P, Davies P (1995) The application of new configurations of coal spirals. In: Proceedings, 12th international coal preparation conference and exhibit, Lexington, Kentucky, 2–4 May, pp 43–52
15. Matthews BW, Fletcher CAJ, Partridge AC, Vasquez S (1999) Computations of curved free surface water flow on spiral concentrators. J Hydraul Eng 125(11):1126–1139
16. Matthews BW, Fletcher CAJ, Partridge TC (1999) Particle flow modeling on spiral concentrators: benefits of dense media for coal processing. In: Second international conference on CFD in the minerals and process industries, CSIRO, Melbourne, Australia, 6–8 Dec, 1999
17. Matthews BW, Fletcher CAJ, Partridge AC (1998) Computational simulation of fluid and dilute particulate flows on spiral concentrators. Appl Math Model 22:965–979
18. Das SK, Godiwalla KM, Panda L, Bhattacharya KK, Singh R, Mehrotra SP (2007) Mathematical modeling of separation characteristics of a coal-washing spiral. Int J Miner Process 84:118–132
19. Doheim MA, Abdel Gawad AF, Mahran GM, Abu-Ali MH, Rizk AM (2008) Computational prediction of water-flow characteristics in spiral separators: part II, the primary and secondary flows. J Eng Sci 34(4):951–961
20. Suchandan K, Kumari P, Bhattacharyya K, Singh R (2010) A mathematical model to characterize separation behavior of a spiral for processing iron ore using a mechanistic approach. In: Proceedings of the XI international seminar on mineral processing technology (MPT-2010), NML Jamshedpur, India, Dec, 2010
21. Doheim MA, Abdel Gawad AF, Mahran GM, Abu-Ali MH, Rizk AM (2008) Computational prediction of water-flow characteristics in spiral separators: part I, flow depth and turbulence intensity. J Eng Sci 34(4):935–950

Mixing Performance Analysis of Serpentine Microchannels with Straight and Curved Bends

Sandeep S. Wangikar, Promod Kumar Patowari, Rahul Dev Misra, Ranjit Gidde, Subhash Jadhav, and Sachin Sonawane

1 Introduction

Microfluidics is a term which deals with transport phenomena at the microscopic scales and techniques and components employed for controlling and actuating the fluids. The microfluidic systems are the fast-growing technology, and the microfluidics study is significant for implementing lab-on-a-chip (LOC). The LOC systems are moreover recognized as micro total analysis systems (μTAS) that can execute maximum stages of chemical and biological processes [1, 2]. Microfluidics has many applications in many different fields including cosmetics, pharmaceuticals, biotechnology, medicine, and also in physical sciences for control systems and heat management. A microchannel is one of the vital components of microfluidic systems. The microchannel is a channel that has a height and width in the order of micrometers (μm). A microchannel which mixes fluid is called a micromixer. The geometries are built into the circuits which are known as microfluidic chips. This technology has been the reason for a good deal of research, as it offers a means for carrying out the key chemical evaluation processes in the biomedical field [1–3].

S. S. Wangikar (✉) · R. Gidde · S. Jadhav · S. Sonawane
Department of Mechanical Engineering, SVERI's College of Engineering, Pandharpur, Maharashtra, India
e-mail: sswangikar@coe.sveri.ac.in

R. Gidde
e-mail: rrgidde@coe.sveri.ac.in

S. Jadhav
e-mail: svjadhav@coe.sveri.ac.in

P. K. Patowari · R. D. Misra
Department of Mechanical Engineering, National Institute of Technology, Silchar, Assam, India

S. Bag et al. (eds.), *Next Generation Materials and Processing Technologies*, Springer Proceedings in Materials 9, https://doi.org/10.1007/978-981-16-0182-8_9

Micromixers can be classified into two main categories as active and passive. Each of these micromixers has different capacity, mixing speed, and operating requirements. For example, an active micromixer requires power input to make mixing possible in the device.

In contrast, a passive micromixer achieves mixing with the applied pressure for fluid motion. As such, some micromixers are more suitable for a particular application than the others. Active micromixer generally provides correct mixing, but their fabrication is cost-intensive, and integration with different devices is difficult. For this reason, passive micromixers are favored in several situations [4–6].

Many researchers report the mixing performance of passive microchannels. The various geometries like wavy structure, curved shape, static micromixers, 3D serpentine, square wave, straight microchannels, spiral-shaped microchannel, serpentine microchannel with cyclic L-shaped units, serpentine microchannel with non-aligned inputs, etc., have been used by different researchers for analyzing the effect of geometry/shape on the mixing performance analysis [7–15]. Many researchers have also studied the microchannel-based on the split and recombine (SAR) process. In SAR, the two fluids to be mixed are split and recombined to optimize the diffusion process. The different passive micromixer configurations developed by various researchers are planar SAR micromixer, micromixer using two-dimensional (2D) modified Tesla structures, two-layer crossing channels (TLCCM), ellipse-like micropillars, P-SAR micromixer with cavities (fan-shaped), modified P-SAR micromixer with dislocation sub-channels, etc. The mixing performance has been observed enhanced due to the SAR process and the subsequent chaotic advection [16–18]. Different researchers report some numerical investigations on mixing behavior of microchannels using different types of obstacles and grooves along the mixing path. [19–24] and reported that recirculation zones are created downstream of these obstructions, which resulted in mixing performance enhancement. Few of the researchers have fabricated the microchannels using different methods like laser machining, photochemical machining, micro-milling, etc. [25–35].

Based on the above reported various studies, it is noted that the microchannel is governed by the two main parameters as pressure drop and mixing index (or mixing length). Still, there is scope for comparative analysis of serpentine microchannels with straight bends and curved bends. Also, the effect of width and height (aspect ratio) on the mixing analysis is impressive. This paper presents the mixing performance analysis with straight and curved bends. COMSOL Multiphysics 5 was used for performing the simulations. The aspect ratio (ratio of channel width to height) was varied as 0.75, 1, and 1.25. The pressure variation (drop) and mixing within straight and curved serpentine microchannel is analyzed.

2 Methodology—Numerical Simulations

2.1 Microchannel Geometry

In the present study, two microchannel configurations with Y-shaped inlet have been considered—a serpentine microchannel with straight bends and a serpentine microchannel with bends. The computational models for the same are developed in COMSOL Multiphysics 5.0 and are presented in Fig. 1a, b for the serpentine microchannel with straight and curved bends, respectively. The dimensions (width and height) of the microchannel considered are 400 μm for both the configurations for aspect ratio 1. The width of the channel is kept constant, and the height is varied for aspect ratio of 0.75 and 1.25. Two different fluids have been fed through two different inlets, namely Inlet 1 and Inlet 2. The fluid velocity (*u* mm/s) for both the inlets has been assumed to be the same.

2.2 Boundary Conditions

The simulations for the developed microchannels have been carried out using COMSOL Multiphysics 5. The physics used for simulations in COMSOL is laminar flow and transport of diluted species. Using suitable boundary conditions, the governing equations, i.e., Eqs. 1–3, have been solved in the software. The boundary

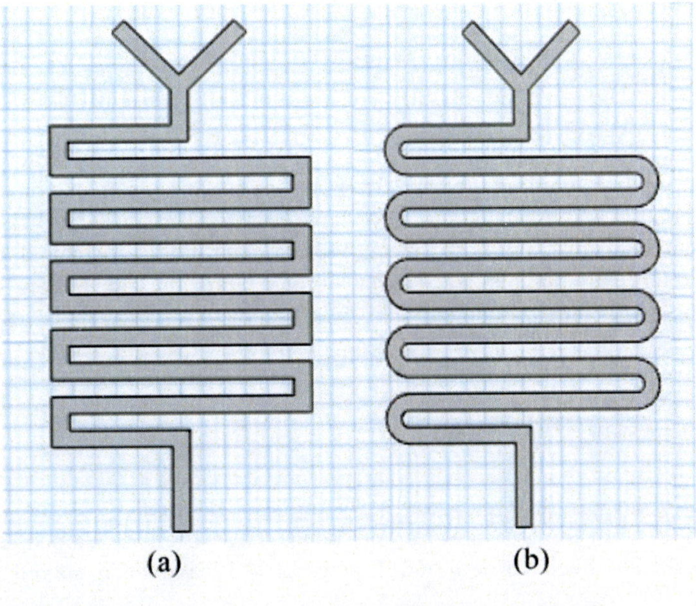

Fig. 1 Serpentine microchannel **a** with straight bends **b** with curved bends

conditions used are the equal velocities at the two inlets, atmospheric pressure at the outlet, symmetry at the interface between the fluids, and no-slip conditions at the microchannel walls. The fluids at the two inlets have been considered as water and ethanol at 25 °C. At the inlet boundaries, the concentrations of fluids have been taken as 10 mol/m^3 and that for fluid 2 as 0 mol/m^3. The diffusion coefficient of ethanol in water has been taken as 1.0×10^{-9} m^2/s. The inlet velocity is varied from 0.5 to 1 mm/s.

For the developed computational models, the steady-state condition for the fluid flow and convection and diffusion of the species have been assumed. The mass and momentum balance for the incompressible and isothermal Newtonian fluids in microchannels are expressed by Navier–Stokes and continuity equations, and the equations are as follows:

$$\nabla.u = 0 \tag{1}$$

$$\nabla(u.\nabla)u = \nabla.\left[-pI + \nabla\left(\nabla u + (\nabla u)^T\right) - 2/3\nabla(\nabla.u)I\right] + F \tag{2}$$

where u = (u, v, w) is the flow velocity field, ρ is the density of the fluid, p is fluid pressure, μ is the dynamic viscosity of the fluid, I is the unit diagonal matrix, and $F = (f_x, f_y, f_z)$ is a volume force affecting the fluid.

Due to the convection and diffusion, the mixing in the flow takes place. The following equation has governed mass transport:

$$\nabla.(-D.\nabla c) + u.\nabla c = R \tag{3}$$

2.3 *Meshing*

For the computational analysis of the microchannel models, the unstructured mesh has been used. For avoiding the effect of increased meshing elements on the quality of the simulation results, the simulations have been carried out with different mesh size (domain elements). For both the configurations of the microchannel, the results for pressure drop are compared at various domain elements. The meshed serpentine microchannel with straight and curved bends is depicted in Fig. 2a, b, respectively.

3 Results and Discussion

Using COMSOL Multiphysics 5.0 software, the 3D models of the serpentine microchannel with straight and curved bends have been developed, and then, simulations have been carried out. Equations 1–3 have been solved by using

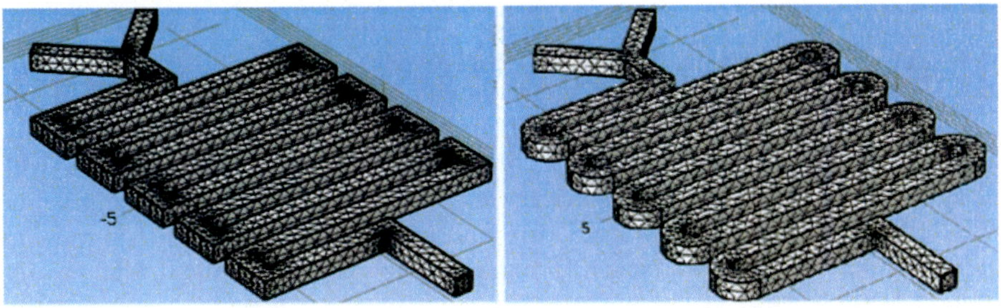

(a) (b)

Fig. 2 Meshing for serpentine microchannel **a** with straight bends **b** with curved bends

considered boundary conditions. Water is used as the primary fluid, while ethanol has been used as the secondary fluid (both at 25 °C).

3.1 Effect on the Pressure Drop (Pa)

The influence of the aspect ratio on the pressure drop is analyzed. The aspect ratio considered for the analysis was 0.75, 1, and 1.25. The velocity of inlet fluids was varied with a velocity of 0.5 mm/s, 0.75, and 1 mm/s. The sample images for the pressure drop measurement for serpentine microchannel with straight bends and curved bends are shown in Fig. 3a, b, respectively.

The pressure drop was recorded, and the effect of aspect ratio on pressure drop for serpentine microchannel with straight bends and curved bends is shown in Fig. 4a, b, respectively. It is observed from Fig. 4 that the pressure drops increase with increase in the aspect ratio from 0.75 to 1.25. Also, the pressure drops increase with increase in velocity from 0.5 to 1 mm/s. The reason behind this is as the aspect

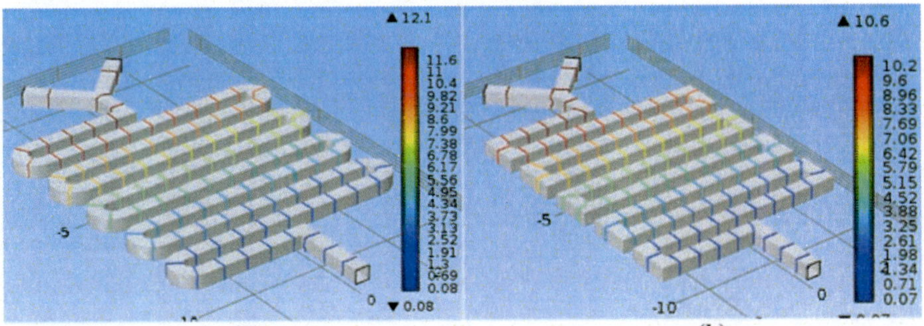

(a) (b)

Fig. 3 Pressure drop for serpentine microchannel at 0.5 mm/s velocity for **a** with straight bends **b** with curved bends ratio increases the cross-sectional area of the microchannel increases. The same fluid will experience more pressure in a lesser area and less pressure in the increased area. Therefore, the pressure drop increases with an increase in aspect ratio. From Figs. 4 and 5, it can also be noted that the pressure drop is more for the serpentine microchannel with curved bends as compared to serpentine microchannel with straight bends. The fluids experience more pressure in curved bend configuration due to its shape, and this leads to increased pressure drop for serpentine microchannel with curved bends as compared to serpentine microchannel with straight bends.

3.2 Effect on the Mixing Length

The term mixing length refers to the distance along the channel where the mixing index is achieved as 1, i.e., the mixing of the two fluids is 100%. The mixing length is recorded in COMSOL Multiphysics 5 software. The sample images for the

Fig. 4 Effect of aspect ratio on pressure drop for serpentine with **a** straight bends **b** curved bends

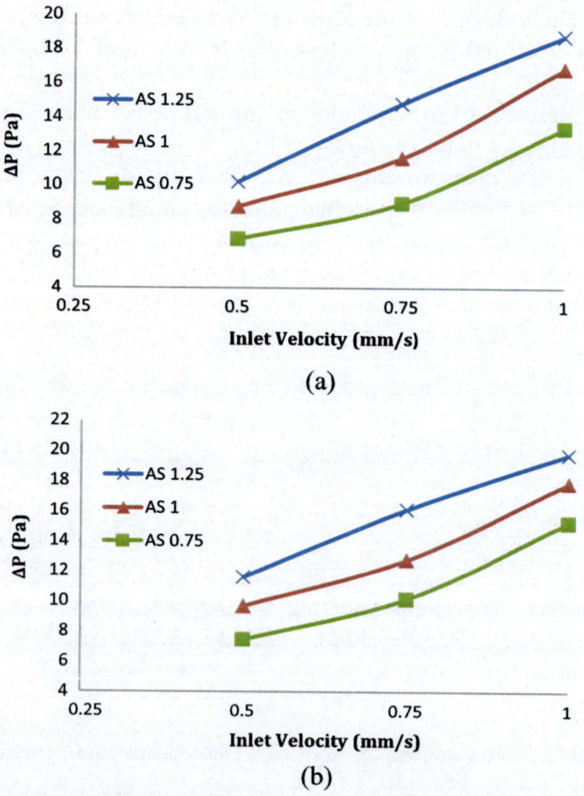

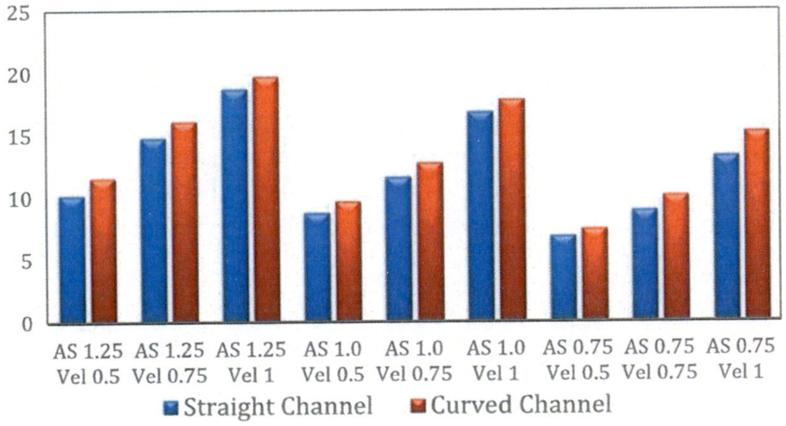

Fig. 5 Comparative pressure drop analysis for serpentine with **a** straight bends **b** curved bends

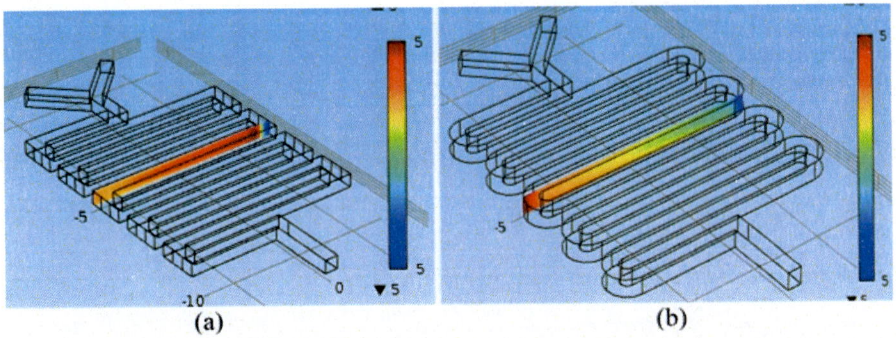

Fig. 6 Sample images for mixing at the cross section of the channel for serpentine with **a** straight bends **b** curved bends

mixture across the cross section for serpentine microchannel with straight bends and curved bends are presented in Fig. 6a, b, respectively.

The effect of aspect ratio on mixing length is presented in Fig. 7a, b for serpentine microchannels with straight bends and curved bends.

From Fig. 7, it can be seen that the mixing length is noted lesser for the microchannels smaller aspect ratio. It increases with increase in aspect ratio from 0.75 to 1.25 for both the considered microchannel configurations. This increase is because the flow is laminar at lesser fluid velocities, and the mixing in the microchannels is because of diffusion. Aspect ratio 0.75 indicates the broader cross section, and the aspect ratio 1.25 showed the smaller the cross section. The more area is available for diffusion in case of the larger cross-sectional area; hence, the lesser mixing length is observed at lower aspect ratio, and the increased mixing lengths are noted for the larger aspect ratio 1.25. Also, the mixing lengths are found

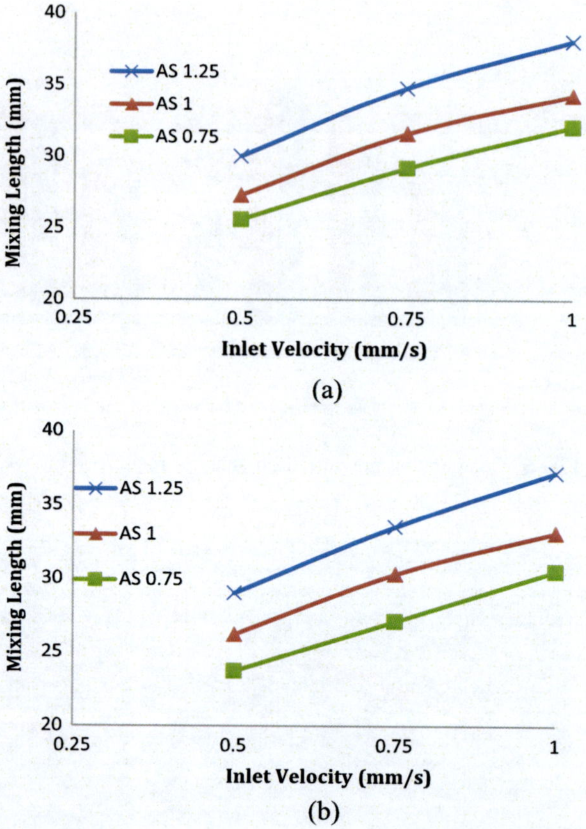

Fig. 7 Effect of aspect ratio on mixing length for serpentine microchannel with **a** straight bends **b** curved bends

lesser for serpentine microchannel with curved bends as compared to that for the serpentine microchannel with straight bends. This smaller mixing length is due to the effect of Dean vortices forming at the curved bends, which enhances the mixing and leads to reduced mixing lengths as compared to that for straight bends.

4 Conclusions

The mixing performance analysis of a serpentine microchannel with straight bends and curved bends has been studied using computational analysis with COMSOL Multiphysics 5.0 software. The aspect ratio considered in the analysis is 0.75, 1, and 1.25. The influence of aspect ratio on the pressure drop and mixing length is investigated. Based on the numerical analysis, the following conclusions are drawn:

- The pressure drop increases proportionally with an increase in the aspect ratio from 0.75 to 1.25.

- The mixing length also increases with an increase in aspect ratio from 0.75 to 1.25.
- The higher-pressure drops are noted for the serpentine microchannel with curved bends as compared to serpentine.
- The mixing lengths are observed lesser for the serpentine microchannel with curved bends as compared to straight bends.

References

1. Lee CY, Wang WT, Liu CC, Fu LM (2016) Passive mixers in microfluidic systems: a review. Chem Eng J 288:146–160
2. Lim YC, Kouzani AZ, Duan W (2010) Lab-on-a-chip: a component view. Microsyst Technol 16(12):1995–2015
3. Whitesides GM (2006) The origins and the future of microfluidics. Nature 442(7101): 368–373
4. Bothe D, Stemich C, Warnecke HJ (2006) Fluid mixing in a T-shaped micro-mixer. Chem Eng Sci 61(9):2950–2958
5. Bahrami M, Yovanovich MM, Culham JR (2005) Pressure drop of fully-developed, laminar flow in microchannels of arbitrary cross-section In: International conference on nanochannels, microchannels, and minichannels, vol 41855, pp 269–280
6. Song H, Wang Y, Pant K (2012) Cross-stream diffusion under pressure-driven flow in microchannels with arbitrary aspect ratios: a phase diagram study using a three-dimensional analytical model. Microfluid Nanofluid 12(1–4):265–267
7. Ansari MA, Kim KY, Kim SM (2010) Numerical study of the effect on mixing of the position of fluid stream interfaces in a rectangular microchannel. Microsyst Technol 16(10):1757–1763
8. Gobby D, Angeli P, Gavriilidis A (2001) Mixing characteristics of T-type microfluidic mixers. J Micromech Microeng 11(2):126
9. Cortes-Quiroz CA, Azarbadegan A, Zangeneh M (2014) Evaluation of flow characteristics that give higher mixing performance in the 3-D T-mixer versus the typical T-mixer. Sens Actuators B: Chem 31(202):1209–1219
10. Lü Y, Zhu S, Wang K, Luo G (2016) Simulation of the mixing process in a straight tube with sudden changed cross-section. Chin J Chem Eng 24(6):711–718
11. Yang ID, Chen YF, Tseng FG, Hsu HT, Chieng CC (2006) Surface tension driven and 3-D vortex enhanced rapid mixing microchamber. J Microelectromech Syst 15(3):659–670
12. Sudarsan AP, Ugaz VM (2006) Fluid mixing in planar spiral microchannels. Lab Chip 6(1): 74–82
13. Hossain S, Kim KY (2015) Mixing performance of a serpentine micromixer with non-aligned inputs. Micromachines 6(7):842–854
14. Gidde RR, Pawar PM, Ronge BP, Shinde AB, Misal ND, Wangikar SS (2019) Flow field analysis of a passive wavy micromixer with CSAR and ESAR elements. Microsyst Technol 25(3):1017–1030
15. Das SS, Tilekar SD, Wangikar SS, Patowari PK (2017) Numerical and experimental study of passive fluids mixing in micro-channels of different configurations. Microsyst Technol 23(12):5977–5988
16. Hong CC, Choi JW, Ahn CH (2004) A novel in-plane passive microfluidic mixer with modified Tesla structures. Lab Chip 4(2):109–113
17. Xia G, Li J, Tian X, Zhou M (2012) Analysis of flow and mixing characteristics of planar asymmetric split-and-recombine (P-SAR) micromixers with fan-shaped cavities. Ind Eng Chem Res 51(22):7816–7827

18. Li J, Xia G, Li Y (2013) Numerical and experimental analyses of planar asymmetric split-and-recombine micromixer with dislocation sub-channels. J Chem Technol Biotechnol 88(9):1757–1765
19. Tran-Minh N, Dong T, Karlsen F (2014) An efficient passive planar micromixer with ellipse-like micropillars for continuous mixing of human blood. Comput Meth Prog Biomed 117(1):20–29
20. Guo L, Xu H, Gong L (2015) Influence of wall roughness models on fluid flow and heat transfer in microchannels. Appl Therm Eng 84:399–408
21. Jain M, Rao A, Nandakumar K (2013) Numerical study on shape optimization of groove micromixers. Microfluid Nanofluid 15(5):689–699
22. Kim DS, Lee SW, Kwon TH, Lee SS (2004) A barrier embedded chaotic micromixer. J Micromech Microeng 14(6):798
23. Wangikar SS, Patowari PK, Misra RD (2018) Numerical and experimental investigations on the performance of a serpentine microchannel with semicircular obstacles. Microsyst Technol 24(8):3307–3320
24. Jadhav SV, Pawar PM, Wangikar SS, Bhostekar NN, Pawar ST (2020) Thermal management materials for advanced heat sinks used in modern microelectronics. IOP Conf Ser: Mater Sci Eng 814(1):012044
25. Wangikar SS, Patowari PK, Misra RD (2017) Effect of process parameters and optimization for photochemical machining of brass and german silver. Mater Manuf Processes 32(15): 1747–1755
26. Wangikar SS, Patowari PK, Misra RD (2016) Parametric optimization for photochemical machining of copper using grey relational method. In: Techno-societal 2016, International conference on advanced technologies for societal applications. Springer, Cham, pp 933–943
27. Wangikar SS, Patowari PK, Misra RD (2018) Parametric optimization for photochemical machining of copper using overall evaluation criteria. Mater Today: Proc 5(2):4736–4742
28. Wangikar SS, Patowari PK, Misra RD, Misal ND (2019) Photochemical machining: a less explored non-conventional machining process. In: Non-conventional machining in modern manufacturing systems. IGI Global, pp 188–201
29. Chavan NV, Bhagwat RM, Gaikwad SS, Shete SS, Kashid DT, Wangikar SS (2019) Fabrication and characterization of microfeatures on PMMA using CO_2 laser machining. Int J Trends Eng Technol 36:29–32
30. Kulkarni HD, Rasal AB, Bidkar OH, Mali VH, Atkale SA, Wangikar SS, Shinde AB (2019) Fabrication of micro-textures on conical shape hydrodynamic journal bearing. Int J Trends Eng Technol 36(1):37–41
31. Raut MA, Kale SS, Pangavkar PV, Shinde SJ, Wangikar SS, Jadhav SV, Kashid DT (2019) Fabrication of micro channel heat sink by using photo chemical machining. Int J New Technol Res 5(4):72–75
32. Jadhav SV, Pawar PM, Shinde AB, Wangikar SS (2020) Performance analysis of elliptical pin fins in the microchannels. In: Techno-societal 2018. Springer, Cham, pp 295–304
33. Bhagwat RM, Gaikwad SS, Shete SS, Chavan NV, Wangikar SS (2020) Study of etchant concentration effect on the edge deviation for photochemical machining of copper. Novyi MIR Res J 5(9):38–44
34. Patil PK, Kulkarni AM, Bansode AA, Patil MK, Mulani AA, Wangikar SS (2020) Fabrication of logos on copper material employing photochemical machining. Novyi MIR Res J 5(7): 70–73
35. Kame MM, Sarvagod MV, Namde PA, Makar SC, Jadhav SV, Wangikar SS (2020) Fabrication of microchannels having different obstacles using photo chemical machining process. Novyi MIR Res J 5(6):27–32

Comparison of Certain Models to Estimate the Best Solar Global Radiation for Jamshedpur, Jharkhand, India

Md. Ahsan, P. Chand, and Kumari Namrata

1 Introduction

For the proper utilization of solar energy in economic and efficient manner, there is need to know the available solar energy data. It is very helpful in order to design and assess the conversion of solar energy system. Many researchers measured solar radiation data of different locations to find out the best performance and efficiency of different thermal devices used by solar energy. Solar energy is a renewable form of energy that helps in reducing the demands of conventional fuels [1]. The solar radiation predictions are being carried out in Jamshedpur considering the weather parameters, sunshine hours as the prime concern [2, 3].

The estimated solar radiation models are established by Angstrom–Prescott. In 2011, the average daily global radiation for seven models was reviewed by Wong and Chow with sunshine hours [4]. Bagheri Tolabi et al. [5] given an estimated solar global radiation depends on Angstrom model. Many researches based on solar radiation model are done by Tiwari and Katiyar [6–8]. Harrouni et al. [9, 10], Maafi [11], and Badescu [12] on its classification.

The data of GSR is very useful for the design and performance of the solar system. This data is not available due to requirement of cost and calibration of the devices used to measure. General methods to determinate the parameter by suitable

Md.Ahsan (✉)
Department of Mechanical Engineering, Al Kabir Polytechnic, Jamshedpur, Jharkhand, India

P. Chand
Department of Mechanical Engineering, NIT Jamshedpur, Jamshedpur, Jharkhand, India

K. Namrata
Department of Electrical Engineering, NIT Jamshedpur, Jamshedpur, Jharkhand, India

S. Bag et al. (eds.), *Next Generation Materials and Processing Technologies*,
Springer Proceedings in Materials 9, https://doi.org/10.1007/978-981-16-0182-8_10

factor which are empirically prescribed for the places where data is not available (Bakirci 2009). The models earlier developed were not for averaged data as they are for hourly records [13]. Temperature-based models, sunshine hour, and cloud-based models are most important factor for findings of global solar radiation [14, 15].

There are various studies that have been published based on the sunshine hour to estimate the monthly mean GSR. In 1924, the regression equation for monthly average daily radiation was suggested by Angstrom for a horizontal surface [16]. Prescott (1940) suggested the ideas to solve the problem of solar radiation using the extraterrestrial radiation and modify it. Later, several models are found to measure monthly mean daily radiation (Mecibah et al. 2014).

2 Climatic Factors of Jamshedpur, Jharkhand

The knowledge of weather condition is in order to utilize solar energy for severe applications. The ideas and methods are provided to find out the global solar radiation for a particular area without accessible data. These equations incorporate the number of sunshine hours, wind speed, number of rainy days, mean sea-level pressure, maximum and minimum temperatures, etc. An analysis and discussion will be done separately to know about the annual and seasonal variation. A basic appraisal and investigation of it will empower us to use these parameters for improving the efficiency of thermal devices depends on thermal energy conversion.

2.1 Sunshine Hours Data for Jamshedpur, Jharkhand

The sunshine data is a numerous factor responsible to find the measured solar radiation for a specific area. This sunshine hours are recorded by the automatic sunshine recorder. The monthly and annual total of bright sunshine hours along with the average daily sunshine hours per day for periods 2010–19 (MNRE Handbook) has been studied [17, 18]. It was observed that a maximum of 324 h is recorded in the month of May, whereas the average of October–November is about 290 h. May and October depict the more values of solar radiation intensity. The lowest mean value of bright sunshine hours occurs in the months of July and August. During these months, it receives monsoon rains, and the sky is mostly overcast attenuating the incident solar radiation.

2.2 *Monthly Mean Daily Global Solar Radiation on Horizontal Surface*

The main objective of gathering solar radiation data is to assess the possibility of application and efficient utilization of this energy source, in order to meet energy demand. With the rapidly growing interest in the area of solar energy, measurements and estimates from developing and underdeveloped countries are becoming readily available. Since it is very typical to cover all the work done regarding solar radiation measurements and estimation all over the world, therefore, many investigators try to correlate relationships between the global or diffuse values of radiation based on meteorological parameters to ensure coverage of maximum number of locations with varying latitudes and climate, located in different continents [19, 20].

In this study, a linear regression model has been produced for assessing monthly average GSR on a flat surface of Jharkhand to be specific city Jamshedpur (22° 48′ N, 86° 11′ E)

The estimations of worldwide sun-based radiation and sunshine hour were estimated at the city Jamshedpur for the period (2010–19) utilizing pyranometer and were gathered from the Solar Radiation Handbook (Solar Energy Centre, MNRE, India Meteorological Department) [21].

The standard methodology is followed to calculate the extraterrestrial and global radiation for the above place. So many findings have been shown that based on the monthly mean value of $\bar{t}$ and GSR.

3 Methodology

Different atmosphere models have been produced for use in foreseeing the month-to-month normal worldwide sun-oriented radiation, the well-known one being the Angstrom–type relapse condition created by Angstrom [22–24]. This relationship between month-to-month day-by-day daylight hours for the global radiation is provided by the following equation [25–27]

$$\frac{\overline{E}_g}{\overline{E}_0} = c + d\left(\frac{\bar{t}}{\overline{T}}\right) \tag{3.1}$$

where $\overline{E}_g$ is monthly average daily global radiation (MJ/m^2-day) falling on an even surface at a specific area [28]. And $\overline{E}_0$ is the month-to-month mean day-by-day radiation (MJ/m^2-day) on a flat surface without climate, $\bar{t}$ is the month-to-month mean every day number of watched daylight hours, $\overline{T}$ is the month-to-month typical step-by-step most extraordinary number of significant stretches of possible light (or day length). The terms c and d are climatologically regression constants [29]. The proportion is regularly $\frac{\bar{t}}{\overline{T}}$ called the level of conceivable daylight hour.

$$\overline{E}_0 = \frac{24}{\pi} I_{sc}\left(1 + 0.033\cos\frac{360n}{365}\right)(\omega_s \sin\phi \sin\delta + \cos\phi\cos\delta\sin\omega_s) \tag{3.2}$$

where n is the quantity of days beginning from January first, I_{sc} is the solar constant with an estimation of 1367 W/m^2, δ is the declination point, ϕ is the scope of the areas, and ω_s is the nightfall hour edges, given as:

$$\delta = 23.45\sin\left[\frac{360}{365}(284 + n)\right] \tag{3.3}$$

$$\omega_s = \cos^{-1}(-\tan\phi\tan\delta) \tag{3.4}$$

The most extreme conceivable daylight span $(\overline{T})$ is given by (Table 1):

$$\overline{T} = 2\omega_s/15 \tag{3.5}$$

The given first-order Angstrom correlation models are to be developed to estimate the values of global solar radiation ($\overline{E}_g$) in city are:

For Jamshedpur

$$\frac{\overline{E}_g}{\overline{E}_0} = 0.2026 + 0.514\left(\frac{\overline{t}}{\overline{T}}\right) \tag{3.6}$$

3.1 Comparison and Validation of Models with Statistical Error

There are numerous boundaries which manage the evaluation and correlation of monthly mean daily sun-oriented radiation estimation models. Here, the factual boundaries like root mean square error (RMSE) and the mean bias error (MBE) assist in finding the mistake or the change of the determined estimation of the deliberate worth. Mean percent error (MPE) and coefficient of correlation (R^2) test the direct connection among measured and estimated values. To improve the outcomes and better correlation, the Nash–Sutcliffe condition (NSE) is additionally chosen as an assessment basis. The t-measurement permits models to be analyzed and simultaneously it is done to decide factual centrality of the anticipated qualities by the models [30–33].

Table 1 Regression constants for selected location

Location	Regression constants		
	c	d	$c + d$
Jamshedpur	0.2026	0.5140	0.7166

$$\mathrm{MBE} = \frac{1}{n}\sum_{1}^{n}\left(E_{i,\mathrm{calc}} - E_{i,\mathrm{meas}}\right) \tag{3.7}$$

$$\mathrm{MPE}(\%) = \frac{1}{n}\sum_{1}^{n}\left(\frac{\left(E_{i,\mathrm{calc}} - E_{i,\mathrm{meas}}\right)}{H_{i,\mathrm{meas}}}\right) \times 100 \tag{3.8}$$

$$\mathrm{RMSE} = \left[\frac{1}{n}\sum_{1}^{n}\left(E_{i,\mathrm{calc}} - E_{i,\mathrm{meas}}\right)^2\right]^{\frac{1}{2}} \tag{3.9}$$

$$\mathrm{NSE} = 1 - \frac{\sum_1^n \left(E_{i,\mathrm{calc}} - E_{i,\mathrm{meas}}\right)^2}{\sum_1^n \left(E_{\mathrm{meas}} - E_{i,\mathrm{meas}}\right)^2} \tag{3.10}$$

where E_{meas} is the measured mean global radiation. A value of NSE closer to 1 is more suitable for model.

$$t = \left[\frac{(n-1)(\mathrm{MBE})^2}{(\mathrm{RMSE})^2 - (\mathrm{MBE})^2}\right]^{\frac{1}{2}} \tag{3.11}$$

3.2 Models Analysis

Rietveld [34] analyzed a few distributed estimations of 'c' and 'd' and noticed that 'c' is connected straightly and 'd' is hyperbolic to the mean estimation of value V so that the condition is accepted to be appropriate any place

$$\frac{\overline{E}_g}{\overline{E}_0} = 0.18 + 0.62(V) \tag{3.12}$$

where $V = \frac{\overline{t}}{T}$.

Ogleman et al. [35] proposed the utilization of a relationship which relates the worldwide sun-based radiation to V in a quadratic structure as:

$$\frac{\overline{E}_g}{\overline{E}_0} = 0.195 + 0.675(V) - 0.142\left(V^2\right) \tag{3.13}$$

A quadratic relationship between the ratio of $\frac{\overline{E}_g}{\overline{E}_0}$ and V is given by Akinoglu and Ecevit.

$$\frac{\overline{E}_g}{\overline{E}_0} = 0.145 + 0.845(V) - 0.280\left(V^2\right) \tag{3.14}$$

Glover and McCulloch endeavored to establish scope reliance with one of the Angstrom–Prescott coefficient and introduced the accompanying

$$\frac{\overline{E}_g}{\overline{E}_0} = 0.29\cos\phi + 0.52(V) \tag{3.15}$$

Gopinathan [36–38] introduced c, d and identified three boundaries, scope, rise, and daylight hours.

$$c = -0.309 + 0.539\cos\phi - 0.0639h + 0.290(V)$$

$$d = 1.527 - 1.027\cos\phi + 0.0926h - 0.359(V)$$

$$\frac{\overline{E}_g}{\overline{E}_0} = 0.32 + 0.42(V) \tag{3.16}$$

Present model

$$\frac{\overline{E}_g}{\overline{E}_0} = 0.211 + 0.489(V) \tag{3.17}$$

4 Results and Discussion

Six empirical models are compared by considering the monthly mean daily solar global radiation on a flat surface with monthly mean sunshine records for city Jamshedpur, Jharkhand is shown in Fig. 2. Further, the models are compared using coefficient of correlation (R^2), root mean square error (RMSE), mean bias error (MBE), and the t-stat. The double axis graph is plotted for monthly average solar radiation in a flat portion for Jamshedpur in Fig. 1 and the scatter plot graph is shown to compare the estimated and measured value of all different models (Fig. 3).

In Table 2, the best coefficient of correlation with R^2 = 0.999 is obtained from the linear regression model while the most minimal correlation coefficients R^2 = 0.971. From the statistical test, the ideal data of MBE and the RMSE plotted in Fig. 4. In correlation with all the models, the least RMSE has 2.52% and the most elevated one with 23.43%, respectively, for Jamshedpur

The results of MBE from all the models show overestimation except present model shows underestimation. Likewise, the estimations of NSE (0.99) as appeared in Fig. 4 and the lowest values of t-marker show the best performance in comparison to all model.

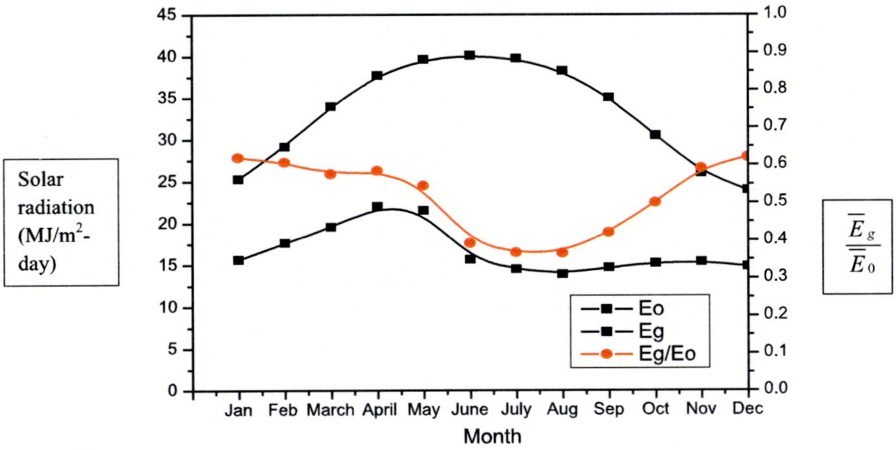

Fig. 1 Monthly average sun-oriented radiation in a flat portion for Jamshedpur

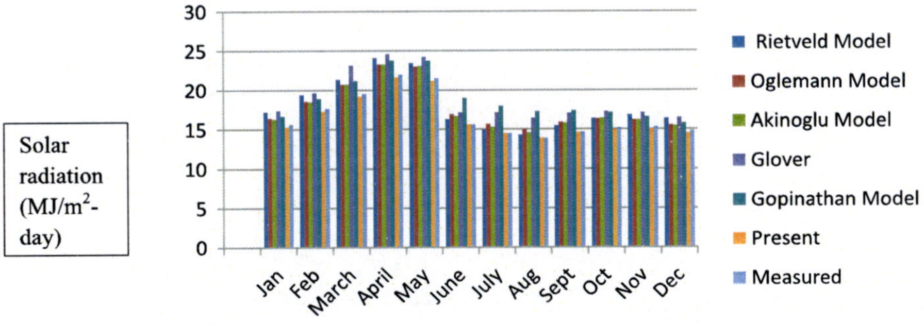

Fig. 2 Measured and estimated monthly average daily global solar radiation $\overline{E}_g$ for the city Jamshedpur in generating various models

5 Conclusions

From the investigation, the possibility of effective use of sunlight accounts extremely brilliant. The target of this examination was to assess different models to estimate month-to-month normal day-by-day global radiation on a flat surface from brilliant daylight hours for Jamshedpur, Jharkhand, and to choose the most suitable model.

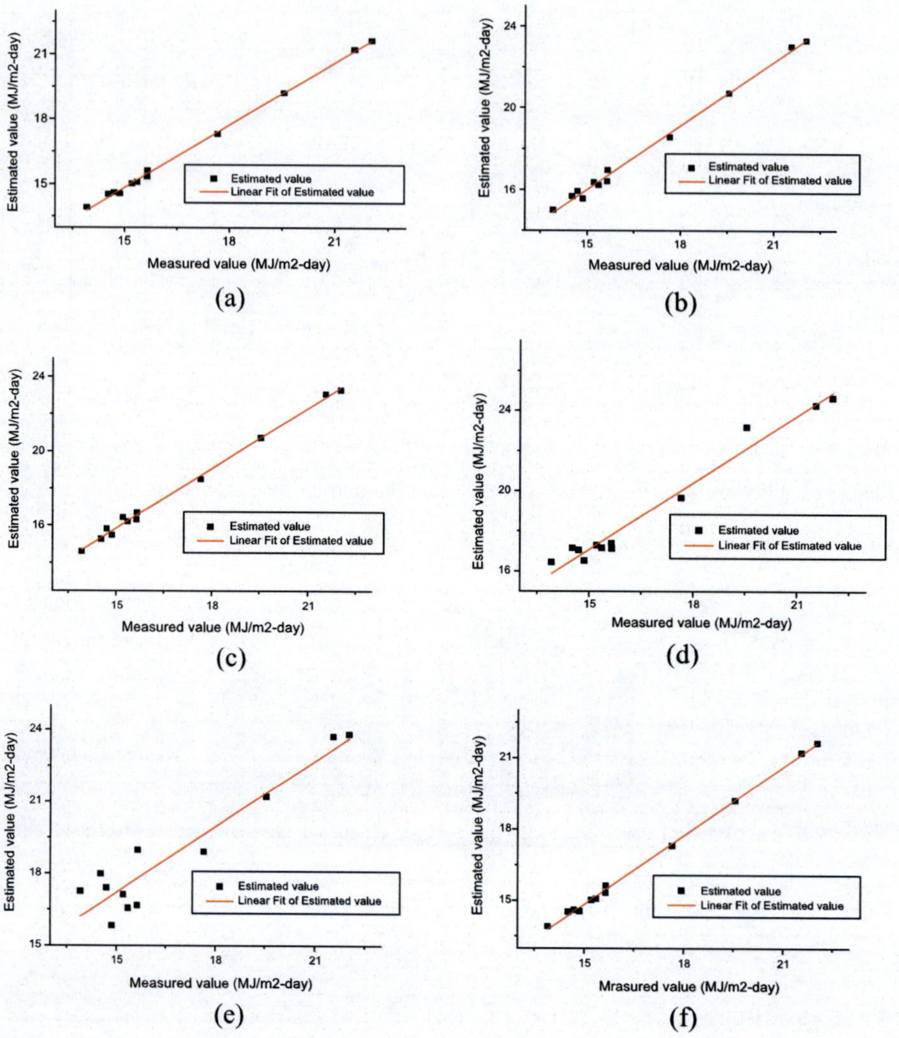

Fig. 3 Measured and estimated monthly mean daily global sun-oriented radiation, $\overline{E}_g$ for the city Jamshedpur in generation of **a** present model **b** Rietveld model **c** Oglemann model **d** Akinoglu model **e** Glover model **f** Gopinathan model

Table 2 Different statistical test for the acceptance of the models for the city Jamshedpur

Statistical parameters	Present	Rietveld	Oglemann	Akinoglu	Glover	Gopinathan
R^2	0.999	0.996	0.998	0.998	0.993	0.971
MBE MJ/m^2-day	−0.208	1.303	1.089	0.989	2.272	2.079
MPE	−0.011	0.078	0.072	0.063	0.151	0.149
RMSE MJ/m^2-day	0.252	1.427	1.111	1.026	2.343	2.262
NSE	0.991	0.720	0.829	0.854	0.242	0.294
MAPE	0.013	0.080	0.073	0.064	0.151	0.149
t-marker	4.878	7.433	10.650	11.956	13.177	7.738

The estimations of month-to-month normal global sun-oriented radiation are determined utilizing the models proposed by Rietveld, Ogleman, Akinoglu, Glover, Gopinathan and present model. The choosen models were contrasted and the current model for assessing monthly GSR for Jamshedpur, based on statistical tests, for example, mean bias error (MBE), root mean square error (RMSE), the mean percent error (MPE), Nash–Sutcliffe condition (NSE,) and the t-marker. The Angstrom model is incredibly prescribed to assess monthly average global solar radiation for Jamshedpur (Jharkhand). Besides, the other new proposed models are additionally being suggested for evaluating the normal day-by-day global sunlight-based radiation for Jamshedpur.

The most extreme normal day-by-day global sun-oriented radiation is gotten in the period of April and May, while it is least in the storm season (July–August) and somewhat moderate in winter (November–January) in all parts. The similar investigation of yearly normal worldwide sun-powered radiation demonstrated that sunlight-based radiation is most extreme at Jamshedpur. The estimations of global solar radiation determined from single recently proposed model for Jharkhand are likewise contrasted. Thus, a recently proposed model is prescribed to find monthly average solar global radiation.

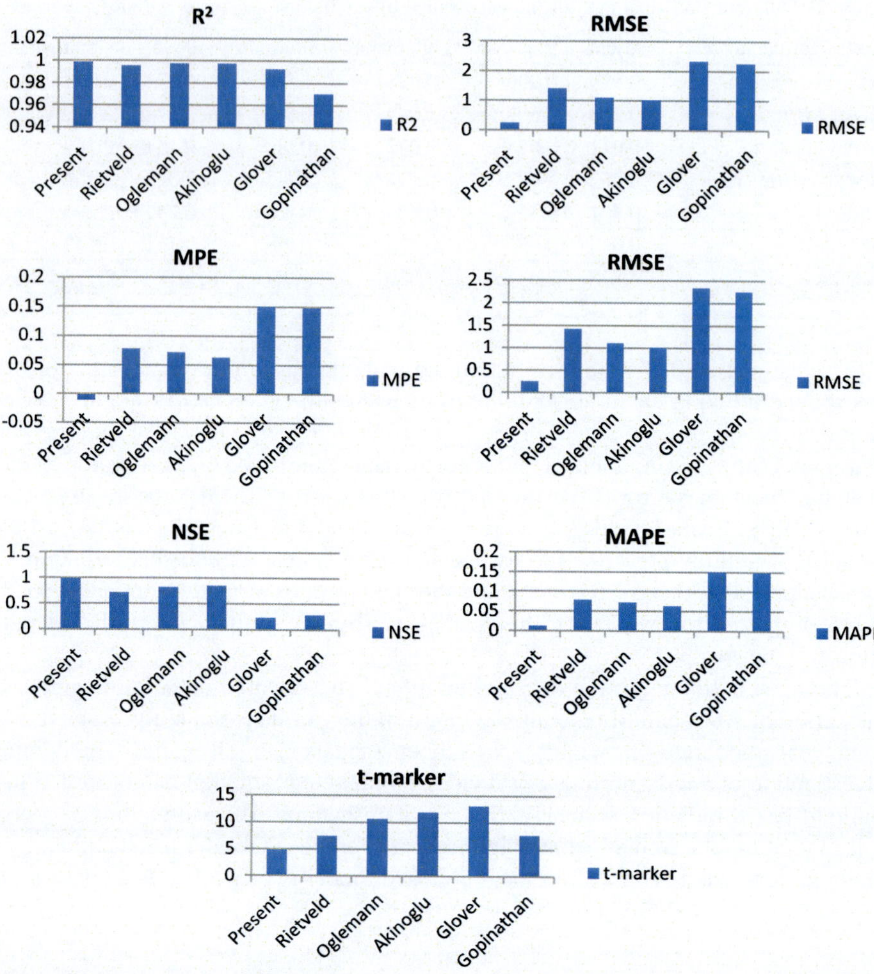

Fig. 4 Statistical error of six model

References

1. Jamil B, Bellos E (2019) Development of empirical models for estimation of global solar radiation exergy in India. J Clean Prod 207:1–16
2. Namrata K, Sharma SP, Seksena SBL (2013) Comparison of different models for estimation of global solar radiation in Jharkhand (India) region. Smart Grid Renew Energy 4:348–352
3. Das A, Park J-K, Park J-H (2015) Estimation of available global solar radiation using sunshine duration over south Korea. J Atmosph Solar Terr Phys 134:22–29
4. Wong LT, Chow WK (2001) Solar radiation model. Appl Energy 69:191–224
5. Bagheri Tolabi H, Moradi MH, Ayob SM, Zandebasiri MR (2013) New technique for global solar radiation prediction using imperialist competitive algorithm. J Basic Appl Sci Res 3(3): 958–964

6. Jamil Ahmad M, Tiwari GN (2011) Solar radiation models—a review. Int J Energy Res 35(4): 271–290
7. Katiyar AK, Pandey CK (2013) A review of solar radiation models—Part I. J Renew Energy. https://doi.org/10.1155/2013/168048
8. Ulgen K, Hepbasli A (2004) Solar radiation models: Part 1: a review. Energy Sources 26:507–520
9. Harrouni, S., & Maafi, A. (2002). Classification des éclairements solaires à l'aide de l'analyse fractale. Revue Internationale des éenergies renouvelables, 5, 107-122.
10. Harrouni, S., Guessoum, A., & Maafi, A. (2005). Classification of daily solar irradiation by fractional analysis of 10-min-means of solar irradiance. Theoretical and applied climatology, 80(1), 27-36.
11. Maafi, A., & Harrouni, S. (2003). Preliminary results of the fractal classification of daily solar irradiances. Solar Energy, 75(1), 53-61.
12. Badescu V (2008) Modeling solar radiation at the earth's surface. Springer-Verlag, Berlin Heidelberg
13. Muneer T, Etxebarria S, Gago EJ (2014) Monthly averaged-hourly solar diffuse radiation model for the UK. Build Serv Eng Res Technol, 573–584
14. Jamil B, Akhtar N (2017) Estimation of diffuse solar radiation in humid-subtropical climatic region of India: comparison of diffuse fraction and diffusion coefficient models. Energy 131:149–164
15. Katiyar AK, Pandey CK (2013) A review of solar radiation models-Part I. J Renew Energy
16. Ahmad L, Kanth RH, Parvaze S, MahdiSS (2017) Measurement of sunshine duration. In: Experimental agrometeorology: a practical manual, pp 37–39
17. Donatelli M, Bellocchi G, Fontana F (2003) RadEst3.00: Software to estimate daily radiation data from commonly available meteorological variables. Euro J Agron 18:363–367
18. Solar radiation handbook, Solar Energy Centre, MNRE
19. Garg HP, Garg SN (1985) Correlation of monthly average daily global, diffuse and beam radiation with bright sunshine hours. Energy Convers Manag 25:409–417
20. Raja IA, Twidell JW (1990) Distribution of global insolation over Pakistan. Sol Energy 44:63–71
21. Quej VH, Almorox J, Ibrakhimov M, Saito L (2016) Empirical models for estimating daily global solar radiation in Yucatán Peninsula, Mexico. Energy Convers Manage, 448–456
22. Doost AK, Akhlaghi M (2014) Estimation and comparison of solar radiation intensity by some models in a region of Iran. J Power Energy Eng 2:345–351
23. Manzano A, Martın ML, Valero F, Armenta C (2015) A single method to estimate the daily global solar radiation from monthly data. Atmosph Res 166:170–182
24. Mousavi SM, Mostafavi ES, Jaafari A et al (2015) Using measured daily meteorological parameters to predict daily solar radiation. Measurement 76:148–155
25. Karkoti I, Das PK, Singh SK (2012) Predicting monthly mean daily diffuse radiation for India. Appl Energy, 412–425
26. Cao F, Li HS, Yang T, Li Y, Zhu TY, Zhao L (2017) Evaluation of diffuse solar radiation models in Northern China: new model establishment and radiation sources comparison. Renew Energy, 708–720
27. Despotovic M, Nedic V, Despotovic D, Cvetanovic S (2016) Evaluation of empirical models for predicting monthly mean horizontal diffuse solar radiation. Renew Sustain Energy Rev, 246–260
28. Almorox Javier (2011) Estimating global solar radiation from common metrological data in Spain. Turkish J Phys 35:53–64
29. Akinoğlu BG, Ecevit A (1990) Construction of a quadratic model using modified Ångstrom coefficients to estimate global solar radiation. Sol Energy, 85–92
30. Angstrom A (1924) Solar and terrestrial radiation. Q J Roy Met Soc 50:121
31. Chakrabarti S, Jamil B, Sakhale CN (2020) Estimation of global solar radiation for the tropical wet climatic region of India: a theory of experimentation approach. Renew Energy 146(Feb 2020):2044–2059

32. Duffie JA, Beckman WA (1994) Solar engineering of thermal process, 2nd edn. John Wiley, New York
33. Spencer JW (1982) A comparison of methods for estimating hourly diffuse solar radiation from global solar radiation. Solar Energy, 19–32
34. Rietveld MR (1978) A new method for estimating the regression coefficients in the formula relating solar radiation to sunshine. Agric Meteorol, 243–252
35. Namrata K, Sharma SP, Seksena SBL (2014) Determining regression constants for calculating global solar radiation at Jharkhand (India) region. In: 2014 International conference on renewable energy research and application (ICRERA), Milwaukee, WI, pp 795–797
36. Gopinathan KK, Soler A (1992) A sunshine dependent global insolation model for latitudes between 60°N and 70°N. Renew Energy 2:401–404
37. Gopinathan KK (1988) A general formula for computing the coefficients of the correlation connecting global solar radiation to sunshine duration. Sol Energy, 499–502
38. Gopinathan KK (1988) A simple method for predicting global solar radiation on a horizontal surface. Sol Wind Technol, 581–593

Processing and Characterization

A Study of Microstructure and Mechanical Properties of Wire Arc Additive Manufactured Component with ER70S6 Alloy Wire Using CMT Process

Yash Khandelwal, Rajneesh Kumar Gupta, K. K. Verma, and Amitava Mandal

1 Introduction

Additive manufacturing has become a common word of mouth and a well-known technology in today's world. As we can see many kinds of research are going on adopting different processes using a variety of materials for manufacturing various products. During this current pandemic time, various products starting from low-value face cover, mask, face shield to high-end components such as mechanical components of ventilators are being made by additive manufacturing technology and are being successfully marketed. Every product, that is, from the smallest to the biggest is developed and produced in large quantities by this technique within a very tight schedule and at a very reasonable cost. This shows that additive manufacturing has a high level of flexibility in the manufacturing components having complex geometries. The product development time and the time gap between product design and commercial production are substantially low in additive manufacturing technology compared to other conventional methods.

Additive manufacturing is also commonly known as 3D printing, which refers to layer-by-layer deposition of raw materials to build the component. According to ASTM standards, additive manufacturing (AM) is defined as the "process of joining materials to make objects from 3D model data, usually layer upon layer, as opposed to subtractive manufacturing methodologies" [1]. According to the type of energy source used in the additive manufacturing process, it has been classified into three main groups: laser-based, electron beam-based, and electric arc welding-based [2].

Y. Khandelwal (✉) · K. K. Verma · A. Mandal
Department of Manufacturing Engineering, National Institute of Foundry and Forge Technology, Ranchi, India

R. K. Gupta
Engineering Division, CSIR-National Metallurgical Laboratory, Jamshedpur, India
e-mail: rkgupta@nmlindia.org

S. Bag et al. (eds.), *Next Generation Materials and Processing Technologies*, Springer Proceedings in Materials 9, https://doi.org/10.1007/978-981-16-0182-8_11

Arc welding-based additive manufacturing process has the advantages of lower cost and higher deposition rates of about 50–130 g/min, in comparison to a laser or electron beam that has a deposition rate of only 2–10 g/min. [3].

Wire arc additive manufacturing (WAAM) is a new advancement in the field of metal additive manufacturing. As the name itself suggests, it is a wire feed and arc-based technology to produce components in a layer-wise manner that uses an electric arc as a heat source and wire as a feedstock material [4]. It is a direct energy deposition technique having a low capital cost and a cost-effective method of producing large and complex shape metallic components by the additive manufacturing method. It has various advantages over other additive manufacturing techniques. The main advantages of WAAM processes are its high deposition rate compared to other AM methods like laser and powder-based technique and electron beam process. WAAM can be used with low and medium complexity parts, and there is no size limit of a part to be manufactured. This technology can make a spark in the development of a new product. The other advantages of this technology are that it has low equipment cost in comparison to other AM techniques such as laser AM and electron beam AM process, also can be build large size component, reduce material wastage, no tool required, near-net-shape components can be produced and most importantly it has very low buy-to-fly ratio [5]. This buy/fly (BTF) ratio is the ratio of the mass of initial workpiece such as billets to one of the finished products. This WAAM technology greatly reduces this BTF ratio and makes a huge reduction in the product cost by saving a large amount of material with a reduction in material wastage. WAAM process also has some limitations and challenges that are mainly residual stress and distortion of the parts, also the surface finish of the near-net-shape component not up to the mark that can be achieved by selective laser melting (SLM) or electron beam melting (EBM) process but these (SLM, EBM) processes are not usable for large size component production and also they are having very fewer deposition rates.

There are commonly three types of WAAM processes that are used based on heat source: gas metal arc welding (GMAW)-based, gas tungsten arc welding (GTAW)-based, and plasma arc welding (PAW)-based. A specific class of WAAM techniques exhibits specific features. The deposition rate of GMAW-based WAAM is 2–3 times higher than the GTAW-based or PAW-based methods. However, the GMAW-based WAAM is less stable and generates more weld fume and spatter due to the electric current acting directly on the feedstock [6]. To reduce fume and eliminate spatter, a variant of the GMAW process has been developed by Fronius. It is called cold metal transfer (CMT) process. This CMT process is a modified MIG/GMAW process variant that relies on controlled short-circuiting/dip transfer mode mechanism. This process works on the principle of push–pull mechanism which retracts the electrode constantly at a very short interval, thus delivers clean and excellent quality weld beads with ultra-lower thermal heat input and spatter-free deposition [7]. It also has another variant where spray mode is mixed up with the dip transfer mode that is called as CMT pulsed mode (CMT-P), further developed advanced versions of these are CMT advanced (CMT-ADV) and CMT pulsed advance (CMT-PADV) processes [8]. This CMT technology has advantages of producing stable arc with high arc deposition rate, accurate digital manipulation in material input, low equipment cost, high precision forming, and easy to popularize the welding process [9].

In the present study, the experiments conducted with CMT MIG robotic welding equipment were carried out by using ER70S6 MS welding wire and S235JR MS support plate (the support plate can be removed after completion). As most of the investigations were conducted on the single-layer deposited wall and the additively manufactured block structure's integrity has not been investigated, this paper presents an investigation on anisotropy behavior of resulted microstructure and mechanical properties including tensile, impact, and hardness of the manufactured solid block component. Finally, the chosen design for the AM part is a solid rectangular block component manufactured by robotic welding on a support plate. The block is subjected to testing of mechanical properties in several directions, hardness measurements, and microscopic investigation of welding defects and grain structure.

2 Experimental Procedure

2.1 Methods and Materials

The experiment was carried out by using a cold metal transfer wire arc additive manufacturing process (CMT-WAAM). The experimental setup mainly consists of a Fronius CMT TPS500i power source that has synergic programs for controlling the current–voltage characteristics and a six-axis KUKA industrial robot KR8R1620 was used as a manipulator/positioning system. Also, included are a wire feeder, robot controller, welding table with clamming devices and CMT welding torch, i.e., mounted on the robot. The welding wire selected for this work was a copper-coated mild steel wire ER70S6 (according to AWS classification) having a diameter of 1.2 mm and a support plate of low carbon structural steel with the dimension of 500 mm length, 350 mm width, and 12 mm thickness and has been used as a substrate to perform experiments. The nominal chemical composition of the wire spool is used is given in Table 1. The shielding gas used was a commercial gas mixture called CORGON®10 that consists of a mixture of 10% CO_2 with 90% argon (M20) with a constant flow rate of 15 L/min. The contact to workpiece distance (CTWD) was kept constant at 12 mm.

Table 1 Nominal chemical composition of the selected ER70S6 welding wire (producer data)

Element	C	Mn	Si	P	S	Fe
wt%	0.09	1.18	0.57	0.012	0.011	Balance

2.2 *Parameter Optimization*

Several experiments were performed to find out the optimum process parameters for the block deposition. Initially starting with a single-layer deposition to multiple layers with a multibead overlapping strategy for deposition has been designed and investigated. The main variable parameters that are taken into account for consideration are the wire feed rate and the welding speed of robot travel speed. A full factorial design experiment was performed to choose the optimum welding parameters for the solid rectangular block deposition based on weld bead profiles.

2.3 *Deposition Process*

After conducting preliminary experiments to achieve the optimized process parameters, the programming codes were generated for the KUKA robot by teaching the coordinates in the teach pendant of the robot. A side layover distance is given for proper overlapping of the weld beads, and a step over a distance of 2 mm is given in the loop program for height increment during the continuous deposition process of the layers to maintain a constant distance between the workpiece and contact tip. Then the deposition of successive layers started on a low carbon steel substrate. The block of dimension 130 mm long, 120 mm wide, and 60 mm high was built at a 4 m/min. wire feed speed (WFS) and at a 0.6 m/min. welding travel speed (TS). The width of the block was achieved by successively overlapping of nearly 20 layers with very integrity and the height of the welded block increases with superimposing of a new layer over the previous one. Thus, a solid metallic block component is developed from ER70S6 MS alloy wire by this WAAM system which is shown in Fig. 1.

2.4 *Testing and Characterization*

The test specimens for tensile test and Charpy impact tests were machined from the deposited weld block in both of the directions parallel as well as perpendicular to the welding direction. Tensile samples were prepared according to ASTM E8 standard as shown in Fig. 3a by using lathe and CNC machines. Tensile tests were carried out by using INSTRON UTM of 35 KN capacity at a strain rate of 0.05/min at ambient temperature. The Charpy impact test specimens of size 55 × 10 × 10 mm were prepared according to standard specimen size as shown in Fig. 3b by using a shaper and grinding machine. The impact tests were performed by using Zwick Roell Charpy impact testing machine at a subzero temperature of −30 °C. The specimens for hardness and microstructure tests were taken from the middle part of the block and then ground and mirror finish by using emery papers from 280

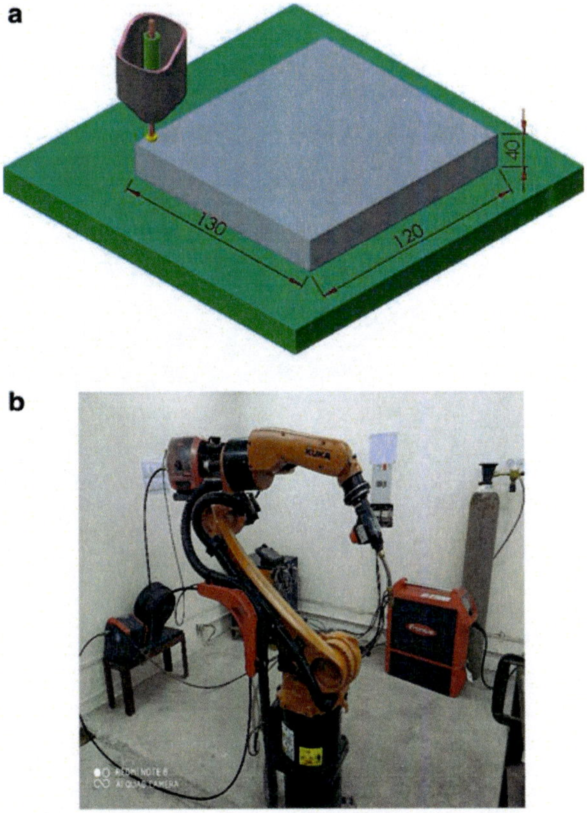

Fig. 1 **a** Deposition process of the solid block component, **b** experimental setup of WAAM equipment

to 2500 grit and then polished with METCO cloth polishing machine. Then these specimens were etched with nital solution to reveal the hardness indentation and the microstructure. A light optical microscope Leica was employed for microstructure analysis. The Vickers hardness was tested by METCO Economet VH 50MD at an applied load of 10 kg for 10 s.

3 Results and Discussion

3.1 Tensile Properties

The tensile properties of the fabricated component produced by the WAAM-CMT process were measured by UTM. The welding parameters of the deposition process are given in Table 2. Tensile testing was performed in two different orientations of direction *X* and *Y*, the *X*-direction belongs to the parallel to deposition of weld beads

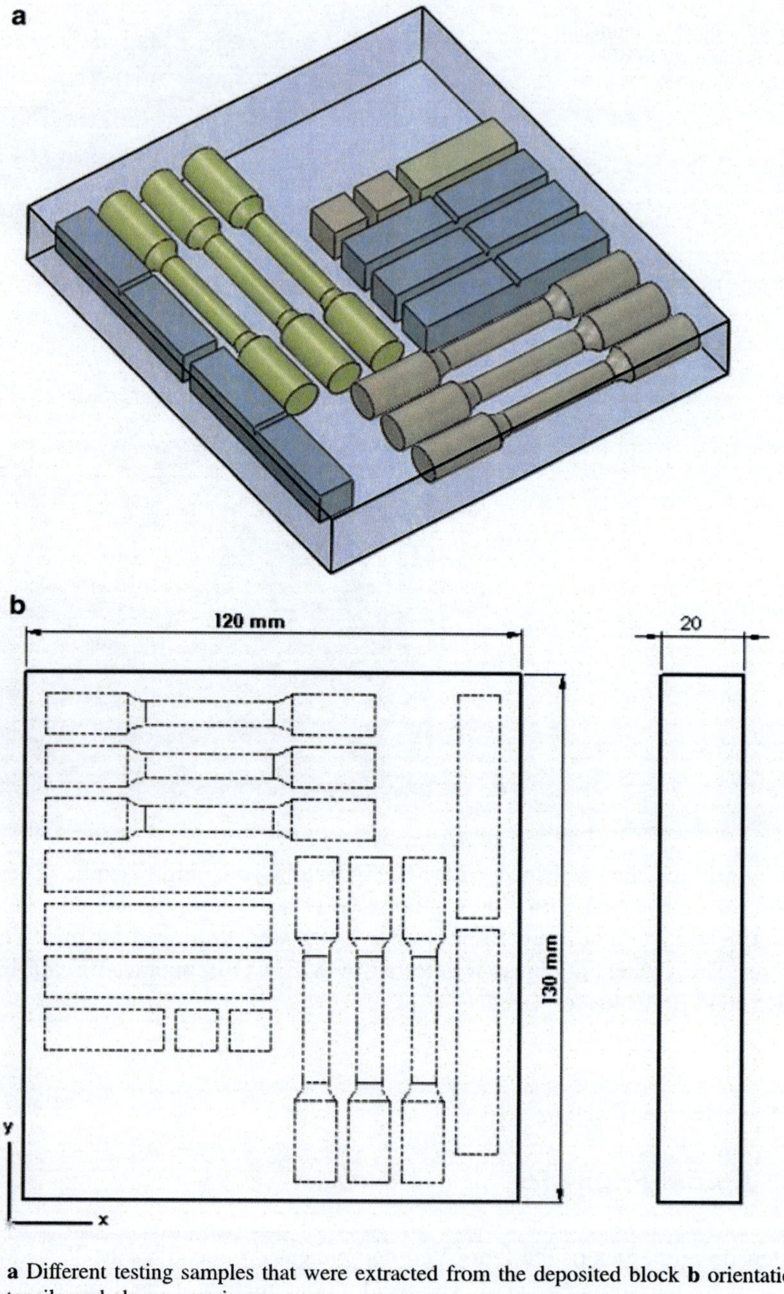

Fig. 2 **a** Different testing samples that were extracted from the deposited block **b** orientation of cut-out tensile and charpy specimens

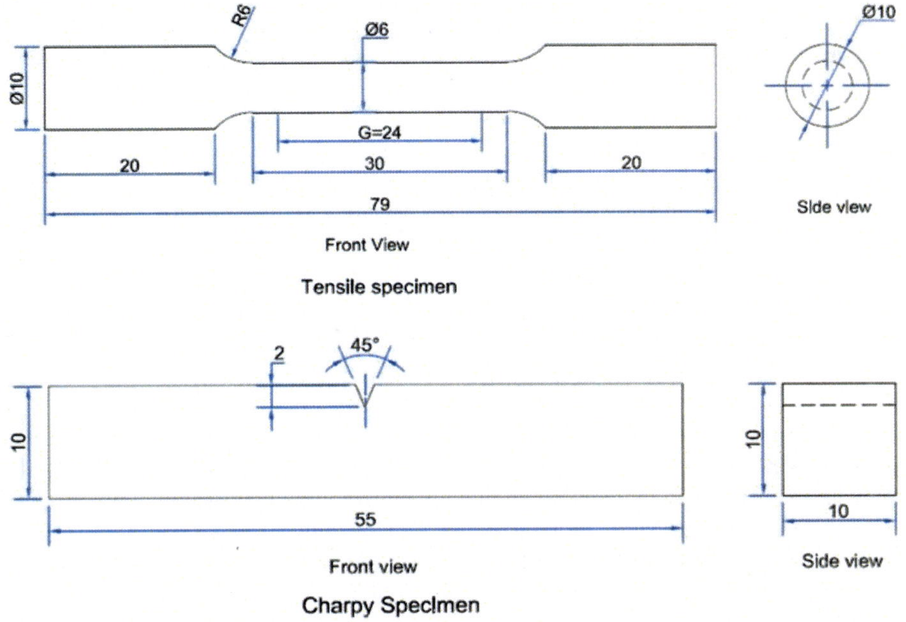

Fig. 3 Dimensions of the tensile and Charpy samples/specimens

Table 2 Welding parameters data

Wire feed rate	4.0 m/min
Welding speed	0.6 m/min
Shielding gas	(Ar + 10% CO_2)
CTWD	12 mm
Current	145 A
Voltage	13.5 V
Gas flow rate	15 l/min
Bypass temp.	70 °C

and *Y* directions show the samples were taken perpendicular to the deposition of weld bead in the horizontal plane. These are illustrated schematically in Fig. 2a, b.

The mean yield and ultimate tensile strengths were 373 MPa and 510 MPa, respectively. The deviation noted is ±5 MPa for both cases. The tensile properties of the samples were found almost uniform in both the directions. Ductility was also found very high with the values of 32% and 73% for mean elongation and reduction in area, respectively. The Young's modulus was measured to be 208 GPa. The result of the tensile test was found as per given in Table 3. From the table, it could be observed that the strength properties have isotropic characteristics with only 4–6 MPa difference in the ultimate tensile strength (UTS), 2–4 MPa difference in the yield strength (YS), and 1–4% difference in elongation (%E) in both parallel and

Table 3 Tensile test result

Sample no.	Avg. gauge dia. (mm)	Area (mm^2)	Yield strength (MPa)	UTS (MPa)	Elongation (%)
X_1	5.94	27.7116746	371.126917	509.8765	35.48
X_2	5.93	27.6184479	378.641073	516.4007	31.25
Y_1	5.95	27.8050585	371.929097	505.777	31.74
Y_2	5.95	27.8050585	369.715228	510.1157	32.39

perpendicular directions. The resulting individual stress–strain graph and the plot/chart are shown in Figs. 4 and 5, respectively. The ultimate tensile strength and % elongation levels obtained in the present work were found to be comparable with the standard requirement for the welding wire used. However, the yield strength values are somewhat lower than the range specified in the standard. Further improvement in the tensile strength can be made possible with heat treatment of the as-welded component. Overall, from the obtained results of tensile properties, it could be seen that ER70S6 alloy parts manufactured by WAAM-CMT have a better performance. The mechanical properties such as tensile strength, yield strength, % elongation, and % reduction in the area are found to be comparable to the published data. Hence, it can be concluded that the components produced using ER70S6 wire will perform as intended.

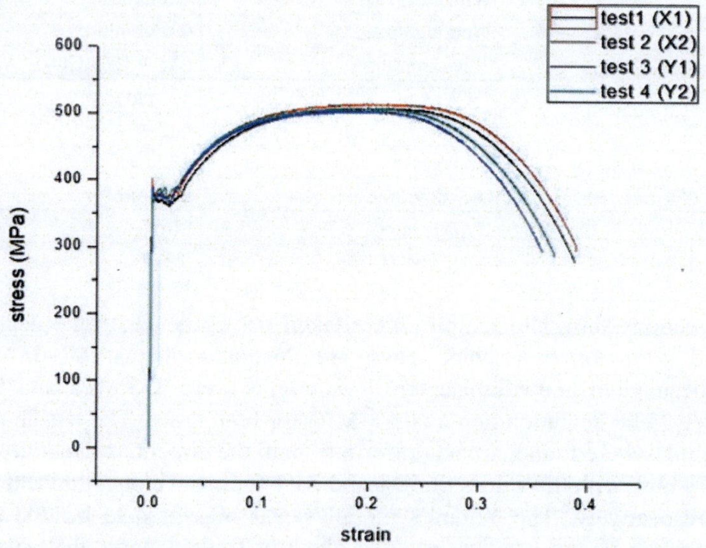

Fig. 4 Individual stress–strain curves. X_1, X_2 are parallel and Y_1, Y_2 are perpendicular to the layer deposition direction

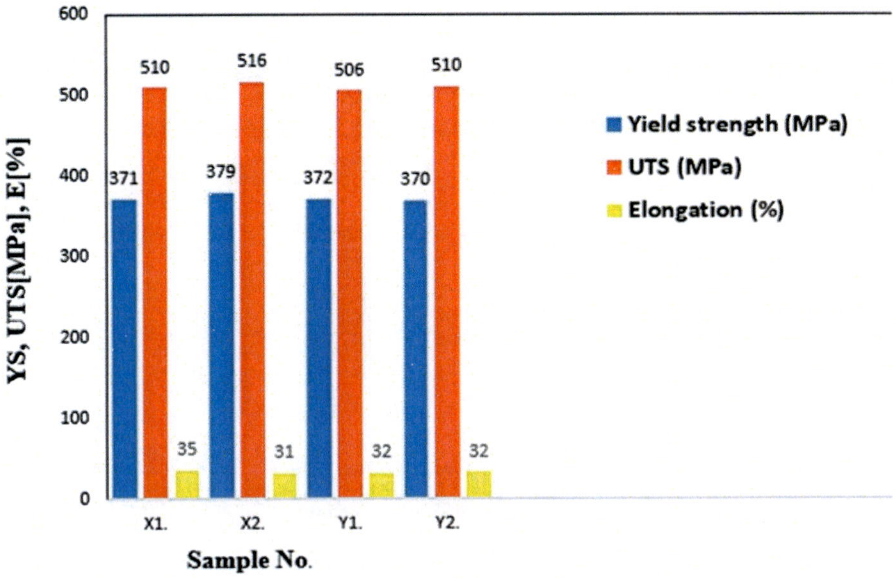

Fig. 5 Mechanical properties of the WAAM fabricated tensile samples

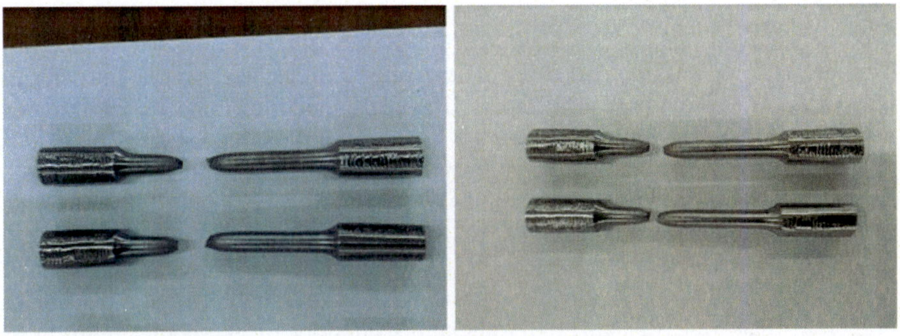

Fig. 6 Tensile test specimens of *X* and *Y* directions after testing

3.2 Hardness

The hardness measurements were carried out by using Vickers hardness tester at a load of 10 kg applied for 10 s. The measured hardness in the horizontal and vertical cross section of the samples is shown in Fig. 7. The Vickers hardness values are between 75 and 80 kg/mm^2 for parallel to the welding direction and between 71 and 74 kg/mm^2 for perpendicular direction.

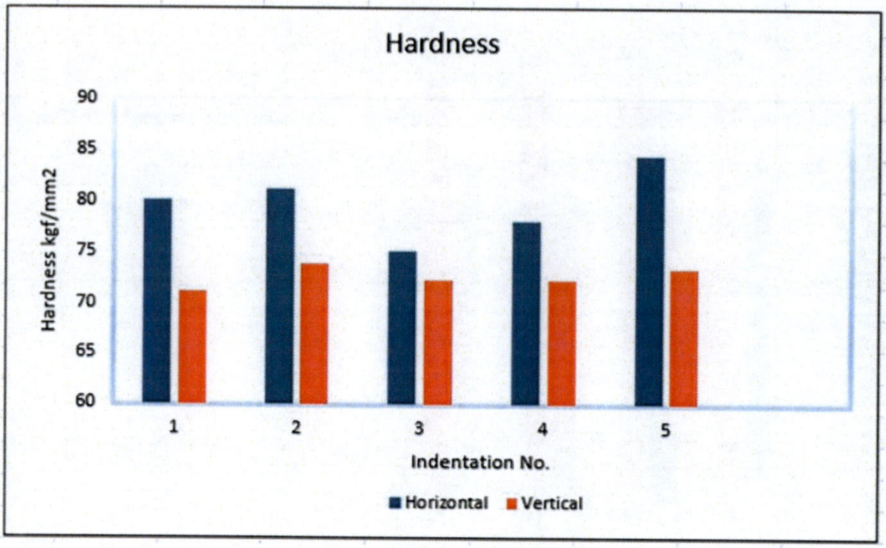

Fig. 7 Horizontal and vertical hardness traverse of the sample

3.3 *Impact Property*

The impact toughness properties of the extracted samples were measured with the help of the Charpy impact instrument. The measurement of impact toughness was taken for the subzero temperature at −30 °C. The measured values were found to be 148 and 170 J in the parallel and perpendicular to the welding direction, respectively. These values of impact toughness are found to be much higher than the required minimum impact strength of 80 J at this subzero temperature as per the AWS requirement for the used ER70S6 welding wire. This indicates that the component fabricated by wire arc additive manufacturing has good mechanical properties that are sustainable and validate its integrity for the applicability in the service areas. The fractured Charpy samples and their fractography image are shown as in Figs. 8 and 9, respectively.

3.4 *Microstructure and Metallography*

The metallurgical examination was also carried out on the developed samples using an optical microscope. The sample for microstructure analysis was prepared by grinding with SiC papers from P280 to P2500 and polished with alumina powder on the cloth polishing machine. The microstructure was subjected to inspection, and a complex pattern of the reheated weld metal part is visible. It is found difficult to

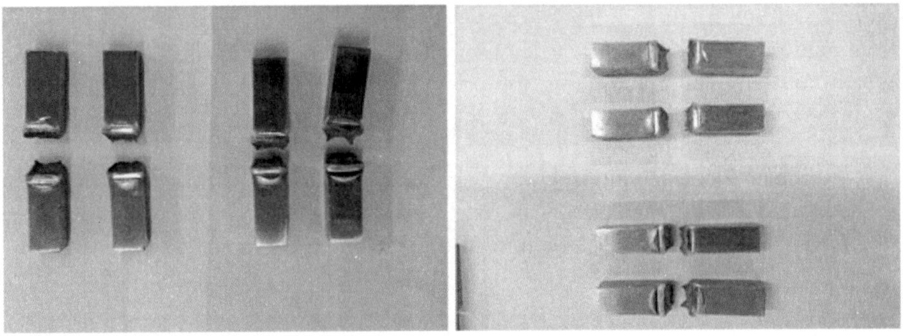

Fig. 8 Fractured Charpy samples of *X* and *Y* directions

Fig. 9 Fractography image of the fractured surface of the Charpy impact specimens

map the different regions in the solid structure component. It is clear after etching with nital acid. The microstructure observed in welded samples is ferrite and bainite which consists inside and along the prior austenite grain boundaries. Most of the grains were found elongated toward the welding direction and characterized by a roughly equiaxed microstructure. The welding wire material is mild carbon steel that justified the ferrite coexists in the equiaxed form with the thin strips of pearlite [10]. The average grain size was found around 9.5 μm as measured by using ImageJ software. The microstructure of the specimen as shown in Fig. 10.

3.5 Macroscopic Morphology and Defects

The macroscopic morphology of the WAAM-CMT solid structure component is shown. As the thick solid block was deposited through successive multilayer multibead deposition, the side surface of the deposited component appeared corrugated. But upon cutting the as-welded deposited solid block, a fully dense surface

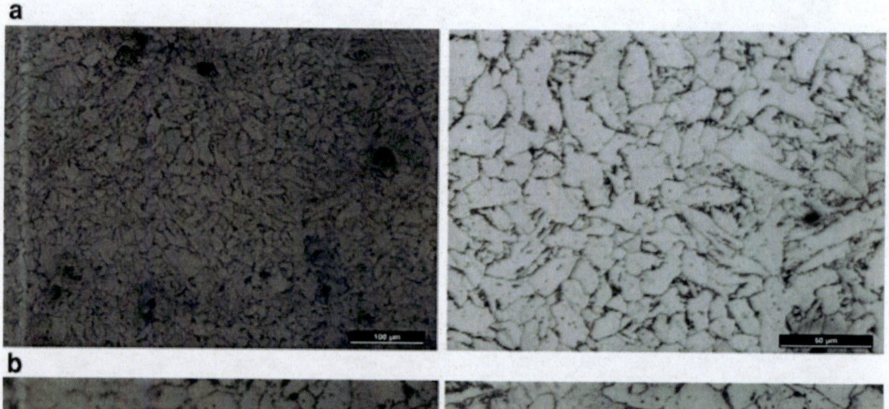

Fig. 10 **a** Microstructure of the specimen at 20X and 50X **b** microstructure of the specimen at 100X

was obtained that shows no sign of any defects and porosity in the fabricated block component. It defines very good integrity of the WAAM process for fabricating large size solid block components and advantage concerning other AM techniques.

4 Conclusions

In this study, the wire arc additive manufacturing (WAAM) technique using a cold metal transfer (CMT) process has been carried out to produce a near-net-shape solid block component of mild steel using ER70S6 copper-coated welding wire. The study of microstructure and mechanical properties of the fabricated component was carried out to analyze the effect of directional variation in the properties of the samples. The following conclusions are taken from the present study:

- The tensile test gave mean values of yield and ultimate tensile strength of 373 and 510 MPa, respectively, which is quite acceptable as compared to the base material.
- The tensile properties of the samples were found almost uniform and isotropic in both the parallel and perpendicular to the directions of welding.
- The Vickers hardness values were found to be between 75 and 80 kg/mm^2 for parallel to the welding direction and between 71 and 74 kg/mm^2 for perpendicular direction.
- Both the strength and percentage of elongation measured are found higher than the required that shows the high ductility of the fabricated component.
- The impact toughness at −30 °C was found to be 148 and 170 J in the parallel and perpendicular direction.
- The toughness obtained is much higher than the required minimum impact strength of 80 J at the subzero temperature as per the AWS requirement for the used ER70S6 welding wire.
- The microstructure of the sample reveals that the microstructure observed in welded samples is ferrite and bainite which consists inside and along the prior austenite grain boundaries and most of the grains were found elongated toward the welding direction.
- The average grain size was found to be around 9.5 μm.
- Based on the observation that no delamination or other defects such as porosity, lack of fusion, etc., have not occurred in the specimens used for mechanical testing, it can be concluded that this process produces the product of high structural integrity.

Thus, wire arc additive manufacturing (WAAM) process based on cold metal transfer (CMT) technology proved to be an economical and cost-effective method of rapid prototyping/additive manufacturing for high-quality metal parts with uniform mechanical properties and fully dense large size complex shape components with solid and hollow both types of structures.

Acknowledgements The authors gratefully acknowledge CSIR-National Metallurgical Laboratory, Jamshedpur for providing material and financial assistance to carry out research work. The authors also thank the Director, CSIR-NML, Jamshedpur for providing infrastructural support to carry out this investigation and for his kind permission to publish this work.

References

1. González J, Rodríguez I, Prado-Cerqueira JL, Diéguez JL, Pereira A (2017) Additive manufacturing with GMAW welding and CMT technology. Procedia Manuf 13:840–847. https://doi.org/10.1016/j.promfg.2017.09.189
2. Prado-Cerqueira JL, Diéguez JL, Camacho AM (2017) Preliminary development of a wire and arc additive manufacturing system (WAAM). Procedia Manuf 13:895–902. https://doi.org/10.1016/j.promfg.2017.09.154
3. Liberini M et al (2017) Selection of optimal process parameters for wire arc additive manufacturing. Procedia CIRP 62:470–474. https://doi.org/10.1016/j.procir.2016.06.124
4. Williams SW, Martina F, Addison AC, Ding J, Pardal G, Colegrove P (2016) Wire + Arc additive manufacturing. Mater Sci Technol 32(7):641–647. https://doi.org/10.1179/1743284715Y.0000000073
5. Cunningham CR, Flynn JM, Shokrani A, Dhokia V, Newman ST (2018) Invited review article: strategies and processes for high quality wire arc additive manufacturing. Addit Manuf 22(June):672–686. https://doi.org/10.1016/j.addma.2018.06.020
6. Wu B et al (2018) A review of the wire arc additive manufacturing of metals: properties, defects and quality improvement. J Manuf Process 35(August):127–139. https://doi.org/10.1016/j.jmapro.2018.08.001
7. Selvi S, Vishvaksenan A, Rajasekar E (2018) Cold metal transfer (CMT) technology—an overview. Def Technol 14(1):28–44. https://doi.org/10.1016/j.dt.2017.08.002
8. Gu J, Cong B, Ding J, Williams SW, Zhai Y (2014) Wire + Arc additive manufacturing of aluminium. In: SFF symposium, Austin Texas, pp 451–458
9. Chen X et al (2018) Cold metal transfer (CMT) based wire and arc additive manufacture (WAAM) system. J Surf Investig 12(6):1278–1284. https://doi.org/10.1134/S102745101901004X
10. American Society for Metals (1988) Metals handbook, 9th edn

Effect of MoS_2 and CeO_2 Powder Addition by Friction Stir Processing on Wear and Corrosion Properties of Al7075 Alloy

Pabitra Maji, Ranit Karmakar, Rahul Kanti Nath, R. K. Bhogendro Meitei, and Subrata Kumar Ghosh

1 Introduction

Aluminium alloys are very extensively used in aerospace and automobile sectors due to their lightweight and good strength. However, due to their poor surface properties like low hardness and wear resistance, more often than not, some reinforcements are added to the alloys to enhance the desired properties for specific applications. Stir casting [1], powder metallurgy [2], laser cladding [3], vapour deposition [4], plasma spraying [5], etc., are some well-established techniques to reinforce secondary particles to metals and metal surfaces. Lately, friction stir processing (FSP) emerged as a new technique for composite fabrication with some notable advantages over other processes such as grain refinement, uniform distribution and environment-friendly technique.

Friction stir processing was derived from friction stir welding (FSW) technique. In this process, a rotating tool is put onto the top surface of the plates to be welded. The constant contact of the tool and material at high speed generates frictional heat, and consequently, the materials to be welded become plasticized. A probe attached with the tool stirs the semi-solid material, and joining takes place. Since development, continuous researches were carried out on this process to know the effects of various materials, parameters and heat flow characteristics in the process [6, 7]. Moreover, with passing time, FSW was used for grain refinement of materials and to enhance uniformity in composites. With the increasing applications of FSW, Mishra et al. [8] used the technique for surface composite development and revealed a wide area in the field of materials processing. Subsequently, several researches were carried out by implementing FSP technique for bulk and surface

P. Maji (✉) · R. K. Nath · R. K. Bhogendro Meitei · S. K. Ghosh (✉)
Department of Mechanical Engineering, NIT Agartala, Agartala, India

R. Karmakar (✉)
Department of Metallurgical and Materials Engineering, IIT Kharagpur, Kharagpur, India

S. Bag et al. (eds.), *Next Generation Materials and Processing Technologies*, Springer Proceedings in Materials 9, https://doi.org/10.1007/978-981-16-0182-8_12

composites fabrication. Heidarpour et al. [9] prepared nano-TiO_2 reinforced copper matrix surface composites by FSP. Balakrishnan et al. [10] fabricated magnesium matrix composite by introducing TiC through FSP technique. Along with these, FSP was found to be successful in fabricating titanium alloys and steel matrix composites [11, 12].

Being the mostly used composite in industrial applications, a significant number of researches are focused on development and application-based characterization of aluminium matrix composites. Dinaharan [13] integrated different particles such as SiC, Al_2O_3, TiC, B_4C and WC. The experimentation suggested that, the particle types did not influence their distribution in metal matrices. A comprehensive review on aluminium matrix composites prepared using FSP was done by Maji et al. [14].

With the necessity of improving more than one aspect of a material, hybrid composites are a lucrative choice in recent researches. Devaraju et al. [15] used mixture of SiC and Al_2O_3 or Gr for fabricating Al6061 matrix composite to achieve high hardness as well as lubricating effect to reduce friction. Similarly, Rejil et al. [16] reinforced Al6360 by TiC and B_4C mixture to attain the aforesaid properties. Palanivel et al. [17] also observed that the mixture of BN and TiB_2 in Al6082 matrix exhibited best wear resistance due to the combining effect of solid lubrication and load carrying by hard particles. Eskandari et al. [18] added a mixture of TiB_2 and Al_2O_3 in Al8026 alloy and observed significant improvement in mechanical properties along with tribological behaviour. Sudhakar et al. [19] observed that, the addition of solid lubricant MoS_2 helped in achieving better ballistic impact resistance of Al7075-B_4C composite. Khan et al. [20] found hybrid composite of B_4C and CNT reinforcement in Al5083 alloy to be beneficial in terms of mechanical properties compared to composites made by individual reinforcements. Hossieni et al. [21] observed addition of CNT and CeO_2 enhanced mechanical properties with retaining corrosion properties of Al5083 alloy.

Addition of MoS_2 was found to be advantageous with respect to wear resistance of metallic alloys. Qu et al. [22] observed lubrication layer during sliding wear of MoS_2 reinforced Ti_6Al_4V composite. Rouhi et al. [23] also identified MoS_2 film over wear tested surfaces of Al–MoS_2 composites. Soleymani et al. also [24] observed reduction in friction coefficient and improvement in wear resistance of Al5083 alloy by reinforcing MoS_2 powder. Ability of CeO_2 in corrosion inhibition was observed by several researchers. Zheng et al. [25] found CeO_2 to be corrosion inhibitor to Mg alloy. Addition of CeO_2 in epoxy coating on carbon steel improved the corrosion performance in NaCl solution [26].

The extensive literature survey indicates that FSP is an efficient way to fabricate aluminium matrix composites. However, very limited studies can be found on improvement of wear and corrosion behaviour of aluminium alloys by reinforcing through FSP technique. Also, from earlier studies, it is well evident that addition of MoS_2 can provide excellent wear resistance and incorporation of CeO_2 yields good corrosion resistance.

Thus, to attain good wear and corrosion properties, in the investigation, aluminium 7075-T651 alloy is reinforced with a mixture of MoS_2 and CeO_2 powder. The effect of particle inclusion on tribological behaviour and corrosion behaviour is examined.

2 Materials and Methods

In this study, Al7075-T651 aluminium alloy rolled plates (Supplied by: Bharat Aerospace Metals, Mumbai) with thickness of 6 mm were used at the base or matrix for this surface composite. The chemical composition of the concerned alloy is given in Table 1. 150 mm × 50 mm rectangular plates were used for the processing. Hybrid composite was made by integrating MoS_2 and CeO_2 particles with the average particle size of 20 μm (Supplied by: Parshwamani Metals, Mumbai) were selected for fabricating the hybrid surface composite using identical proportions of both. A longitudinal groove (width and depth of 2 mm) was milled on alloy plate prior to fabrication for sitting of the powder reinforcements. Filling of the powder mixture was followed by covering of the groove by a pinnless friction stir tool (with 1000 rpm and 60 mm/min traverse speed) of shoulder diameter 16 mm to hinder the splashing of powder. Covering and the fabrication of FSP was performed in a 10 t friction stir welding machine set-up, which is shown in Fig. 1a, b. The final processing was performed by a concave shoulder FSP tool made of high carbon high chromium steel. The detailed specification of this tool is illustrated in Fig. 2a. It was indigenously made, heat treated and oil quenched followed by normalized. The rotational speed and transverse speed of the tool were kept constant at 1500 rpm and 60 mm/min, respectively, selected from trial experiment. For superior scattering of particles in the matrix, two FSP passes were employed. The second pass was performed exactly to the opposite direction of the first pass with 100% overlap after subsequent cooling after first pass. A realistic top view of the processed plate levelled with different details is shown in Fig. 2b.

The dry sliding wear properties of the composite and unreinforced alloy were examined in the ball on disc tribo testing machine. 10 × 10 mm^2 square wear test sample was cut from the middle of the stir zone of the FSPed specimen (Fig. 2b) using wire EDM as disc. The samples were tested under the conditions of vertical load—4.9 N, sliding speed 0.126 m/s, sliding distance—150.79 m and 6 mm tungsten carbide (WC) ball was used for this experiment. The material loss as well as the specific wear rate was calculated thereafter. The wear tracks of the samples were examined in field emission scanning electron microscope (FE-SEM) to know

Table 1 Chemical composition of as supplied Al7075-T651 alloy

Element	Zn	Mg	Cu	Ti	Mn	Fe	Si	Al
wt%	5.6	2.4	1.5	0.02	0.02	0.26	0.08	Balance

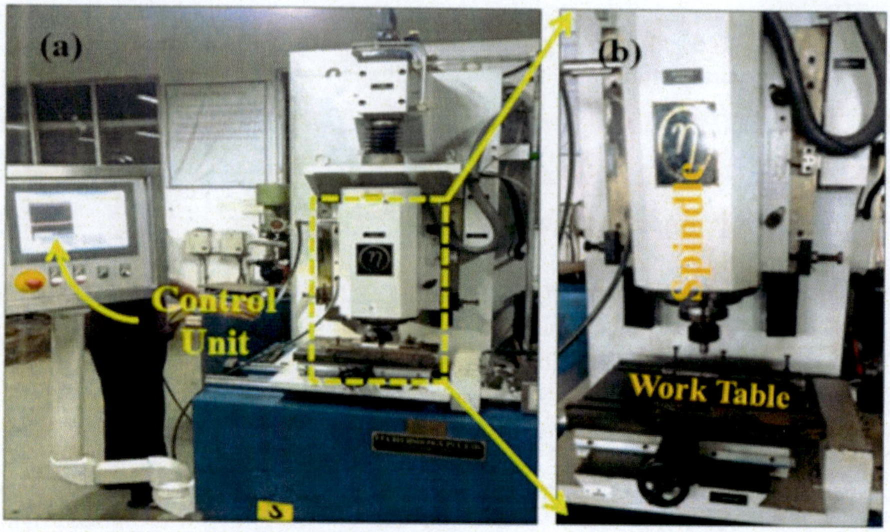

Fig. 1 **a**, **b** Friction stir processing machine set-up

the cause of wear. Electrochemical corrosion behaviour of the composite specimen (Fig. 2b) and unreinforced alloy was performed in a potential dynamic polarization equipment in 3.5% NaCl solution. Prior to the polarization test, samples were epoxy coated barring one surface to be exposed and were immersed in the NaCl solution for a half an hour duration. The corrosion potential (E_{corr}) as well as the corrosion current density (I_{corr}) was calculated by means of Tafel extrapolation method with a scanning from −1.5 to 1 V (versus SEC) at speed of 1 mV/s.

3 Results and Discussion

Figure 3a, b represents the scanning electron microscopic (SEM) images of the cross-sectional microstructure of the hybrid composite in low magnification and high magnification, respectively. Any significant agglomeration cannot be identified in the SEM images. From highly magnified image, it is evident that the reinforcing particles, i.e. MoS_2 (black particles) and CeO_2 (white particles), are almost evenly distributed throughout the composite surface. In spite of having different density of the reinforcement particles, the solid state nature of the process and proper stirring of the particles by tool pin is responsible for obtaining uniform distribution. It is interesting to note that, the particles size in the composite is much lower than the size of the supplied particle. The extreme shearing of the particles by the rotating pin induced fracture in the micro-particles. As a result, the fragmented particles can be observed in the composite. No pore and void can be observed in the cross-sectional images. The energy dispersive spectroscopy (EDS) analysis of the

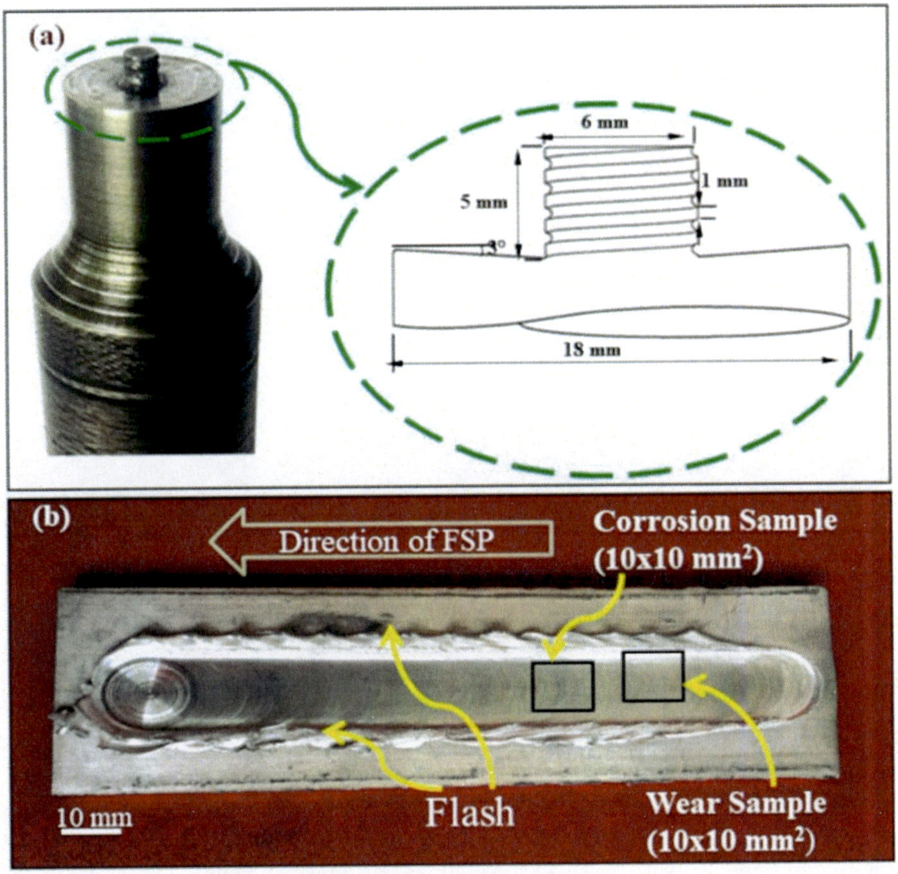

Fig. 2 **a** Threaded cylindrical FSP tool, **b** plate after FSP

cross section (Fig. 3a) reveals the presence of major alloying elements of the Al7075 alloy such as Al, Zn, Mg and Cu. Also, the elements of reinforcing particles, i.e. Mo, S, Ce and O can be observed in the elemental analysis of the cross section. This suggests successful fabrication of MoS_2 and CeO_2 reinforced Al7075 matrix composite.

Prior to wear test the specimens were mechanically polished up to 0.001 average surface roughness. The material loss in wear was measured by measuring the mass of the specimens before and after subjected to dry sliding wear. Then, the volume loss of specimens was measured by using predetermined density such as volume loss = mass loss/density of specimen. The specific wear loss was calculated as specific wear rate = volume loss/(load applied × sliding distance). The specific wear loss provides an assumption on possibility of material erosion in a particular sliding encounter [27]. The specific wear rate of the prepared composite and as supplied alloy is shown in Fig. 4a. The specific wear rate of 7.95×10^{-4} mm^3/N-m

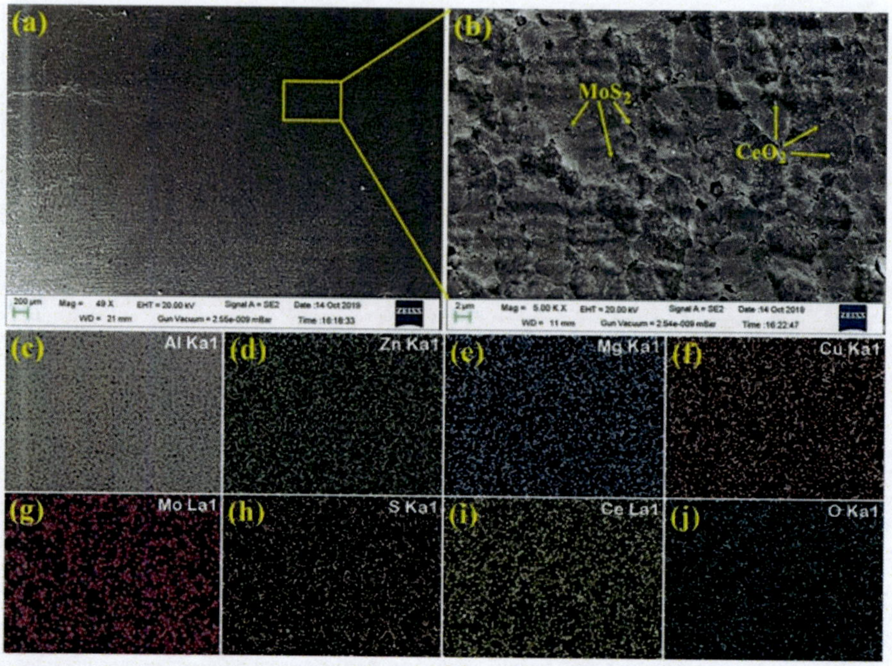

Fig. 3 **a** Cross-sectional SEM image of the FSPed composite, **b** enlarged view, **c–j** EDS area mapping of the cross section

of the unprocessed alloy was reduced to 3.98×10^{-4} mm^3/N-m in case of the hybrid composite. In a certain sliding condition, the material loss in wear depends on mainly three aspects: (a) surface roughness of the specimen, (b) hardness of the material and (c) coefficient of friction (COF) between test specimen and counter body. The values of average surface roughness of the specimens were kept similar. The measured average microhardness of the unreinforced alloy was $175HV_{0.1}$, whereas the measured microhardness of the fabricated composite was $171HV_{0.1}$. Therefore, the coefficient of friction was most influential in the difference in material loss of the tested specimens. The variation of coefficient of friction of the specimens with sliding distance is shown in Fig. 4b. After the initial increase, the unreinforced alloy shows less variation compared to the hybrid composite. This may be due to uniformity of wear occurrence in the aluminium alloy. In case of hybrid composite, after the initial increase, the COF decreases. After a slight stability, the COF increases, decreases and finally increases in current span of sliding distance. The initial increase in COF in both the specimens is attributed to the interlocking between asperities of test surfaces and the counter body. The decrease afterwards in the composite is due to reduction in normal force by unlocking of the asperities. The de-bonded particles got trapped in between the mating faces and transformed the mode of wear to three body wear from two body wear [28]. However, after the trapped particles were removed, due to successive locking and unlocking asperities, the normal load and consequently the COF were increased, decreased and increased as observed in Fig. 4b. Also, it is clearly evident form the figure that the average COF of the aluminium alloy is decreased from 0.659 to 0.405 by inclusion of the micro-powders. This is due to solid lubricant property of MoS_2. The reduction of COF is responsible for obtaining lower wear rate in the composite compared to the unprocessed alloy.

The SEM images of the worn surfaces are shown in Fig. 5. In case of aluminium alloy, material removal in large flakes can be observed. Also, a white coloured phase distinct from aluminium alloy is seen. To identify the different phase, spot EDS analysis was performed on that. The EDS result demonstrates that, the new phase is Al and O_2-rich phase. This observation suggests that oxidation happened in the unreinforced alloy during sliding contact with WC ball. The rubbing action of WC ball generated heat in the aluminium alloy. The soft alloy became plasticized by the frictional heat and adhered with the WC ball. Then, the adhered materials were removed by further movement of the ball. With more continuous contact, the amount of heat generated in the alloy was increased. At high temperature, the alloy reacted with oxygen and thick oxide layer formed. With further sliding contact, due to internal stress and brittle nature, the oxides became unstable and broke after reaching critical thickness. As a result, rough delaminated surface can be observed in the SEM micrographs of aluminium alloy (Fig. 5a). In the hybrid composite, parallel groove like surface is observed in the SEM image of the specimen subjected to sliding wear. Along with the parallel grooves, some flake removal is also found in the worn surface. The presence of reinforcing particles reduced the direct contact of aluminium matrix with the WC ball. Rather, the particles acted as load carrying medium. As observed in Fig. 4b, with progression in sliding contact, the particles came out and stuck between ball and matrix plate. The cutting action of the comparatively harder reinforcing particles generated grooves on the aluminium matrix surface. However, due to low load carrying capacity, the composite also got heated simultaneously. Consequently, the flake structured material removal due to sticking of plasticized material to WC ball can be observed in the worn surface

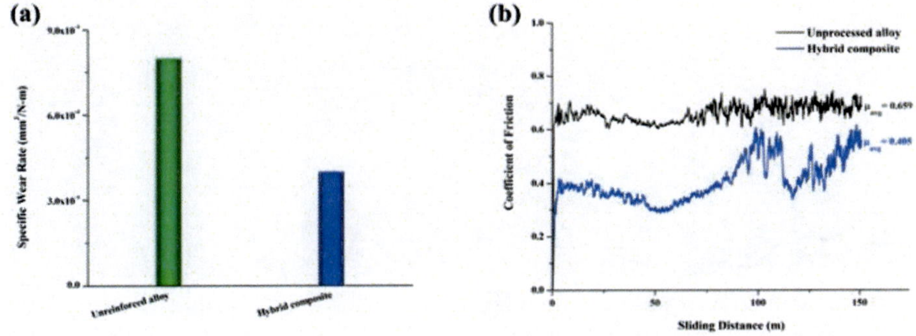

Fig. 4 **a** Specific wear rate and **b** COF curves of the as supplied aluminium alloy and the FSPed hybrid MMC

(Fig. 5b). Due to the low COF between the hybrid composite and WC ball owing to the lubricating property of MoS_2 reinforcement particle (Fig. 4b), the intensity of heat generated in the composite was lower than that of in the unreinforced alloy. Consequently, the severity of adhesive wear was reduced, and oxidation did not initiate. The change of wear mechanism is responsible for achieving better wear resistance in the hybrid composite.

A saturated calomel electrode (SCE) was employed as the reference electrode for the corrosion test, while the auxiliary electrode was made of platinum. The exposed surfaces of both the specimens were polished up to average roughness of 0.001 to eliminate the influence of surface roughness from the corrosion behaviour as higher roughness results in higher corrosion rate [5]. The potentiodynamic polarization curves of Al7075 alloy and the fabricated hybrid composite are illustrated in Fig. 6a. For the aluminium alloy, the anodic current increases rapidly then somewhat stabilizes. The rapid increase in corrosion current suggests onset of rapid corrosion. The stabilization of corrosion current is maybe due to hindrance of rapid corrosion by a layer formed during initial chemical reaction. In case of the hybrid composite, the anodic current does not increase rapidly, rather the corrosion potential increases. This suggests onset of passivation in the hybrid composite. The presence of anti-corrosive element CeO_2 in the processed composite stalled the corrosion propagation and enforced passivation. The corrosion current density (I_{corr}) and corrosion potential (E_{corr}) of the specimens were measured using Tafel's extrapolation method and shown in Fig. 6b. The corrosion potential (E_{corr}) represents the onset of reduction from oxidation, whereas the corrosion current density (I_{corr}) is directly related to corrosion rate such as corrosion rate = $(0.13 \times I_{corr} \times EW)/d$, where EW is equivalent weight of the specimen and d is density of the specimen [29]. From Fig. 6b, it can be observed that for the composite, the E_{corr} is higher, and I_{corr} is lower compared to the aluminium alloy. The corrosion potential is increased from −863.962 to −672.192 V, and the corrosion current density is decreased from 2×10^{-6} A/cm^2 to 8.18×10^{-7} A/cm^2. The presence of CeO_2 is

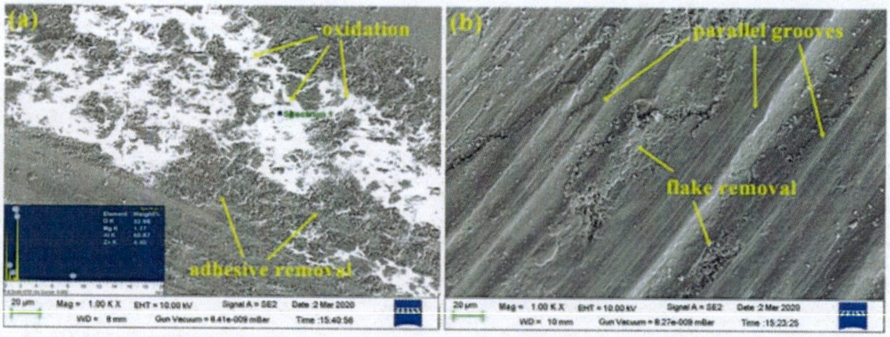

Fig. 5 SEM images of wear abraded surfaces of **a** as supplied Al7075 alloy, **b** hybrid composite

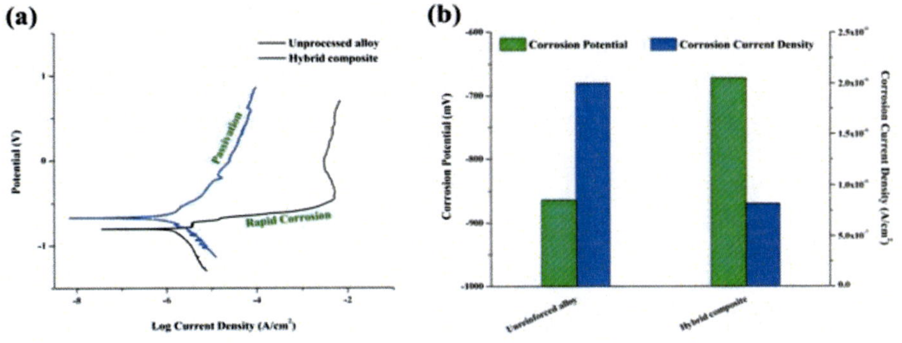

Fig. 6 **a** Polarization curves of the as supplied aluminium alloy and hybrid composite, **b** corresponding corrosion potential and current density

responsible for delaying the onset of corrosion in the composite. This observation yields that the addition of hybrid power mixture is beneficial in terms of corrosion resistance of the Al7075 alloy.

4 Conclusions

A hybrid composite was made by reinforcing Al7075-T651 alloy with a mixture of MoS_2 and CeO_2 powder through friction stir processing technique. From this investigation, the following outcomes can be concluded:

(a) The microstructure of the hybrid composite reveals pore free and nearly uniform distribution of MoS_2 and CeO_2 particles.
(b) Specific wear rate of the hybrid composite is reduced compared to the aluminium alloy due to decrease in the coefficient of friction and presence of particles.
(c) The reduction of COF is mainly attributed to the existence of solid lubricant MoS_2 in the composite.
(d) Severe adhesive and oxidation wear is observed in aluminium alloy. Due to the decrease in friction coefficient, the severity of adhesive wear is reduced, and onset of abrasive wear is observed in the composite.
(e) The composite exhibits passivation in corrosive medium compared due to hindrance of corrosion propagation by CeO_2 particles present in the composite.
(f) The corrosion potential is increased, and corrosion current density is decreased in the composite compared to the unreinforced aluminium alloy. This indicates better corrosion resistance in the composite.

This study is focused on the effect of addition of MoS_2 and CeO_2 in aluminium alloy. However, the FSP parameters also play an important role in the properties of the fabricated composite which may be investigated. Mechanical properties such as tensile strength, fatigue strength, bending strength of the fabricated composite are needed to be examined for using it in industrial applications. Also, the wear and corrosion performances reported in this study were performed in room temperature (~25 °C). Behaviour of the composite in high temperature and very low temperature may be examined.

References

1. Ravi B, Balu Naik BB, Udaya Prakash J (2015) Characterization of aluminium matrix composites (AA6061/B4C) fabricated by stir casting technique. Mater Today Proc 2:2984–2990. https://doi.org/10.1016/j.matpr.2015.07.282
2. Chen HS, Wang WX, Nie HH, Li YL, Wu QC, Zhang P (2015) Microstructure and dynamic compression properties of PM Al6061/B4C composite. Acta Metall Sin (English Lett) 28:1214–1221. https://doi.org/10.1007/s40195-015-0315-8
3. Karmakar R, Ghosh SK (2020) Effect of composition and process parameter on mechanical properties of composite coating by laser cladding: an overview. In: Lecture Notes in Mechanical Engineering, pp 387–395
4. Beach DB, Blum SE, LeGoues FK (1989) Chemical vapor deposition of aluminum from trimethylamine-alane. J Vac Sci Technol A Vac Surf Film 7:3117–3118. https://doi.org/10.1116/1.576322
5. Kumari R, Majumdar JD (2017) Studies on corrosion resistance and bio-activity of plasma spray deposited hydroxylapatite (HA) based TiO_2 and ZrO_2 dispersed composite coatings on titanium alloy (Ti_6Al_4V) and the same after post spray heat treatment. Appl Surf Sci 420:935–943. https://doi.org/10.1016/j.apsusc.2017.05.208
6. Yaduwanshi DK, Bag S, Pal S (2015) Heat transfer analyses in friction stir welding of aluminium alloy. Proc Inst Mech Eng Part B J Eng Manuf 229:1722–1733. https://doi.org/10.1177/0954405414539297
7. Nath RK, Maji P, Barma JD (2019) Development of a self-heated friction stir welding tool for welding of polypropylene sheets. J Brazilian Soc Mech Sci Eng 41:1–13. https://doi.org/10.1007/s40430-019-2059-2
8. Mishra RS, Ma ZY, Charit I (2003) Friction stir processing: a novel technique for fabrication of surface composite. Mater Sci Eng, A 341:307–310. https://doi.org/10.1016/S0921-5093(02)00199-5
9. Heidarpour A, Mazaheri Y, Roknian M, Ghasemi S (2019) Development of Cu–TiO_2 surface nanocomposite: effect of pass number on microstructure, mechanical properties, tribological and corrosion behavior. J Alloys Compd 783:886–897. https://doi.org/10.1016/j.jallcom.2018.12.382
10. Balakrishnan M, Dinaharan I, Palanivel R, Sivaprakasam R (2015) Synthesize of AZ31/TiC magnesium matrix composites using friction stir processing. J Magnes Alloy 3:76–78. https://doi.org/10.1016/j.jma.2014.12.007
11. Li B, Shen Y, Luo L, Hu W (2013) Fabrication of TiCp/Ti_6Al_4V surface composite via friction stir processing (FSP): process optimization, particle dispersion-refinement behavior and hardening mechanism. Mater Sci Eng, A 574:75–85. https://doi.org/10.1016/j.msea.2013.03.019
12. Ghasemi-Kahrizsangi A, Kashani-Bozorg SF (2012) Microstructure and mechanical properties of steel/TiC nano-composite surface layer produced by friction stir processing. Surf Coatings Technol 209:15–22. https://doi.org/10.1016/j.surfcoat.2012.08.005
13. Dinaharan I (2016) Influence of ceramic particulate type on microstructure and tensile strength of aluminum matrix composites produced using friction stir processing. J Asian Ceram Soc 4:209–218. https://doi.org/10.1016/j.jascer.2016.04.002

14. Maji P, Ghosh SK, Nath RK, Karmakar R (2020) Microstructural, mechanical and wear characteristics of aluminum matrix composites fabricated by friction stir processing. J Brazilian Soc Mech Sci Eng 42. https://doi.org/10.1007/s40430-020-02279-5
15. Devaraju A, Kumar A, Kotiveerachari B (2013) Influence of addition of Grp/Al_2O_3p with SiCp on wear properties of aluminum alloy 6061-T6 hybrid composites via friction stir processing. Trans Nonferrous Met Soc China (English Ed) 23:1275–1280. https://doi.org/10.1016/S1003-6326(13)62593-5
16. Rejil CM, Dinaharan I, Vijay SJ, Murugan N (2012) Microstructure and sliding wear behavior of AA6360/(TiC+B4C) hybrid surface composite layer synthesized by friction stir processing on aluminum substrate. Mater Sci Eng, A 552:336–344. https://doi.org/10.1016/j.msea.2012.05.049
17. Palanivel R, Dinaharan I, Laubscher RF, Davim JP (2016) Influence of boron nitride nanoparticles on microstructure and wear behavior of AA6082/TiB_2 hybrid aluminum composites synthesized by friction stir processing. Mater Des 106:195–204. https://doi.org/10.1016/j.matdes.2016.05.127
18. Eskandari H, Taheri R, Khodabakhshi F (2016) Friction-stir processing of an AA8026-TiB_2–Al_2O_3 hybrid nanocomposite: Microstructural developments and mechanical properties. Mater Sci Eng, A 660:84–96. https://doi.org/10.1016/j.msea.2016.02.081
19. Sudhakar I, Madhu V, Madhusudhan Reddy G, Srinivasa Rao K (2015) Enhancement of wear and ballistic resistance of armour grade AA7075 aluminium alloy using friction stir processing. Def Technol 11:10–17. https://doi.org/10.1016/j.dt.2014.08.003
20. Khan M, Rehman A, Aziz T, Naveed K, Ahmad I, Subhani T (2017) Cold formability of friction stir processed aluminum composites containing carbon nanotubes and boron carbide particles. Mater Sci Eng, A 696:552–557. https://doi.org/10.1016/j.msea.2017.04.074
21. Hosseini SA, Ranjbar K, Dehmolaei R, Amirani AR (2015) Fabrication of Al5083 surface composites reinforced by CNTs and cerium oxide nano particles via friction stir processing. J Alloys Compd 622:725–733. https://doi.org/10.1016/j.jallcom.2014.10.158
22. Qu CC, Li J, Juan YF, Shao JZ, Song R, Bai LL, Chen JL (2019) Effects of the content of MoS_2 on microstructural evolution and wear behaviors of the laser-clad coatings. Surf Coatings Technol 357:811–821. https://doi.org/10.1016/j.surfcoat.2018.10.100
23. Rouhi M, Moazami-Goudarzi M, Ardestani M (2019) Comparison of effect of SiC and MoS_2 on wear behavior of Al matrix composites. Trans Nonferrous Met Soc China (English Ed) 29:1169–1183. https://doi.org/10.1016/S1003-6326(19)65025-9
24. Soleymani S, Abdollah-zadeh A, Alidokht SA (2012) Microstructural and tribological properties of Al5083 based surface hybrid composite produced by friction stir processing. Wear 278–279:41–47. https://doi.org/10.1016/j.wear.2012.01.009
25. Zheng Z, Zhao MC, Tan L, Zhao YC, Xie B, Yin D, Yang K, Atrens A (2020) Corrosion behavior of a self-sealing coating containing CeO_2 particles on pure Mg produced by micro-arc oxidation. Surf Coatings Technol 386:125456. https://doi.org/10.1016/j.surfcoat.2020.125456
26. Lei Y, Qiu Z, Tan N, Du H, Li D, Liu J, Liu T, Zhang W, Chang X (2020) Polyaniline/CeO_2 nanocomposites as corrosion inhibitors for improving the corrosive performance of epoxy coating on carbon steel in 3.5% NaCl solution. Prog Org Coatings 139:105430. https://doi.org/10.1016/j.porgcoat.2019.105430
27. Maji P, Dube RK, Basu B (2009) Enhancement of wear resistance of copper with tungsten addition ($\leq$20 wt%) by powder metallurgy route. J Tribol 131:1–9. https://doi.org/10.1115/1.3204776
28. Kumari R, Majumdar JD (2018) Wear behavior of plasma spray deposited and post heat-treated hydroxyapatite (HA)-based composite coating on titanium alloy (Ti-6Al-4 V) substrate. Metall Mater Trans A Phys Metall Mater Sci 49:3122–3132. https://doi.org/10.1007/s11661-018-4626-z
29. Popov BN (2015) Basics of Corrosion Measurements. Corros Eng 181–237. https://doi.org/10.1016/b978-0-444-62722-3.00005-7

Effect of Current Pulsation on Weld Microstructure During Micro-Plasma Arc Welding of Inconel 718

A. K. Sahu and Swarup Bag

1 Introduction

Superalloy 718 (also known as Inconel 718) is the most widely recognized Ni–Fe-based alloy which exhibits excellent mechanical and oxidation resistance property up to the temperature of 650 °C [1]. Hence, the alloy is found to be desirable for numerous high-temperature applications even in the extreme environments, such as gas turbine disks, casings, shafts, and liners in aerospace and nuclear power generation industries. The superalloy shows exceptional fusion welding quality owing to the sluggish precipitation of alloy's principal strengthening phase γ''(BCT, Ni_3Nb), which provides resistance to strain age cracking and solidification cracking [2, 3]. The primary challenge involved with the fabrication of superalloy 718 is the segregation of Niobium in the interdendritic zone and precipitation of various secondary phases (i.e., NbC and Laves) which promotes the microfissuring or liquation cracks in the heat-affected and fusion zone (FZ) [4–7]. Hence, the presence of Laves phase which is a Nb enriched intermetallic phase (represented as, Fe_2Nb) in the interdendritic zones during solidification has deleterious impact on the mechanical properties like tensile strength, hardness, ductility, fatigue, and creep rupture strength [1, 8]. Precipitation of Laves and different carbides not merely provides micro-crack initiation sites in the FZ and HAZ but also consume a substantial amount of favorable strengthening alloying elements out of the metal matrix.

Laves phase is an unavoidable intermetallic phase present mainly in the interdendritic region of fusion welded component [9]. Hence, reduction in volume percentage of Laves phase by using various advanced welding technique has resulted in weld quality improvement. The amount of heat input (J/mm) during

A. K. Sahu (✉) · S. Bag
Department of Mechanical Engineering, Indian Institute of Technology Guwahati, Guwahati, Assam 781039, India
e-mail: ajit.sahu@iitg.ac.in

S. Bag et al. (eds.), *Next Generation Materials and Processing Technologies*, Springer Proceedings in Materials 9, https://doi.org/10.1007/978-981-16-0182-8_13

welding plays a pivotal role for amount of segregation and precipitation of Laves phase. Low heat input during welding assures a steep thermal gradient during solidification of weld pool which hinders the segregation tendency of the solutes like Nb, Mo, and Ti in the interdendritic zones and leads to a refined microstructure in the weld zone. Many studies have focused on various advanced welding techniques like laser beam welding (LBW) and electron beam welding (EBM) to achieve low heat input and high cooling rate to control the segregation tendency and to further improve the mechanical strength of Inconel 718 welded structure [1, 2, 10]. Apart from the heat input, techniques like current pulsation and beam oscillation were also used to control the Nb segregation and Laves phase formation [10–12]. The application of current pulsation during welding concludes in remarkable grain refinement of the solidified structure. The microstructure refinement during solidification occurs mainly due to the periodic change in the weld pool temperature. The cyclic temperature variation enhances the fluid flow inside the weld pool and facilitates to break the long continuous chain of Laves particle during the solidification. Ram et al. [11] reported the application of current pulsation during GTA welding leads to contraction in Laves phase volume percentage and enhancement in the weld mechanical characteristics. Ram et al. [13] witnessed very fine and discrete weld morphology and lower segregation of principal alloying elements in the interdendritic region with the pulsed Nd-YAG laser welding as compared to the GTA welding technique. Radhakrishna and Rao [10] used electron beam (EB) with circular oscillation technique to weld the Inconel 718 and found significant reduction in segregation of Nb and Laves phase precipitation in the weld zone. Till date, very minimal work is available on arc welding of superalloy 718 with current pulsation mode as compared to LBW and EB advanced welding techniques. Hence, it is worthwhile in the current study to analyze the cost-efficient micro-plasma arc welding (μPAW) in constant and pulsed current mode to weld the solution treated superalloy 718 sheets and highlight the possibility in reducing weld segregation and precipitation of Laves phase. The primary aim of the present work is to investigate the impact of current pulsation on microstructure and mechanical characteristics of the welded joints.

2 Experimental Details

As received superalloy 718 rolled sheets of 0.7 mm thickness are undergone solution treatments at 980 °C for 1 h followed by water quenching to obtain a homogeneous microstructure. The solution treated sheets are cut into 90 × 55 mm coupons, and the welding direction is kept in normal to the sheet rolling direction. In order to figure out the elemental composition (wt%), energy dispersive X-ray (EDX) investigation is conducted on the base material and the result is shown in Fig. 1a. In reference to the EDX probe, the parent material composition is presented in Table 1. The optical microstructure of superalloy 718 parent material is demonstrated in Fig. 1b. The base material wrought austenite consisting of

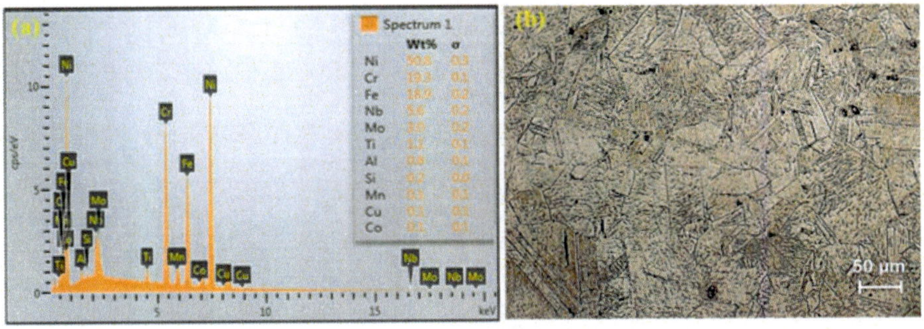

Fig. 1 Showing **a** EDX elemental analysis and **b** optical microstructure of the base material

Table 1 Chemical composition (wt%) of superalloy 718

Elements	Ni	Cr	Nb	Mo	Ti	Al
Composition (wt%)	50.8	19.3	5.6	3.0	1.1	0.8
Elements	Si	Mn	Cu	Co		Fe
Composition (wt%)	0.2	0.1	0.1	0.1		Bal.

equiaxed-type grains with an average grain diameter of 63 μm (ASTM 5–6) are observed in the micrograph. Discrete needle shape δ phase and randomly distributed MC (M: Nb, Ti) carbides are also evident in the microstructure. The welding surfaces are thoroughly cleaned before the welding process to remove the oxides and different impurities present on the surface. Sheets are tightened in a fixture made up of copper to curb the deformation and achieve high cooling rate throughout the welding process. The welding torch of micro-plasma setup is held stationary, and the workpiece is enabled to move with a consistent velocity. Commercial pure argon gas of 99% purity is employed as plasma gas in addition to shielding gas to cover the weld pool from atmosphere. Butt configured welded joints are produced through constant current (CC) and pulsed current (PC) method using a micro-plasma arc welding machine with DCEN power supply. The plasma and shielding gas flow speed are kept at 0.4 and 5 L/min, respectively, along with a nozzle standoff gap of 2 mm. The remaining welding variables adopted for producing defect less welds by constant and pulsed current conditions are presented in the Table 2, usually acquired by broad range of welding trials with keeping the heat input (J/mm) constant in both the condition.

The welded specimens are cut normal to the welding direction for microstructural analysis and polished with emery paper followed by velvet cloth polishing with the diamond paste (0.5–1 μm) to obtain mirror finish. The polished samples are dip for few seconds in a mixture of HCl, HF, and HNO_3 in the proportion of 4:2:1, respectively, to obtain the microstructure. The microstructural investigation is performed through the optical microscope and scanning electron microscope

Table 2 Welding parameters employed for CC (58.5 J/mm) and PC (59 J/mm) weld

Constant current welding	
Current, A	15
Welding speed, mm s^{-1}	4
Voltage, V	24
Pulse current welding	
Peak current, A	18
Base current, A	5.4
Welding speed, mm s^{-1}	4
Pulse frequency, Hz	6
Pulse-on time	30% of cycle time
Voltage, V	24

(SEM). The volume percentage of Laves intermetallic phase is evaluated through the image analysis technique on the SEM micrographs, taken from fusion boundary to weld center for both the welding conditions. X-ray diffraction (XRD) technique is carried out on CC and PC weld zone to reveal the composition and presence of various intermetallic phases along with the main austenite matrix. Micro-hardness measurements are conducted across the weld bead with an indentation spacing of 0.2 mm for each specimen, using Vickers hardness tester at a load of 500 g and a dwell time of 20 s. The hardness is evaluated in the center portion of the welded joint. Tensile testing of the welded samples and base material is undertaken at room temperature to analyze the proof stress, tensile strength, and degree of elongation. For every condition, three sub-size specimens are prepared according to usual ASTM E8M subsize standard. The tests are performed with a computer controlled servo-hydraulic universal tensile testing setup with extensometer attachment and a crosshead speed of 0.5 mm/min.

3 Results and Discussion

3.1 Microstructure

Figure 2 depicts the top and cross-sectional view of the weld bead obtained by CC and PC mode of welding. In both the cases, full penetration is achieved with a weld bead cap width of 1.9 mm and 2.7 mm and root width of 1.4 mm and 1.6 mm for CC and PC weld, respectively. Higher weld bead width in PC mode is owing to the increased heat input over the application of peak current (i.e., 18 A) during the pulse-on period as compared to CC mode of welding (i.e., 15 A). No significant difference in the weld surface geometry is observed except overlapped type weld pool due to the current pulsation in the PC mode as compared to wavy nature of weld pool in CC mode of welding is witnessed. From the optical microscopic

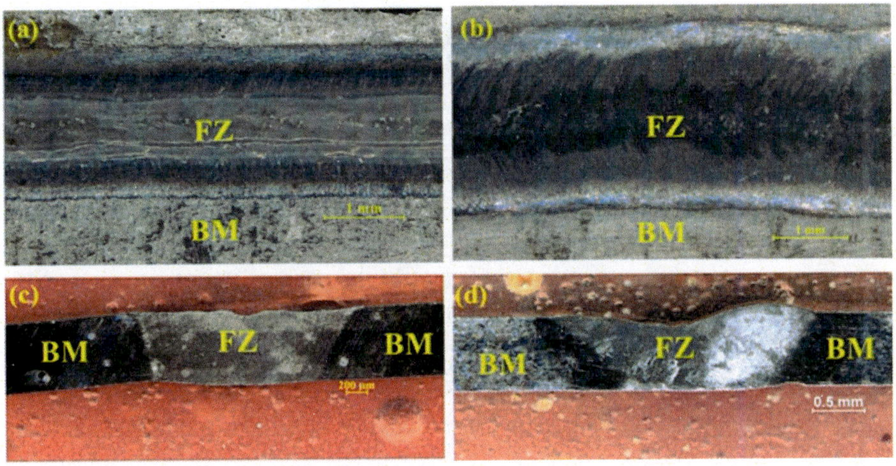

Fig. 2 Top and transverse view of weld bead produced by **a**, **c** constant current (CC) and **b**, **d** pulse current (PC) mode of welding

inspection, it is found that the acquired weld beads are independent from any kind of weld surface imperfections.

Figure 3 describes the microscopic view of weld fusion zone and fusion boundary for PC and CC conditions. Inconel 718 is a highly alloyed material; hence, the mode of solidification is of dendritic type. Columnar dendritic chain developed from the solid–liquid interface (fusion boundary) is seen in both the welding conditions. This is owing to the higher cooling rate near the fusion boundary [13]. The columnar dendritic growth near to the weld interface of PC weld results in fewer as compared to larger and continuous dendrites of CC weld, thus producing coarser and interconnected Laves particles in CC mode of welding [14]. The welds are observed to be free from microfissuring in the FZ an HAZ, and an extremely constricted heat-affected zone (HAZ) is witnessed in both the welding conditions.

The higher magnified micrographs of the weld interior reveal the mode of solidification dendrites for the PC and CC welds. The PC weld is witnessed to be fine and of equiaxed type, whereas the microstructure is observed to be coarser and columnar type for CC mode weld (Fig. 4). The variation in the fusion zone microstructural morphology is mainly due to the current pulsation effect during the welding [15, 16]. The solidification morphology depends upon the ratio of temperature gradient (G) and solidification growth rate (R). Lower value of (G/R) produces equiaxed dendritic structure. The current pulsation reduces thermal gradient in the weld pool by enhancing the fluid flow due to cyclic variation of temperature during the welding process [11]. In both the welding condition, solidification growth rate would be same due to equal welding speed [17]. Hence, at constant (R) reduction in (G) will reduce the (G/R) for the PC condition weld, resulting in equiaxed type of microstructure. Due to current pulsation, the cooling

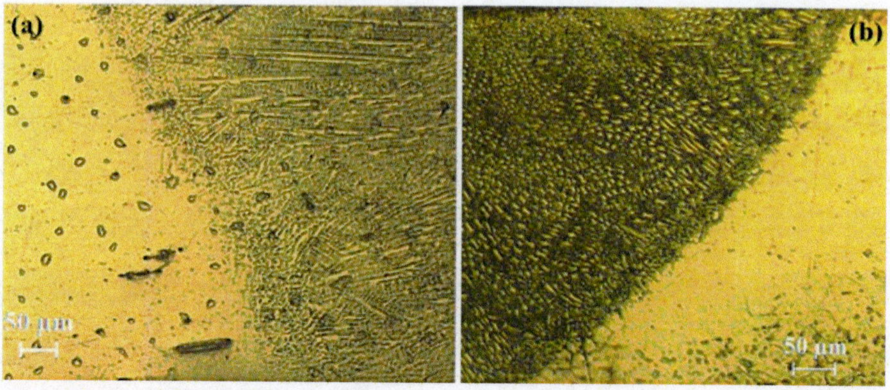

Fig. 3 Showing fusion boundary and heat affected zone of **a** CC and **b** PC welding condition

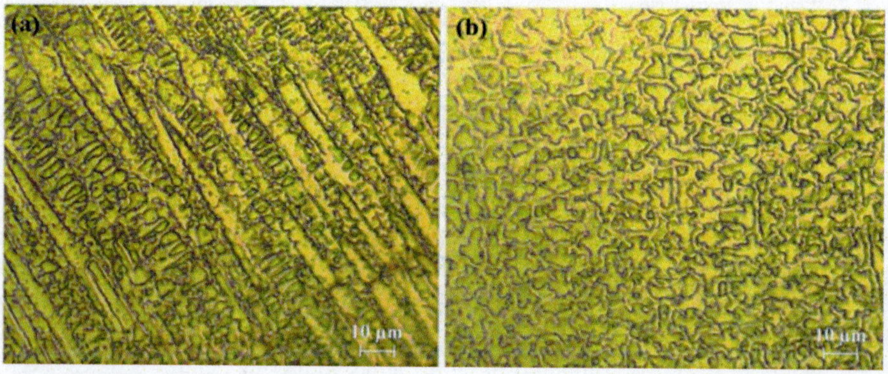

Fig. 4 Microstructure of the weld interior for **a** CC and **b** PC mode welding

rate also increases and results in decline of weld dendritic arm spacing and scale of the deleterious intermetallic phases [11].

The SEM analysis of the weld zone interior is shown in Fig. 5. At the weld interior, numbers of bright irregular-shaped Laves particles are observed in the interdendritic region. The Laves particles in CC weld are witnessed to be interconnect columnar type and moderately coarser against to the equiaxed dendrites of the PC weld. The current pulsation has refined the microstructure due to the temperature variation during the welding process. Through image analysis technique the average volume fraction (%) of interdendritic Laves phase in the weld interior is calculated as 15.5 ± 1.3% and 6.9 ± 1.1% for CC and PC weld, respectively. The continuous temperature variation during pulsed welding causes remelting and breaking off of the long continuous dendritic chain at the solidification front as a result reducing the segregation and volume percentage of intermetallic phase.

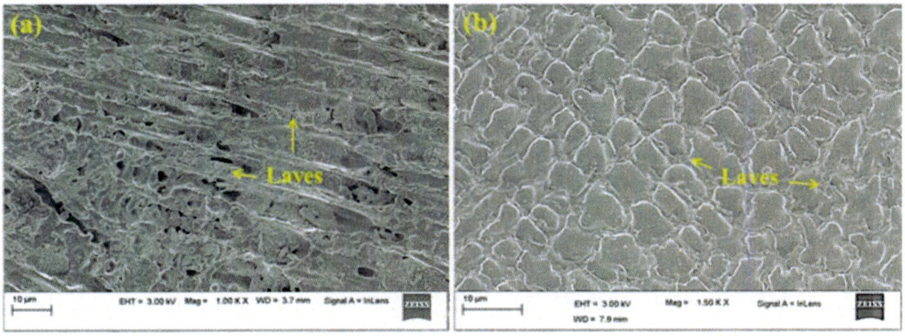

Fig. 5 SEM microstructure of weld interior for **a** CC mode and **b** PC mode weld

Hence, the amount (volume %) of Laves phase is higher in the columnar dendritic regions of CC weld as compared to the equiaxed dendrites of the PC mode weld [11].

3.2 *XRD Analysis*

The diffraction (XRD) peaks corresponding to the CC and PC weld zones are shown in Fig. 6. The main Fe–Ni austenite (γ) matrix dominating in (111) plane is witnessed from the sharp diffraction peak. The presence of diffraction peaks corresponding to various intermetallic phases such as Laves (Fe_2Nb), NbC, and TiC dominating in the direction (110) (112), (111) (220), and (222), respectively, are observed from the diffraction plot. The diffraction peaks corresponding to CC weld

Fig. 6 XRD analysis of the CC and PC weld zone

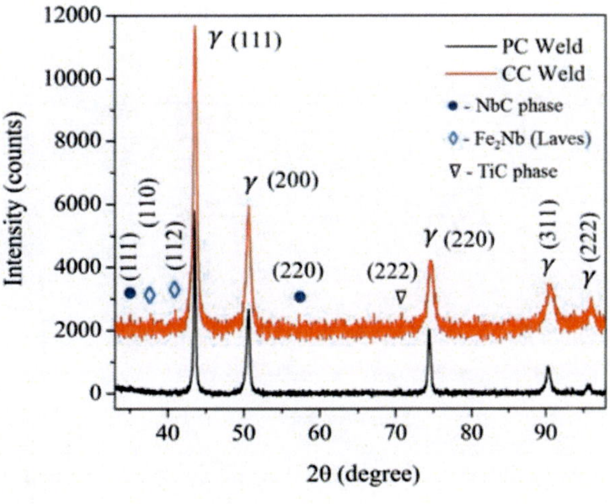

are found to be sharp and strong as compared to the peaks of PC weld. The small and weak peaks of PC weld will reduce the normalized intensity ratio (NIR) with respect to the CC weld, which signifies reduction in relative quantity of different intermetallic phases in the weld zone [18]. Hence, current pulsation during the welding hinders the precipitation of various secondary intermetallic phases in the solidified weld zone.

3.3 Mechanical Properties

The effect of segregation and formation of intermetallic phases on the weld mechanical properties are investigated. The micro-hardness test was carried out across the weld bead for both the cases, and the hardness profile is depicted in Fig. 7a. The base material hardness is found to be 215 ± 6 HV in the solution treated condition. The hardness of the CC weld falls in the fusion zone against the base material which is mainly due to segregation of principal strengthening alloying element. The development of intermetallic phases absorbs a substantial amount of principal alloying elements (i.e., Nb, Mo, and Ti) from the matrix, therefore reducing the hardness of the weld. The maximum fusion zone hardness in CC weld is found to be 206.2 HV. The presence of continuous Laves phase in the weld microstructure of CC weld, absorbs ample amount of Nb that were initially present in the base material, making accessible very less quantity of Nb for the precipitation of strengthening phases. For PC weld, the micro-hardness value is witnessed to be marginally higher than the CC weld zone. The maximum hardness of PC weld is found to be 222.2 HV. The hardness variation is primarily due to the refined microstructure of the PC fusion zone which leads to reduction of Laves phase formation, hence making ample amount of alloying element available in the metal matrix for the precipitation of strengthening phases. The hardness value of the HAZ drops in both the cases is attributed to the grain coarsening in the region due to generation of high temperature in the course of welding process.

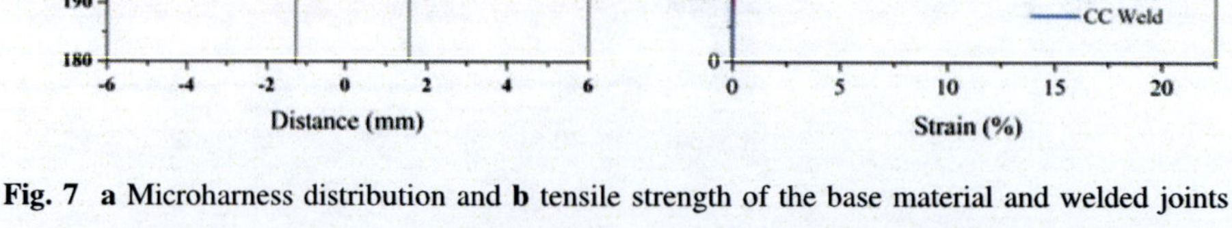

Fig. 7 **a** Microharness distribution and **b** tensile strength of the base material and welded joints

Table 3 Tensile test results of welded samples

	Base material	PC weld metal	CC weld metal
Yield stress (MPa)	595 ± 11	485 ± 8	499 ± 10
UTS (MPa)	880 ± 13	803 ± 9	762 ± 11
Elongation, %	21.0 ± 1.1	19.0 ± 0.9	10.0 ± 0.8

The base material and weld metal tensile properties for PC and CC mode weld are given in Table 3, and the corresponding stress–strain plot is shown in Fig. 7b. The tensile strength of the base material is witnessed to be superior to both the CC and PC weld metals, and the properties of CC welds are found inferior to the PC weld material. All of the welded specimens are failed in the fusion zone area, hence making the FZ as the weakest region in the component due to the presence of intermetallic phases. The joint efficiency of the welded component with respect to the BM is found to be 91.2% and 86.5% for PC and CC condition, respectively. The differences in tensile characteristics are owing to the development of intermetallic Laves phase during the weld solidification. The Laves phase not just consumes essential alloying element from the metal matrix in addition also provides desirable location for crack initiation and growth in the fusion zone during the loading [13]. As a result, the tensile ductility of the CC weld is found to be less than 47% against the base material and around 52% of the PC weld. Hence, a significantly improvement in the tensile properties is witnessed by applying current pulsation during the welding process.

4 Conclusion

In the present work, micro-plasma arc welding of superalloy 718 in constant current (CC) and pulsed current (PC) mode has been undertaken and its effect on weld microstructure and mechanical behavior has been studied. The findings of the present work are as follows:

- Due to current pulsation, significant refinement of fusion zone microstructure is witnessed.
- In PC weld equiaxed dendrites are dominating in the weld interior, whereas in CC, columnar dendrites are dominating.
- Grain refinement and equiaxed dendrites reduce the amount of brittle intermetallic Laves phase more than 50% in the PC weld.
- Microstructural refinement due to current pulsing led to substantial enhancement in the micro-hardness and tensile characteristics of the welded joint.
- The tensile ductility improves by 47% by the application of current pulsation as compared to CC mode of welding.

Acknowledgements The authors are thankful to the department of Mechanical Engineering and Central Instrument Facility of IIT Guwahati for providing financial and experimental facilities to carry out the present study.

References

1. Radhakrishna C, Rao KP (1994) Studies on creep/stress rupture behaviour of superalloy 718 weldments used in gas turbine applications. Mater High Temp 12:323–327. https://doi.org/10.1080/09603409.1994.11752536
2. Gobbi S, Zhang L, Norris J, Richter KH, Loreau JH (1996) High powder CO_2 and NdYAG laser welding of wrought Inconel 718. J Mater Process Technol 56:333–345. https://doi.org/10.1016/0924-0136(95)01847-6
3. Sahu AK, Bag S (2020) Effect of pre-weld solution treatment on mechanical properties and microstructure of micro-plasma arc welded Inconel 718. In: Shunmugam MS, Kanthababu M (eds) Advances in additive manufacturing and joining, pp 373–383. Springer, Singapore. https://doi.org/10.1007/978-981-32-9433-2_33
4. Thompson RG (1988) Microfissuring of alloy 718 in the weld heat-affected zone. JOM 40:44–48. https://doi.org/10.1007/BF03258151
5. Radhakrishnan B, Thompson RG (1992) A model for the formation and solidification of grain boundary liquid in the heat-affected zone (HAZ) of welds. Metall Mat Trans A 23:1783–1799. https://doi.org/10.1007/BF02804371
6. Cieslak MJ, Headley TJ, Knorovsky GA, Romig AD, Kollie T (1990) A comparison of the solidification behavior of INCOLOY 909 and INCONEL 718. Metall Mat Trans A 21:479–488. https://doi.org/10.1007/BF02782428
7. Sahu AK, Bag S (2019) Finite element modelling and experimental verification of dissimilar joining between Inconel 718 and SS 316L by micro-plasma arc welding. In: Narayanan RG, Joshi SN, Dixit US (eds) Advances in computational methods in manufacturing, pp 231–243. Springer, Singapore. https://doi.org/10.1007/978-981-32-9072-3_20
8. Hong JK, Park JH, Park NK, Eom IS, Kim MB, Kang CY (2008) Microstructures and mechanical properties of Inconel 718 welds by CO_2 laser welding. J Mater Process Technol 201:515–520. https://doi.org/10.1016/j.jmatprotec.2007.11.224
9. Cao X, Rivaux B, Jahazi M, Cuddy J, Birur A (2009) Effect of pre- and post-weld heat treatment on metallurgical and tensile properties of Inconel 718 alloy butt joints welded using 4 kW Nd:YAG laser. J Mater Sci 44:4557–4571. https://doi.org/10.1007/s10853-009-3691-5
10. Radhakrishna C, Prasad Rao K (1997) The formation and control of laves phase in superalloy 718 welds. J Mater Sci 32:1977–1984. https://doi.org/10.1023/A:1018541915113
11. Ram GDJ, Reddy AV, Rao KP, Reddy GM (2004) Control of laves phase in Inconel 718 GTA welds with current pulsing. Sci Technol Weld Joining 9:390–398. https://doi.org/10.1179/136217104225021788
12. Kumar B, Bag S, Paul CP, Das CR, Ravikumar R, Bindra KS (2020) Influence of the mode of laser welding parameters on microstructural morphology in thin sheet Ti6Al4V alloy. Opt Laser Technol 131:106456. https://doi.org/10.1016/j.optlastec.2020.106456
13. Janaki Ram GD, Venugopal Reddy A, Prasad Rao K, Reddy GM, Sarin Sundar JK (2005) Microstructure and tensile properties of Inconel 718 pulsed Nd-YAG laser welds. J Mater Process Technol 167:73–82. https://doi.org/10.1016/j.jmatprotec.2004.09.081
14. Sahu AK, Bag S, Huang K (2020) Mitigation of micro-cracks in dissimilar welding of Inconel 718 and austenitic stainless steel. Philos Mag Lett 0:1–10. https://doi.org/10.1080/09500839.2020.1774674
15. Ye X, Hua X, Wang M, Lou S (2015) Controlling hot cracking in Ni-based Inconel-718 superalloy cast sheets during tungsten inert gas welding. J Mater Process Technol 222:381–390. https://doi.org/10.1016/j.jmatprotec.2015.03.031

16. Sahu AK, Bag S (2020) Probe pulse conditions and solidification parameters for the dissimilar welding of Inconel 718 and AISI 316L stainless steel. Metall Mat Trans A 51:2192–2208. https://doi.org/10.1007/s11661-020-05705-4
17. Manikandan SGK, Sivakumar D, Rao KP, Kamaraj M (2014) Effect of weld cooling rate on laves phase formation in Inconel 718 fusion zone. J Mater Process Technol 214:358–364. https://doi.org/10.1016/j.jmatprotec.2013.09.006
18. Peelamedu RD, Roy R, Agrawal DK (2002) Microwave-induced reaction sintering of $NiAl_2O_4$. Mater Lett 55:234–240. https://doi.org/10.1016/S0167-577X(01)00653-X

Effect on Microstructure and Mechanical Property of 4043 Wire Arc Additively Manufactured Aluminum Alloy with Different Process Parameters

Atosh Kumar Sinha, Sameer Ranjan, and Krishna Priya Yagati

1 Introduction

Unlike well-established traditional manufacturing processes such as casting, forging, and machining [1], additive manufactured (AM) products can be made with less restriction. AM process can create highly complex and customized geometry directly by the deposition of material in the layer-by-layer form by employing a 3D model designed by CAD software [2], so it is referred to as 3-D printing [3]. Wire arc additive manufacturing (WAAM) is a new additive manufacturing technology in which components can be formed by feeding wire via layer-by-layer deposition on the predefined path by using an electrical arc as a power source. Various material such as aluminum and its alloy [4–6], chromium and its alloy [7–9], nickel and its alloy [10–12], steel and its alloy [13–15], titanium and its alloy [16–18], and superalloys [19–21] can be processed by WAAM techniques with high deposition rate. Demands of WAAM products in automobiles, aerospace, aircraft, and shipbuilding industries [22] are increasing owing to various advantages such as reduced material wastage, shaving load times, offer a low cost of production, and able to produce large metallic parts.

Gas tungsten arc welding (GTAW)-based WAAM technology has been reported in various literature [4, 23–25]. Microstructure and mechanical properties in different regions of Inconel 625 WAAM samples prepared by the GTAW process had investigated [23]. GTAW process based on Fe_3Al WAAM-based iron aluminate alloy had prepared, and the microstructure and mechanical properties of the components were evaluated [24]. By varying traveling speed during GTAW-based WAAM technology on 2219 aluminum alloy, the microstructure and mechanical

A. K. Sinha (✉) · S. Ranjan · K. P. Yagati (✉)
Department of Metallurgical and Materials Engineering, NIT Durgapur, Durgapur, India
e-mail: aks.19mm1103@phd.nitdgp.ac.in

K. P. Yagati
e-mail: kp.yagati@mme.nitdgp.ac.in

S. Bag et al. (eds.), *Next Generation Materials and Processing Technologies*,
Springer Proceedings in Materials 9, https://doi.org/10.1007/978-981-16-0182-8_14

properties of the WAAM samples were studied [25]. WAAM samples deposited with the help of the GTAW process by feeding two wires 2319 Al alloy and 5087 Al alloy together and properties of the samples were investigated [4]. Till now, out of different series of aluminum alloy, mostly Al-Cu (2XXX) [26, 27] and Al-Mg (5XXX) [28–30] have been carried out by various researchers.

In this paper, WAAM of 4043 aluminum alloy on 1060-H16 substrate had built by the GTAW process. The macrostructure, microstructure, and microhardness of different samples had investigated. Double-layer and triple-layer samples were prepared by 2-min and the 5-min time interval between two layers, and changes of the different properties had been recorded. The effect on macrostructure, microstructure, and hardness had studied by partially melting the double-layer and triple-layer WAAM samples. To study the cooling effect on 4043 WAAM aluminum alloy, copper tube and molding sand were used, and the deposit had done surroundings the copper tube and molding sand.

2 Experimental Details

In this study, 1.6 mm diameter 4043 aluminum alloy wire was used to build up WAAM samples on the 150 mm × 150 mm × 8 mm 1060-H16 substrate, and the chemical composition of the wire and the substrate is listed in Table 1. The substrate was washed initially with alkaline water and dried in air and then followed by mechanical cleaning, and acetone was used to degrease the substrate immediately before deposition. The schematic diagram of the experimental setup is shown in Fig. 1. KEMPPI MasterTIG machine was chosen to fabricate layer-by-layer 4043 WAAM samples with 2T contact pulsed mode DC. The process parameters like—deposition mode (direct current electrode negative-2T-contact pulse-electrode negative), travel speed (2.5 mm/sec), shielding gas flow rate (10 lit/min) of pure argon (99.99%), deposit current (125A), and deposition voltage (11.2 V)—were kept constant, whereas time interval between two deposited layers and number of deposited layers were the variable parameters. The process parameters are listed in Table 2. Double- and triple-layer samples were prepared, shown in Fig. 2a, b, with 2 min and 5-min intervals, and then, the samples were re-melted at high speed. The copper tube of square cross section filled with molding sand has been used as a core to build a three-layer four-walled 4043 Al alloy WAAM component as shown in the following Fig. 2c–d.

Table 1 Chemical composition of the 4043-filler wire and 1060-H16 substrate

Elements	Si (wt %)	Fe (wt %)	Cu (wt %)	Mn (wt%)	Mg (wt%)	Zn (wt %)	Ti (wt %)	Vn (wt%)	Al (wt %)
4043	5	≤ 0.40	≤ 0.05	≤ 0.05	0.05	≤ 0.10	0.20	–	Bal.
1060-H16	0.25	≤ 0.35	≤ 0.05	≤ 0.03	≤ 0.03	≤ 0.05	0.03	0.05	Bal.

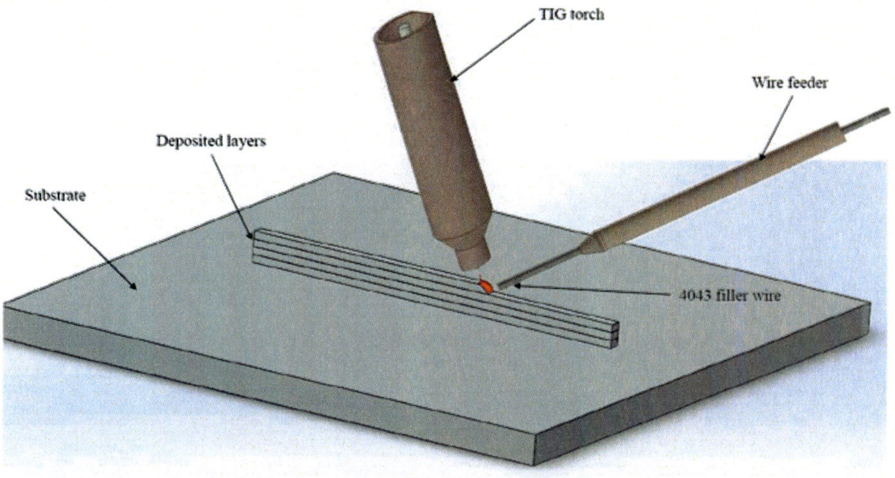

Fig. 1 Schematic diagram of the WAAM setup

Table 2 Process parameters used to fabricate 4043 Al alloy WAAM samples

Parameter	Values
Type of welding current	DC
Travel speed	2.5 mm/sec
Argon	99.99% purity
Shielding gas flow rate	10 lit/min
Current intensity	125 A
Tungsten electrode diameter	3.2 mm
Interlayer time interval	2, 5 min
Number of layers	Double (2) and Triple (3)

The physical properties such as deposit width, deposit height, and deposit angle have been measured on the fabricated WAAM samples. The deposit width and height are measured from the surface of the substrate where layer-by-layer deposit has built up. It is shown in Fig. 3—deposit angle is the angle between the substrate and the deposited layers shown in Fig. 3. It can also be termed as a wetting/bead angle. Microstructure and microhardness had performed on the longitudinal section of the WAAM specimens. The grain structure formed after the deposited samples had been studied by this process. The deposited samples were grounded by 400, 600, 800, 1200, 1500, and 2000 abrasive papers, and final mirror finish polish was obtained by 6, 3, and 1 μm alumina paste. Leica's polarized optical microscope had operated to examine the microstructural characteristics of the prepared WAAM samples under several conditions. Vickers hardness testing machine was employed to calculate the microhardness along with the deposit samples with 500 gf load for 10 s dwell time.

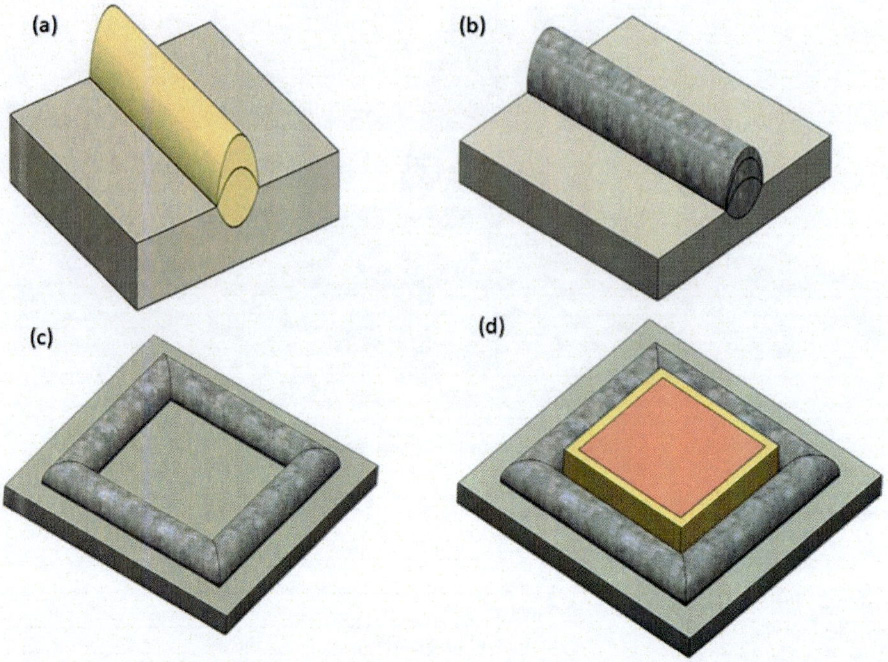

Fig. 2 Schematic diagram of WAAM samples **a** double-layer deposit, **b** triple-layer deposit, **c** four-wall triple-layer without copper tube and molding sand, **d** four-wall triple layer with copper tube and molding sand

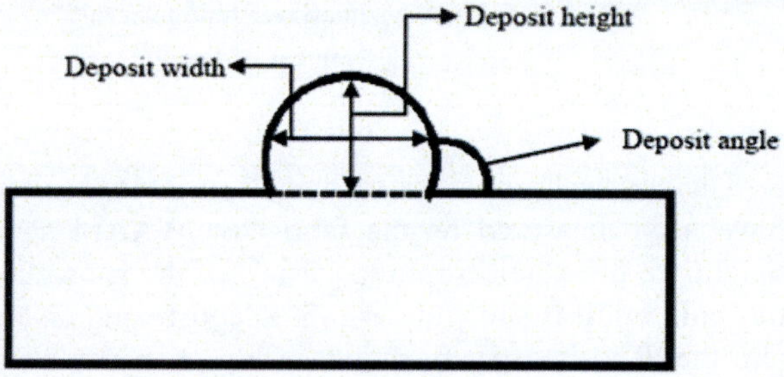

Fig. 3 Deposit width, deposit height, and deposit angle of the WAAM samples

3 Result and Discussion

3.1 Deposit Profile

Deposit profile refers to deposit width, height, and deposit angle. Figure 4a, b illustrates the macrostructures of double- and triple-layer built up, and the bead profile measurement is listed in Table 3.

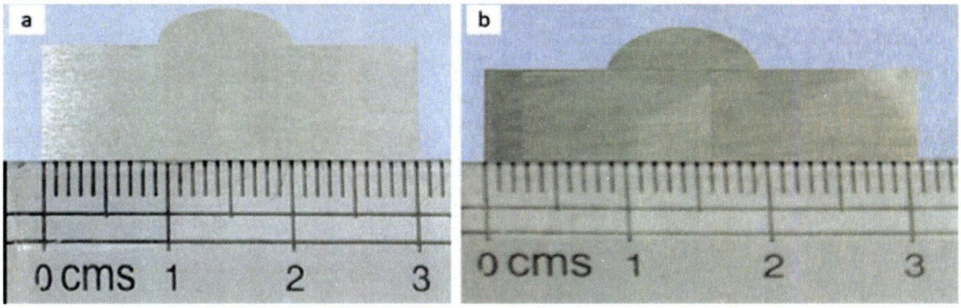

Fig. 4 Images of 4043 aluminum alloy WAAM deposit with a 2-min time interval **a** double layer and **b** triple layer

It is noticed that the deposit width, deposit height, and deposit angle of double deposit are around 8.4 mm, 3 mm, and 65°, respectively. The three layers built up recorded the deposit width, deposit height, and deposit angle of 9.2 mm, 3.8 mm, and 64°, respectively. The recorded deposit profile revealed that the deposit width increased from 8.4 mm for the double layer to 9.2 mm for triple-layer samples and the bead height increased from 3 mm for the double layer to 3.8 mm for triple-layer deposit. Therefore, increment in the number of layers found to increase the deposit width to a large extent compared to bead height and deposit angle. The extended fluid flow caused due to the high fluidity of the aluminum alloy might result in shooting up of deposit height compared to deposit width.

Figure 5a, b shows the macrostructures of the double- and three-layered samples obtained at a time interval of 5 min between the two successive deposits. The deposit width, deposit height, and deposit angle of the double layer WAAM samples at 2 min and 5 min time intervals are 8.4 mm, 3 mm, and 65°, and 8 mm, 3.6 mm, and 64°, respectively. It had observed that, in double-layered built-up, with an increase in the time interval, the deposit width reduced from 8.4 mm to 8 mm, and deposit height increased from 3 mm to 3.6 mm. Similarly, in three-layered deposits with an increase in the time interval, deposit width is reduced from 9.2 mm to 8.6 mm and bead height increased from 3.8 mm to 4.5 mm. The increment in the time interval between the layer deposition facilitates the cooling down of the previous layer. Hence, the temperature of the previous layer is lower in 5-min interval samples compared to a 2-min interval WAAM sample. The high temperature during the new layer deposition might have aided in an increase in bead width rather than bead height. Both double and triple layer samples have recorded a similar trend concerning the time gap between two layers.

Table 3 Deposit width, deposit height, and deposit angle of 4043 WAAM samples

Types of Samples	Double layer with 2 min interval	Triple layer with 2 min interval	Double layer with 5 min interval	Triple layer with 5 min interval	Re-melted double layer	Re-melted triple layer	Four-wall triple layer with copper tube and molding sand	Four-wall triple layer without copper tube and molding sand
Width (mm)	8.4	9.2	8	8.6	8	8.6	8.2	8.6
Height (mm)	3	3.8	3.6	4.5	3.4	4.3	5.1	4.6
Angle (°)	65	64	84	82	84	82	80	70

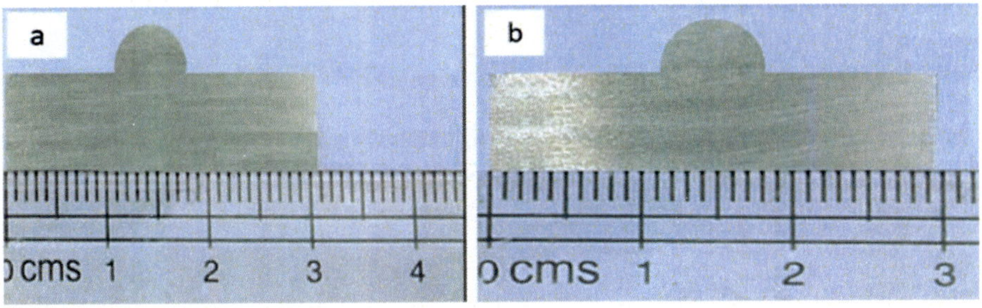

Fig. 5 Images of 4043 aluminum alloy WAAM deposit with a 5-min time interval **a** double layer and **b** triple layer

Figure 6a, b shows the macrostructure of the re-melted double- and three-layered deposits, and the measurements are listed in Table 3. For a double-layered built-up, the deposit width, deposit height, and deposit angle of the re-melted layers are 8 mm, 3.4 mm, and 84°, respectively, and the deposit width, deposit height, and deposit angle of the triple-layer re-melted samples are 8.6 mm, 4.3 mm, and 82°, respectively. Figure 6a, b illustrates that the bead has become slightly flat on the top in both doubles- and three-layered built-up. The re-melting of the layer surface might have resulted in flattening of the bead surface, keeping the bead width and bead angle constant.

Figure 7a, b shows the macrostructure of the triple-layer four-wall samples with copper tube and molding sand and without copper tube and molding sand, respectively, and the deposit profile measurements are listed in Table 3. For the four-wall triple-layer built-up by copper tube and molding sand WAAM samples, the deposit width, deposit height, and deposit angle are 8.2 mm, 5.1 mm, and 80°, respectively. For normally deposited four-wall triple-layered built up, the deposit width, deposit height, and deposit angle are 8.6 mm, 4.6 mm, and 70°, respectively.

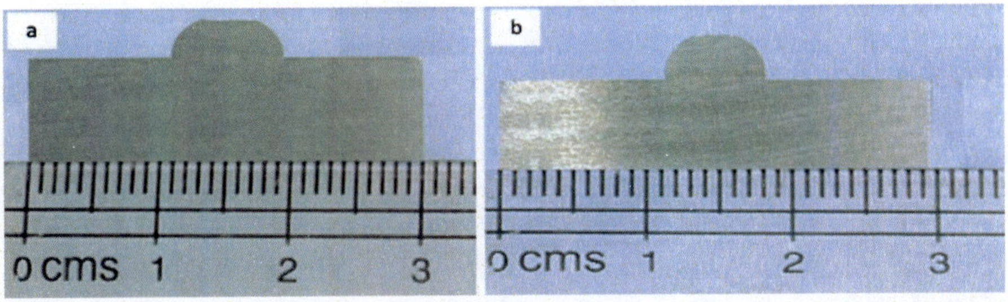

Fig. 6 Macrostructure images of 4043 aluminum alloy WAAM deposit with a 5-min time interval **a** Re-melted double layer and **b** Re-melted triple layer

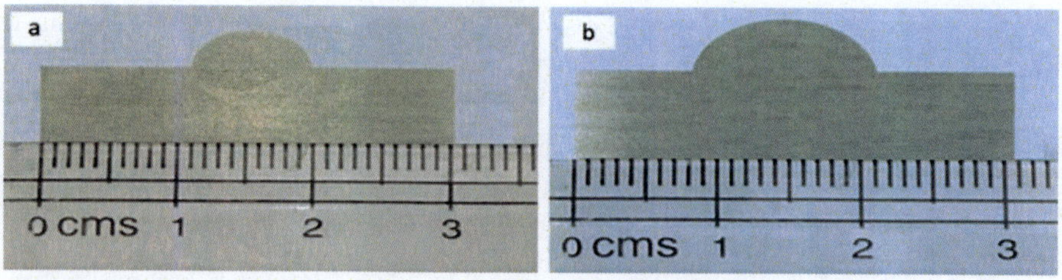

Fig. 7 Macrostructure images of 4043 aluminum alloy WAAM deposit **a** four-wall triple layer with copper tube and molding sand and **b** four-wall triple layer without copper tube and molding sand

3.2 Microstructure and Microhardness

The microstructures obtained in layer 1, layer 2 of the double-layered deposit made with a 2 min time interval between the deposits are shown in Fig. 8a, b. It is observed that both the layers show the α-Al dendritic microstructure with variation in dendrite size. It is observed that the α-Al dendrites in layer 1 are coarser than in layer 2. This is because, in a double-layered deposit, layer 1 experiences longer non-uniform temperature cycles compared to layer 2 resulting in coarser grains. This is also supported by the respective hardness testing, in which layer 2 recorded a high hardness of 75 ± 5 $HV_{0.05}$ compared to layer 1 which is shown in Fig. 8c. Similarly, both the layers of double-layered deposit made at a time interval of 5 min also showed the presence of dendritic microstructure as shown in Fig. 9a, b. However, it is also noticed that the deposits obtained at a time interval of 5 min showed the finer α-Al dendritic structure in both layers 1 and 2 compared to the deposits fabricated at a time interval of 2 min. This may be due to the availability of sufficient time for cooling in case of deposits made at 5 min time interval. Therefore, at high time intervals, there will be a sufficient amount of time for the earlier deposited layer to cool down facilitating the formation of fine dendritic structure. Figure 10a illustrates the microstructures of the re-melted double-layered deposits. The microstructure shows fine α-Al dendrites in both the layers compared to non-re-melted deposits. This is because re-melting at high processing speeds enables faster cooling rates facilitating the formation of finer dendrites throughout the deposited layers which is also supported by the obtained hardness values of the respective re-melted deposits. This is also proved by the obtained hardness data as shown in Fig. 10b. In the microstructures, some black dots have been observed which are debris formed during polishing which might not be removed during the cleaning step.

The microstructures of the triple-layered deposits are shown in Fig. 11a–d. The figures illustrate α-Al dendritic solidified structure at all the layers in the triple-layer WAAM samples. Similar to the double-layered deposits, finer dendrites are obtained in deposits made at longer time intervals and in re-melted deposits. Under similar processing conditions, the triple-layer deposits recorded a slight increase in

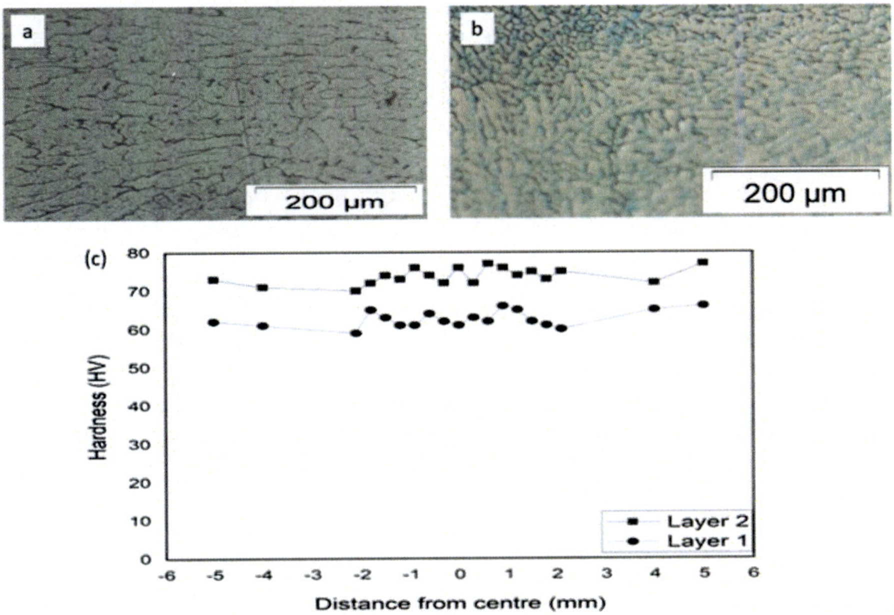

Fig. 8 Microstructure images of **a** layer 1 and **b** layer 2 for double-layer deposit and microhardness of **c** double-layer deposit of 2 min time interval samples

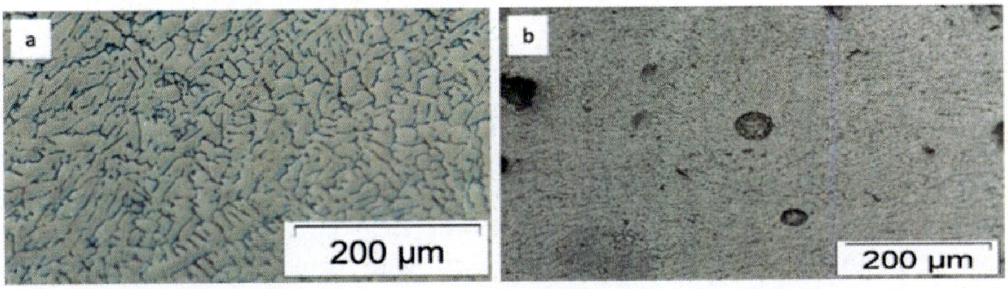

Fig. 9 Microstructure images of **a** layer 1 and **b** layer 2 for double-layer deposit with 5-min time interval samples

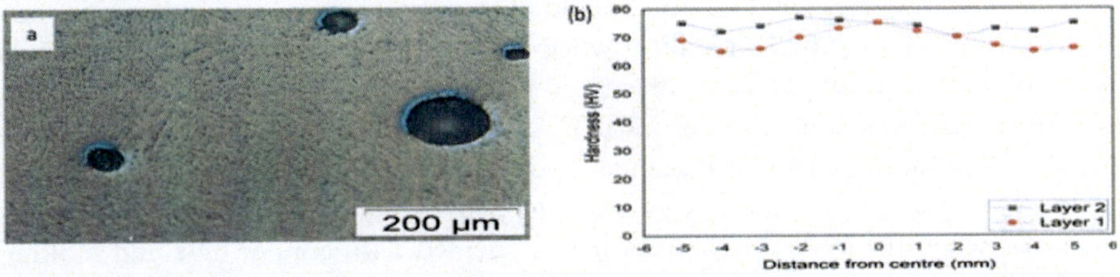

Fig. 10 **a** Microstructure and **b** microhardness images of double-layer re-melted samples

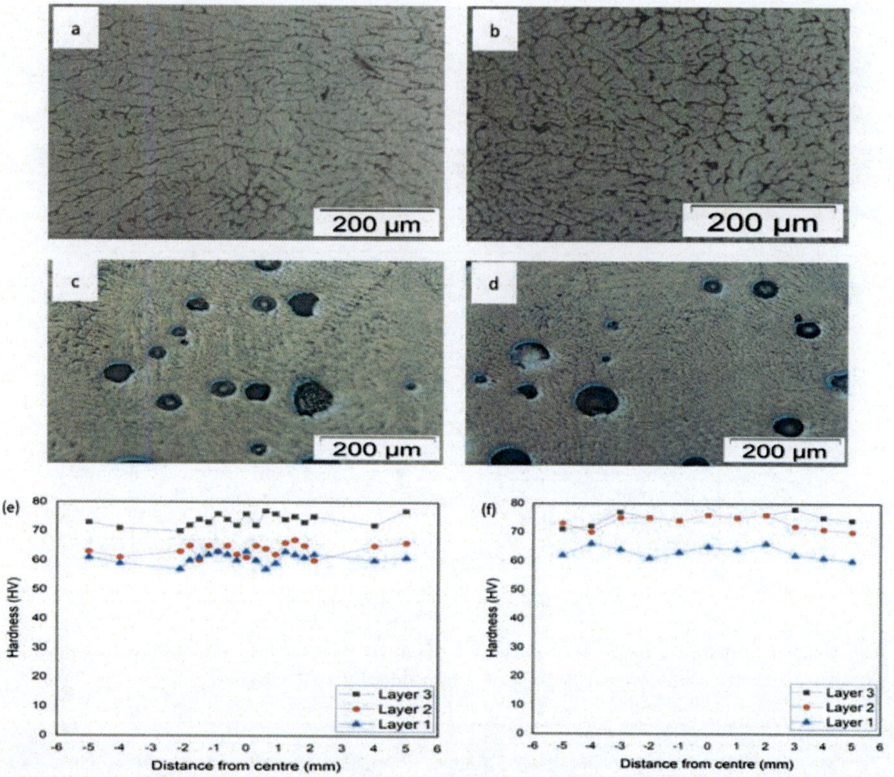

Fig. 11 **a–c** Microstructure of layers 1, 2, and 3, respectively, for three-layer deposit, **d** microstructure of three-layered re-melted components, **e–f** hardness of three layers and three-layer re-melted deposit at different positions

dendrite size compared to a double-layered deposit. This might be due to longer heat exposure and accumulation associated with triple-layered deposits. The respective hardness testing of the deposits records the hardness values around 74 ± 5 $HV_{0.05}$ of layer 1, 65 ± 5 $HV_{0.05}$ layer 2, and 60 ± 5 $HV_{0.05}$ for triple-layered and 75 ± 5 $HV_{0.05}$ to 65 ± 5 $HV_{0.05}$ re-melted triple-layer deposit which is shown in Fig. 11e and f.

The microstructures and microhardness of four-wall triple layer with copper tube and molding sand samples have shown from Fig. 12a–c and without copper tube and molding sand samples have illustrated from Fig. 13a–c. From the images of the microstructure, it can be observed that WAAM samples prepared under the copper tube and molding sand had obtained finer α-Al dendritic structure compared to normally deposited samples. This may happen owing to the faster cooling rate occurred when the WAAM samples had fabricated with copper tube and molding sand condition. The average hardness of WAAM samples prepared by copper tube and molding sand was recorded 74 ± 5 $HV_{0.05}$, whereas the average hardness of

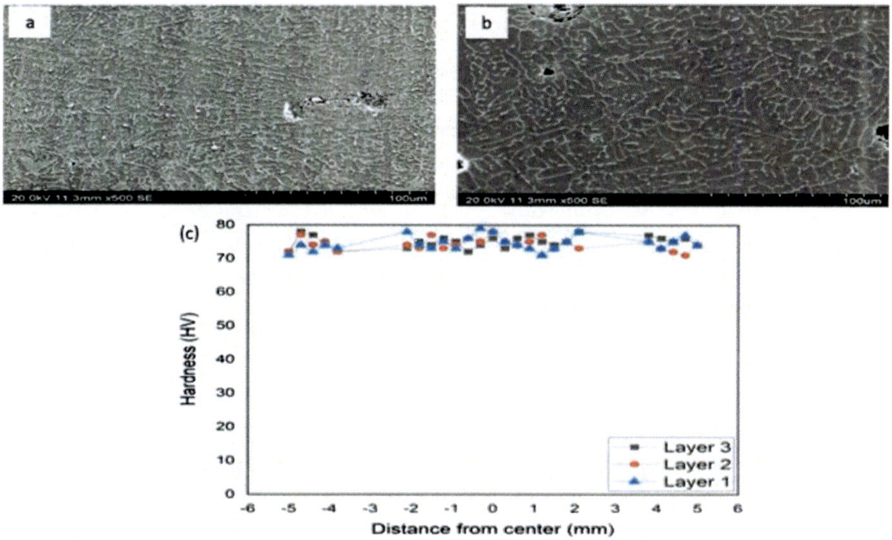

Fig. 12 Microstructure images of **a** Layer 3 and **b** Layer 1 and **c** microhardness of triple-layer four-wall WAAM samples prepared under copper tube and molding sand condition

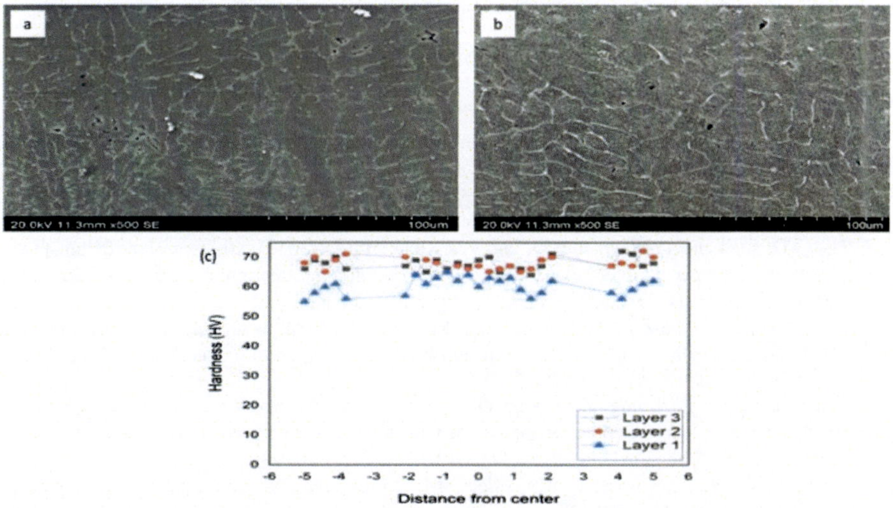

Fig. 13 Microstructure images of **a** Layer 3 and **b** Layer 1 and **c** microhardness of triple-layer four-wall WAAM samples prepared under normal condition

the normally deposited samples had been recorded in the range for third and second layer is 68 ± 5 $HV_{0.05}$ and 1st layer is 60 ± 5 $HV_{0.05.}$ Since the obtained hardness data had proved that samples fabricated under the copper tube and molding sand condition had a finer grain structure.

4 Conclusion

By employing the GTAW process, double and triple layers of 4043 WAAM samples were successfully deposited. From the study, the following major conclusions have been observed:

(i) With 2-min and 5-min time interval samples for double- and triple-layer deposit, in both cases, 5-min time interval samples had shown better microstructural characteristics due to sufficient time was available to cool down the samples.
(ii) Double- and triple-layer re-melted samples had obtained finer α-Al grains compared to normally deposited samples.
(iii) The microhardness of the double and triple-layer re-melted samples had shown almost uniform hardness in both cases. In the case of the time gap, a 5-min interlayer time gapping sample had recorded better microhardness compared to 2-min time interval samples.
(iv) Four-wall triple-layer 4043 Al alloy WAAM samples which are prepared under copper tube with molding had shown better properties compared to normally deposited samples. This may be happened due to a faster cooling rate occurred in the presence of a copper tube.

References

1. Lathabai S (2018) Additive manufacturing of aluminium-based alloys and composites. Elsevier Ltd.
2. ASTM International (2013) F2792-12a—standard terminology for additive manufacturing technologies. Rapid Manufacture Association 10–12. https://doi.org/10.1520/F2792-12A.2
3. Ngo TD, Kashani A, Imbalzano G, Nguyen KTQ, Hui D (2018) Additive manufacturing (3D printing): a review of materials, methods, applications and challenges. Compos Part B Eng 143:172–196. https://doi.org/10.1016/j.compositesb.2018.02.012
4. Qi Z, Cong B, Qi B, Zhao G, Ding J (2018) Properties of wire + arc additively manufactured 2024 aluminum alloy with different solution treatment temperature. Mater Lett 230:275–278. https://doi.org/10.1016/j.matlet.2018.07.144
5. Wang L, Suo Y, Liang Z, Wang D, Wang Q (2019) Effect of titanium powder on microstructure and mechanical properties of wire + arc additively manufactured Al–Mg alloy. Mater Lett 241:231–234. https://doi.org/10.1016/j.matlet.2019.01.117
6. Ortega AG, Corona Galvan L, Salem M, Moussaoui K, Segonds S, Rouquette S, Deschaux-Beaume F (2019) Characterisation of 4043 aluminium alloy deposits obtained by wire and arc additive manufacturing using a Cold Metal Transfer process. Sci Technol Weld Join 24:538–547. https://doi.org/10.1080/13621718.2018.1564986
7. Ge J, Lin J, Fu H, Lei Y, Xiao R (2018) A spatial periodicity of microstructural evolution and anti-indentation properties of wire-arc additive manufacturing 2Cr13 thin-wall part. Mater Des 160:218–228. https://doi.org/10.1016/j.matdes.2018.09.021

8. Zhang X, Zhou Q, Wang K, Peng Y, Ding J, Kong J, Williams S (2019) Study on microstructure and tensile properties of high nitrogen Cr–Mn steel processed by CMT wire and arc additive manufacturing. Mater Des 166:107611. https://doi.org/10.1016/j.matdes.2019.107611
9. Hu Z, Qin X, Shao T (2017) Welding thermal simulation and metallurgical characteristics analysis in WAAM for 5CrNiMo hot forging die remanufacturing. Procedia Eng 207:2203–2208. https://doi.org/10.1016/j.proeng.2017.10.982
10. Abe T, Mori D, Sonoya K, Nakamura M, Sasahara H (2019) Control of the chemical composition distribution in deposited metal by wire and arc-based additive manufacturing. Precis Eng 55:231–239. https://doi.org/10.1016/j.precisioneng.2018.09.0100.2
11. Wang J, Pan Z, Yang G, Han J, Chen X, Li H (2019) Location dependence of microstructure, phase transformation temperature and mechanical properties on Ni-rich NiTi alloy fabricated by wire arc additive manufacturing. Mater Sci Eng, A 749:218–222. https://doi.org/10.1016/j.msea.2019.02.029
12. Marenych OO, Kostryzhev AG, Pan Z, Li H, van Duin S (2019) Comparative effect of Mn/Ti solute atoms and TiC/Ni 3 (Al, Ti) nano-particles on work hardening behaviour in Ni–Cu alloys fabricated by wire arc additive manufacturing. Mater Sci Eng, A 753:262–275. https://doi.org/10.1016/j.msea.2019.03.040
13. Dirisu P, Supriyo G, Martina F, Xu X, Williams S (2020) Wire plus arc additive manufactured functional steel surfaces enhanced by rolling. Int J Fatigue 130:105237. https://doi.org/10.1016/j.ijfatigue.2019.105237
14. Caballero A, Ding J, Ganguly S, Williams S (2019) Wire + Arc Additive Manufacture of 17-4 PH stainless steel: effect of different processing conditions on microstructure, hardness, and tensile strength. J Mater Process Technol 268:54–62. https://doi.org/10.1016/j.jmatprotec.2019.01.007
15. Li M, Lu T, Dai J, Jia X, Gu X, Dai T (2020) Microstructure and mechanical properties of 308L stainless steel fabricated by laminar plasma additive manufacturing. Mater Sci Eng, A 770:138523. https://doi.org/10.1016/j.msea.2019.138523
16. Yin B, Ma H, Wang J, Fang K, Zhao H, Liu Y (2017) Effect of CaF_2 addition on macro/microstructures and mechanical properties of wire and arc additive manufactured Ti-6Al-4 V components. Mater Lett 190:64–66. https://doi.org/10.1016/j.matlet.2016.12.128
17. Wang J, Pan Z, Cuiuri D, Li H (2019) Phase constituent control and correlated properties of titanium aluminide intermetallic alloys through dual-wire arc additive manufacturing. Mater Lett 242:111–114. https://doi.org/10.1016/j.matlet.2019.01.112
18. Ding J, Colegrove P, Martina F, Williams S, Wiktorowicz R, Palt MR (2015) Development of a laminar flow local shielding device for wire + arc additive manufacture. J Mater Process Technol 226:99–105. https://doi.org/10.1016/j.jmatprotec.2015.07.005
19. Xu X, Ganguly S, Ding J, Seow CE, Williams S (2018) Enhancing mechanical properties of wire + arc additively manufactured INCONEL 718 superalloy through in-process thermo-mechanical processing. Mater Des 160:1042–1051. https://doi.org/10.1016/j.matdes.2018.10.038
20. Tanvir ANM, Ahsan MRU, Ji C, Hawkins W, Bates B, Kim DB (2019) Heat treatment effects on Inconel 625 components fabricated by wire + arc additive manufacturing (WAAM)—part 1: microstructural characterization. Int J Adv Manuf Technol 3785–3798. https://doi.org/10.1007/s00170-019-03828-6
21. Xu X, Ding J, Ganguly S, Williams S (2019) Investigation of process factors affecting mechanical properties of INCONEL 718 superalloy in wire + arc additive manufacture process. J Mater Process Technol 265:201–209. https://doi.org/10.1016/j.jmatprotec.2018.10.023
22. Ding D, Pan Z, Cuiuri D, Li H (2015) Wire-feed additive manufacturing of metal components: technologies, developments and future interests. Int J Adv Manuf Technol 81:465–481. https://doi.org/10.1007/s00170-015-7077-3

23. Wang JF, Sun QJ, Wang H, Liu JP, Feng JC (2016) Effect of location on microstructure and mechanical properties of additive layer manufactured Inconel 625 using gas tungsten arc welding. Mater Sci Eng, A 676:395–405. https://doi.org/10.1016/j.msea.2016.09.015
24. Shen C, Pan Z, Cuiuri D, Dong B, Li H (2016) In-depth study of the mechanical properties for Fe_3Al based iron aluminide fabricated using the wire-arc additive manufacturing process. Mater Sci Eng, A 669:118–126. https://doi.org/10.1016/j.msea.2016.05.047
25. Zhou Y, Lin X, Kang N, Huang W, Wang J, Wang Z (2020) Influence of travel speed on microstructure and mechanical properties of wire + arc additively manufactured 2219 aluminum alloy. J Mater Sci Technol 37:143–153. https://doi.org/10.1016/j.jmst.2019.06.016
26. Hönnige JR, Colegrove PA, Ganguly S, Eimer E, Kabra S, Williams S (2018) Control of residual stress and distortion in aluminium wire + arc additive manufacture with rolling. Addit Manuf 22:775–783. https://doi.org/10.1016/j.addma.2018.06.015
27. Gu J, Ding J, Williams SW, Gu H, Bai J, Zhai Y, Ma P (2016) The strengthening effect of inter-layer cold working and post-deposition heat treatment on the additively manufactured Al-6.3Cu alloy. Mater Sci Eng A 651:18–26. https://doi.org/10.1016/j.msea.2015.10.101
28. Gu J, Wang X, Bai J, Ding J, Williams S, Zhai Y, Liu K (2018) Deformation microstructures and strengthening mechanisms for the wire + arc additively manufactured Al–Mg4.5Mn alloy with inter-layer rolling. Mater Sci Eng A 712:292–301. https://doi.org/10.1016/j.msea.2017.11.113
29. Su C, Chen X, Gao C, Wang Y (2019) Effect of heat input on microstructure and mechanical properties of Al–Mg alloys fabricated by WAAM. Appl Surf Sci 486:431–440. https://doi.org/10.1016/j.apsusc.2019.04.255
30. Zhang B, Zhang L, Wang C, Wang Z, Gao Q (2019) Microstructure and properties of Al alloy ER5183 deposited by variable polarity cold metal transfer. J Mater Process Technol 267:167–176. https://doi.org/10.1016/j.jmatprotec.2018.12.011

Influence of WEDM Parameters for Estimating the Surface Integrity of Laser Additive Manufactured Hybrid Material

Nehem Tudu, Mayuri Baruah, Shashi Bhushan Prasad, and Christ Prakash Paul

1 Introduction

In this developing technology world, aerospace industry is also growing and trying to cope up. To keep with the momentum, aerospace industries are modifying aircrafts and its components to improve its efficiency and productivity. To meet the requirement for the modification of the component, manufacturer demands better quality of the materials with better capability, lifespan, mechanical and chemical properties [3]. To fabricate aeronautical component, materials characteristics such as fatigue and fracture toughness are need to be minimized.

To fabricate different components of aircraft, there are different manufacturing processes using which a manufacturer can give shape to the components in the desired shape. In present day, there are varieties of modern machining process which are very much reliable to implement for the production of components using aerospace materials. As conventional manufacturing process are not economical, the precision is very low [1]. In any type of manufacturing process, machining plays a vital role during fabrication and it is the crucial consideration in case of finishing of any kind of component. Amid all the modern machining processes, this article is mainly focused in wire EDM process as it is considered as one of the best machining processes for hard materials.

N. Tudu (✉) · M. Baruah · S. B. Prasad
Department of Production and Industrial Engineering, National Institute of Technology Jamshedpur, Jamshedpur, Jharkhand 831014, India
e-mail: nehemiah.tudu@gmail.com

C. P. Paul
Laser Development and Industrial Applications Division, Raja Ramanna Centre for Advanced Technology, Indore Madhya Pradesh, 452013, India

C. P. Paul
Homi Bhabha National Institute, Anushaktinagar, Mumbai, Maharashtra 400094, India

S. Bag et al. (eds.), *Next Generation Materials and Processing Technologies*, Springer Proceedings in Materials 9, https://doi.org/10.1007/978-981-16-0182-8_15

Wire EDM is very much useful for the machining of electrically conductive and difficult to cut materials as it is a potential electro-thermal process [11]. Demand of materials with high impact resistance, toughness and hardness in aerospace industries is increasing day by day. Many researchers are focusing on the non-conventional machining techniques rather than the conventional methods as the complexity issues occur during machining of these materials such as rapid tool wear [18]. In comparison to the conventional technique of machining, non-conventional technique has various advantages such as machining with higher precision and accuracy, better productivity with lesser machining cost, high surface quality, etc. [37]. To fabricate aerospace component, one of the best techniques and economic technique is wire EDM as it can manufacture complex geometry with very small tolerances by removing the redundant materials. Along with this, it can machine any materials with high hardness or toughness without any distortion of the materials [12].

1.1 *Materials for Aerospace Industries*

In common words, aerospace material can be defined as the materials used to manufacture the aircraft component which can sustain the load applied during the operation [27]. To manufacture any component, component design and material selection play a very important role. For selection of material and designing the component, consideration of various aspects is very crucial as it is very much dependent on several factors. Material cost, manufacturing cost, fuel consumption, engine efficiency, performance of operation, maintenance, safety, reliability, life span of aircraft, recycling, etc., are the major factors to be considered. Characterization analysis of any materials such as study of composition, atomic bonding, crystal structure, defect structure, macro and micro structure, etc., can help to imply the control of materials properties like density, stiffness, strength, fatigue, toughness, corrosion, etc.

There are large varieties of materials available for applications in manufacturing of the aeronautical component. To choose the appropriate material for any component, it is very important to identify the suitable material properties. As per the demand of aeronautical industries, desired properties defined are lightweighted, high stiffness, high strength, good durability, high structural efficiency, damage tolerance, etc. Along with the mentioned properties, cost effectiveness and ease of manufacturing also play important role in material selection. From environmental point of view, it is very essential to consider the renewability and sustainability of the materials. Materials mainly used by the manufacturer are steel, aluminium, titanium, magnesium, superalloys, composites, functionally graded materials and others. In other materials for the application in electrical wiring, copper is mainly used, for electronic devices semiconductor and for seat and other furnishing synthetic fibres are used. Heat insulation tiles used in rockets and space craft are made of magnesium, fibre-metal laminated, metal matrix composites, woods and ceramic, and these are radiation absorption materials and can be used for aircraft in defence.

1.1.1 Steel Alloy

Most commonly used materials for the manufacturing of structural component is steel, but in aircraft use of this alloy is limited due to high density which is about twice of titanium alloy and thrice of aluminium alloy. Along with high density, it has also poor resistance of corrosion and embrittlement which cause crack and also limits the application. To manufacture the aircraft components, Fe (Iron) is being alloyed followed by heat treatment, resulting in superior strength as compared to Al alloy (thrice) and Ti alloy (twice). Inspite of several limitation, steel alloy is still preferable due to high elastic in nature, good resistivity to fatigue failure and high fracture toughness, and it also possesses safety critical structural component because of its high strength.

1.1.2 Aluminium Alloy

In early days, aluminium alloy was preferred for most of the aircraft structure, and in 1920s and 1930s, it replaced wood. In most of the airlines, 70–80% of the structural weight is conserved by aluminium alloy and more than 50% in helicopters and defence aircraft. However, in recent years, it is replaced by composites material but importance of aluminium alloy in manufacturing will remain. Lightweight, good strength, stiffness and toughness, ease to manufacture and low cost keep the aluminium alloy still in demand. But high corrosive in nature and high fatigue failure limit the use of aluminium alloy. These properties can be controlled by alloying with appropriate constituents and proper heat treatment. Few examples of applications of the material are upper wing skin of the aircraft which is manufactured of high strength aluminium alloy to support high bending loads. Other aluminium alloy is used at lower wing skin for high fatigue resistance.

1.1.3 Titanium Alloy

To manufacture airframe and jet engine in the aircraft, titanium alloy is preferred due to its good resistivity to corrosion, lightweight and good structural properties, and even at high temperature, it can retain its mechanical properties. As compared to aluminium alloy, titanium alloy has better structural properties but it is more expensive and heavier. Its use is recommended in heavy loaded structure component and which occupies minimum space, for example, landing gear, etc. In commercial aircraft, use of titanium alloy is lesser than 10% but fighter aircraft has greater use for achieving greater strength.

1.1.4 Magnesium Alloy

It is one of the lightest metals in nature and is most preferable material for manufacturing of lightweighted aircraft. In early days, use of this material was more but composites and aluminium alloy have replaced the use of the material in recent days. Nowadays, use of magnesium alloys reduced to less than 2% due to its high cost, low strength and stiffness as compared to aluminium alloy. These materials require high for maintenance and repairmen due to its poor resistance of corrosion. These reasons limit the use of the material in non-gas turbine engine component, gear box and its housing, etc.

1.1.5 Superalloys

Superalloys are the group of metallic alloys of Ni, Fe–Ni and Co which is highly recommended for the application in the components of jet engine as it possesses excellent properties of heat resistance and is capable of retaining its toughness, strength, stiffness and stability of structural dimension at very high temperature. Its resistivity against oxidation and corrosion is very good and even in the work environment of high temperature [45]. Ni based alloy is considered as one of the most important superalloys as it contains Cr, Fe, Ti, Co and other constituents which enhances its properties. Components of aircraft where superalloys used are combustion chamber, high pressure turbine blades, thrust reverser, etc.

1.1.6 Composites

Composite materials are fabricated as required properties of materials according to the demand of application. Composite materials mostly used for aerospace components are aluminium alloy and carbon fibre matrix. Application of carbon fibre composite for aircraft and helicopter for airframe component is mostly as structural materials with aluminium alloy. Composites are usually stronger than aluminium alloy but are more expensive. Types of composites used as materials for aerospace components are carbon fibre composite, glass fibre, metal matrix composite, etc. For examples, E-glass (type of glass fibre) is used for small passenger AC parts, high strength carbon fibre in antenna dishes, satellite, missiles, etc. [23].

1.1.7 Functionally Graded Materials (FGM)

For the manufacturing of aerospace components, FGM are amongst the important materials as it can be fabricated as thermal barrier which can operate in extreme environmental condition [6]. Functionally graded material is one of the best options for the application where there is extreme change in the environment, as these material possess different properties in different part [4]. These materials are used as heat shielding materials, and for these properties, this is found to be very promising material for their use in nuclear plant components. Major advantages of functionally graded material are variation of components within the materials which helps to show different properties or materials behaviour throughout the material [20]. Major applications of functionally graded materials in aerospace components are landing gear, turbine wheel, engine casing, wings, rocket nozzle, etc.

1.1.8 Others

In early days, wood was being used as major materials for the manufacturing of aerospace components, as it is easily available, very cheap and light in weight. But it absorbs moistures, warps and decays in very short period of time which results in requirement of regular maintenance and regular repairing. Later, it is modified as the laminated plywood containing thin sheet of wood grained at different angle-oriented timber. As compared to normal wood, laminated plywood possesses good strength and toughness. Availability of good quality of timber is not found in every country, which restricts the use of these materials and also it lacks stiffness and strength compared to other materials used, although it is still in use for the fabrication of gliders, ultralight and piston-driven aircraft.

Few other materials are also there which are being used as the materials for the manufacturing of aerospace component. Amongst that fibre-metal laminated (FML) materials are also being used. Fibre-metal laminated materials are light-weighted with layers of thin bonded metal and fibre polymer composites. These combination makes the material light in weight, more fatigue resistant, impact strength and damage tolerance are enhanced and higher in strength. Glass reinforced aluminium (GLARE®) is one of the FML materials which are commonly used as aerospace materials which contains thin layer of aluminium alloy with layer of fibre polymer composite.

Amongst all the materials discussed for the application in the manufacturing of aerospace components, functionally graded material is found to have a very wide range of application. As the microstructure of the materials varies gradually within, comprise variation in properties which makes the material to be useful for the fabrication of aerospace component. There are many different methodologies by which functionally graded material can be fabricated, amongst which additive manufacturing is found to be the most promising technique. As using this technique, material can be fabricated by varying the material composition [6].

In this article, few works or study presented by researchers all over the world is reviewed for better understanding of machining parameters effect on the materials properties. According to the literatures, many researchers experimentally analysed the input parameter effect wire EDM on the machined sample. During machining by wire EDM removal of material takes place by high temperature spark generation due to high voltage difference between workpiece and wire electrode which melts and evaporates. Solidification of molten materials which does not evaporates, causes formation of recast layer and overlapping discharge crater [46]. Effect of pulse duration on kerf width, heat affected zone (HAZ) thickness, surface roughness, cutting speed and material removal rate (MRR) is analysed experimentally comparison is done for different composition of materials.

1.2 Background

In this section, literature reveals that machining using wire EDM is mainly due to electrical erosion of the material by the high temperature generated by spark voltage [21]. Heat generation due to rise in temperature during the development of spark between workpiece and wire electrode, material starts melting from the workpiece surface. Molten materials then solidifies due to contamination with the dielectric fluid and retains at surface as recast layer [46]. Generation of heat during the process causes the formation of HAZ at machined surface, these recast layers and HAZ possess poor material properties as compared to the base material. Hence, optimization of parameter becomes important in order to reduce these effects and to obtain better result.

1.2.1 Optimization of Parameters

Many researchers used various techniques of optimization to obtain optimum result for different input and output parameters. Fard, et al. studied on SiC reinforced aluminium metal matrix composite (AMMC), machined by dry wire EDM [11]. Their study includes the effect of input factors on cutting speed and surface finish. For optimization, they adopted adaptive neuro-fuzzy inference system (ANFIS) and artificial bee colony (ABC) algorithm and compared to find the best optimized result. They found that there is significant impact of peak current and pulse duration on surface finish and cutting speed. Similar study is done by Liao, et al. where they studied the effects of pulse width, pulse interval, feed rate and spark voltage on surface roughness and optimized the parameters using Taguchi method and F-Test [18].

Rajyalakshimi and Ramaiah optimized multiple parameters of WEDM processed Inconel 825 by Taguchi GRA method [31]. They optimized process parameters using L36 Taguchi's orthogonal array to achieve enhanced MRR, surface finish and spark gap. They considered spark voltage gap, pulse interval, pulse duration, corner servo voltage, wire tension, wire feed rate, servo feed and dielectric flow rate. Selvakumar et al. studied and optimized the parameters of wire EDM of 5083 aluminium alloy [34]. They designed their experiment using Taguchi's L9 orthogonal array and considered wire tension, peak current pulse interval and pulse duration as process parameter. They optimized cutting speed and surface finish using pareto optimality approach, and analysis of variance (ANOVA) was performed to classify significance level of the process parameters. Caydas and Ay also used ANOVA for significance level identification on wire cut Inconel 718 [5]. They developed empirical model using regression analysis to investigate the effect of pulse width, peak current and injection pressure on kerf width, surface finish and thickness of recast layer.

Majumder et al. analysed the response variable of wire EDM, i.e. machining time and surface finish for Inconel 800 using response surface methodology (RSM) [22]. Optimization of the parameters was compared amongst principal component analysis (PCA), grey rational analysis (GRA) and hybrid of PCA and GRA named hybrid PCA-GRA and found that hybrid PCA-GRA shows lesser error. Saha and Mondal also made similar work on nanostructured hardfacing material (NanoCarb 110) machined using WEDM with coated brass wire electrode with different material of brass and zinc [32]. Durairaj et al. optimized wire EDM input parameters for SS304 using Taguchi method of single objective and multi-objective grey rational grade [10]. They considered wire feed, gap voltage, pulse duration and pulse interval as input factor for their studies with 4 level. Shandaliya et al. also considered similar parameters with 3 level for their studies and optimize using response surface methodology on wire EDM machined surface of aluminium metal matrix composite (AMMC) reinforced with SiC particles [35].

1.2.2 Survey on Process Parameters Effects

EDM is used to manufacture aeronautical component as it fulfils the requirement of aircraft sector for the components [2]. But thermal effect due to the machining causes metallurgical change and reduces fatigue cycle life which restricts to use the process for selected components. They compared the effect on fatigue life of superalloy Inconel 718 and grounded sample machined by wire EDM. It has been informed that the fatigue life of WEDM machined component is reduced significantly as compared to the ground material. Rajurkar and Wang developed in situ spark frequency monitor to detect thermal load on wire electrode during WEDM process [30]. They modelled equations for the mechanism of wire electrode breakage due to thermal load during the machining. Also, optimization for MRR and surface finish was done and developed relationship between them.

MRR, machining efficiency and accuracy are very much affected by wire electrode recycled through drum, and hysteresis occurs when achieved MRR does not match with servo feed speed [7]. They modified the thermal model according to the previously available and the equations for latent heat of evaporation. Model which is modified are found to evaluate more precise result than that of previous one. They observed that at optimal servo feed rate, faster cutting speed can be achieved with narrow kerf width.

Imprecision machining is caused due to bending of wire which is further caused due to wire lag and gap force [33]. They investigated on cylindrical workpiece and found that accuracy level depends on wire lagging and gap force, wire lagging is more for workpiece with smaller radius and gap force is less for the job with more height. Mahapatra and Patnaik optimized the input parameters like flow of dielectric, wire tension, wire speed, frequency of pulse, pulse width and discharge current [21]. For optimization, they maximized the MRR value and surface finish and minimized the kerf width.

To manufacture any product or component, surface integrity became one of the major concern for the customer for which many researchers presented in their studies [39]. They performed experimental investigation on EN-31 die steel considering discharge gap voltage, peak current and pulse duration as input factor. They did analysis of MRR and surface finish of machined workpiece in EDM and also studied heat affected zone. Their study concludes that Copper as electrode results in good MRR, whereas for better surface finish brass electrode is recommended. Newton et al. studied on Inconel 718 investigated the effect of input factor on recast layer formation by wire EDM process [29]. They found that recast layer thickness is directly dependent on energy per spark, pulse width and peak current. Surface roughness is found to directly dependent on diameter of wire electrode and energy per spark. Recast layer fabricated on surface possess, in-plane tensile residual stress, and in comparison, to base material elastic modulus and hardness are lower. Izquierdo et al. modelled to predict layer of heat affected zone (HAZ) which can estimate thermal damage and can help to achieve better surface [15]. To validate the model, they machined Inconel 718 in various parameters.

Sharma et al. studied the effect of wire electrode material on surface quality and the productivity on wire EDM processed Inconel 706 [38]. They used three different types of wire as electrode (diffused wire, zinc coated wire and hard-brass wire) for their investigation and studied cutting speed, surface quality like surface finish, recast layer, residual stress, metallurgical and microstructural study. Amongst these three wires, zinc coated wire is found to have best productivity but best surface quality is achieved using hard-brass wire. In addition to this, hard-brass wire possesses lower tensile residual stress whereas diffused wire shows average effect on surface quality and productivity. Similar study is performed by Kumar et al. on Inconel 625 and they compared the effect of zinc coated electrode wire (brass wire) to uncoated brass wire [16]. Manjaiah et al. presented similar study for TiNiCu shape memory alloy (SMA) [24]. They found that in comparison to uncoated brass wire, zinc coated wire improves surface finish and MRR. Liu et al. analysed white layer and HAZ of wire cut Nitinol SMA in TEM, XRD and EBSD [19]. They spotted crystalline structure of white layer instead of being amorphous structure. Najm studied the effect of pulse width, peak current and wire tension on HAZ, white layer and surface finish of wire cut high speed steel [28]. He found that surface roughness is significantly affected by wire tension, whereas HAZ and white layer are affected by pulse width and peak current.

Machining characteristics comparison is done by Li et al. on Inconel 718 by different types of EDM process (viz. wire EDM and die sinking EDM) using electrode with different material [17]. They found that machining with Cu–SiC electrode shows better machining characteristics than that of Cu electrode. Significant reduction in electrode wear is observed in case of Cu–SiC and also shows improved MRR and surface finish. Unune et al. investigated the effect of input factors on MRR and kerf width of machined Inconel 718 by micro-WEDM [45]. They assisted vibration of low frequency in the machining process, and the parameters considered for the experiment were feed rate, gap voltage, frequency of vibration and capacitance as control factor. Kerf width and MRR are found to be significantly affected by capacitance, and they observed improvement in machining performance by assisting low-frequency vibration on micro-WEDM. Goyal et al. studied the machinability of machined Inconel 625 by cryogenic treated WEDM [13]. They compared the machining process of WEDM by diffused and cryogenic treated diffused wire, considered peak current, pulse width, pulse interval, wire feed and wire tension as parameter. They found that cryogenic treatment on wire electrode improves the performance in term of improvement in MRR and wear of electrode.

Tonday and Tigga, experimented to studied the effect of input factors on surface properties and morphology on Inconel 718 [42] and Inconel 825 [41] and Inconel 625 [40]. For optimization the input factors they used response surface methodology (RSM), and significant factors are determined from signal to noise ratio (S/N) by analysis using ANOVA. Chowdhury et al. analysed the effect on corner radius of voltage, wire tension and dielectric pressure and optimised using ANOVA on stainless steel 304 [8]. They used dry wire EDM technique in which gas is used as dielectric medium.

Tosun et al. obtained mathematical modelling for the variation of wire electrode crater size due to the variation in parameters [43]. They analysed the effect of pulse duration, OCV, dielectric flush pressure and feed rate of the wire on AISI 4140 steel using wire electrode of brass. They found that if pulse duration is increased, wire crater erosion also increased and similar trend is observed with wire speed and open circuit voltage, but in case of dielectric flush pressure, it is opposite. Mouralova et al. analysed surface layer of wire EDM machined surface of 16MnCr5 steel in different parameters [26]. Morphology of surface detected the presence of crater due the WEDM spark generation during the machining process. At highest speed, no major crater is detected but few deep craters are found along with recast deposition. Microstructure analysis revealed that 70% of machined surface are covered with mixture of workpiece and electrode materials deposition. Sharma et al. reported that pulse duration, pulse interval and servo voltage significantly influences surface finish and MRR on WEDM of Inconel 706 [36]. At lower duration of pulse and higher servo voltage, at machined surface, there is significantly reduction in microvoids and microglobules.

Durairaj et al. studied the input factors such as pulse width, gap voltage, pulse interval and wire feed effect on kerf width and surface finish on wire EDM machined surface of SS304 [10]. They observed pulse width to be the most significant factors for kerf width and surface roughness. Better surface finish and less kerf width can be observed when pulse on time is reduced. Similar study is done by Shandaliya et al. on AMMC reinforced with SiC particles, and similar effect is observed [35]. Ugrasen et. also observed similar result in their study done on HCHCr surface which is machined by wire EDM process [44].

From various literature surveys, it is observed that peak current, pulse width and wire electrode materials are the significant factor for surface quality of workpiece machined by WEDM process. Amongst these factors, pulse width is most significant on MRR and surface finish [42]. In this article, WEDM process was performed on two different materials which are fabricated by additive manufacturing of laser assisted directed energy deposition process. Considering process parameters, surface quality is studied for the different materials and compared to analyse the effect. Combination of Inconel 625 and SS304L is preferable where high strength and highly corrosion resistance is necessary [4]. This study mainly concentrates on comparison of the effect of input factors on the output variables in pure Inconel 625 and the combination of Inconel 625 and SS304L.

2 Experimental Investigation

In this section, the effect of input factors on various output response is studied by investigating experimentally. For experimentation, selection of variable parameters and constant factors is done on the basis of the survey of the work of researchers. Post experimentation results are analysed and compared to understand the effects. Analysed results are later optimized to find the optimal parameters to machine the materials with optimum output.

2.1 Design of Experiment

For experimental investigation, selection of input factors is very important that affect the output parameters to optimize the process for best results. From various literatures survey and experimentation performed by the researchers all over the world, it is found that pulse width is most significant factor on the machined surface. Major effects found on the wire EDM processed surface of workpiece are surface roughness, recast layer and heat affected zone.

As per the literature survey, for experimentation chosen input parameters are pulse width with 3 level, pulse interval, peak current and spark voltage at single level. The investigation contains study effect of pulse width on surface finish, cutting speed, MRR, kerf width and HAZ layer thickness. Comparison is done on the machining of different materials which are made by deposition of metallic powder using additive manufacturing process.

2.2 Materials and Methodology

Materials used for the experimentation were fabricated on stainless steel (SS304L) substrate by laser assisted directed energy deposition method as shown in Fig. 1a, b. For deposition process, the powder used was Inconel 625 and SS304L, and composition of the materials are shown in Table 1. Powders are mixed properly for

Fig. 1 Fabricated sample by laser assisted directed energy deposition method, **a** 50% 304L + 50% Inconel 625, **b** 100% Inconel 625

the deposition process in different composition, one with 50% Inconel 625 and 50% SS304L, and another is 100% Inconel 625. Powders are fabricated as block, layer by layer in continuous wave (CW) 2 kW Ytterbium laser system (YLS) fibre laser on SS304L substrate. Then fabricated block is then machined using wire EDM machining process at different process parameters. Parameters selected for deposition of powder material on the substrate are shown in Table 2.

Experimentation is performed on Electra Hitech Job Master D-zire wire EDM machine using brass wire electrode of 0.25 mm diameter as shown in Fig. 2a. Process parameters for experimental work are shown in Table 3, and all other constant parameters are shown in Table 4. Level of input parameters is selected based on literature and trials prior to experiment. To measure MRR and kerf width, machining process is done upto 2 mm depth from the surface of the workpiece, and varying cutting speed is recorded during the machining process. Methodology for the experimental is explained through the flowchart as shown in Fig. 2b.

3 Results and Discussion

After the experiment, output responses for specimens are recorded after the test was performed in different equipment. Analysis of obtained output response is important to understand the effect of the input factors on it. In this study, variation of surface roughness, kerf width, cutting speed, MRR and HAZ thickness is considered as output response. Three different values of pulse on time for two different additive manufactured materials are considered as input parameter. It is observed that pulse width and different material composition have significant effect on selected output response. Output responses are measured from different tests and reported in Table 5.

Table 1 Composition distribution of Inconel 625 (IN625) and SS304L

	Fe	Ni	Mn	C	Cr	Mo	Ti	Al	Si	P	S
IN625	9.2	62.2	–	0.1	22.5	4.7	0.36	0.4	0.5	–	–
SS304L	Bal	10	1.5	0.03	19	–	–	–	0.7	0.04	0.03

Table 2 Process parameter for deposition of Inconel 625 and SS304L by laser additive manufacturing process

Parameters	Values
Laser power (kW)	1300
Scanning speed (m/min)	0.5
Nozzle distance (mm)	12
Feed rate (rpm)	4
Gas flow rate (lpm)	6 at 2 kg pressure
Wavelength (nm)	1070

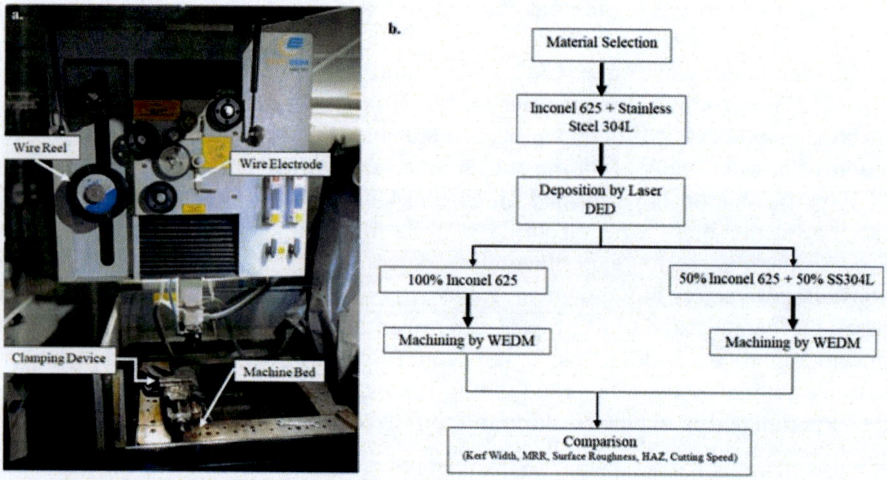

Fig. 2 **a** Experimental setup for wire EDM process at Anita enterprise, **b** flowchart of the experimentation

Table 3 Experimental process parameters

Material	Sample No.	Pulse duration (μs)
50% Inconel 625 + 50% SS304L	1	2
	2	5
	3	8
100% Inconel 625	4	2
	5	5
	6	8

Table 4 Parameters with single level for wire EDM process

Parameter	Value
Wire diameter	0.25 mm
Pulse interval	57 μs
Peak current	1 A
Servo voltage	20 V
Wire material	Brass
Wire tension	1000 N/mm^2
Depth of cut (*d*)	2 mm

Table 5 Obtained results from experimentation

Sample No.	Surface roughness (μm)	Kerf width (k_f) (mm)	Cutting speed (v) (mm/min)	Thickness of HAZ (μm)
1	1.4587	0.286	0.5	22.60938
2	1.6328	0.286	0.8	24.37747
3	2.1038	0.287	0.8	28.06525
4	1.6009	0.285	0.5	28.21311
5	1.8521	0.287	0.7	29.85204
6	2.0364	0.303	0.8	31.74013

3.1 Analysis of Surface Roughness

Machining in wire EDM process occurs mainly due to melting of materials because of high temperature generated during spark generation. As spark occurs only during pulse duration, so due to discontinuous pulse duration machining irregularity caused which produce high surface roughness of the materials. It is been observed from literature and work done by various researchers that surface roughness increases with increasing pulse duration, as increase of pulse duration results in machining irregularity which causes roughness on surface.

Obtained samples were analysed using various analysis technique on the measuring equipment such as surface roughness tester, profile projector and optical microscope to determine the output results. Surface roughness is measured on Zeiss Acoretech Surfcom 1500SD3 surface roughness tester. Response recorded in Table 4 reveals the reduction in surface finish with increasing pulse duration for both the materials. But pure Inconel 625 shows more surface roughness than that of the mix of Inconel 625 and SS304L. As Inconel 625 is much harder than that of SS304L which causes uneven material erosion from the job during machining process.

Measured data are plotted using origin with pulse duration in X-axis and surface roughness in Y-axis for 100% Inconel 625 and 50% Inconel 625 + 50% SS304L in Fig. 3. It can be clearly seen that there is gradual increment in the surface roughness

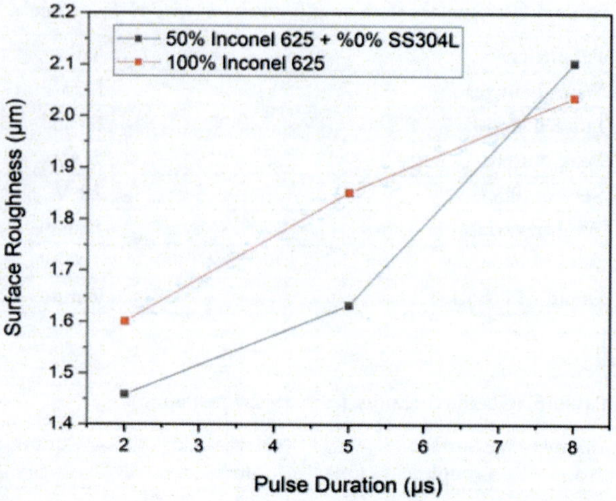

Fig. 3 Surface roughness measured after WEDM on different pulse duration

of machined 100% Inconel 625, and for other material, there is sudden increase between 5 and 8 μs. From the graph, it can be concluded that presence of SS304L in Inconel 625 causes drastic increment in surface roughness when increasing the pulse duration while machining.

3.2 Analysis of Kerf Width

Kerf width can be defined as the width obtained during the machining process due to high temperature rise by the generation spark between workpiece and wire electrode. Spark generation occurs during pulse duration due to occurrence of voltage difference in the gap between workpiece and wire electrode. As pulse duration increases, generation of sparks also increases which gives resulting in the more melting of the workpiece which results in increase of kerf width.

Profile projector, Mitutoyo-PJ-A3000 is used for measuring the kerf width by focusing on the tranverse section of the sample. Width is measured at five different points in X-axis by measuring the point of the both edge of the width at same point in Y-axis. Average value has been taken from the points measured from the profile projector and plotted in the origin as shown in Fig. 4. From the result plotted in the origin shows that pulse duration is not a significant factor kerf width for 50% Inconel 625 + 50% SS304L. But there is drastic increment between 5 μs and 8 μs pulse duration for the 100% Inconel 625 fabricated machined block. It can be concluded that presence of SS304L in Inconel 625 reduces the influence of pulse duration in kerf width.

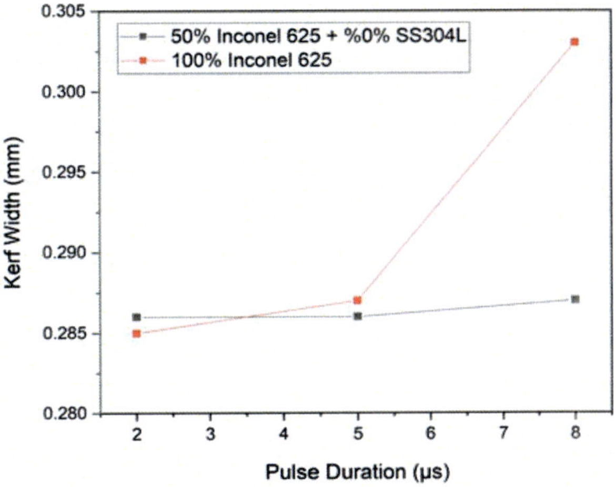

Fig. 4 Kerf width measured after WEDM on different pulse duration

3.3 Analysis of Material Removal Rate (MRR)

Melting of workpiece due to the highly rise in temperature during pulse duration which results in spark generation and removes the material. As discussed earlier, increase in pulse duration causes in increase of spark generation duration which results more removal of material. From literature, it is found that removal rate of material is dependent on kerf width. As kerf width increases with pulse duration, MRR also increases.

MRR is calculated from the kerf width measured in profile projector to determine the cutting speed using Eq. 1. Where *MRR* is material removal rate k_f is kerf width, t is depth of cut which is kept constant as 2 mm and v is cutting speed.

$$MRR = k_f \times t \times v \quad (1)$$

Using Eq. 1 MRR is calculated and recorded the values in Table 6. Calculated value of MRR is plotted against pulse duration in origin as shown in Fig. 5 to understand and compared the effects of both the materials. From plotted graph, it can be clearly seen that there is constant increment in MRR with pulse duration for pure Inconel 625 fabricated block. But in mix of SS304L and Inconel 625, no such trend is found, so it can be concluded that presence of SS304L in Inconel 625 reduces the machinability of the material. The reason for this may be difference in specific heat capacity, thermal conductivity and electrical resistivity which plays vital role in wire EDM process.

Table 6 MRR calculated using Eq. 1

Sample no.	Kerf width (k_f) (mm)	Depth of cut (t) (mm)	Cutting speed (v) (mm/min)	MRR (mm^3/min)
1	0.286	2	0.5	0.286
2	0.286	2	0.8	0.4576
3	0.287	2	0.8	0.4592
4	0.285	2	0.5	0.285
5	0.287	2	0.7	0.4018
6	0.303	2	0.8	0.4848

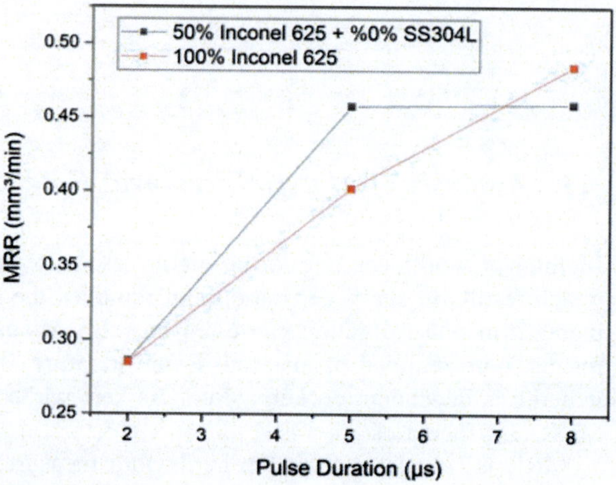

Fig. 5 Material removal rate (MRR) measured after WEDM on different pulse duration

3.4 *Analysis of Heat Affected Zone (HAZ) Thickness*

HAZ can be defined as layer which forms during the machining of workpiece due to the rise of high temperature by spark generation. When material melts due to the spark generation, erosion of molten material it affects the materials which form a layer. From literature, it can be observed that thickness of HAZ increases with increasing temperature.

Heat affected zone (HAZ) is measured from the macrograph obtained from the optical microscope (25X–1000X) Axio Cam ERc5s as shown in Figs. 7 and 8. In macrograph, a thin layer of HAZ is visible which is measured in microscope. Thickness of HAZ is measured through the microscope, and data are plotted in origin as shown in Fig. 6.

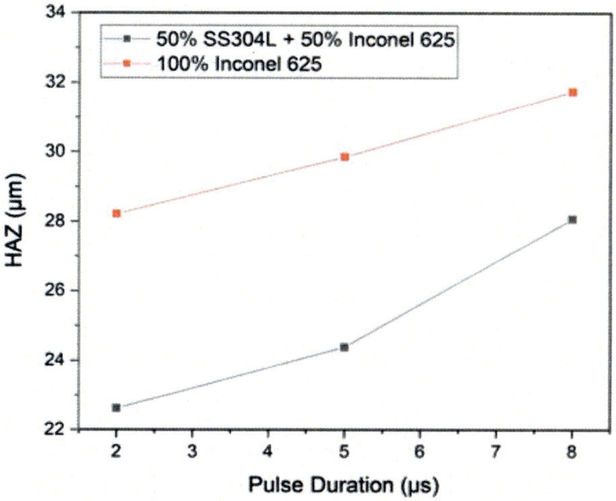

Fig. 6 Heat affected zone (HAZ) recorded after WEDM on different pulse duration

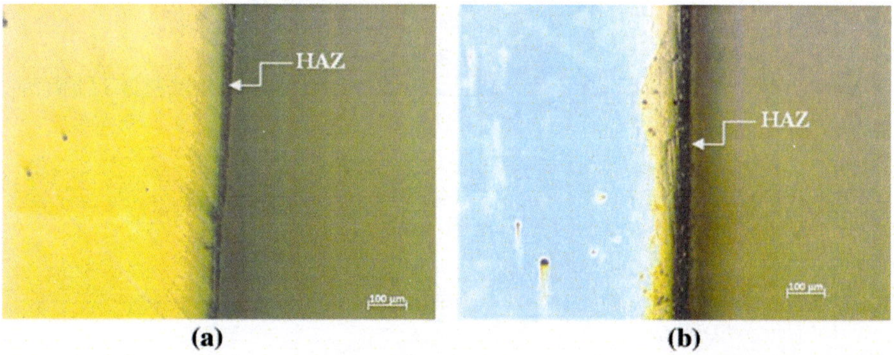

Fig. 7 Macrograph of the samples at 100X observed to measure HAZ thickness **a** 50% SS304L + 50% Inconel 625 samples machined at 2 μs, and **b** 100% Inconel 625 samples machined at 5 μs

3.5 Analysis of Cutting Speed

Cutting speed varies with the variation of parameters and variation in the composition of materials and is obtained during the machining process. Recorded data are plotted against pulse duration for both the materials to compare the effect in origin as shown in Fig. 9. From plotted data of cutting speed, it can be easily predicted that when pulse duration is increased, cutting speed is also increases. Amongst two materials, the mix of 50% SS304L with 50% Inconel 625 does not show variation when pulse duration is increased from 5 μs to 8 μs but drastic increase in speed can be visible between 2 and 5 μs.

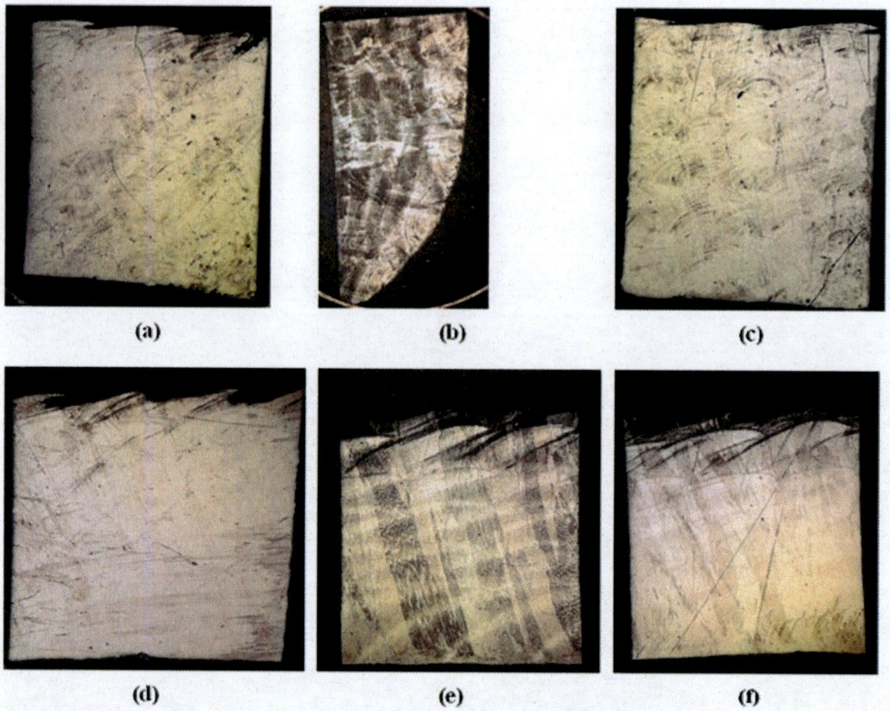

Fig. 8 Macrograph of the samples at 25X observed **a**, **b** and **c** are 50% SS304L + 50% Inconel 625 samples machined at 2 μs, 5 μs and 8 μs, respectively, and **d**, **e** and **f** are 100% Inconel 625 samples machined at 2 μs, 5 μs and 8 μs, respectively

Fig. 9 Cutting speed recorded after WEDM on different pulse duration

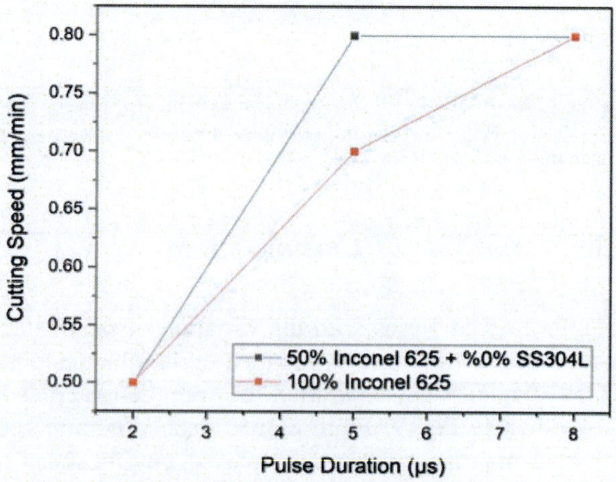

4 Future Research Directions

Study presented by various researchers on machining by wire EDM involves optimization of effect of input factors on different output like surface finish, MRR, kerf width, HAZ and cutting velocity. Different researchers performed on different materials, but as discussed in previous section, this study was done on the additive manufactured material with different volume fraction of Inconel 625 and SS304L. Similarly, various researchers suggested various methodologies to improve the quality of the surface machined by wire EDM. Few researcher suggested to study the surface morphology and topology which shows the proper information of surface quality [25]. They optimized the parameter which leads to defect free surface or surface with very less crack with precise machining. Hence, this study can be carried forward with the comparison for different volume fraction and analyse the recast layer.

Few researchers reported that coating of electrode improves the surface topology and the metallurgy of the surface [24]. Li et al. reported that alloying or reinforcement of electrode material improves the machining efficiency [17]. As discussed by Unune and Mali adhesion between electrode-workpiece causes reduction in MRR and surface finish [45]. This effect can be reduced by assisting low-frequency vibration in the machining process. Recast layer developed on machined workpiece during the machining by wire EDM process, and the surface roughness can be reduced by wire electro discharge chemical machining (WEDCM) [46]. In WEDCM, wire electrochemical machining is used with the EDM process. Goyal et al. reported that cryogenic treatment of electrode wire improves the performance [13].

For designing and fabrication of any components of product in manufacturing industries, it is very important to study the influence of input factors on the materials properties. To understand the effects, designer or engineers need to investigate or study the process either by experimentation or by theoretical. But performing experimental investigation becomes more difficult as it consumes more time and becomes costly. So, there is very much importance of performing simulation or develop mathematical modelling to analyse the effects. Few researchers have worked in this direction [15].

Few researcher performed simulation in the machining by wire EDM when cutting the corner surface [14]. They simulated to optimize the machining error at the corner during the rough cut in wire EDM. They analysed the wire path during the process by providing numerical control (NC) path and input parameter as simulation input and calculated the wire vibration and influence of discharge are on modelled geometry. Such studies can lead to achieve precise machining which can improve productivity and the product quality. Effect of different parameters on wire electrode crater generation can be analysed by simulation as done by Das and Joshi [9].

5 Conclusions

Materials required for the manufacturing of aerospace components should have some specific properties such as high resistivity to corrosion, high strength, light in weight, etc. All materials do not possess all the properties desired for aerospace body; for this reason, different materials are used in different parts of the aerospace body which are selected based on the properties required in the particular parts. For fabricating the component materials need to go through the manufacturing process, and amongst all the manufacturing process, machining of aerospace materials is discussed in this article. Article also provides an extensive review related to the various materials used in different parts of the aircrafts.

- Literature survey reveals that the superalloys and fabricated materials (viz. Composites and Functionally Graded Materials) are more preferable as these materials can be manufactured according to the desired properties.
- To study input parameter's effect on the surface integrity of the aerospace material, wire EDM process is chosen as it is very difficult to machine the material used in presentday by conventional machining process.
- For the study, materials chosen for experimentation are 100% Inconel 625 and 50% SS304L + 50% Inconel 625 which were fabricated by laser assisted directed energy deposition method.
- For machining, wire EDM machining process is chosen as it is considered as one of the important machining processes in aerospace industries. Wire EDM though costly, is a high precision and effective process and is being utilized in various industries for cutting as well as finishing purpose. But this process produces some unavoidable defects such as formation of recast layer, HAZ and may slightly increase the surface roughness if the parameters are not optimized properly. Thus, proper investigation of the process will prove to be helpful in optimizing the process for obtaining better result.
- This article highlights the effect of pulse duration on different composition of material during wire EDM process. From the study, it can be concluded that pulse duration directly affects the surface roughness, MRR, kerf width, cutting speed and HAZ. All the effect shows increasing trend with increasing pulse duration. After studying the input factor effect for both the materials, it can be observed that mix of steel and Inconel (50% SS304L + 50% Inconel 625) shows better properties than that of 100% Inconel 625.

Study can be carry forwarded by considering more input process parameters and more variation in volume fraction of composition of the materials.

Acknowledgements The authors acknowledge all the member of the Laser Additive Manufacturing Laboratory of Raja Ramanna Centre of Advanced Technology, Indore, Madhya Pradesh, India for technical support in deposition of the additive manufactured sample. Also acknowledge to Anita Enterprise, Industrial area, Adityapur, Jamshedpur, Jharkhand, India for providing setup for the experimentation.

References

1. Alting L, Boothroyd G (2020) Nontraditional manufacturing processes. In: Manufacturing engineering processes. CRC Press, pp 363–416
2. Ayesta I, Izquierdo B, Flaño O, Sánchez JA, Albizuri J, Avilés R (2016) Influence of the WEDM process on the fatigue behavior of Inconel 718. Int J Fatigue 92:220–233. https://doi.org/10.1016/j.ijfatigue.2016.07.011
3. Cantor B, Goringe MJ (2001) Aerospace materials. IOP Publishing, London
4. Carroll BE, Otis RA, Borgonia JP, Suh JO, Dillon RP, Shapiro AA, Hofmann DC, Liu ZK, Beese AM (2016) Functionally graded material of 304L stainless steel and inconel 625 fabricated by directed energy deposition: characterization and thermodynamic modeling. Acta Mater 108:46–54. https://doi.org/10.1016/j.actamat.2016.02.019
5. Caydas U, Ay M (2016) WEDM cutting of Inconel 718 nickel-based superalloy: effects of cutting paramters on the cutting quality. Mater Technol 50:117–125. https://doi.org/10.17222/mit.2015.026
6. Chakraverty S, Pradhan KK (2016) Origin and basics of functionally graded structural members. In: Vibration of functionally graded beams and plates. Elsevier, pp 7–18
7. Chen Z, Zhang G, Han F, Zhang Y, Rong Y (2018) Determination of the optimal servo feed speed by thermal model during multi-pulse discharge process of WEDM. Int J Mech Sci 142–143:359–369. https://doi.org/10.1016/j.ijmecsci.2018.05.006
8. Chowdhury AGK, Ali MY, Banu A (2020) Analysis of corner radius in dry micro WEDM. Int J Mech Eng Robot Res 9:158–162. https://doi.org/10.18178/ijmerr.9.1.158-162
9. Das S, Joshi SN (2020) Thermal modeling and simulation of crater generation on wire electrode during wire EDM operation, pp 121–135
10. Durairaj M, Sudharsun D, Swamynathan N (2013) Analysis of process parameters in wire EDM with stainless steel using single objective Taguchi method and multi objective grey relational grade. Procedia Eng 64:868–877. https://doi.org/10.1016/j.proeng.2013.09.163
11. Fard RK, Afza RA, Teimouri R (2013) Experimental investigation, intelligent modeling and multi-characteristics optimization of dry WEDM process of Al–SiC metal matrix composite. J Manuf Process 15:483–494. https://doi.org/10.1016/j.jmapro.2013.09.002
12. García Navas V, Ferreres I, Marañón JA, Garcia-Rosales C, Gil Sevillano J (2008) Electro-discharge machining (EDM) versus hard turning and grinding—comparison of residual stresses and surface integrity generated in AISI O1 tool steel. J Mater Process Technol 195:186–194. https://doi.org/10.1016/j.jmatprotec.2007.04.131
13. Goyal A, Pandey A, Sharma P (2017) Machinability of Inconel 625 aerospace material using cryogenically treated WEDM. Solid State Phenom 266:38–42. https://doi.org/10.4028/www.scientific.net/SSP.266.38
14. Han F, Zhang J, Soichiro I (2007) Corner error simulation of rough cutting in wire EDM. Precis Eng 31:331–336. https://doi.org/10.1016/j.precisioneng.2007.01.005
15. Izquierdo B, Plaza S, Sánchez JA, Pombo I, Ortega N (2012) Numerical prediction of heat affected layer in the EDM of aeronautical alloys. Appl Surf Sci 259:780–790. https://doi.org/10.1016/j.apsusc.2012.07.124
16. Kumar A, Mohanty CP, Bhuyan RK, Shaik AM (2020) Performance analysis and optimization of process parameters in WEDM for Inconel 625 using TLBO couple with FIS. Springer, Singapore
17. Li L, Li ZY, Wei XT, Cheng X (2015) Machining characteristics of Inconel 718 by sinking-EDM and wire-EDM machining characteristics of Inconel 718 by sinking-EDM. 6914. https://doi.org/10.1080/10426914.2014.973579
18. Liao YS, Huang JT, Chen YH (2004) A study to achieve a fine surface finish in wire-EDM 149:165–171. https://doi.org/10.1016/j.jmatprotec.2003.10.034
19. Liu JF, Guo YB, Butler TM, Weaver ML (2016) Crystallography, compositions, and properties of white layer by wire electrical discharge machining of nitinol shape memory alloy. JMADE 109:1–9. https://doi.org/10.1016/j.matdes.2016.07.063

20. Loh GH, Pei E, Harrison D, Monzón MD (2018) An overview of functionally graded additive manufacturing. Addit. Manuf. 23:34–44
21. Mahapatra SS, Patnaik A (2007) Optimization of wire electrical discharge machining (WEDM) process parameters using Taguchi method. Int J Adv Manuf Technol 34:911–925. https://doi.org/10.1007/s00170-006-0672-6
22. Majumder H, Paul TR, Dey V, Dutta P, Saha A (2017) Use of PCA-grey analysis and RSM to model cutting time and surface finish of Inconel 800 during wire electro discharge cutting. Measurement 107:19–30. https://doi.org/10.1016/j.measurement.2017.05.007
23. Mangalgiri PD (1999) Composite materials for aerospace applications. Bull Mater Sci 22:657–664. https://doi.org/10.1007/BF02749982
24. Manjaiah M, Narendranath S, Basavarajappa S, Gaitonde VN (2015) Effect of electrode material in wire electro discharge machining characteristics of Ti50Ni50-xCux shape memory alloy. Precis Eng 41:68–77. https://doi.org/10.1016/j.precisioneng.2015.01.008
25. Mouralova K, Kovar J, Klakurkova L, Bednar J, Benes L, Zahradnicek R (2018) Analysis of surface morphology and topography of pure aluminium machined using WEDM. Measurement 114:169–176. https://doi.org/10.1016/j.measurement.2017.09.040
26. Mouralova K, Matousek R, Kovar J, Mach J, Klakurkova L, Bednar J (2016) Analyzing the surface layer after WEDM depending on the parameters of a machine for the 16MnCr5 steel. Measurement 94:771–779. https://doi.org/10.1016/j.measurement.2016.09.028
27. Mouritz AP (2012) Introduction to aerospace materials. Woodhead Publishing Limited
28. Najm VN (2018) Experimental investigation of wire EDM process parameters on heat affected. Eng Technol J 36:46–54
29. Newton TR, Melkote SN, Watkins TR, Trejo RM, Reister L (2009) Investigation of the effect of process parameters on the formation and characteristics of recast layer in wire-EDM of Inconel 718. Mater Sci Eng, A 513–514:208–215. https://doi.org/10.1016/j.msea.2009.01.061
30. Rajurkar KP, Wang WM (1993) Thermal modeling and on-line monitoring of wire-EDM. J Mater Process Technol 38:417–430
31. Rajyalakshmi G, Ramaiah PV (2013) Multiple process parameter optimization of wire electrical discharge machining on Inconel 825 using Taguchi grey relational analysis. https://doi.org/10.1007/s00170-013-5081-z
32. Saha A, Mondal CS (2016) Multi-objective optimization in WEDM process of nanostructured hardfacing materials through hybrid techniques. Measurement 94:46–59. https://doi.org/10.1016/j.measurement.2016.07.087
33. Sarkar S, Sekh M, Mitra S, Bhattacharyya B (2011) A novel method of determination of wire lag for enhanced profile accuracy in WEDM. Precis Eng 35:339–347. https://doi.org/10.1016/j.precisioneng.2011.01.001
34. Selvakumar G, Sornalatha G, Sarkar S, Mitra S (2014) Experimental investigation and multi-objective optimization of wire electrical discharge machining (WEDM) of 5083 aluminum alloy. Trans Nonferrous Met Soc China 24:373–379. https://doi.org/10.1016/S1003-6326(14)63071-5
35. Shandilya P, Jain PK, Jain NK (2012) Parametric optimization during wire electrical discharge machining using response surface methodology. Procedia Eng 38:2371–2377. https://doi.org/10.1016/j.proeng.2012.06.283
36. Sharma P, Chakradhar D, Narendranath S (2015) Evaluation of WEDM performance characteristics of Inconel 706 for turbine disk application. Mater Des 88:558–566. https://doi.org/10.1016/j.matdes.2015.09.036
37. Sharma P, Chakradhar D, Narendranath S (2016) Effect of wire diameter on surface integrity of wire electrical discharge machined Inconel 706 for gas turbine application. J Manuf Process 24:170–178. https://doi.org/10.1016/j.jmapro.2016.09.001
38. Sharma P, Chakradhar D, Narendranath S (2016) Effect of wire material on productivity and surface integrity of WEDM-processed Inconel 706 for aircraft application. J Mater Eng Perform. https://doi.org/10.1007/s11665-016-2216-z

39. Vb T, Soni JS, Rao ML (2016) Analysis and evaluation of heat affected zones (HAZ) of the work piece surface machined using different electrode by die-sinking EDM of EN-31 die steel. Int J Eng Sci 5:37–42
40. Tonday HR, Tigga AM (2016) Analysis of effects of cutting parameters of wire electrical discharge machining on material removal rate and surface integrity. IOP Conf Ser Mater Sci Eng 115:012013. https://doi.org/10.1088/1757-899X/115/1/012013
41. Tonday HR, Tigga AM (2017) Evaluation of the influence of wire electrical discharge machining parameters on material removal rate and surface characteristics in cutting of Inconel 825. Mater Today Proc 4:9865–9869. https://doi.org/10.1016/j.matpr.2017.06.283
42. Tonday HR, Tigga AM (2019) An empirical evaluation and optimization of performance parameters of wire electrical discharge machining in cutting of Inconel 718. Measurement 140:185–196. https://doi.org/10.1016/j.measurement.2019.04.003
43. Tosun N, Cogun C, Pihtili H (2003) The effect of cutting parameters on wire crater sizes in wire. Int J Adv Manuf Technol 21:857–865
44. Ugrasen G, Ravindra HV, Prakash GVN, Prasad YNT (2015) Optimization of process parameters in wire EDM of HCHCr material using Taguchi' s technique. Mater Today Proc 2:2443–2452. https://doi.org/10.1016/j.matpr.2015.07.185
45. Unune DR, Mali HS (2017) Experimental investigation on low-frequency vibration assisted micro-WEDM of Inconel 718. Eng Sci Technol an Int J 20:222–231. https://doi.org/10.1016/j.jestch.2016.06.010
46. Wu X, Li S, Jia Z, Xin B, Yin X (2019) Using WECM to remove the recast layer and reduce the surface roughness of WEDM surface. J Mater Process Technol 268:140–148. https://doi.org/10.1016/j.jmatprotec.2019.01.016

Effect of Vibrations on Solidification Behavior and Mechanical Properties of Shielded Metal Arc Weld

Pravin Kumar Singh, Shashi Bhushan Prasad, and D. Patel

1 Introduction

The application of vibrations during welding leads to the detachment of grains from the dendrites and works as new nucleation sites [1]. The forced vibrations during welding as well affect the temperature distribution, which is cause of generation of nucleation [2, 3]. The application of vibration technique is expected for improving the weld qualities. The key goal of vibratory welding is to improve the grain structure and reduction in the residual stress [4–8]. Researchers have found that the application of ultrasonic vibration is very beneficial for stress relief and applicable for small thickness specimen. As per the result, the stress can be reduced by forty percent by using the ultrasound vibration during welding [9–11].

There are various methods used to apply the vibratory technique during the welding operations. Table 1 shows various methods used by the researchers for the vibratory welding technique. According to the literature, it has been discovered that no application of vibration into the weld pool has been reported by any of the researchers. So following objectives have been decided:

(i) To design a vibratory welding setup which is capable to stir the molten weldpool during the welding process and
(ii) To study the effect of vibration on metal coating process.

P. K. Singh (✉)
Mechanical and Automation Engineering Department, AMITY University Ranchi, Jharkhand, India
e-mail: pksingh@rnc.amity.edu

S. B. Prasad · D. Patel
Production Engineering Department, NIT Jamshedpur, Jharkhand, India

S. Bag et al. (eds.), *Next Generation Materials and Processing Technologies*, Springer Proceedings in Materials 9, https://doi.org/10.1007/978-981-16-0182-8_16

Table 1 Literature on vibratory setups

S.no.	Material used	Vibratory technique	Frequency produced	Process	Reference
1	AISI 310	Electromagnetic	0–40 Hz	GTAW	[12]
2	MS	Vibratory table	80–400 Hz	SMAW	[2]
3	A1 Alloy (1085,2214)	Electromagnetic	50 Hz	Casting process	[13]
4	Nickel alloy (690)	Vibratory table	58 Hz	GTAW	[14]
5	MS	Vibratory table	25 Hz	MEG	[5]
6	D6AC D406A	Vibratory table	2.5 Hz	MIG	[7]
7	Niomol 490K	Vibratory table	–	SAW	[15]
8	al-6xn	Ultrasonic	20 kHz	SMAW	[16]
9	A-105	Vibratory table	54–59 ms	SAW	[3]
10	A-105	Vibratory table	54–59 ms	SAW	[17]
11	Superalloy 800	Electromagnetic	–	GTAW	[18]
12	SiC_0/6061 Al	Ultrasound	50 kHz	PAW	[19]
13	Al alloy	Vibratory table	100–3000 Hz	GTAW	[11]
14	MS	Vibratory table	–	SMAW	[20]
15	Al alloy	Wave guide	20 kHz	MIG and TIG	[21]
16	304-SS	Vibratory table	375 Hz	GTAW	[9]
17	304-SS	Vibratory table	150–350 Hz	TIG	[8]
18	AISI 304	Horn plus tool	429 Hz	FSW	[10]
19	AISI 304	Vibratory table	60.9 Hz	TIG	[22]
20	AZ31 Mg alloy	Vibratory table	15 kHz	TIG	[8]
21	MS	Vibration transfers into the molten weldpool during welding operation	80–300 Hz	SMAW	PW*

2 Methodology

This experiment is a comparative study between SMAW and Vibratory SMAW (V-SMAW) of various weld bead using the SMAW process under the same welding parameters. The coding of the weld bead used in present study is shown in Table 2.

Table 2 Various material combinations, welding techniques, and specimen codes used in the present work

Welding technique used	Filler material and diameter	Base material	Specimen code	Welding current	Industrial relevance
Bead on plate	Mild steel (E-3016, Φ3.2)	Mild steel (6 mm)	$WB_{MS/MS}$	130A	General structural work
	Austenitic stainless steel (AISI 316L) (Φ3.2 mm)	Mild steel (6 mm)	$WB_{SS/MS}$	110A	Corrosion resistant
	Stellite (Co-6) (Φ3.2 mm)	Mild steel (6 mm)	$WB_{ST/MS}$	130A	Hot forging dies

2.1 Design of Vibratory Setup

With a motivation to improve the weld structure properties vibratory setup has been designed. The schematic figure of the present setup is delineated in Fig. 1. A rod having diameter of 3 mm made up of Zirconium–tungsten alloy is mounted with a vibratory setup. The one side of rod is merged into the weldpool and transfers vibrations from vibratory setup into the weldpool. The schematic diagram of vibratory setup is presented in Fig. 1. A motor with a balance load on its drive shaft is used to make a vibration during the process. A centripetal force comes into action due to rotation of asymmetric mass resulting in displacement of the motor, this repeated displacement produces vibration. Vibratory setup works over the range of voltage. Motor can produce maximum 300 Hz of frequency.

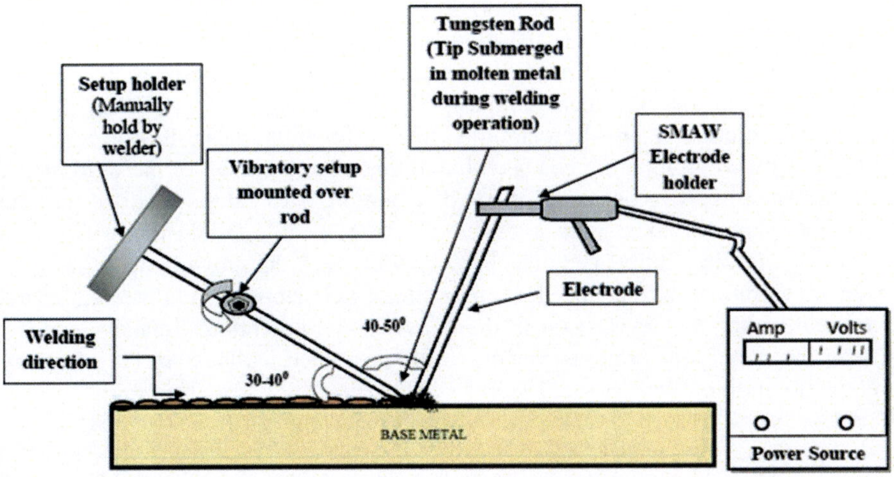

Fig. 1 Vibration setup used in present investigation

SMAW operation was used for bead on plate technique, and it is a process in which coalescence is produced by electric arc forms between consumable filler metal and base metal. The filler metal is coated by the flux which is helpful to prevent the base metal from atmospheric contaminations during welding. For investigation, the mild steel is used as a baseplate, stainless steel (AISI 316), mild steel electrode (E-3016), and Stellite (Co-6) is used as electrode to construct weld bead. Two weld beads are generated for the comparative study, one is by simple process without use of vibratory called as conventional welding (C-SMAW) and another was under the vibratory conditions, vibratory welding (V-SMAW). On macrolevel, the weld bead structure was studied and the difference in bead geometry, like ripple formation, reinforcement, and weld width structure was noted (Fig. 2).

3 Result and Discussion

3.1 Microhardness Testing

For microhardness testing, the specimens were prepared using standard procedures like belt grinding, polishing using fine grades of emery paper up to 1500 grit size. This helped in removing coarse and fine oxide layers as well as scratches on the surfaces that were to be analyzed. Microhardness tester (Make: Omnitech, Capacity: 1000 grams) was used to measure microhardness at various zones of interest in different weldments. A load of 500 grams and a dwell time of 20 s were used for these studies. The hardness value was measured along the center line, and the mid-thickness of the weld joints, i.e., the distance between the two measuring points was 1 mm, each point was measured three times to inquire about its average value. Due to various intensity of vibration at various zones of the weld bead, the hardness distribution is uneven. Therefore, hardness was measured at various zones like heat-affected zone, weld metal zone, and base material. Microhardness across different zones of the weldment samples $WB_{MS/MS}$, $WB_{SS/MS}$,, and $WB_{ST/MS}$ was measured in transverse direction and the same is shown graphical form (Figs. 3, 4, and 5). As shown in figure, it has been found that microhardness of the weld metal/ zone has been improved because of use of vibration. This can be ascribed into the way that when vibrations are actuated into the weld pool during welding, a wave-like unsettling influence is made which will break the growing dendrites. The result of vibrations is that the developing dendrites cannot arrive at their actual dendritic lengths as they can do in the conventional condition. Smaller dendritic size in the weld pool prompts moderately higher microhardness in the vibratory weldpool condition.

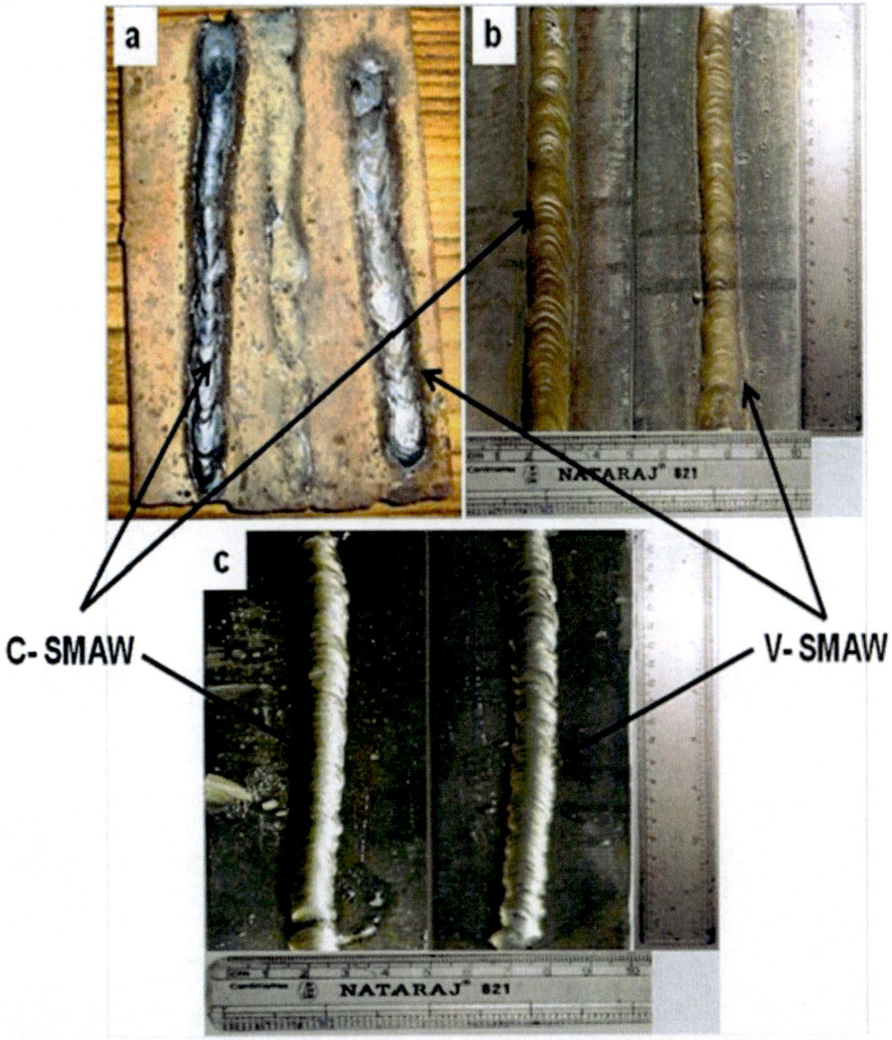

Fig. 2 Weld bead structure

3.2 Microstructural Study

The initial results were found to be encouraging as the weld metal microhardness and tensile properties have been improved in almost all the specimens to a significant extent due to application of vibration, which indicates that setting up of auxiliary mechanical vibrations of low amplitude and high frequency into the weld pool could be used as a means for influencing the weld microstructure and hence mechanical properties favorably increased. As per the result obtained from hardness

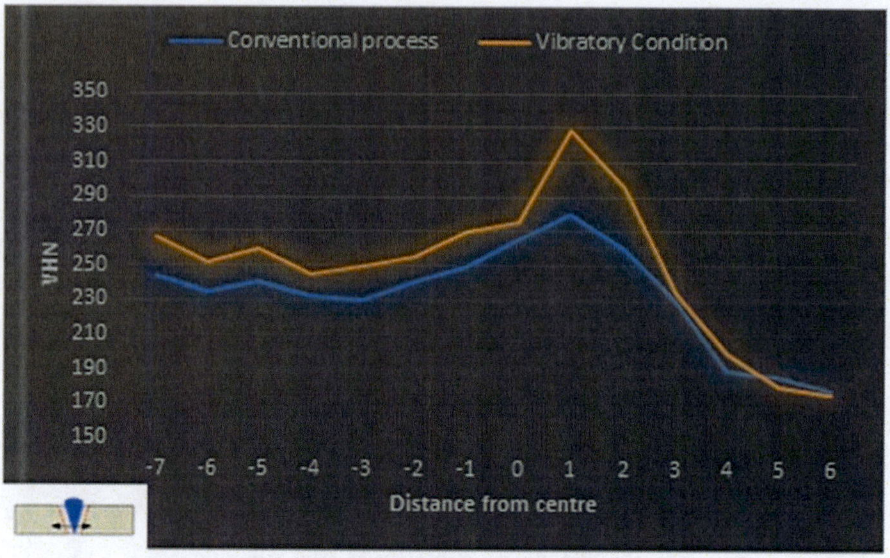

Fig. 3 Microhardness plot of $WB_{SS/MS}$ (AISI 316 SS on mild steel) specimen

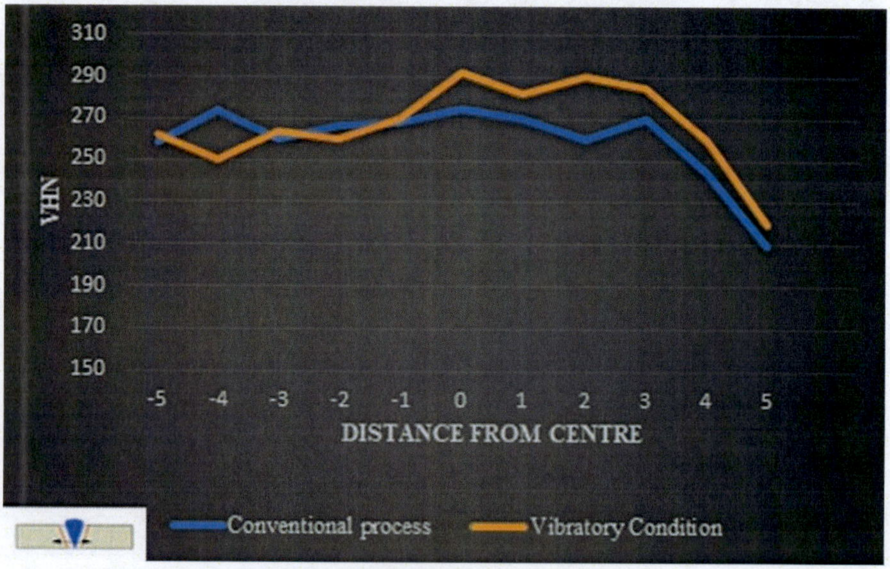

Fig. 4 Microhardness plot of $WB_{MS/MS}$ specimen

test, microstructural behavior of vibratory welding was observed. The microstructure figures show the drastic changes in the grain size due to application of vibrations. Figures 6, 7 and 8 showing the microstructure figures (C-SMAW:

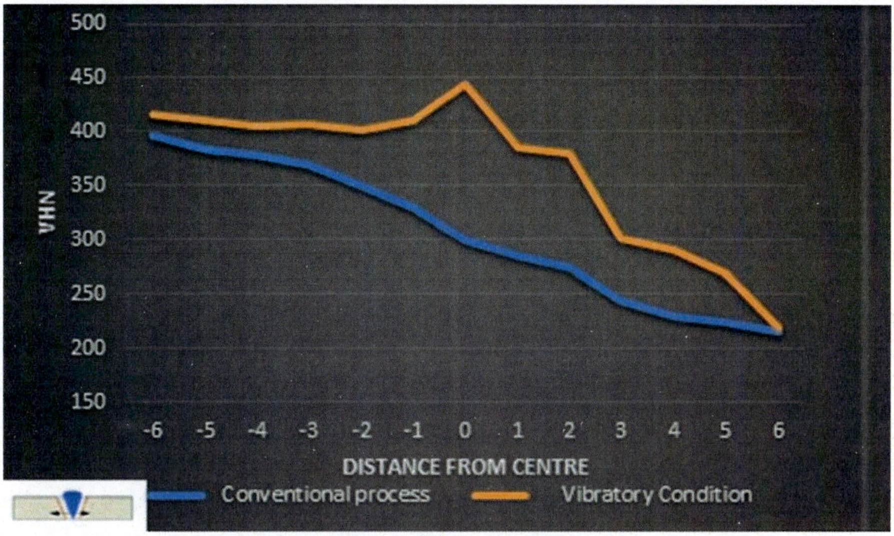

Fig. 5 Microhardness plot test of $WB_{ST/MS}$ (stellite on mild steel) specimen

Conventional SMAW Process, V-SMAW: Vibratory welding process), it has been observed that under the vibratory condition, fine-grain structures have been formed compared to the conventional SMAW weld bead. The dendrites are not able to grow to their fullest extent during vibratory conditions as these dendrites would have grown up during conventional SMAW condition.

Applied vibration fragmentizes the growing dendrites, these dendrites act as a new nucleus site for further solidification process. Vibration techniques during welding stir the molten metal before its solidification. This type of interruption increases the cooling rate of the molten weld zone and helps to reach the super-cooling temperature.

Above discussion concluded that the applied Vibration:

- breaks the growing dendrites,
- increases the cooling rate,
- helps to prevent the newborn nucleus from remelting,
- increase the number of grains and
- finally produces the fine-grain structure.

3.3 Tensile Testing

The tensile specimens were prepared in accordance with ASTM E-08 standards. Tensile specimens were tested on a universal testing machine, (Make: FIE, Capacity: 600 kN). Observations that were recorded from this test include yield

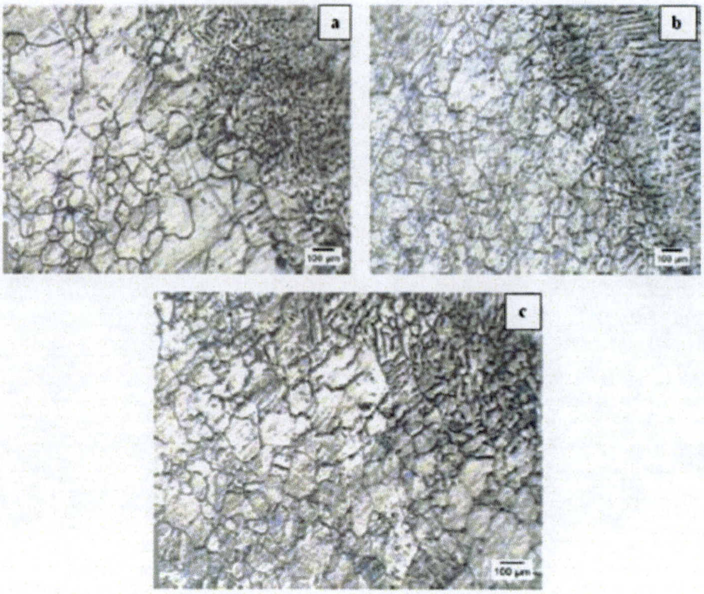

Fig. 6 Microstructure of weld bead (SS over MS plate)

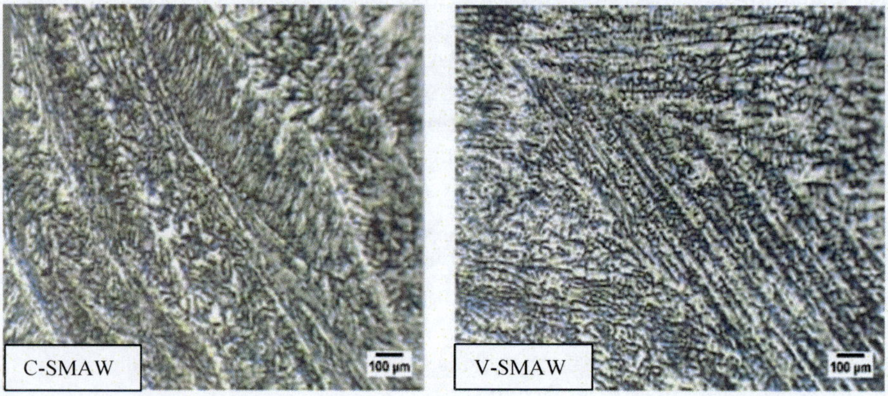

Fig. 7 Microstructure of weld bead (MS over MS plate)

strength (YS), ultimate tensile strength (UTS), and percentage elongation. The displacement rate was 0.5 mm/min. The elongation was measured with the extensometer (FIE makes) GL-50 mm. The tensile tests of weld bead structures are presented in Figs. 9 and 10. For investigation, three specimens were tested on UTM machine and average value has been taken for study. As per the result obtained,

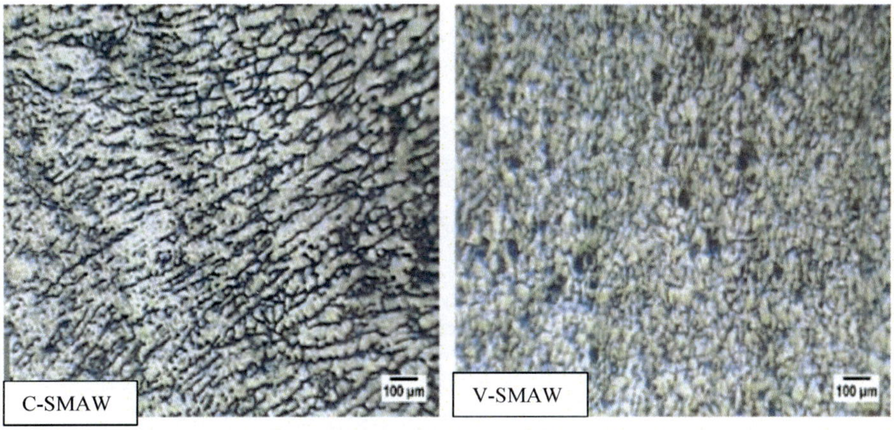

Fig. 8 Microstructure of weld bead (ST over MS plate)

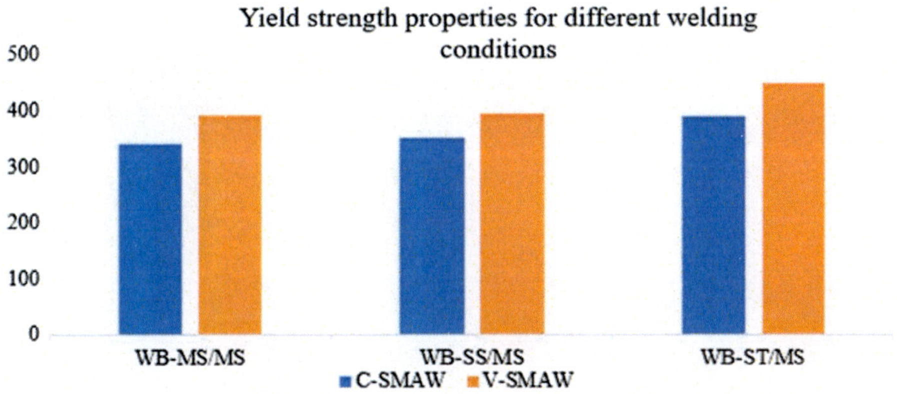

Fig. 9 Comparative study of yield strength

a drastic improvement has been found in yield strength and tensile strength of weld bead. The fine-grain structure might be the reason for improvement in strength. Maximum 15% of improvement has been found in weld bead for the case ST over MS in yield strength and UTS. To check the ductility of vibratory welding, the percentage elongation has been measured. The result shows that the percentage elongation is increased by twice as compare to the conventional welding condition. Figure 11 is showing the percentage elongation of $WB_{SS/MS}$.

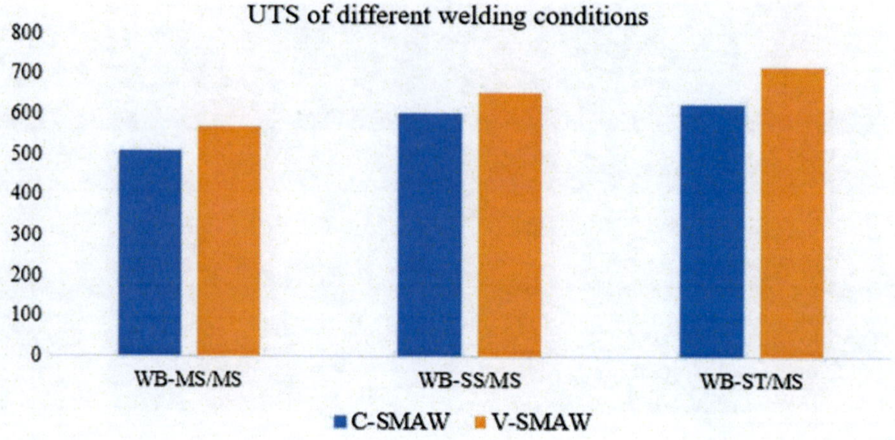

Fig. 10 Comparative study of UTS

Fig. 11 Comparative study of percentage elongation

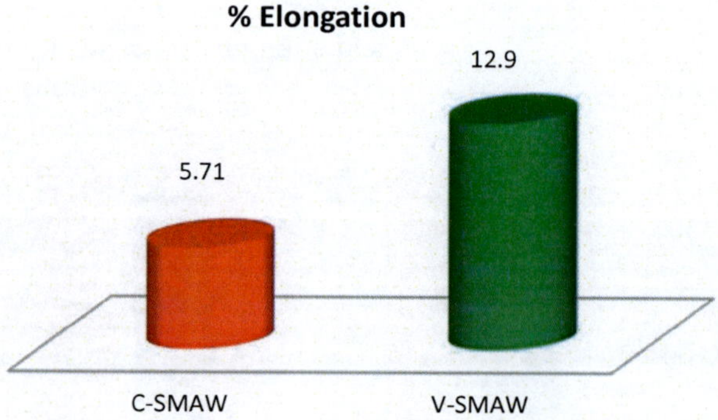

4 Conclusion

The auxiliary mechanical vibrator could successfully transmit vibrations directly into the weld molten pool. These vibrations did not allow the grains to grow, due to which fine microstructures are formed. With increase in frequency of the applied vibration, the hardness values increase.

The microstructure properties are hugely changed by the application of vibrations. The external vibrations disturbed the solidification mechanism of the weld zone due to which fine grains and steeper HAZ form. Weld bead under the resonant frequency condition has improved grains as compared to the other ones.

References

1. Kou S, Le Y (1986) Nucleation mechanism and grain refining of weld metal. Welding J 471
2. Davies GJ, Garland JG (1975) Solidification structures and properties of fusion welding. Int Metal Rev 20:83–106
3. Hsieh C-C, Lai C-H, Weite Wu (2013) Effect of vibration on microstructure and mechanical properties of 304 Stainless steel GTA welds. Met Mater Int 19:835–844
4. Lu Q, Chen L, Ni C (2007) Improving weld quality by vibratory weld conditioning. Mater Sci Eng A 457:246–253
5. Xu J, Chen L, Ni C (2007) Effect of vibratory weld conditioning on residual stress and distortion in multipass girth butt welded pipes. Int J Pressure Vessel Piping 84:298–303
6. Wang J-S, Hsieh C-C, Lin C-M, Chen E-C, Kuo C-W, Xu W (2014) The effect of residual stress relaxation by the vibratory stress relief technique on the textures of grains in AA 6061 Aluminum alloy. Mater Sci Eng A 605:98–107
7. Shalvandi M, Hojjat Y, Abdullah A, Asadi A (2012) Influence of ultrasonic stress relief on Stainless steel 316 specimens: a comparison with thermal stress relief. Mater Des 46:713–723
8. Mostafapour, Gholizadeh V (2014) Experimental investigation of the effect of vibration on mechanical properties of 304 stainless steel welded parts. Int J Adv Manuf Technol 70:1113–1124
9. Balasubramanian K, Balusamykeshavan VD (2011) Studies on the effect of vibration on hot cracking and grain size in AA7075 Aluminum alloy welding. Int J Eng Sci Technol 3(1)
10. Tiwari SP, Shanker (1993) Effect of longitudinal vibration on mechanical properties of mild steel weldments. Proc Instn Mech Engrs Part B J Eng Manuf 207:173–177
11. Govindarao P, Srinivasarao P, Gopalakrishna A, Sarkar MMM (2012) Affect of vibrations on butt welded joints. IJMER 2:2766–2770
12. Krajewski A, Włosiński W, Chmielewski T, Kołodziejczak P (2012) Ultrasonic vibration assisted arc-welding of aluminum alloys. Bulletin of the polish academy of sciences and technical sciences 6
13. Dehmolaei R, Shamanian M, Kermanpur A (2008) Effect of electromagnetic vibration on the unmixed zone formation in 25 Cr- 35Ni heat resistant steel/Alloy 800 dissimilar welds. Mater Charact 59:1814–1817
14. Cui Y, Xu CL (2006) Effect of ultrasonic vibration on unmixed zone formation. Scripta Mater 55:957–958
15. Pucko B, Gliha V (2005) Charpy toughness of vibrated microstructure. Original scientific paper-Izvorni Znanstveni Rad. Metalurgija 44:103–106
16. Weite Wu (2000) Influence of vibration frequency on solidification of weldments. Scripta matter 42:661–665
17. Vives Ch (1993) Effect of electromagnetic vibration on the microstructure of continuously cast alloys. Mater Sci Eng A 173:169–172
18. Malinowaski-BRODNICKA M, Den G, Wink WJ (1990) Effect of magnetic field on GTA welds in austenitic stainless steel. Welding Research Supplement
19. Shi L, Wu CS, Liu XC (2015) Modeling the effects of ultrasonic vibration on friction stir welding. J Mater Process Technol 222:91–102
20. Singh PK, Deepak Kumar S, Patel D, Prasad SB (2017) Optimization of vibratory welding process parameters using response surface methodology. J Mech Sci Technol 31(5):2487–2495
21. Singh PK, Patel D, Prasad SB (2018) Investigation on the effect of auxiliary vibrations on microstructure and mechanical properties of SMAW butt welded joints. Indian J Eng Mater Sci NISCAIR 25:155–162
22. Singh PK (2019) Investigation on the effect of mechanical vibration in mild steel weldpool. Manuf Rev 6:13

Investigating Laser Surface Texturing on SS 304 for Self-cleaning Applications

Ambar Choubey, Sabir Ali, Sunil Yadav, C. S. Mandloi, Christ Prakash Paul, and K. S. Bindra

1 Introduction

In laser surface texturing, various geometric structures are generated using laser ablation processes for a specific applications [1, 2]. The advantage of laser texturing is the ability to produce high precision structures of micro to nanoscale on the surface. The process is highly localized, automated, repeated, dry, non-contact type and can be performed under normal ambient conditions. The process is also eco-friendly because no chemicals or toxic wastes are used and can process nearly all types of materials. For laser surface texturing of metals like stainless steel, an infrared pulsed laser is a preferred choice, because of minimum thermal distortion in the material. When metallic surface ablated using ns laser, initially it shows hydrophobicity and then after exposing some time with ambient air surface converted shows super-hydrophobicity. Super-hydrophobic surfaces have proven cost-effective solutions for anti-fouling coatings, anti-icing, anti-corrosion, and self-cleaning surface and have attracted research interest in recent times due to the variety of biological, scientific, and industrial applications [3, 4].

Most of the researchers reported works for the generation of the super-hydrophobic surface on metals use costly ultrafast pico or femto second duration lasers [5, 6]. However, for industrial applications, there is a need to make the generation of super-hydrophobic surface by a cost-effective technique using compact and low price ns fiber lasers. Hence, in the present experiment, a ns fiber laser is deployed for the generation of super-hydrophobic surfaces with good

A. Choubey (✉) · S. Ali · S. Yadav · C. S. Mandloi · C. P. Paul · K. S. Bindra
Laser Technology Division, Raja Ramanna Centre for Advanced Technology,
Indore 452013, India
e-mail: ambarchoubey@rrcat.gov.in

S. Yadav · C. P. Paul · K. S. Bindra
BARC Training School Complex, Homi Bhabha National Institute, Anushakti Nagar,
Mumbai 400 078, India

S. Bag et al. (eds.), *Next Generation Materials and Processing Technologies*,
Springer Proceedings in Materials 9, https://doi.org/10.1007/978-981-16-0182-8_17

scalability, ruggedness and durability. The super-hydrophobic surface can also be generated using different types of low surface energy coatings on the surface [7, 8]. However, in terms of life and durability in a harsh environment, the laser-textured super-hydrophobic surfaces are superior as compared to that with coated surfaces. The actual life for a coated self-cleaned super-hydrophobic surfaces generally varies from few weeks to few months, but the requirement is larger usually in the 5–10 years range, especially for applications in harsh and outdoor environments. Some of the examples of the potential applications are reduction of dust accumulation by self-cleaning on the solar cell surfaces in desert regions and prevention of bio fouling underwater on a ship hull. Hence, laser-based surface texturing is the best solution for the generation of the self-cleaned super-hydrophobic surface.

In the present work, the laser texturing of SS304 stainless steel is carried out for the generation of super-hydrophobic surface and the subsequent application for self-cleaning. The major focus of the work is to optimize laser process parameters investigating the effect of laser power, laser fluence, and laser track separation on super-hydrophobicity. ANOVA is used for the optimization of laser process parameters and wettability of the surface.

2 Theory and Analytical Model

There are many examples of hydrophobic surfaces in nature. For example, lotus rose leaves and some insect wings like butterfly having natural super-hydrophobicity. The structure on their surface has minimum surface energy and minimum contact between water drop and surface. The water drop stays on the surface with minimum contact area, the effect of low surface energy and surface tension leads to form a spherical shape of the water drop with minimum contact area.

The contact angle of these drops is around 150° or more on such surfaces. Because of the minimum contact area with surface, the water drop easily rolls or slides from the surface (depends on surface friction) when it turns by a few degrees (<10°) [9]. The hydrophobicity nature of the surface is measured by the contact angle of water droplets on the surface. The contact angle θ is given by Young's equation as follows [9].

$$\cos\theta = \frac{\gamma_{SG} - \gamma_{SL}}{\gamma_{LG}} \quad (1)$$

where γ_{SG}, γ_{SL}, and γ_{LG} are the surface tension of the fluid with the solid represented by S, liquid by L and gas by G. Due to contact angle hysteresis [10], a droplet can be pinned on inclined surfaces, as shown in Fig. 1. Sliding angle (SA) α is defined as the angle between the tilted substrate and horizontal plane when a droplet starts to move down the surface due to gravity. The slide angle (SA) is used to quantify the slippery nature of a surface and its self-cleaning characteristics. The contact angle hysteresis (CAH) ($\cos\theta_{rec} - \cos\theta_{adv}$) is related to the sliding angle as [10]

$$mg \sin \alpha / w = \gamma_{LG}(\cos \theta_{rec} - \cos \theta_{adv}) \quad (2)$$

where m is the droplet mass, g is the gravity, and w is the droplet width contact with the surface. The lower the value of CAH, the contact area within the liquid/solid interface reduced and shows hydrophobicity, and self-cleaning and slippery nature of the surface were observed. So, the smallest value of CAH with the minimum surface contact area of fluid drop on the solid surface is responsible for the easy sliding features as shown in Fig. 1.

When a liquid drop falls on a textured surface, it can sit on the surface either with minimum contact with the surface or flow between the grooves of the textured surface. It can be explained with two different model Wenzel and Cassie-Baxter model [11]. In Wenzel state, the drop penetrates the gap of the grooves of the texture surface resulting in an increased contact area fraction of liquid with a solid surface. In Wenzel model, the movement of the water drop is restricted because of larger contact area with the surface. On the other hand, in Cassie-Baxter state, the water drop has a minimum surface contact area on the texture surface because, and the air gets entrapped in the grooves of the surface as shown in Fig. 2. This effect generates spherical shape of the droplet with minimum contact area, and drops easily can roll on the surface. For self-cleaning application, Cassie-Baxter state is more favorable.

The contamination on the surface is attached with strong force. If the capillary force of the water drop is more than the adhesion force of contamination, it generates friction forces, results in the contamination, and removes easily during the sliding of the water drop. This leads to a self-cleaning effect on the surface. In most of the cases, friction forces (self-cleaning forces) are in the range of micro-Newton.

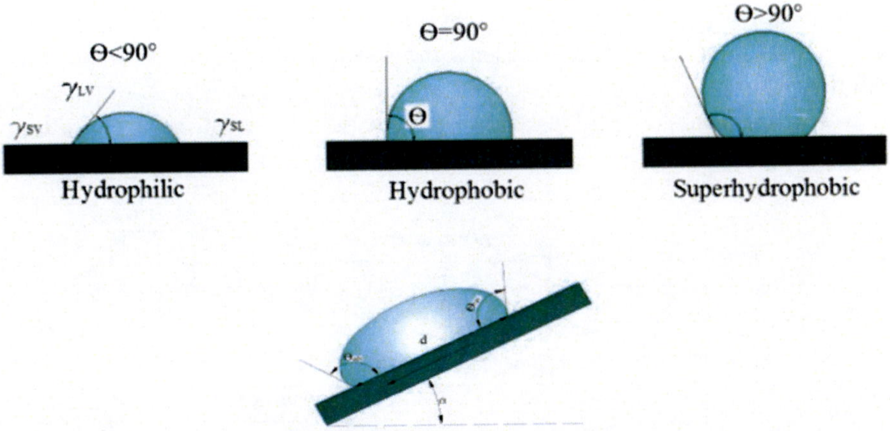

Fig. 1 Different state of water drop and corresponding contact angle on the surface

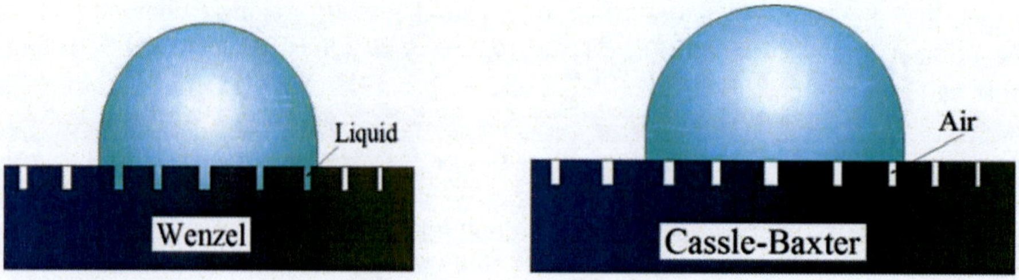

Fig. 2 Difference in Wenzel and Cassie-Baxter model

If the textured surfaces exposed to the ambient air it becomes more contaminated with time. These deposited particles are originated from nature (e.g., microorganisms and pollen). For example, deposition of contamination on the solar cell surface in ambient atmosphere required self-cleaning in quick time span to regain their efficiency.

3 Experimental Work

The texturing experiments are performed on the SS 304 samples with a dimension of 50 mm × 50 mm. The samples were polished manually followed by ultrasonic cleaning using water and surface cleaning using acetone. A 100 ns pulsed fiber laser system (Make: IPG photonics, Model: YLPN-1-1X130-100-M) with variation of 1–20 W average power and repetition rate in the range of the 30–150 kHz is deployed for the laser texturing experiment. The laser system is having a Galvano scanning system for X–Y movement of the laser beam and beam delivered through *f*-theta focusing lens (focal length $f = 163$ mm). The scan speed is in the range of 100–800 mm/s. The minimum focal spot diameter was 60 μm, and its variation is in the range from 60 to 800 μm for setting the different fluence level during texturing on the surface. Figure 3 presents the schematic arrangement of the laser texturing setup.

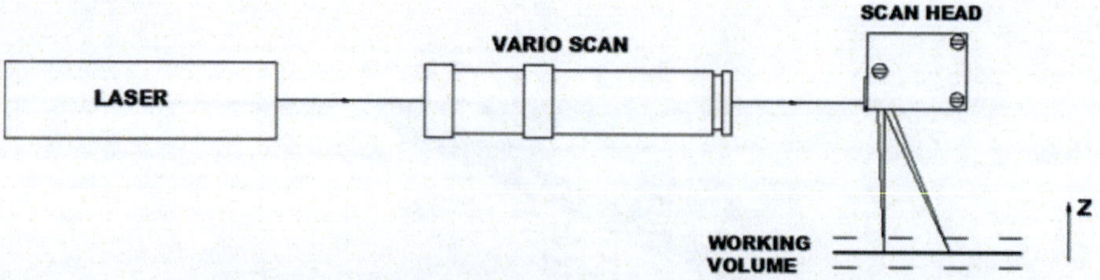

Fig. 3 Schematic arrangement of laser texturing setup

The laser beam is scanned on the substrate surface, along the *x* (0°)-axis and then along the *y* (90°)-axis achieving the final laser-textured grid-patterned structure. The number of laser scan passes *N* varies in the range from 1 to 10. The line spacing (*S*) between two laser scanning lines set 100 μm and the corresponding spot overlapping shown in Fig. 4a, b. All the experiments were carried out under open atmosphere. The spatial profile of the laser beam is Gaussian. Taguchi method is deployed to find out the effects of laser process parameters on the surface texturing [12]. The process parameters under investigation are (1) laser fluence (2) scanning speed and (3) line spacing or hatch spacing each set at three levels. Hatch spacing is directly related with the spot overlapping and spot size of the laser beam according to relation, hatch spacing = (1-overlapping) × spot diameter. The laser average power varies in the range from (10 to 20 W), spot diameter varies (70–800 μm), and pulse duration (100 ns) was fixed during the different sets of experiment. Taguchi's L9 (3^3) orthogonal array and three levels are selected for the optimization of parameters during the experiments. Table 1 presents the laser texturing process parameters and their corresponding level used in the experiment under report. Surface profile of the laser-textured SS304 surfaces is analyzed using a confocal microscope (Make: Leica, Model: TCS SP8). Measurements of the contact angle are performed using Goniometer (Make: Ossila, Model: L2004A1) with a 25 μL distilled water droplet. The apparent contact angle (APCA) is recorded everyday under normal condition. The fresh SS304 laser-textured sample was treated as reference.

4 Results and Analysis

Laser texturing experiments are carried out by varying laser fluence (*F*), scanning speed (*V*), and line spacing (*h*) as shown in Table 1. ANOVA is performed to estimate the effect of *F*, *V*, and *h* on contact angle in ANOVA input and response as

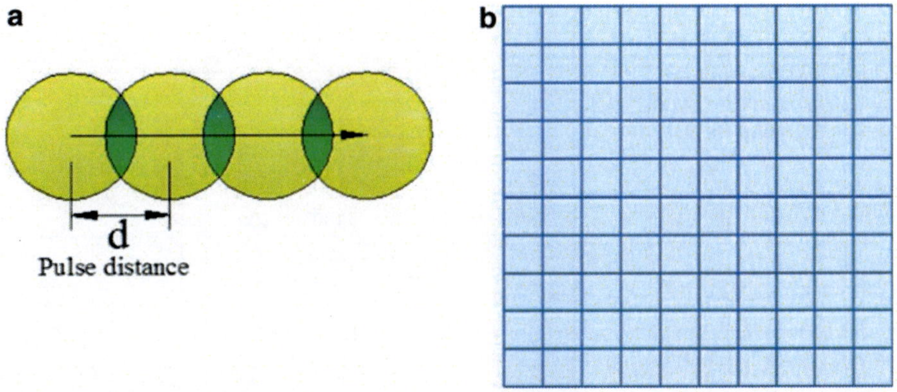

Fig. 4 **a** Hatch spacing and spot overlapping. **b** X–Y scan pattern for texturing

Table 1 Laser texturing process parameters and their corresponding levels values

S. No.	Process parameter	Symbol	Unit	Levels		
				1	2	3
1	Laser fluence	*F*	J/cm^2	3	5	7
2	Scanning speed	*V*	mm/s	200	300	400
3	Line spacing	*h*	μm	30	50	100

given in Tables 2 and 3, respectively. From the ANOVA analysis, in the present experiment, it is observed that the contact angle is mainly governed by laser fluence having minimum *P*-value in Table 3 with a contribution percentage of 46.04%. Laser fluence is followed by line spacing and scanning speed with a contribution of 26.72% and 16.61%, respectively. *P*-value greater than 0.05 indicates weak correlation between the contact angle and process parameter. However, the satisfactory *R* square value (89.42%) is obtained from ANOVA response. *R* square indicates the degree of closeness between a measured response variable and predicated response variables by ANOVA.

A further effect of process parameter, i.e., laser fluence (*F*), scanning speed (*V*), and line spacing (*h*) is studied by means of mean plot as shown in Fig. 5. It is observed that as laser fluence increases, contact angle increases up to a maximum value of contact angle at 7 J/cm^2, and afterward contact angle decreases with laser fluence. Similarly, as scanning speed increases, contact angle increases up to scanning speed of 300 mm/s, afterward contact angle decreases with scanning speed. Decrease in contact angle can be attributed to changes in surface profile model from Cassie-Baxter to Wenzel as scan speed increases as shown in Fig. 2. Further, as line spacing increases, contact angle decrease initially and then increases. Reduced value of contact angle at lower values of line spacing is attributed to merging of laser tracks leading to disappearing of texturing pattern. However, at higher values of line space, measured value of contact angle increases because of non-merging of laser track and clear texture pattern.

Further, optimized process parameter is identified by using the *S*/*N* ratio plot as shown in Fig. 6. Criteria chosen for an optimized parameter are higher contact

Table 2 L9 orthogonal array, nine rows present the set of parameters for the experiments, and the value of contact angle measured immediately after laser texturing

Column sample ID	L (parameters)			Contact angle measured
	F	*V*	*h*	
20200701	3	200	30	45°
20200702	3	300	50	48°
20200703	3	400	100	68°
20200704	5	200	50	85°
20200705	5	300	100	110°
20200706	5	400	30	75°
20200707	7	200	100	82°
20200708	7	300	30	94°
20200709	7	400	50	44°

Table 3 Analysis of variance for contact angle and super-hydrophobic surface

Source	DF	Adj SS	Adj MS	F-value	P-value
Laser fluence (J/cm^2)	2	1984.7	992.3	4.35	0.187
Scanning speed (mm/s)	2	716.7	358.3	1.57	0.389
Line spacing (μm)	2	1152.7	576.3	2.53	0.283
Residual error	2	456.0	228.0		
Total	8	4310.00			

angle as it is required for generation of super-hydrophobic surface [13]. From *S/N* plot, it is observed that optimum values of process parameters of level 1, level 2, and level 3 of laser fluence, scanning speed, and line spacing, respectively, give the highest contact angle 110°. Thus, identified optimum parameter for surface texturing is 5 J/cm^2, 300 mm/s, and 100 μm.

The laser scanning speed v is related to the percentage laser spot overlapping (n), laser spot diameter S, and pulse repetition rate f with the relation [14]

$$v = S \times (1 - n) \times f \tag{3}$$

The surface profile of textured surfaces affected by laser fluence, scanning speed, and the laser track separation. For a small line separation of 10 μm, the laser track overlapping is high, so the multi-directional structure is not well observed. In contrast, for a separation in the range of 50–100 μm, these structures are clearly observed as shown in Fig. 7a, b. As given in Table 2, the value of contact angle is measured with Goniometer for different sets of laser processing parameters. For sample Id 20200705, values of contact angle are the maximum (~110°), and the

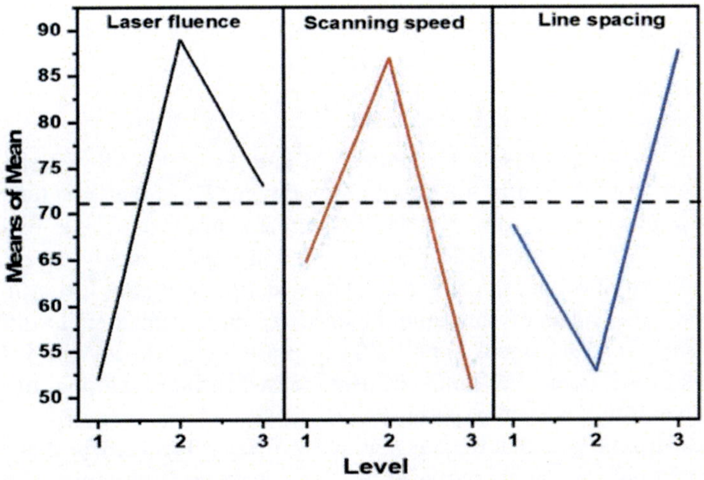

Fig. 5 Means of mean plot for variations in contact angle

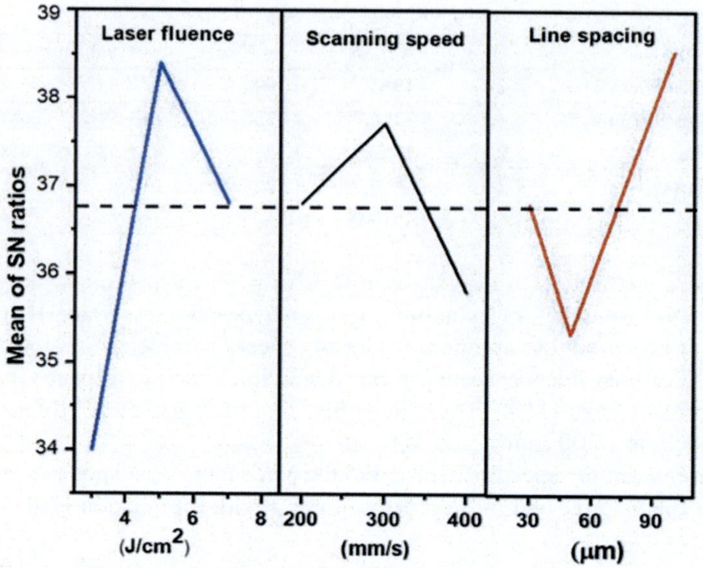

Fig. 6 Means of SN ratio for variations in contact angle

process parameters are the optimum parameters. Table 3 presents the statistical analysis of process parameters using ANOVA and shows that laser fluence is (J/cm^2) having the minimum *P*-value and is the most significant parameters during the texturing followed the line spacing and scanning speed. The ablation depth for *N* number of pulses is given by [14]

$$d = N \times (1/\alpha)\ln(F/F_{\text{th}}) \tag{4}$$

and ablated mass is

$$m_a = N(\rho A d) \tag{5}$$

where $1/\alpha$ is the effective penetration depth of laser energy in SS material having a calculated value of 20 nm in the present experiment, F_{th} is the laser fluence at the threshold, ρ is the density, and A is the textured area of the surface as explained by Tam et al. [15]. It was observed that the ablation depth strongly depends on the fluence level. At threshold fluence, F_{th} ablation started, and after increasing fluence and crossing the value of saturation fluence F_s further ablation is difficult and materials surface is damaged. For SS 304 in present experiment, the value of F_{th} varies from 3 to 8 J/cm^2. For optimum damage free surface texturing, fluence level 5 J/cm^2 is set in the present experiment.

It is clear from Fig. 8 that there is a window of laser fluence, where ablation was achieved without damaging the surface. For single laser scanning, the ablation depth was ~80 nm and can be enhanced by an increasing number of laser scanning

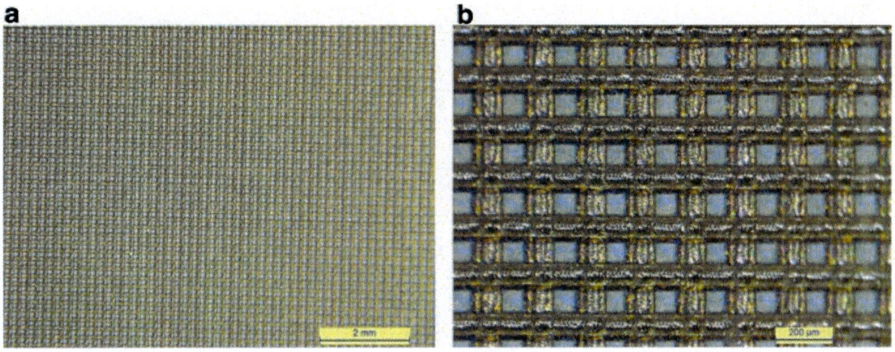

Fig. 7 **a** Microscopic images of the grid-patterned after laser texturing and high-magnified image are shown in **(b)**, respectively, for 100 μm line spacing. Many microscaled grid-shaped channels were formed on the SS 304 surface

Fig. 8 Comparison of calculated and measured ablation depth variation with fluence, after 8 J/cm^2 fluence level ablation depth start saturates

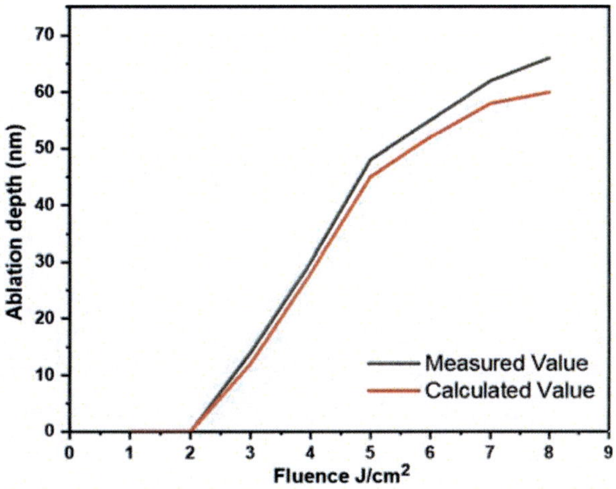

passes *N*. The ablation depth was measured using confocal microscope. Contact angle and super-hydrophobic behavior of the surface depend on the line spacing and exposure time of laser-textured surface. It shows clearly that after 10 days exposure, contact angle or super-hydrophobicity is maximum for 100 μm line or hatch spacing. The ablated mass during laser texturing is collected by a suction system to avoid re-deposition of ejected debris on the textured surface. After performing several experiments and surface analysis, the optimum values of process parameters were 100 ns pulse duration, 15 W average power, 300 mm/s speed, 20 kHz repetition rate, 5 J/cm^2 fluence level, 100 μm lines spacing, and $N = 5$ number of laser passes in *x*- and *y*-axis. The textured groove width was ~180 μm and depth of about 8–10 μm and center-to-center distance of ~100 μm. Laser-textured surface displayed enhanced hydrophobicity with respect to untreated

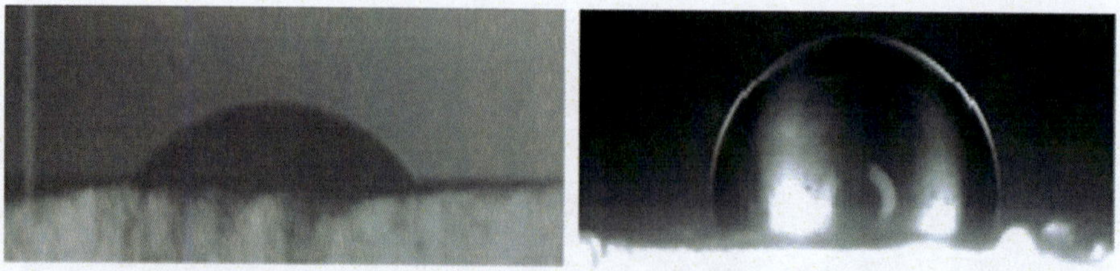

Fig. 9 Image of water drop on surface of untreated and laser-textured surface of SS 304

specimens, as shown in Fig. 9. The measured contact angle of untreated and laser-textured surfaces are 40° and 110°, respectively, measured with Goniometer. After 10 days exposure with air ambient, this 110° contact angle value reaches to 156^0, and surface becomes super hydrophobic. The possible reason is for generation of super-hydrophobic surface, and after 10 days it was the gradual conversion of contact angle related to the absorbed airborne contaminants from air moisture [16]. To remove particles from a super-hydrophobic surface, the capillary force F_C acting on the particle at the rear side of the drop needs to overcome the adhesion force between the particle and the surface, F_{Adh}. Then the surfaces intentionally are contaminated with particles of sizes (80 nm to 1 mm) and polarities (hydrophobic/ hydrophilic) [16].

It was observed that for an optimized grid pattern, the surfaces were cleaned by using 15–30 water drops of 25 ml volume. These drops rolled off on the laser-textured surface inclined at 15° angle and cleaned almost all of the contamination. This laser-textured super-hydrophobic surface is also tested by depositing contamination of dust and carbon, and when the water drop shower is applied for only 5 s, it cleans the surface completely by self-cleaning process with 98% cleaning efficiency, as compared to the normal contaminated surface takes 120 s with only 40% cleaning efficiency. The self-cleaning efficiency is given by [17]

$$\eta = 1 - \left(N_f/N_i\right) \tag{6}$$

Here, N_i and N_f are the respective surface densities of the contaminated particulates on the surface before and after self-cleaning. Analysis software for materials science, i.e., Avizo Fire from Visualization Sciences Group, attached to the Leica microscope has been used to count the initial and the final number of particles. The microscopic image of self-cleaning laser-textured super-hydrophobic surface (lower portion of sample) and unclean non-textured, normal surface (upper portion of the sample) is shown in Fig. 10.

Fig. 10 Image shows difference in self-cleaned textured (lower part) and non-textured surface (upper part)

5 Conclusion

In this study, the laser-textured surfaces on SS 304 with square grid shape were generated successfully using selective ns duration laser-based texturing process. Initially, the as-prepared SS 304 laser-textured surface showed super-hydrophilicity. However, after the ambient air exposure for 10 days, the surfaces experienced gradual wettability conversion to steady super-hydrophobicity. The effect of different laser process parameters on super-hydrophobicity during laser texturing is also performed. ANOVA is used for optimization of process parameters. Laser fluence was the most important parameters for the generation of super-hydrophobic surface on SS 304. There is an optimum window of fluence between 3 and 8 J/cm^2 for good quality texturing on SS 304 surface. Development of laser texturing-based self-cleaning technique is more convenient in terms of durability, efficiency, cost effectiveness, and feasibility. Laser-textured self-cleaning surfaces can have a broad range of applications from self-cleaning of machine parts for food industry, food packaging, home appliances, shipping, aerospace, and medical instrumentation, for reducing the growth of corrosions, scaling bacteria on surface, and enhancing the life of the parts.

Acknowledgements Authors thankfully acknowledge the valuable technical support of Mr. Anil Adbol, Mr. Sokhen Tudu, and Mr. Lalit of Laser Additive Manufacturing Laboratory during the work.

References

1. Chu F, Xiaomin W (2016) Fabrication and condensation characteristics of metallic super hydrophobic surface with hierarchical micro-nano structures. Appl Surf Sci 371:322–328. https://doi.org/10.1016/j.apsusc.2016.02.208
2. Etsion I (2005) State of the art in laser surface texturing. J Tribo Trans ASME 127:248–253
3. Chen C, Yang S, Liu L, Xie H, Liu H, Zhu L, Xiaoliang X (2017) A green one-step fabrication of superhydrophobic metallic surfaces of aluminum and zinc. J Alloys Compd 711:506–513. https://doi.org/10.1016/j.jallcom.2017.04.050
4. Etsion I (2010) Laser surface texturing and applications. In: Nikas GK (ed) Recent developments in wear prevention, friction and lubrication. Research Signpost, Kerala (India), pp 137–158
5. Bonse J, Kirner SV, Griepentrog M, Spaltmann D, Krüger J (2018) Femto second laser texturing of surfaces for tribological applications. Materials 11:801. https://doi.org/10.3390/ma11050801
6. Schille J, Ullmann F, Schneider L, Graefensteiner M, Schiefer S, Gerlach M, Leidich E, Exner H (2015) Experimental study on laser surface texturing for friction coefficient enhancement. J. Laser Micro Nanoeng 10:245–253
7. Zhang Z, Bo G, Men X, Li Y (2016) Mechanically durable, superhydrophobic coatings prepared bydual-layer method for anti-corrosion and self-cleaning. Colloids Surf 490:182–188. https://doi.org/10.1016/j.colsurfa.2015.11.049
8. Huang W-H, Lin C-S (2014) Robust superhydrophobic transparent coatings fabricated by a low temperature sol–gel process. Appl Surf Sci 305:702–709. https://doi.org/10.1016/j.apsusc.2014.03.179
9. Zhang JH, Sheng XL, Jiang L (2009) The dewetting properties of lotus leaves. Langmuir 25:1371–1376
10. Geyer F, D'Acunzi M, Sharifi-Aghili A, Saal A, Gao N, Kaltbeitzel A, Sloot T-F, Berger R, Butt H-J, Vollmer D (2020) When and how self-cleaning of superhydrophobic surfaces works. Sci Adv 6:1–11
11. Lafuma A, Quéré D (2003) Superhydrophobic states. Nat Mater 2:457–460
12. Chatterjee S, Sankar Mahapatra S, Bharadwaj V, Choubey A, Upadhyay BN, Bindra KS (2018) Drilling of micro-holes on titanium alloy using pulsed Nd:YAG laser: parametric appraisal and prediction of performance characteristics. Proc I Mech E Part B: J Eng Manuf 1–18. https://doi.org/10.1177/0954405418805604
13. Yilbas BS, Keles O, Yagmur Toprakli A (2017) Surface engineering towards self-cleaning applications: laser textured silicon surface. Procedia En 184:716–724. https://doi.org/10.1016/j.proeng.2017.04.147
14. Choubey A, Vishwakarma SC, Vachhani DM, Singh R, Misra P, Jain RK, Arya R, Upadhyaya BN, Oak SM (2014) Study and development of 22 kW peak power fiber coupled short pulse Nd:YAG laser for cleaning applications. Optics Lasers Eng 62:69–79. http://dx.doi.org/10.1016/j.optlaseng.2014.04.012
15. Tam AC, Leung WP, Zapka W, Ziemlich W (1992) Laser cleaning techniques for removal of surface particulates. J Appl Phys 71:3515–3523
16. Bhushan B, Jung YC (2011) Natural and biomimetic artificial surfaces for superhydrophobicity,:self-cleaning, low adhesion, and drag reduction. Prog Mater Sci 56:1–108
17. Singh A, Choubey A, Modi MH, Upadhyaya BN, Oak SM, Lodha GS, Deb SK (2013) Cleaning of carbon layer from the gold films using a pulsed Nd:YAG laser. Appl Surf Sci 283:612–616

Critical Assessment of Metallurgical and Mechanical Characteristics of Pulse Laser Welded α + β Phase Ti-Alloy

Bikash Kumar and Swarup Bag

1 Introduction

Dual-phase Ti-alloy is of utmost interest for extensive application at room as well as moderately elevated temperature, especially due to its appealing properties such as lightweight and high strength-to-weight ratio, good compatibility with body fluid, resistance against corrosion and oxidation, and superior mechanical properties. In principle, this alloy is comprised of α-phase and β-phase stabilized by 6 vol.% of Al and 4 vol. % of V, respectively. Reversible transformation (depicted in Fig. 1) is an exceptional attribute of Ti-6Al-4V alloy, which might be the reason for stable and improved mechanical properties at elevated temperature. Initially, Ti-6Al-4V alloy exists as mixture of α-phase and β-phase hence termed as $\alpha + \beta$ alloy. As the metal heated beyond 940 K, dissolution of α-phase starts and transforms into β-phase up to 1273 K which is known as β-transus temperature. Upon further heating up to solidus temperature, i.e., 1877 K, weld metal exhibits fully β-phase (BCC crystal structure). In the range of solidus and liquidus temperature, weld metal renders β as well as liquid phase. On the contrary, molten metal (liquid phase) again transforms into β-phase during cooling cycle at β-transus temperature. Afterwards, whether it would follows diffusional mechanism of transformation or non-diffusional, it depends on the cooling rate followed beyond β-transus temperature. If attained cooling rate is higher than critical cooling rate (i.e., 410 K/s) non-diffusion α'-martensitic transformation takes place which results in acicular α'-martensite. Otherwise, diffusional growth of α-phase comes into existence which leads to the formation of lamellar structured α-phase. Final microstructure of mechanically processed material plays a decisive role to determine the overall engineering properties of the structure. In addition, microstructure and morphology of

B. Kumar (✉) · S. Bag
Department of Mechanical Engineering, Indian Institute of Technology Guwahati, Assam 781039, India
e-mail: bikash.kumar@iitg.ac.in

S. Bag et al. (eds.), *Next Generation Materials and Processing Technologies*, Springer Proceedings in Materials 9, https://doi.org/10.1007/978-981-16-0182-8_18

Ti-6Al-4 V alloy possibly can be controlled through the cooling rate. From the perspective of fusion welding, shielding system also contributes to the alteration of weld quality. Shielding environments surrounding the weldment–heat source interplay region are perhaps the crucial facet that can severely affect the quality of welded joints. Inert gases for instance argon, helium, or combination of both is usually placed to protect the molten zone [1–3]. One of the authors suggested that combination of Ar + He in varied volume ratio can minimize the loss parameters which leads to growth in penetration depth in the course of laser welding [1]. Shielding gas applied to inhibits contamination from external detrimental factors that promote defects, i.e. porosity and crack development. Thus, other author endorsed isolated chamber to furnish shielding environment in the weld zone and to minimize the generation of porosity as well [4, 5]. Proper shielding environment favourably changes the quality of weld joint since Ti-alloy is susceptible towards the formation of oxides, nitrides and carbides that can severely deteriorates the overall service life of welded structure. Hence, it is vital to pay attention towards the metallurgical–mechanical facets during welding for stable and reliable welded joints of Ti-6Al-4V alloy.

Welding provides an efficient and affordable technique of fabricating large and small structural assembly by utilizing this Ti-alloy. Several researchers have applied different welding processes such as gas tungsten arc welding (GTAW) [6], gas metal arc welding (GMAW) [7], ultrasonic welding [8, 9], linear friction welding [10], laser beam welding and electron beam welding to analyse the performance of process parameters on microstructural evolution, and thermal–mechanical–metallurgical behaviour of welded structure. El-Batahgy and DebRoy [6] did comparative study between laser and GTA welded Ti-6Al-4V alloy. It is revealed that laser fabricated samples render relatively low distortion and smaller molten zone than GTA welded specimens. Apart from that, higher amount of acicular α'-martensite is observed in laser welded samples which is prominent cause of high hardness owing to high cooling rate. Pardal et al. [7] indicated that the performance of GMA welded specimens is found to be less productive owing to high spatter generation, wandering of arc and subsequent wavy nature of molten pool. However, combination of laser welding with arc process makes it happen to benefit from advantageous characteristics of each process [8]. Zhao et al. [9] convey that the application of ultrasonic vibration can efficiently prohibits the growth of columnar crystal within the molten zone which enhances the tensile properties of welded joints. Linear

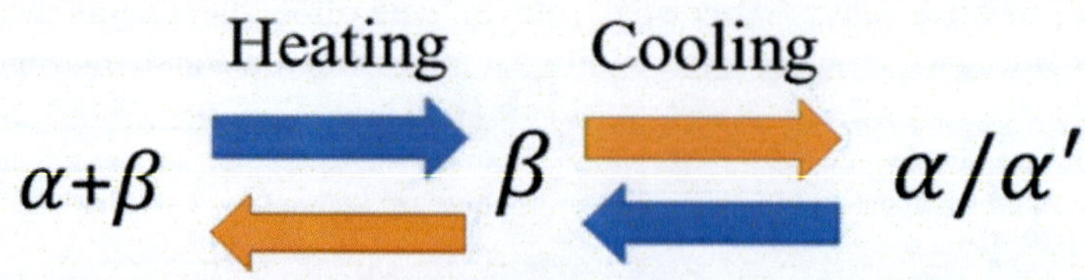

Fig. 1 Reversible transformation by $\alpha + \beta$ alloy

friction welding is solid-state joining technique in which thermo-mechanically affected region exhibits recrystallization of Widemanstatten $\alpha + \beta$ phase within prior β-grain boundary [10].

Pulse Nd:YAG laser is characterized by high peak power that is significantly larger than average power of laser beam, allowing melting and solidification to take place in a fraction of second [11]. The lasers have potential of pulse shaping at frequency of up to several kHz and with pulse duration ranging from 0.5 to 20 ms [12]. Akman et al. [5] implemented Nd:YAG pulsed laser for the fabrication of Ti-alloy joints and stated that ratio of pulse energy and pulse duration is critical factor to determine penetration depth. It is also emphasized that transformation of phase from α to β through heat treatment and coarsening of grain can deteriorates the mechanical properties. Hence, low hardness is observed at large heat input. Baruah and Bag [13] investigated the influence of pulsation effect during simulation of heat source and reported that sophistication can be improved through pulsation since it facilitates elucidation of thermo-mechanical behaviour pattern of laser welded joints. Furthermore, Kumar et al. [14] interpreted the welded joints fabricated at continuous and pulsed mode. At optimized minimum heat input in pulse welded samples, the acicular shape α'-martensite within fine prior β grain boundary results extraordinary tensile strength even more than base metal. Few researchers elucidated that the risk of porosity formation and undercut during welding of Ti-alloy owing to entrapment of different gases can be minimized through the utility of pulsation effect [15, 16]. In addition, Kumar and Bag [17] studied the metallurgical transformation behaviour of pulse laser welded thin sheets of Ti-alloy and concluded that relative stability of phase transformation during heating cycle can trigger the state of stress pattern during cooling cycle through manipulating martensitic fraction. Therefore, it convinces that it is important to pay attention towards the optimization pulse parameters during welding to own desired structural properties, microstructural and metallurgical features, and improvement in the weld characteristics.

Nowadays, finite element-based mathematical tool promotes the accurate measurement of thermal data at difficult-to-access location and to trace other intricate characteristics. Several researchers developed different heat source model to precisely improve the qualitative and quantitative prediction of thermal history [18, 19]. A lot of literature is available which focuses on mathematical representation of heat source where geometrical dimension is primarily mapped with molten pool geometry [20–22]. Hence, the selection of appropriate heat source model is critical for the adequate prediction of time-temperature field in conduction-based heat transfer phenomena.

The present study extensively contributes to the core insight into parallel interaction between thermal–metallurgical and associated mechanical performance of pulse laser welded thin joints of Ti-6Al-4V alloy. The investigation is carried out on laser welding and evolved bead analysis, XRD-based metallurgical analysis, development of heat transfer model and its validation with experimental result, and microstructural–mechanical interaction. Interrelation between thermal history and assisted phase transformation is briefly explored. In addition, surface discoloration methodology is applied to track the influence of effectiveness of applied shielding environment of molten zone. FESEM-based microstructural attributes are also studied to trace the morphological variation at weld zone.

2 Materials and Methodology

As received $\alpha + \beta$ alloy (Ti-6Al-4V alloy is also known as dual-phase Ti-alloy, i.e., $\alpha + \beta$ alloy) of dimension $120 \times 100 \times 0.5\text{mm}^3$ is used as weldments for the fabrication of welded joints. The pulse Nd: YAG laser system of capacity 2 kW is used to perform the weld phenomena [23]. The pulse laser system is comprised of four major sub-components, i.e. laser head, 2-axis automated table, cooling system, CNC controller retrofitted with laser system. Here, laser head is in static position that can only move in Z-direction to control the stand-off distance. Hence, only 2-axis automated table is used for the movement of the workpiece. CNC controller imposes the laser beam by controlling the process parameters as per given input by the user. External shielding environment is provided around the molten pool to inhibit the contamination as much as possible. Another shielding pipe is connected to the laser nozzle system to cool the nozzle chamber so that discrepancy on the lens can be avoided. In the present study pulse parameter, i.e., defocusing distance, is considered as effective controlling parameter during welding. Hence, two different conditions, i.e., focal point on the top surface ($f_p \rightarrow 0$) and the bottom surface ($f_p \rightarrow -0.5$) of workpiece (as pictorial representation is demonstrated in Fig. 2), are considered for the investigation. It indicates that the minimum laser spot size can be accomplished when the workpiece surface lies of the focal point, whereas relatively broadened laser spot can be obtained if workpiece surface deviates from its focal point due to convergence and divergence of the laser beam. Corresponding process parameters on which welded joints are fabricated are illustrated in Table 1. Constant scanning speed of 800 mm/min is used for all weld fabrication.

Conventional sample preparation procedure has been performed on the surface of weldment to remove the contaminated layer of oxides, grease or oil. The edge of weldments that has to be welded is cleaned with grit paper. Furthermore, after fabrication of welded joints, every sample is cross-sectioned normal to the welding direction at middle of the plate to get samples for microstructural analysis, XRD analysis, FESEM analysis and evaluation of tensile properties. Microhardness testing is performed on the weld interface by using Vicker's microhardness tester. Furthermore, tensile testing is performed to estimate the joint efficiency and mechanical behaviour of welded joints. The sample of cross section $5 \times 5\,\text{mm}^2$ is used to carry out X-ray diffraction analysis to trace the phase constituents present over the weld interface specially at fusion zone. Cross-sectioned samples are crossed through different stage of polishing with several grade of grit paper followed by velvet cloth polishing to reveal the microstructure and highly magnified FESEM images.

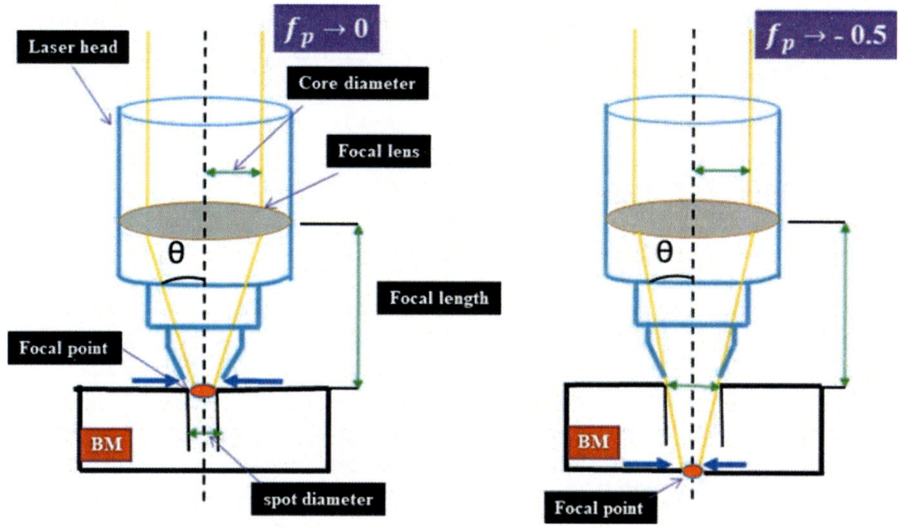

Fig. 2 Schematic of welding phenomena at different focal distances

Table 1 Process parameters used for the fabrication of welded joints

Sample	Peak power (W)	Pulse frequency (Hz)	Pulse energy (J)	Pulse width (ms)	Average power (W)	Defocus (mm)	Heat input (J/mm)
Pulse mode							
S1	800	20	6.4	8	128	0	9.6
S2	800	20	6.4	8	128	0.5	9.6
S3	1000	20	12	10	160	0	15
S4	1000	20	12	10	160	0.5	15

3 Theoretical Background

A conduction-based three-dimensional heat transfer model is established to estimate temperature history at different zones of weld interface. Half of the geometrical domain is designed to perform the thermal analysis owing to the symmetry of fabricated joints. Fourier based nonlinear heat conduction equation is solved for each node during movement of heat source to calculate the attained temperature. Non-uniform meshing is implemented over the geometrical domain, i.e., very fine mesh at weld zone and HAZ, coarse mesh at nearby region and subsequent more coarse mesh at farther away location. Temperature-dependent thermo-physical properties and some constant properties utilized for thermal modelling are considered from the literature [23, 24]. Double ellipsoidal heat source model is applied for the generation of heat flux to substitute the representation of laser beam, since, it can predict accurate weld profile and temperature history [25]. For the generation and movement of heat flux, FORTRAN-based in-house developed DFLUX subroutine is written which incorporated the effect of pulsation as well.

4 Results and Discussions

4.1 Laser Welding and Bead Analysis

Figure 3 illustrates the bead (top view) evolved during laser welding under various weld conditions in pulse mode. Level of contamination and whether the applied shielding system working well or not during fabrication of welded joints can be estimated through the appearance of bead colour or surface discoloration technique is an important feature of Ti-6Al-4V alloy. The presence of silver colour bead for S1 (800 W) and S2 (800 W) reflects the application of better shielding system and minimum level of contamination [14]. However, transformation of silver into reddish straw for case S3 or into reddish brown for weld metal S4 confirms relatively higher contamination as compared with S1 and S2. It could be due to sudden growth (double time) of heat input from 9.6 J/mm → 18 J/mm for S1 → S4. However, Ti-alloy is prone to contamination, i.e., carbide, nitride and oxide formation at elevated temperature. Among various parameters, pulse width (pulse duration) and defocus distance are considered to track its impact on weld joint. Figure 3 (top view) and 4 (cross-sectional view) show that the fusion width increases from 1.504 mm → 1.617 mm upon increasing pulse duration from 8 ms → 10 ms. It increases from 1.504 mm → 1.571 mm with increasing defocusing position from top (~0 mm) to bottom surface (~0.5 mm) at a pulse width of 8 ms and 1.617 mm → 1.690. The threshold energy is the minimum energy required to melt the weldments for the fusion to form a successful joint. It is expressed as a function of power density and interaction time. The time for which laser source is in contact with workpiece is known as interaction time. As the pulse duration increases from S1 to S3, the interaction time also increases during pulse ON time, by virtue of which volume of molten metal increases and enhances the fusion width. The other parameter is the average power density that can be expressed as inversely proportional to the area of laser spot. Defocus distance dramatically influences the bead geometry owing to the variation of power density. As the defocus position increases from S1 → S2 or S3 → S4, i.e., 0 mm → 0.5 mm, the laser spot area on the surface of welded metal increases due to convergence of laser beam as shown in Fig. 2. Therefore, same laser power has to be distributed on large spot area, hence, laser power density decreases and fusion width increases as reported in Fig. 4. It can be inferred that upon increasing defocus position for same laser power, local cooling rate enhances extensively at the molten surface. Furthermore, Fig. 4 depicts that partial penetration is observed for the samples S1

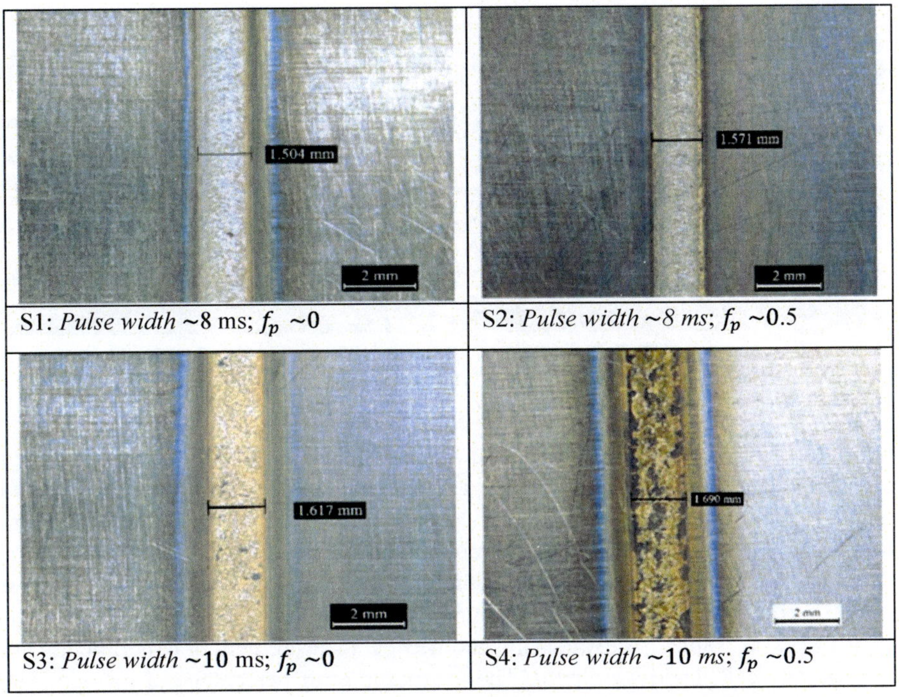

Fig. 3 Demonstration of bead appearance (top view) under various weld conditions

and S2 at laser power of 800 W and pulse duration of 8 ms, owing to insufficient heat input. However, full depth of penetration is obtained for the weld metal S3 and S4 welded at laser power 1000 W and pulse duration of 10 ms. Hence, the cases S3 and S4 are regarded as optimized weld condition from successful weld joints perspective.

4.2 Comparison Between Experimental and Numerical Bead Profile

Figure 5 portrays the validation of three-dimensional thermal model by comparing numerical result with experimentally measured weld bead profile. Figure 5a represents the spatial and temporal variation of temperature distribution at the middle of the plate in pulse ON condition. The arrow represents the velocity direction of heat flux. Figure 5b depicts the magnified image of Fig. 5a. The calculated weld bead profile and different zones are characterized by different colour contour. Fusion zone (FZ) and fully transformed-heat-affected zone (FT-HAZ) are represented by grey and yellow band, whereas mushy zone which is surrounded by FZ

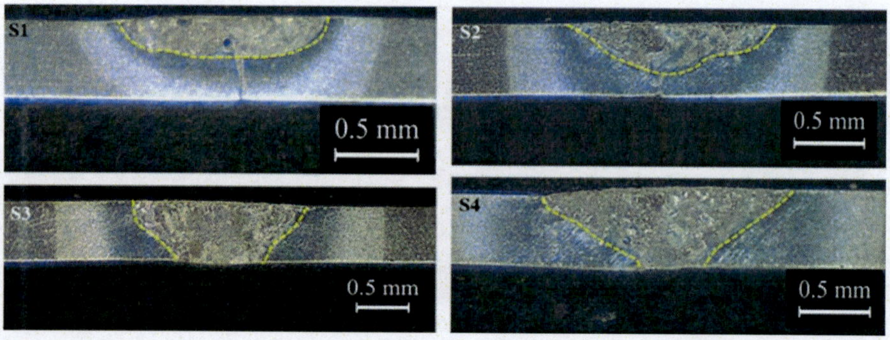

Fig. 4 Cross-sectional view of weld bead profile at different process conditions

and FT-HAZ, which is depicted as red band. Moreover, partially transformed-heat affected zone (PT-HAZ) is symbolized as green and another band. The comparative study reveals good and satisfactory correspondence between experimental and numerical outcomes. The maximum temperature obtained by the case S1 and S4 is 2289 K and 2772 K, respectively, which is found to be sufficiently more than melting point. However, exhibited temperature range corresponding to HAZ is 1659–1292 K. Table 2 illustrates the geometrical measurement of experimental and numerically calculated weld pool. The maximum error percentage of FZ and HAZ is found to be 12% and 8%, respectively, and shows the robustness of the developed model. Based on the visual inspection, it is clarified that the aspects ratio (*w*/*h*) of weld zone is more than 1 and the corresponding laser power is under 1 kW. Hence, all the fabricated joints are categorized as conduction mode rather than keyhole mode of welding.

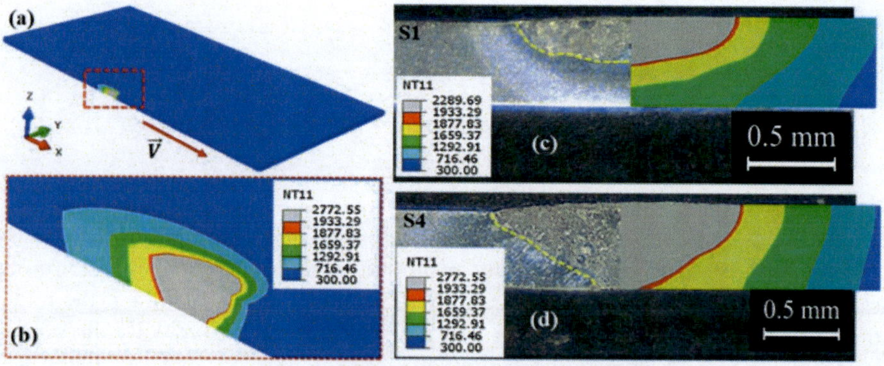

Fig. 5 Validation of heat transfer model: **a** 3D distribution of temperature at the middle of the plate at pulse ON condition, **b** magnified image of temperature distribution; comparison of experimental (left) and numerically calculated (right) temperature profile for the cases **c** S1, and **d** S4

Table 2 Quantitative analysis of bead profile and associated error in numerical model

Sample	FZ			HAZ		
	Exp.	Num.	Err.%	Exp.	Num.	Err.%
S1	0.72	0.65	10	0.35	0.33	6
S2	0.76	-		0.40	-	
S3	0.80	-		0.63	-	
S4	0.85	0.76	12	0.71	0.66	8

4.3 Interrelation Between Thermal Cycle and Phase Constituents

Figure 6a depicts the interrelation between thermal cycle and solid-state phase transformation that occurs within thermally agitated region during and after welding. Temperature associated with α-dissolution, β-transus, martensitic start (Ms) and finish (Mf) is recognized as critical temperature, since metallurgical diversification has been observed beyond these limits. Figure 6 shows that the point '1' lies on room temperature at which Ti-alloy exist as dual-phase alloy, i.e., composed of α-phase and β-phase. As the temperature rises beyond 940 K, α-dissolution takes place and begins to transform into β-phase. Furthermore, at β-transus temperature (i.e. 1273 K) single phase, i.e. β-phase retained in the alloy. As the temperature grows up to peak value due to pulsation effect (pulse ON condition), only liquid phase exist in the molten pool designated as point '2'. Pulsation effect is confirmed by the zig-zag path because pulse on and off phenomena possessed for time duration of millisecond. Additionally, point '3' lies within temperature range of solidus and liquidus limit which indicates the presence of liquid phase and solid (β) phase during cooling cycle. Upon further cooling, columnar β-phase emerges beyond solidus temperature (1877 K). However, as the metal attains temperature less than β-transus, $\beta \rightarrow \alpha$ transformation takes place. As a result, growth of α-phase occurs within prior β-phase grain boundaries. Since the present study is associated with pulse-assisted process, hence high cooling rate, i.e., $\geq$420 K/s, is expected during solidification phenomena. At the temperature beyond martensitic start value (1053 K), α'-martensitic lath formation occurs, and martensitic transformation finishes at Mf. Nevertheless, the mixture of $\alpha + \alpha' + \beta$ phase exists after solidification at ambient temperature with various morphologies.

Figure 6b demonstrates the X-ray diffraction pattern revealed for base metal of Ti-6Al-4V alloy. It signifies that plane (110) and (002) corresponds to β-phase and the planes $(1\ 0\ \bar{1}\ 0)$, $(1\ 0\ \bar{1}\ 1)$, $(1\ 1\ \bar{2}\ 2)$, $(1\ 0\ \bar{1}\ 2)$, $(2\ 0\ \bar{2}\ 1)$ corresponds to α/α' possessed by the base metal. Crystallographic dimensions of α and α' are approximately same; therefore, it is somewhat difficult to distinguish between them. Figure 6c illustrates the diffraction pattern of fusion zone under different weld conditions (800 and 1000 W). Peak broadening and peak shifting characteristics are noticed at weld zone of both samples, i.e. S1 and S4. This may be due to the retained lattice micro-strain within the molten metal. It is indicated that peaks

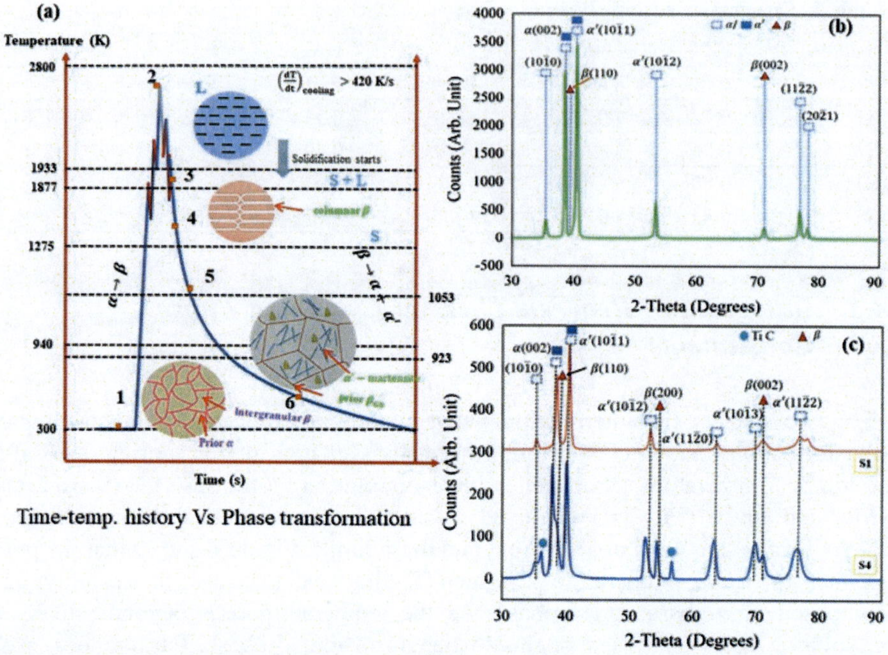

Fig. 6 **a** Correlation b/w thermal history and phase transformation; illustration of XRD pattern of **b** base metal and **c** weld metal

corresponding to every plane exhibit higher intensity at higher heat input, i.e., 18 J/mm rather than 9.6 J/mm. Enough heat input and lower cooling rate attained at the highest heat input might be the predominant cause for the presence of intense peak. Higher heat input allows the metal to get transformed more and favours the growth of martensitic lath significantly. Relatively higher intensity of plane corresponds to α'-phase confirms the existence of higher volume fraction of martensite. In addition to this, some peaks correspond to TiC which is also noticed for higher heat input, i.e., for sample S4. It signifies that the contamination is also prevalent at higher heat input rather than lower heat input as also concluded by surface discoloration technique.

4.4 Compositional Analysis

Figure 7a depicts the EDS scanning at fusion zone of welded sample. It shows the compositional variation of α-stabilizing elements, i.e., Al and β-stabilizer, i.e., V, across the scanned section in terms of intensity counts. It is obvious that there is no segregation or agglomeration of α- or β-stabilizer is observed throughout the scanned area. However, enriched amount of α-stabilizer is noticed at core region of

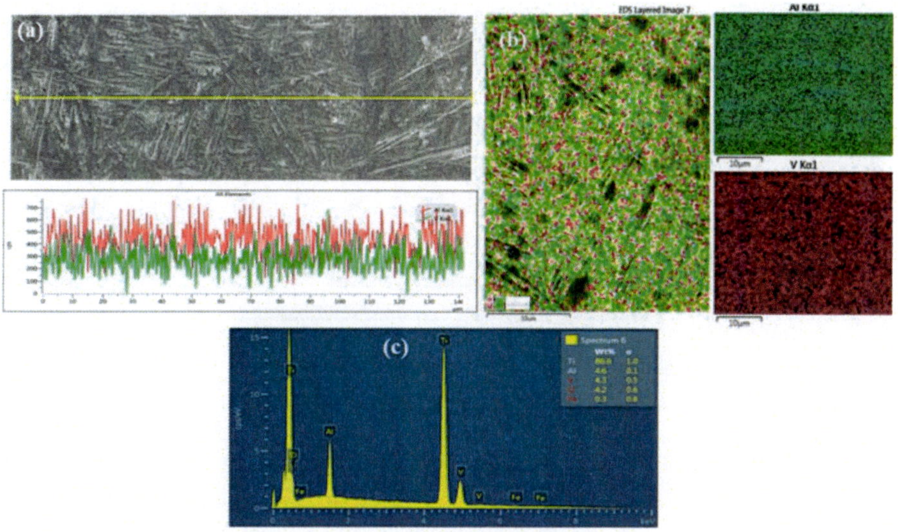

Fig. 7 EDS analysis of **a** line scanning, and **b** elemental area of weld zone; **c** compositional analysis by EDS

α'-lath, whereas lean composition of β-stabilizer is obtained on those locations. Furthermore, Fig. 7b represents EDS scanning over elemental area of weld metal S4. It reveals that the dispersion of α-stabilizer (green) is severe and more prominent than that of β-stabilizer (pink). It confirms that $\beta \rightarrow \alpha/\alpha'$ transformation is prevalent after melting and re-solidification phenomena during pulse welding. Moreover, retained β-phase contributes minimum amount in the molten zone. Dark spot in scanned area signifies the segregation of oxide and other detrimental compounds which formed or entrapped within molten area during welding. Figure 7d manifests that the wt% composition of oxygen is significantly higher, i.e., 4.3% at scanned area of dark spot.

4.5 Microstructural Evolution

Variation in microstructural morphology plays a decisive role to trigger the final mechanical properties of Ti-alloy welded structure. Figure 8 shows the evolved microstructure at different zones of welded specimens S1 and S4. Figure 8a signifies dramatic difference in microstructure at FZ, FT-HAZ and PT-HAZ owing to non-uniform temperature gradients during welding. Solidification of molten metal emphasizes epitaxial growth of columnar β-grain boundaries and emergence of α-phase within prior β-grain boundary. Fusion zone of case S4 is characterized by hexagonal grey blocky structured martensite composed of thick α'-lath and surrounded by coarsed prior β-grain boundary as shown in Fig. 8b. Attainment of low

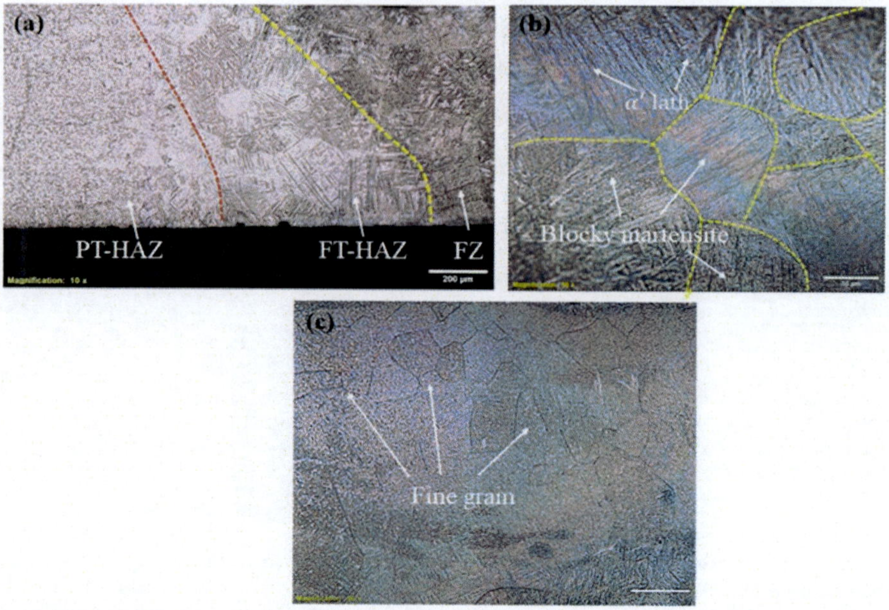

Fig. 8 Microstructural pattern: **a** depiction of microstructure at different zones, **b** grain size demonstration for weld metal (S4) and **c** grain size demonstration for weld metal (S1)

heat input at farther away region, i.e., FT-HAZ and PT-HAZ, inhibits the columnar growth of prior β-grain boundary; however, fine acicular structured martensite is revealed at FT-HAZ. Furthermore, PT-HAZ attains insufficient heat input due to which it is unable to transform into fully β-phase. Nevertheless, small-sized (fine) grain with pre-mature grain boundary is observed for the case S1 or lowest heat input as demonstrated in Fig. 8c. It is because the minimum heat input exhibits lowest solidification time, hence, high cooling rate. Rapid cooling does not permit the grain or grain boundary to grow more due to which it retained as small and fine in domain of molten zone. On the contrary, higher heat input favours significant grain growth.

4.6 Mechanical Performance

Figure 9 demonstrates the mechanical performance of fabricated joints at various welding conditions. Ultimate tensile strength (UTS), elongation, yield strength are the crucial factors that decide the joints strength and joint efficiency. Figure 9a elucidates the engineering stress and strain curve estimated using universal tensile testing (UTM) machine. It is obvious that UTS ($\sim$900 MPa) and elongation (14.5%) of base metal are significantly higher than other welded joints. It conveys that welded joints are somewhat weaker than base metal. In addition, welded joints

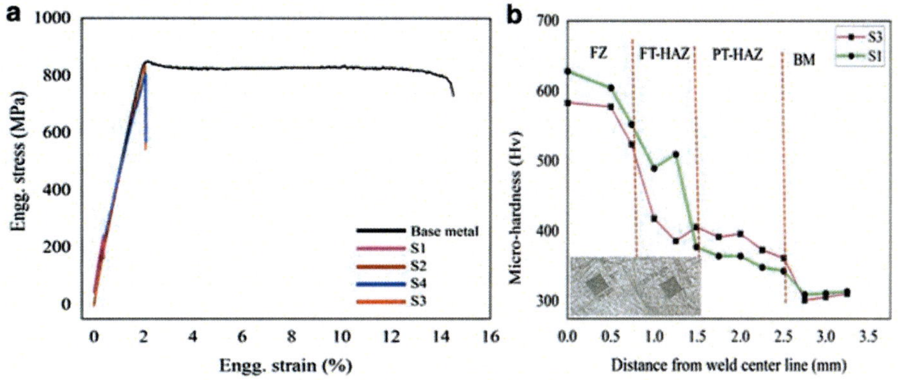

Fig. 9 **a** Depiction of engineering stress versus strain at different weld conditions; **b** distribution of hardness across the weld interface

fabricated at heat input of 18 J/mm (samples S3 and S4) renders UTS of approximately 820 and 810 MPa, respectively, whereas 220 and 210 MPa for specimens S1 and S2. Although weld metal S1 and S2 fabricated at lowest heat input and achieved needle-shaped acicular structured morphology within fine prior β-grain boundary, attained minimum contamination (as confirmed by discolouration technique) exhibited inferior strength and ductility properties. Insufficient depth of penetration due to application of less heat input might be the causes of poor mechanical properties. Furthermore, it is clarified that elongation percentage of each sample is very less. It might be due to the generation of brittle martensitic phase at the weld zone. Figure 9b signifies the hardness distribution across the weld interface for the specimens S1 (9.6 J/mm) and S3 (18 J/mm). It shows that hardness is remarkably higher at fusion zone for both the cases; however, distinguishable variation in magnitude is observed. Martensite exhibits relatively higher hardness among all the phases and fusion zone of each sample is comprised of martensitic phase. Hence, fusion zone exhibited higher hardness than another zone. Attainment of higher cooling rate at lowest heat input promotes needle-shaped acicular α'-martensite which is the hardest morphology accordingly weld metal S1 possessed higher hardness. Moreover, grey blocky structured α'-martensite within coarsened prior β-grain boundary attributed at relatively higher heat input is relatively soft phase as a result sample S3 emphasized relatively lower hardness at weld zone.

5 Conclusions

The laser welding of 0.5-mm-thin sheet Ti-6Al-4 V alloy is performed at pulse duration of 8 ms and 10 ms by considering focusing and defocusing distances. A finite element-based thermal model is established to analyse the pulse welding.

Interrelation between thermal history and associated phase transformation is predicted. The following conclusion is drawn from the present investigation.

- Surface discoloration methodology can qualitatively predict the level of contamination within weld zone only by virtue of visual inspection.
- Lowest heat input of 9.6 J/mm reveals insufficient depth of penetration; however, attained minimum contamination, whereas higher heat input of 18 J/mm, achieves full depth of penetration with higher level of contamination.
- Elemental scanning analysis shows enriched α-stabilizing elements throughout the weld zone. It indicates that diffusionless transformation is prevalent due to which α'-martensite is found to be distributed uniformly.
- Different plane corresponds to α/α' phase, and transformed β-phase is observed in fusion zone. Grey blocky structured martensite with thicker α'-lath is observed for pulse duration of 10 ms, whereas, fine grain with acicular α'-martensite is obtained for heat input of 9.6 J/mm.
- Ultimate tensile strength (UTS) of 820 MPa is obtained, when focal point is considered at top surface of weld metal (i.e. $f_p \rightarrow 0$). While UTS of 810 MPa is perceived, when focal point is shifted to bottom surface (i.e. $f_p \rightarrow -0.5$) for pulse duration of 10 ms. However, very low UTS, i.e. in the range of 200 MPa, is attained for the pulse duration of 8 ms by virtue of insufficient penetration.
- The weld metal fabricated at the lowest heat input achieves needle-shaped acicular morphology within fine prior β-grain boundary. Despite it attained minimum contamination, exhibited inferior engineering properties. Insufficient depth of penetration due to low heat input is the dominating factor for poor mechanical properties.
- Presence of acicular α'-martensite favours the enhancement of hardness within weld zone. Grey blocky structured martensite accomplishes at higher heat input found to be relatively soft.

Acknowledgements The authors thankfully acknowledge the Department of Mechanical Engineering (IIT Guwahati, Assam, India), Central Instrument Facility (CIF) and Central workshop for offering the experimental facility of welding and equipments for further analysis.

References

1. Motlagh NH, Parvin P, Jandaghi M, Torkamany MJ (2013) The influence of different volume ratios of He and Ar in shielding gas mixture on the power waste parameters for Nd: YAG and CO_2 laser welding. Opt Laser Technol 54:191–198
2. Wang H, Shi Y, Gong S, Duan A (2007) Effect of assist gas flow on the gas shielding during laser deep penetration welding. J Mater Process Technol 184(1–3):379–385
3. Amaya JS, Amaya-Vázquez MR, Botana FJ (2013) Laser welding of light metal alloys: aluminium and titanium alloys. In: Handbook of laser welding technologies, pp 215–254. Woodhead Publishing

4. Reed CB, Natesan K, Xu Z, Smith DL (2000) The effect of laser welding process parameters on the mechanical and microstructural properties of V–4Cr–4Ti structural materials. J Nucl Mater 283:1206–1209
5. Akman ERHAN, Demir A, Canel TİMUR, Sınmazçelik T (2009) Laser welding of Ti6Al4 V titanium alloys. J Mater Process Technol 209(8):3705–3713
6. El-Batahgy AM, DebRoy T (2014) Nd-YAG laser beam and GTA welding of Ti-6Al-4 V alloy. Int J Eng Tech Res 2(12):43–50
7. Pardal G, Martina F, Williams S (2019) Laser stabilization of GMAW additive manufacturing of Ti-6Al-4 V components. J Mater Process Technol 272:1–8
8. Casavola C, Pappalettere C, Pluvinage G (2011) Fatigue resistance of titanium laser and hybrid welded joints. Mater Des 32(5):3127–3135
9. Zhao G, Wang Z, Hu S, Duan S, Chen Y (2020) Effect of ultrasonic vibration of molten pool on microstructure and mechanical properties of Ti-6Al-4 V joints prepared via CMT + P welding. J Manuf Process 52:193–202
10. Wanjara P, Jahazi M (2005) Linear friction welding of Ti-6Al-4V: processing, microstructure, and mechanical-property inter-relationships. Metall Mater Trans A 36(8):2149–2164
11. Zhou J, Tsai HL (2007) Porosity formation and prevention in pulsed laser welding. J Heat Transfer 129(8):1014–1024
12. Chmelíčková H, Šebestová H. Pulsed laser welding
13. Baruah M, Bag S (2017) Influence of pulsation in thermo-mechanical analysis on laser micro-welding of Ti6Al4V alloy. Opt Laser Technol 90:40–51
14. Kumar B, Bag S, Paul CP, Das CR, Ravikumar R, Bindra KS (2020) Influence of the mode of laser welding parameters on microstructural morphology in thin sheet Ti_6Al_4V alloy. Opt Laser Technol 131:106456
15. Gao XL, Zhang LJ, Liu J, Zhang JX (2014) Porosity and microstructure in pulsed Nd:YAG laser welded Ti_6Al_4V sheet. J Mater Process Technol 214(7):1316–1325
16. Blackburn J (2012) Laser welding of metals for aerospace and other applications. In: Welding and joining of aerospace materials, pp 75–108. Woodhead Publishing
17. Kumar B, Bag S (2019) Phase transformation effect in distortion and residual stress of thin-sheet laser welded Ti-alloy. Opt Lasers Eng 122:209–224
18. Trivedi A, Suman A, De A (2006) Integrating finite element based heat transfer analysis with multivariate optimization for efficient weld pool modeling. ISIJ Int 46(2):267–275
19. Bag S, De A (2008) Development of a three-dimensional heat-transfer model for the gas tungsten arc welding process using the finite element method coupled with a genetic algorithm–based identification of uncertain input parameters. Metall Mater Trans A 39 (11):2698–2710
20. Kumar B, Nawani R, Bag S (2020) Effect of mechanical constraints on thermo-mechanical behaviour of laser-welded dissimilar joints. In: Advances in simulation, product design and development, pp 107–119. Springer, Singapore
21. Zhan X, Mi G, Zhang Q, Wei Y, Ou W (2017) The hourglass-like heat source model and its application for laser beam welding of 6 mm thickness 1060 steel. Int J Adv Manuf Technol 88 (9–12):2537–2546
22. Yadaiah N, Bag S (2014) Development of egg-configuration heat source model in numerical simulation of autogenous fusion welding process. Int J Therm Sci 86:125–138
23. Kumar B, Kebede D, Bag S (2018) Microstructure evolution in thin sheet laser welding of titanium alloy. Int J Mechatro Manuf Syst 11(2–3):203–229
24. Baruah M, Bag S (2016) Microstructural influence on mechanical properties in plasma microwelding of Ti_6Al_4V alloy. J Mater Eng Perform 25(11):4718–4728
25. Sahu AK, Bag S (2020) Probe pulse conditions and solidification parameters for the dissimilar welding of Inconel 718 and AISI 316L stainless steel. Metall Mater Trans A, 1–17

Characterization as a Tool for Failure Investigation: Tata Steel Experience

Kaushal Kishore, Arthita Dey, Sanjay Kushwaha, Suman Mukhopadhyay, Manashi Adhikary, and Anup Kumar

1 Introduction

Failure is often defined as the inability of a component or a system to efficiently and effectively perform its intended function(s). Failures can be relatively simpler or extremely complex owing to the extent of availability of background information, operating parameters, state of the failed component, environmental conditions, etc. Over the years, analysis of some of the historic failures such as de Havilland's comet [1, 2], sinking of Titanic [3], Great Molasses Flood of Boston [4, 5] and the Chernobyl nuclear explosion in Ukraine [6] have shaped the way critical failures are analysed even today and recommendations are made to improve the reliability and durability at the component, assembly and system levels. In the past few decades, many guidelines are set in the form of handbooks for failure analysis of components illustrated in the form of numerous case studies [7–9]. However, in some way or the other, failures vary from industry-to-industry, even if the basic science remains the same. Integrated steel plants consist of components experiencing a wide range of stresses and environment varying from temperature up to 1600 °C in Linz-Downawitz convertor [10], severe thermal cycling in corrosive atmosphere like the hood tubes of steelmaking vessel and boiler tubes of a power house [11, 12], shafts, pins, gear boxes and bearings operating at or around room temperatures but under different state of mechanical stresses [13–15], highly corrosive atmosphere of coke and by-product recovery plants [16] to rolling mills where situations are entirely different and dynamic [17]. Therefore, the present work is aimed at presenting a methodology for failure investigations of components in steel plant.

K. Kishore (✉) · A. Dey · S. Kushwaha · S. Mukhopadhyay · M. Adhikary · A. Kumar
Scientific Services, Tata Steel Limited, Jamshedpur-831001, India
e-mail: kaushal.kishore3@tatasteel.com

S. Bag et al. (eds.), *Next Generation Materials and Processing Technologies*, Springer Proceedings in Materials 9, https://doi.org/10.1007/978-981-16-0182-8_19

2 Methodology for Failure Investigations

Figure 1 shows the common chronology of steps adopted when a critical component fails in a steel plant.

The first important step in failure investigation which can easily be determined during visual observation is the identification of the broad category of failure, that is, whether that component fractured, worn out or corroded. This would direct the investigation for collection of operating parameters and the use of characterization techniques. It is essential to find out the service life that component lasted before the failure. In certain cases, it is observed that components are used beyond their design life and as such failure is not surprising. However, if the failure is premature, it needs to be viewed in a different light. On-site observations including the location

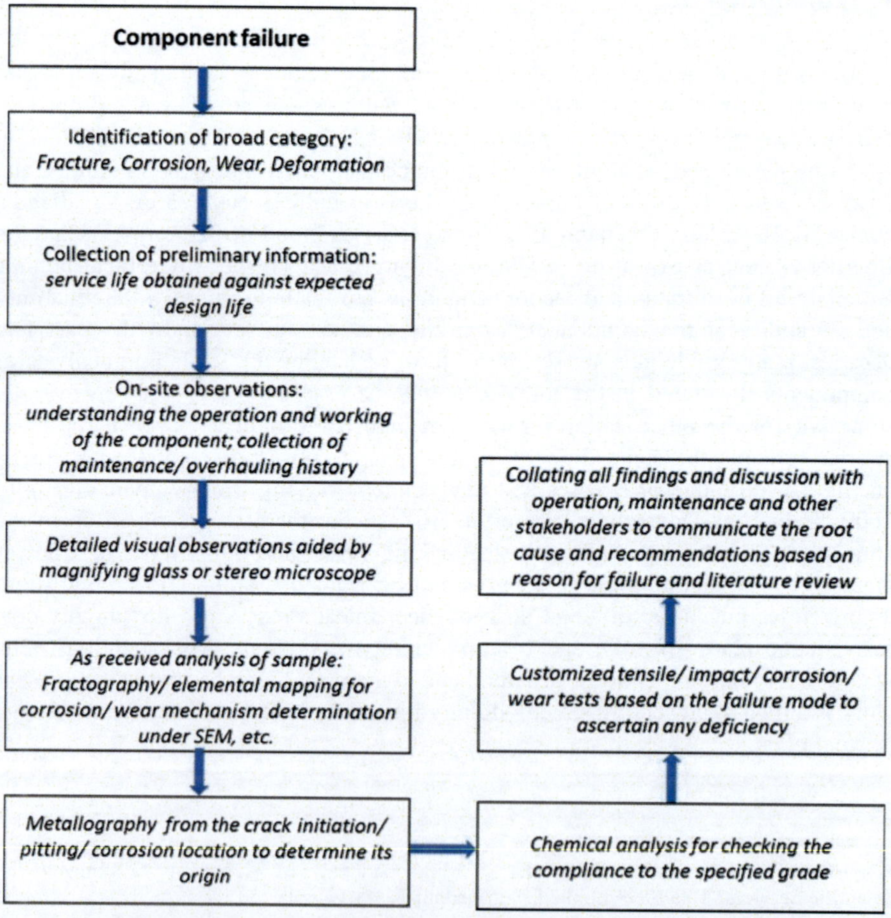

Fig. 1 Typical chronology followed for failure analysis of a component in a steel plant

of failed component and its detailed function in the system or assembly help in understanding the type of stresses and environment it could experience during its service life. If the component fractured, then it becomes essential to look at the load or current data (if available) over its service life and check if prior to the failure, there was an overload. An operational overload can inevitably result in brittle or ductile fracture depending up on the state of stress, inherent ductility of the material, strain rate and operating temperature [18]. However, if the loading history does not show any signature of overload and the component still fractured, then it is likely that it is a progressive failure, which can be low or high cycle fatigue, stress corrosion cracking, etc. For corrosion and wear related failures, it is essential to analyse the environment, fluid or mass flow rate, location of failure with respect to any abrupt cross-sectional change, contaminants like sulphur, chlorine, etc., which can accelerate the rate of corrosion in the presence of moisture, pH of the medium and so on [19]. It should be noted that special attention should be paid on overhauling history of the failed component. Faulty assembly in terms of greater clearance, misalignment, poor lubrication, hammering, etc., have led to failures which are otherwise difficult to decode without an overhauling history.

Once adequate detail of operational data, maintenance practices and history is collected, careful visual observations are made using unaided eyes as well as with the help of magnifying glass or stereomicroscope. This is intended for macrofractography and identification the of damage (crack, corrosion pit or localized wear) initiation for subsequent investigations. In case of fracture, macroscopic features such as beach marks, ratchet marks and colour contrast would indicate fatigue mode of failure. Visual image of a brittle fracture would show fresh and shiny appearance with no gross plastic deformation. These are often characterized by chevron marks whose origin would also indicate the crack initiation point. In contrast, a ductile fracture would be characterized by dull and woody appearance with gross plastic deformation.

Visual observation is often followed by fractography (in case of fracture) under scanning electron microscope. It can confirm the mode of fracture: ductile, brittle and fatigue. Further detail about the size, shape and orientation of dimples would throw light on the loading condition and relative ductility of the material. Likewise, the presence of cleavage planes or rock candy structure will suggest whether the brittle fracture is transgranular or intergranular. Fatigue fracture surfaces, if preserved are usually associated with striation marks. However, an absence of striation marks should not lead to a conclusion that the component has not failed under fatigue mode. Less ductile material with limited slip systems often do not leave striation marks when cracks propagate under cyclic loading. Furthermore, the corrosive or oxidizing environment under which the component is placed may leave a layer of corrosion or oxidation products over the fracture surface which may not reveal striations even after extensive cleaning. In such cases, the role of visual and stereoscopic observations to detect signatures of fatigue such as ratchet marks, beach marks and multiple planes of crack propagation becomes pivotal. It should be noted that in addition to ascertaining the failure mode and mechanism, carefully fractography under scanning electron microscope can also suggest reason for crack initiation such as the presence of hard inclusions or entrapment like mould powder near the crack initiation site. The rate of fatigue crack growth and the number of stress reversals can also be estimated and its detail is discussed elsewhere [13, 20].

Once the mode of failure and possible crack origin is established, metallography is carried out to observe general steel cleanliness and inclusion ratings. Inclusions as the root cause of failure of components is discussed in detail [21]. Metallography at the crack initiation location also reveals if the presence of decarburization, pre-existing cracks, hard phases like untempered martensite or cementite network contributed to the failure. Furthermore, the adequacy of heat treatment (quenching and tempering, normalizing or temper embrittlement) can be determined by microstructural analysis in conjugation with mechanical properties evaluation like hardness, tensile and impact tests. Likewise, in the case of corrosion and related failures, whether there is severe crack branching or unrelated blind cracks would indicate for stress corrosion cracking and hydrogen induced cracking, respectively [22].

Once all the information is collected, it is important to link the sequence of events to establish the mode of failure, its reason and corrective measures. Here, few case studies of failure analysis in a steel plant are presented with a thrust on the importance of different characterization techniques.

3 Results and Discussion (Case Studies)

3.1 Failure Analysis of 2 Ton Electric Fork Lift in a Cold Rolling Mill

2 Ton electric fork lift model EV-20 is used to run in ARP section at CRM as shown in Fig. 2a. It is used to handle red oxide gunny bag. Approximately 32 gunny bags are handled each day by the fork lift. The fork lift was under operation for last 20 years and was dedicated for this operation. As can be seen from Fig. 2b–c, there were two roller plates which were welded to the frame of the machine at two different locations. Further two forks mounted on the frame handle the gunny bags. Thus, it formed the whole roller bracket. Among these two roller plates, one plate broke from both weld locations as shown in Fig. 2d–e. Discussion with plant suggested no evidence of overloading.

Fracture surfaces showed distinct and well-defined beach marks confirming fatigue mode of failure (Ref. Fig. 3). This is consistent with the fact that no overload was recorded as per the plant professionals. The origin of the fatigue crack as found out by tracing the centre of the beach marks was around the root of the welding.

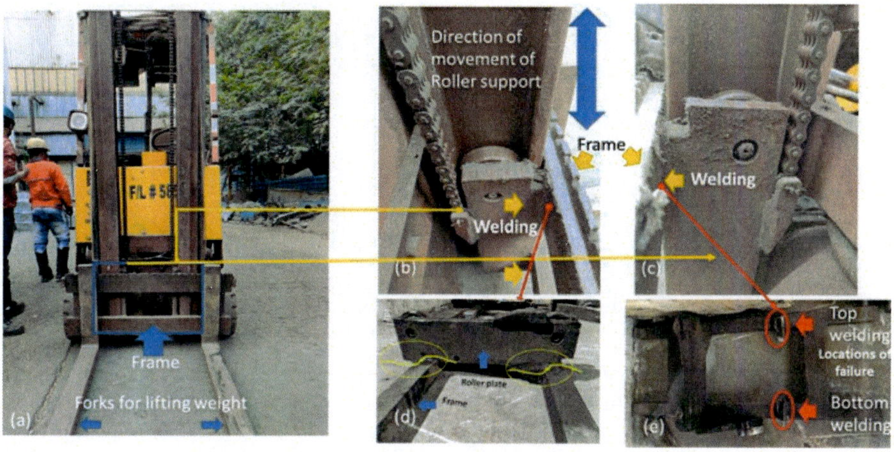

Fig. 2 On-site visual images showing the application of the fork lift and the location of failure from the weld

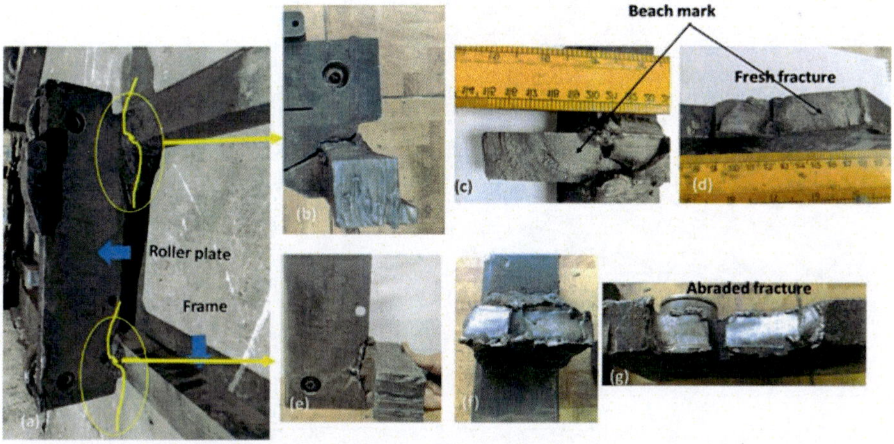

Fig. 3 **a** Failed roller plate; **b–d** closer views of the fracture surface or top welding locations; **e–g** closer view of the fracture surface of bottom welding location

SEM fractography shown in Fig. 4a revealed the presence of weld pool at the fatigue crack initiation location. This is followed by distinct striation marks in crack propagation region confirming progressive failure by fatigue mechanism.

Hardness survey across the weld revealed variation (Fig. 5) of weld hardness from roller plate, which seemed to cause the initiation of fatigue crack from the weld and propagation of the same in the direction of roller plate finally resulting in the failure of the same. There was a weld undercut which would have led to ineffective stress transfer and can contribute to the fatigue failure. No crack was

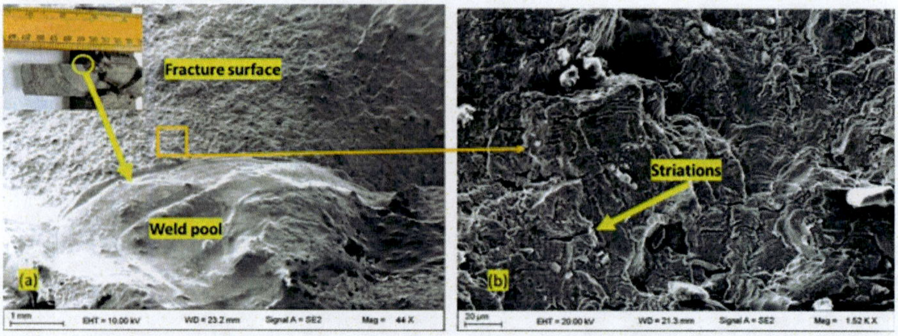

Fig. 4 SEM fractography near the fatigue initiation zone: **a** weld pool present near the location of fatigue crack initiation, **b** higher magnification image showing striation marks in the crack propagation region

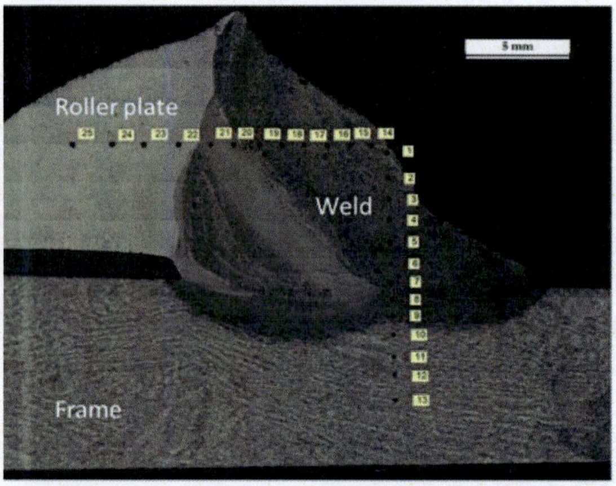

		Location	Hardness in HV/10
Weld to Frame	WELD	1	204
		2	216
		3	218
		4	215
		5	222
		6	223
		7	238
		8	301
		9	256
	HAZ	10	223
	MATRIX	11	217
		12	221
		13	218
Weld to roller plate	WELD	14	212
		15	215
		16	221
		17	206
		18	198
		19	193
		20	158
		21	132
	HAZ	22	125
	MATRIX	23	127
		24	126
		25	129

Fig. 5 Macrostructure of the weld and corresponding hardness values

observed to propagate from weld towards frame as hardness of the weld and frame material were compatible to each other. It was also important to note that the fork gave a life of 20 years, this indicates that service stress was very low, that is why the plate sustained the long life in spite of being presence of inherent weld notch and undercut.

Based on these characterization techniques, it can therefore be concluded that the fork lift failed from the welded plates in progressive fatigue mode as revealed by beach marks and striations. Crack initiation occurred at the weld pool and the presence of undercut there would have aggravated the situation.

Following recommendations are provided for extension of service life:

(i) Ultrasonic testing of the weld joints before the installation and during service life at predefined intervals.
(ii) Only skilled welder should be allowed for fabrication of such critical components as it can lead to serious safety issues.

3.2 *Failure Analysis of Bend Pulley Shaft of INBA Slag Granulation System of a Blast Furnace*

Granulation of blast furnace slag is an important step in converting otherwise waste into useful resource and raw material for cement industries. Conveyor belt carries the granulated slag to the silos. A bend pulley of the conveyor belt failed after 2.5 years as against a design life of 5 years. Visual image of the failed shaft is shown in Fig. 6a. It failed from the step portion. Fracture surface shown in Fig. 6b, c revealed multiple ratchet marks all along the circumference of the shaft. This indicates that the shaft failed in torsional fatigue mode which is consistent with the loading condition of the shaft.

In order to ascertain the reason of premature fatigue failure, detailed metallurgical analysis was carried out. Table 1 shows the summary of chemical analysis of the shaft relative to the specified steel grade.

Chemistry of the failed shaft showed higher carbon and phosphorus than the specification. This means there was some issue with oxidation potential or blowing time during the steelmaking stage. The presence of higher carbon and phosphorus can induce brittleness in the material.

Figure 7a, b shows the unetched micrographs of the failed shaft. Multiple non-metallic inclusions were observed near the fracture contour. Such severe sulphide inclusions can contribute to fatigue crack initiation. Furthermore, inclusion rating of the failed shaft summarized in Table 2 shows that sulphide inclusions of 2.5 severity was present in thin series, whereas beyond ratable limit due to excessive thickness was present in thick series. This confirms improper steel cleanliness. Etched microstructures revealed the presence of pro-eutectoid α-ferrite and pearlite phases. It appears that the crack as indicated by the fracture contour propagated along the pro-eutectoid α-ferrite/pearlite interface as shown in Fig. 7c, d. It is well established and demonstrated that for rotating components under cyclic loading condition, an adequately quenched and tempered martensitic microstructure is superior compared to the ferrite–pearlite microstructure in terms of fatigue life.

Based on these metallurgical characterizations, it can be concluded that shaft failed in torsional fatigue mode due to a combination of metallurgical deficiencies. Higher phosphorus, extremely severe sulphide inclusions and the presence of ferrite–pearlite microstructure are all undesirable from the standpoint of fatigue failure.

Following recommendations are provided for extension of service life:

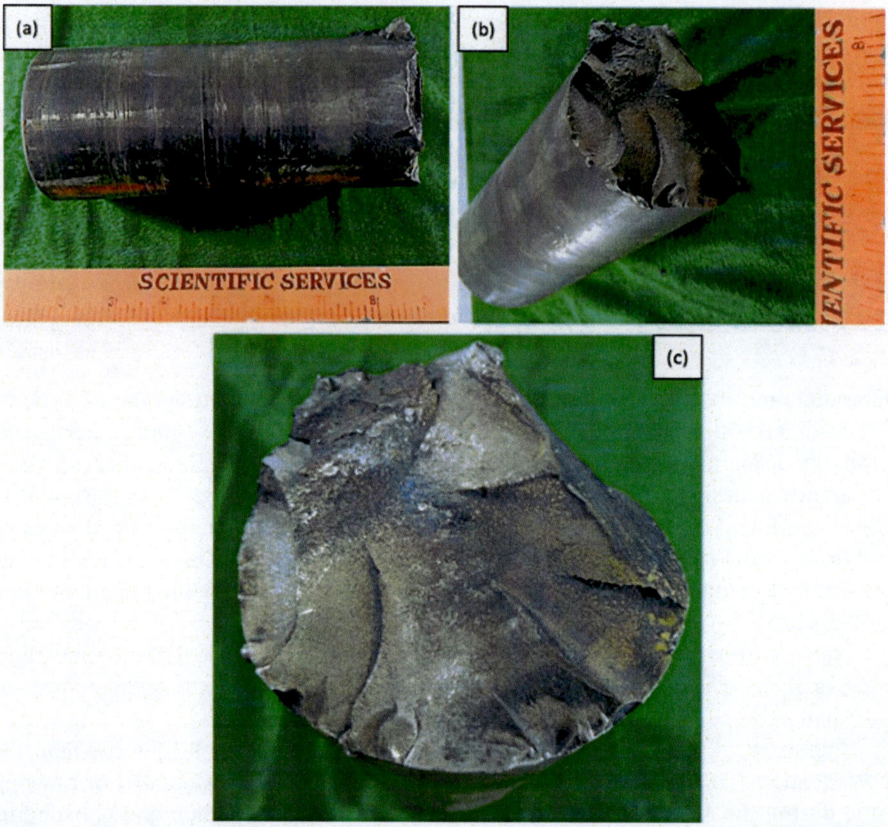

Fig. 6 Visual images of failed bend pulley: **a** bend pulley along its length, **b** angular view of the shaft, **c** fracture surface revealing multiple ratchet marks

Table 1 Summary of chemical analysis of the failed shaft

Sample	C	Mn	S	P	Si
Shaft	0.48	0.67	0.038	0.09	0.268
Spec IS 1570 C40 steel (As per drawing)	0.37–0.44	0.50–0.80	0.045 max	0.045 max	0.40 max

(i) Better steel cleanliness is recommended to reduce the length of sulphide inclusions. These inclusions act as stress raisers and reduce the fatigue life of components.
(ii) Adequate quenching and tempering heat treatment should be carried out for the shaft as tempered martensitic microstructure have superior fatigue strength compared to ferrite–pearlite microstructure.
(iii) Shot peening may be carried out to induce compressive residual stresses to further improve the fatigue life.

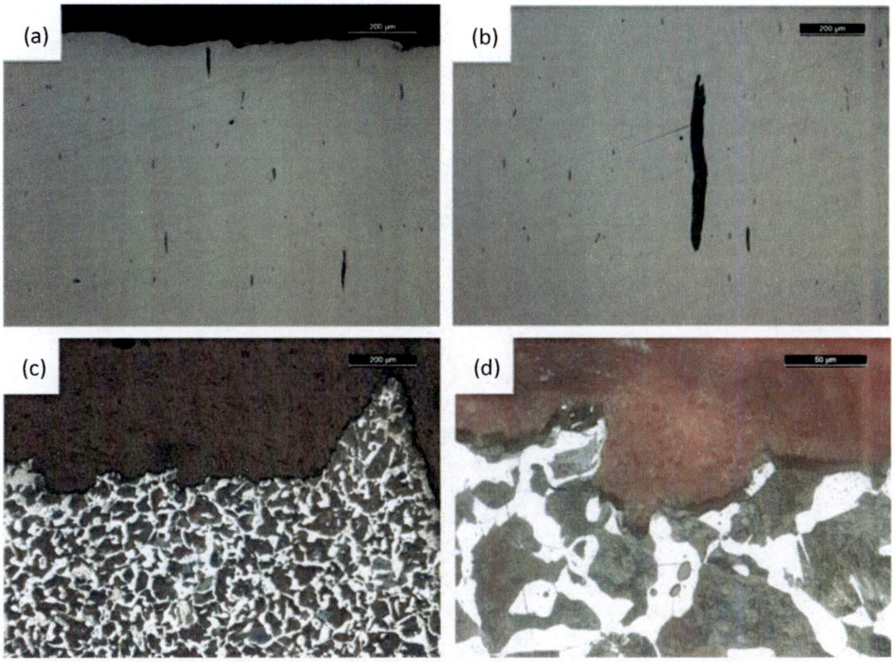

Fig. 7 Optical micrographs showing **a** unetched microstructure near the fracture contour, **b** unetched micrograph at the quarter-thickness position, **c** etched microstructure near the fracture contour, **d** higher magnification etched micrograph

Table 2 Inclusion rating as per ASTM E-45

A (Thin/heavy)	B (Thin/heavy)	C (Thin/heavy)	D (Thin/heavy)
3.0/*	0.0/0.0	0.0/0.0	0.5/0.0

*Not ratable due to excessive length or thickness

3.3 Failure Analysis of Heat Exchanger of a Batch Annealing Furnace

During cooling cycle of the batch annealing furnace (BAF), H_2 gas from BAF at a temperature of around 600 °C is passed through the heat exchanger (HE) to cool down around 50 °C. Industrial cooling water (ICW) is used for cooling the gas. Thereafter, it is recirculated to BAF through blower. The schematic of heat exchanger as used in plant is shown in Fig. 8a, b; the cover under which the heat exchanger is put. As shown in Fig. 8c, H_2 gas is coming from BAF through the ducts marked by orange coloured arrows and going to the blower through the tube as shown in Fig. 8d. Water flowing through the tube is marked by blue coloured arrows. As per the drawing, there were 207 tubes of SA 179 steel grade covered by fins. Fins are used for better heat transfer efficiency. All the tubes'

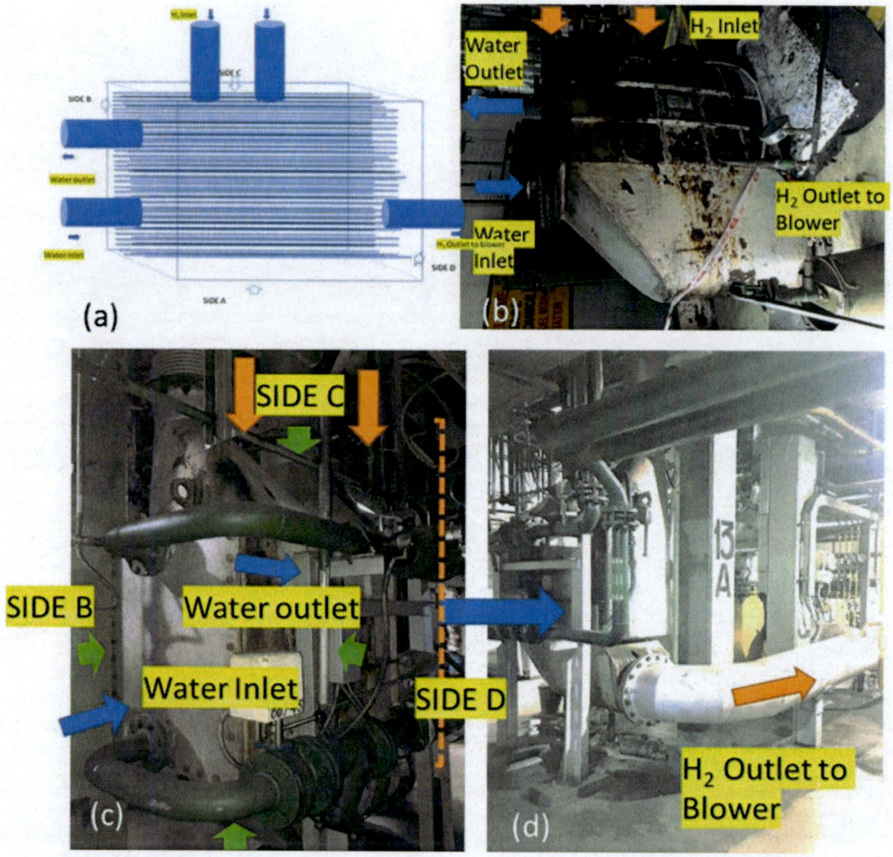

Fig. 8 **a** Schematic of heat exchanger, **b** cover of the heat exchanger; **c–d** different water line arrangement and H_2 outlet at the site

specification covers seamless cold drawn low carbon steel pipe. Both fin and tubes are supplied in hot dip galvanized. Outside surface of the tubes are expected to be in sand blasted condition and primer zinc silicate dry coating layer to be applied. It was also mentioned in the drawing that two coats of aluminium paint suitable for 500 °C must be applied on the outer surface and inside should thoroughly be cleaned and rust preventive oil shall be applied. Heat exchanger of BAF bypass cooling got punctured during annealing cycle. The tube was on side A facing the bottom of heat exchanger as shown in Fig. 9a, b. In addition to the leakage, the tubes at side C facing H_2 gas were found to be white rusted as depicted in Fig. 9d. There were several other tubes which were found rusted on different sides of the heat exchanger (Ref. Fig. 9 c and e). Several samples from this heat exchanger were collected for metallurgical analysis as shown in Fig. 10. The tube which was leaked was cut and found chocked or blocked (Ref. Fig. 11). The same also revealed cracking of tube due to corrosion from inner side (Fig. 12).

Fig. 9 Failed heat exchanger tube assembly from different sides **a** and **b** showing SIDE A; **c**, **d** and **e** showing SIDE B, C and D, respectively; **f** shows the arrangement of water distribution

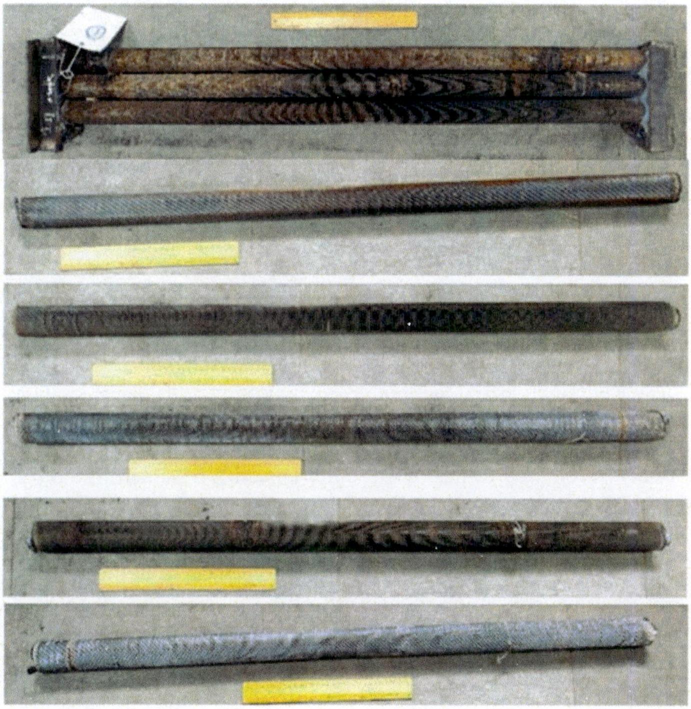

Fig. 10 Sample referred for failure analysis

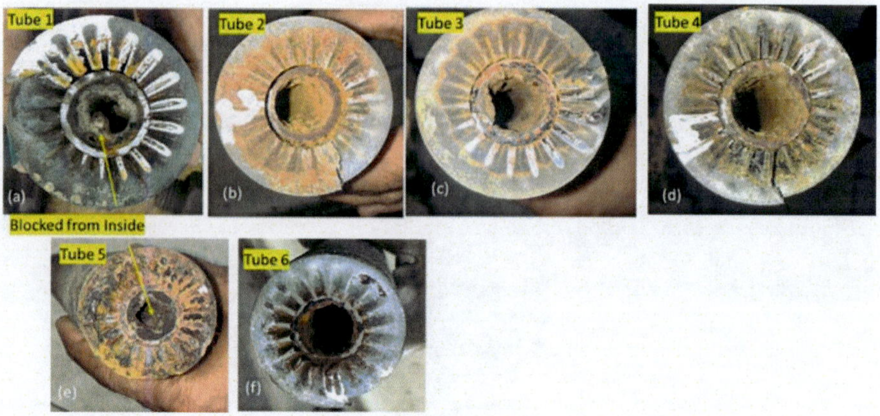

Fig. 11 Tubes were chocked/blocked

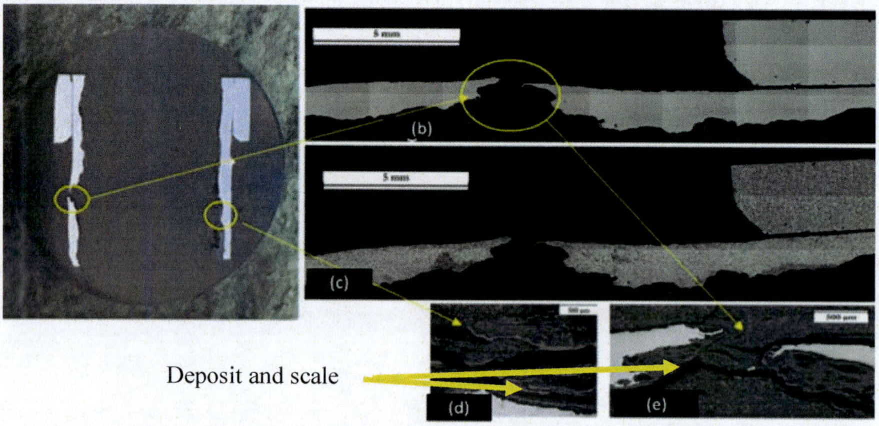

Fig. 12 Microstructure from leaked location revealing deposit at the defect location

Similar observation was found in the other rusted tubes, with the only difference that there was deposition inside, but those tubes were not fully blocked. This indicated that deposition happened over a period of time. Chemical analysis of the tube and fins conformed to the specification. White rusting confirmed the presence of galvanized coating. But since most of the tube was corroded, therefore coating was not visible. EDS analysis (Figs. 13 and 14) and chemical analysis of deposits revealed presence of oxides of Ca, Si, Mg, Na, Al along with Cl in certain locations. The water analysis also confirmed the source of these elements to be from the water in the circuit. From the above analysis, it seems that the tubes are failing due to deposit corrosion. Literature survey shows that, deposit corrosion is typically related to minute volumes of inactive solution from lap joints, surface deposits or

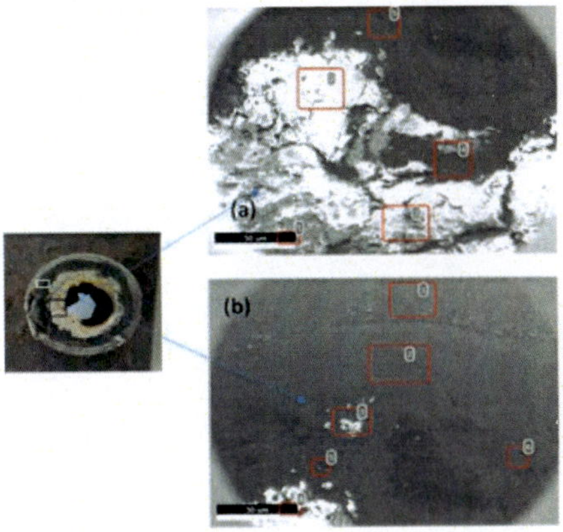

	O	Na	Mg	Al	Si	P	Nb	Ca	Fe	S
1	49.12	2.5	19.76	0.76	12.91	2.2	3.06	9.68		
2	50.69	1.45	15.5	0.44	11.02	2.07	2.2	16.63		
3	47.56	2.28	16.04	0.78	10.37		3.97	14.89	4.11	
4	51.7	1	18.41	0.18	11.46	0.58	0.61	14.87	1.19	
5	38.67	1.49	2.35	0.43	2.62	1.06	2.06	20.12	29.52	1.68

	O	Si	Cl	Fe	Al	Mg	P	Nb	S	Ca	Na
1	9.76	0.45	1.17	88.62							
2	26.52	1.82		71.21	0.45						
3	45.84	5.06		4.66	0.66	5.26	1.47	2.63	10.44	23.98	
4	47.76	11.56		2.55	0.82	14.45	2.3	2.3		16.25	1.99
5	45.76	7.02			1.14	7.01	2.99		10.02	23.52	2.53
6	28.3	0.58		64.29	0.15		0.54			3.26	2.89

Fig. 13 EDS analysis of the deposit

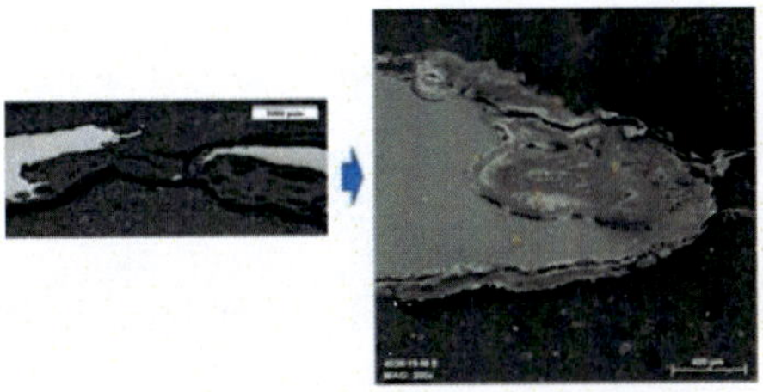

Normalized mass concentration [%]

Spectrum	Oxygen	Silicon	Phosphorus	Sulfur	Calcium	Chromium	Manganese	Iron	Zinc
1	26.19	0.97	0.25	0.44	0.09	0.12		71.54	0.41
2	34.17	4.97	0.45	0.35	0.38	0.25	0.15	59.27	
3	41.15	3.13	0.34	1.06	0.17	0.35	0.20	53.61	
4	29.69	0.60		0.50		0.22	0.08	68.90	
5		0.26					0.36	99.38	

Fig. 14 EDS analysis of the leakage location

crevice. Deposit corrosion is also known as crevice or under-deposit corrosion [23]. This type of corrosion is one of the most severe types of corrosion, especially affecting piping systems. It is very localized and aggressive, leading to deep penetration of the metal surface with less corrosion involved in the adjacent areas. Surface deposits are the major reason why corrosion factors and other forms of initiating mechanisms choose individual sites [24]. A deposit over a bare unprotected metal spot worsens and accelerates the corrosion. The deposit restricts the free access of oxygen [air] under the deposit and creates differential aeration corrosion cells which multiplies corrosion rate as shown schematically in Fig. 15.

Following recommendations were given to avoid premature failure:

(i) Cleaning of tube after a certain interval by purging water/nitrogen gas at high pressure could be beneficial to remove blockage or deposit. As the failure is attributed to under-deposit corrosion, the best preventive measure is to avoid the accumulation of corrosion products.

(ii) Pulsed injections can also be made by introducing a gas (compressed air or nitrogen) at very high speed (3 ms^{-1}) periodically at short intervals of time (5, 10 and 60 min). The effect is to increase the turbulent regime, creating

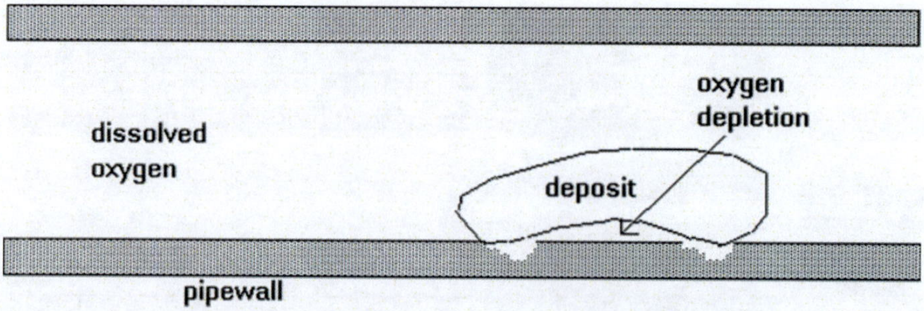

Fig. 15 Schematic explaining the mechanism of under-deposit corrosion

pressure fluctuations inside the tube and higher shear stresses on the surface. As a consequence, a greater detachment of deposit can be achieved than that obtained by simply increasing the speed of water flow.

4 Concluding Remarks

Careful sequence of characterization of failed components when combines with adequate background information about operation and maintenance practice helps in root cause analysis of critical components in steel plants. Suggested methodology can be helpful guide for future failure analysis and life cycle extension.

References

1. Cohen, BL, Farren W, Duncan WS, Wheeler AH (1955) Report of the court of Inquiry into the accidents to comet G-AL YP on 10 January, 1954 and Comet G-AL Y Y on 8 April, 1954. HMSO, London
2. Barter S, Sharp PK, Clark G (1994) Eng Fail Anal 1:255–266
3. Felkins K, Leighly HP, Jr, Jankovic A (1998) The royal mail ship titanic: did a metallurgical failure cause a night to remember? JOM, 50(1):12–18
4. https://www.bostonglobe.com/metro/2015/01/14/nearly-century-later-new-insight-into-cause-great-molasses-flood/CNqLYc0T58kNo3MxP872iM/story.html. Last Accessed on 14 Aug 2020
5. Erik H (2010) The boston molasses disaster: causes of the molasses tank explosion. Bright Hub. Last Accessed on 10 Aug 2020)
6. https://ukrns.org/en/publications/item/1272-low-safety-culture-of-the-entire-system-the-cause-of-the-chernobyl-accident. Last Accessed on 14 Aug 2020
7. Failure Analysis and Prevention, ASM Handbook, Volume 11
8. Affonsso L (2007) Machinery failure analysis handbook
9. Makhlouf A, Aliofkhazraei M (2018) Handbook of materials failure analysis

10. Kishore K, Singh R, Nirmal B, Yadav A, Adhikary M, Kumar A (2020) In situ failure investigation of weld cracking in a linz-donawitz vessel. J Fail Anal Prev 20(2):555–562
11. Kishore K et al (2019) Movable hood tube in LD convertor: failure analysis and coating solution. Eng Fail Anal 105:25–39
12. Pal U et al (2019) Failure analysis of boiler economizer tubes at power house. Eng Fail Anal 104:1203–1210
13. Kishore K, Adhikary M (2019) Metallurgical investigation and life cycle assessment of a piston rod of thin slab caster. J Fail Anal Prev 19(5):1407–1419
14. Kishore K et al (2020) Failure analysis of a gearbox of a conveyor belt. J Fail Anal Prev 1–7
15. Kishore K et al (2018) Metallurgical analysis of premature failure of a phosphor bronze worm wheel. J Fail Anal Prev 18(6):1321–1326
16. Mukhopadhyay S et al (2015) Failure analysis of the liner plate of CDQ chamber made of AISI 310. J Fail Anal Prev 15(5)
17. Kishore K et al (2019) Root cause failure analysis of pinch roll bearing at hot strip mill. J Fail Anal Prev 19(1):219–229
18. Dieter GE. Mechanical metallurgy, 3rd edn
19. Fontana MG (1986) Corrosion engineering. Mcgraw-Hill, 3rd edn
20. DeVries PH, Ruth KT, Dennies DP (2010) Counting on fatigue: striations and their measure. J Fail Anal Prev 10:120–137
21. Zerbst U et al (2019) Defects as a root cause of fatigue failure of metallic components. II: non-metallic inclusions. Eng Fail Anal 98:228–239
22. Raja VS (2011) Tetsuo, stress corrosion cracking theory and practice. Woodhead Publishing Limited
23. https://www.corrosionpedia.com/definition/375/deposit-corrosion. Last Accessed on 10 Aug 2020
24. Corrosion behavior of pipeline carbon steel under different iron oxide deposits in the district heating system Yong-Sang Kim and Jung-Gu Kim; MDPI, Received: 21 April 2017; Accepted: 17 May 2017; Published: 19 May 2017

Assessment of Bio-dielectric *Calophyllum inophyllum* (Polanga) Oil in Electro-discharge Machining: A Step Toward Sustainable Machining

Bikash Chandra Behera, Matruprasad Rout, and Arpan Kumar Mondal

1 Introduction

The electro-discharge machining (EDM) process is a non-conventional machining process in which electrically conductive material is machined. The mechanism of material removal in EDM is due to the electron bombardment on the workpiece in repeated intervals. In this process, material is machined in the form of tiny debris. This process is mainly used to manufacture forging dies, press tool and molds for injection molding. It is also employed for producing intricate and irregular shaped profiles. Small holes on carbide and hardened steel can be machined by EDM. There are different varieties of EDM processes such as die shrink EDM, micro-EDM and wire EDM [1]. The hydrocarbon-based oils are generally used for the die shrink EDM, but in wire EDM and micro-EDM deionized water is used. These dielectric fluids work as a medium in which controlled electrical discharge occurred. It also helps to carry away debris and heat generated by the discharge. The dielectric fluid significantly affects the productivity, quality and cost of the machining and when hydrocarbon-based dielectric fluid is used health hazards and environmental impact becomes two other important aspects for selecting the dielectric fluid [2].

B. C. Behera
Department of Mechanical Engineering, C. V. Raman Global University, Bhubaneswar 752054, India

M. Rout (✉)
Department of Production Engineering, National Institute of Technology, Tiruchirappalli 620015, India
e-mail: matruprasadrout87@gmail.com

A. K. Mondal
Department of Mechanical Engineering, National Institute of Technical Teachers Training and Research, Kolkata 700106, India
e-mail: arpan@nitttrkol.ac.in

S. Bag et al. (eds.), *Next Generation Materials and Processing Technologies*, Springer Proceedings in Materials 9, https://doi.org/10.1007/978-981-16-0182-8_20

The manufacturing industries have been considered as one of the main sources of environmental hazards [3]. The minimization of environmental hazards is the prime focus after implementation of the ISO 14000 environmental management system standard. The main requirements of ISO 14000 standard are identifying the source of pollution in the manufacturing industries. Hence, today's researchers are thinking about eco-friendly manufacturing, i.e., green manufacturing. In EDM process, hydrocarbon-based dielectric fluid is the main source of pollution. The complete replacement of the hydrocarbon-based dielectric fluid has not yet been developed. Zia et al. [4] reported that the environmental concern is due to the emission from and disposal of the hydrocarbon dielectric. However, the main health concern is connected to fire explosion, inhalation of fume and contamination of dielectric on the skin. Kou and Han [5] experimentally investigated the EDM of Ti–6Al–4V using water-based dielectric for their study and reported that the environmental pollution can be minimized by using water-based dielectric. The environmental pollution and health hazards can also be minimized by using vegetable oil as a dielectric fluid. Higher operational safety due to low risk of vegetable-based dielectric catching fire makes it as safer fluids to use. The hydrocarbon-based dielectric like kerosene has higher emission of pollutant into atmosphere making it less environment friendly, but vegetable oil emits less pollutant and hence environment friendly. As discussed, vegetable oil is environment friendly, so they provide better personnel health. The vegetable oils have low manufacturing cost and also have higher breakdown voltage, hence more efficient than conventional oils. Valaki et al. [6] reported the practical viability of jatropha curcas oil-based bio-dielectric for EDM. They suggested that the jatropha curcas oil can be used as an alternative to hydrocarbon-based bio-dielectric. Das et al. [7] experimentally investigated the technical feasibility of jatropha, canola and neem oil as the bio-dielectric and reported that all the bio-dielectrics ensure better MRR and better surface finish. However, jatropha and canola are remarkable. Mishra and Routara [8] have conducted EDM of EN 24. They used polanga bio-dielectric (PBD) and hydrocarbon-based dielectric for their study and reported that the performance of PBD is remarkable in terms of MRR and eco-friendly aspect. Das et al. [9] experimentally studied efficacy of neem oil as a dielectric with EDM of Ti–6Al–4 V. They reported an improvement in material removal rate by 22 % with neem oil as the dielectric. The better surface integrity is also achieved in neem dielectric environment.

Though works have been reported on PBD as dielectric, more investigation is needed to establish PBD as a bio-dielectric. In this study, EDM of P20 steel is examined using PBD and kerosene as dielectric fluids.

2 Preparation of Bio-dielectric

The PBD has been chosen as the bio-dielectric fluid to study the EDM process. Since the dielectric fluids have different properties, their operating efficiency differs from each other. The raw polanga oil might get pre-breakdown during EDM process as it has high viscosity. So, to make the polanga oil suitable for the dielectric fluid, multiple refining stages have been followed to decrease the viscosity and achieve other desirable properties (Fig. 1). In general, three stages have been used to minimize the viscosity of the raw polanga oil. These steps are degumming and esterification which is followed by transesterification. After transesterification water washing, heating and steering have been conducted to convert it to the desired dielectric fluid. The characterization of the dielectric was done based on the ASTM standard. The raw polanga oil and the processed polanga oil, i.e., PBD are shown in Fig. 2. The different fluid properties, viz. viscosity, thermal conductivity, specific heat, etc., have been recorded for this study.

3 Experimental Methods

The experiments have been performed by using a CNC electric discharge machine classified as (die-sinking type) MIC 432CS (Make: ECOWIN, Taiwan) machine whose polarization on the electrode is considered as negative and the workpiece as positive. The schematic diagram of the setup is shown in Fig. 3. The process parameters used for the experiments are given in Table 1. Two types of dielectrics, viz. kerosene and PBD are used to perform the experiments. Different properties of

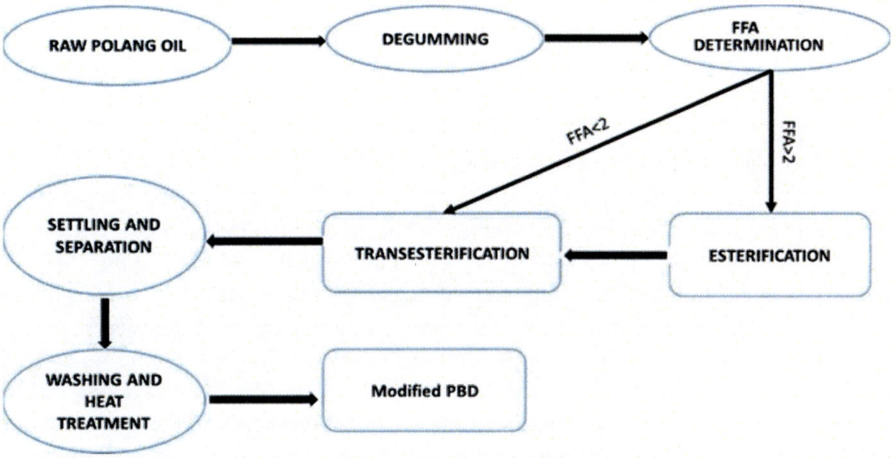

Fig. 1 Various purifying process used to produce suitable dielectric from raw polanga oil (FFA stands for free fatty acid test)

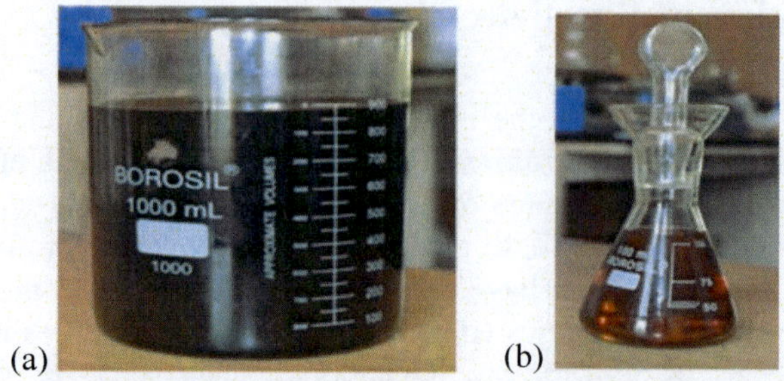

Fig. 2 **a** Raw polanga oil and **b** processed polanga oil

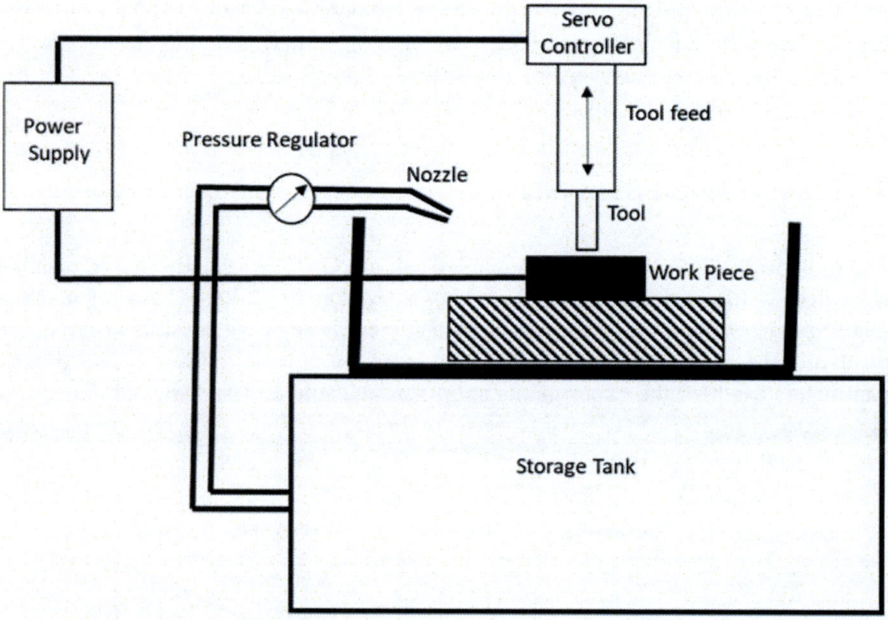

Fig. 3 Line diagram of the experimental setup

these two dielectrics are given in Table 2. The material removal rate (MRR) and surface roughness (R_a) have been recorded to analyze the efficacy of the PBD. In this experiment, P20 plastic mold steel of size 220 mm × 203 mm × 15 mm has been used. The physical properties and chemical composition of the workpiece are presented in Tables 3 and 4, respectively. The workpieces before and after machining are shown in Fig. 4a, b, respectively. The 99% pure copper rod of 20 mm diameter (Fig. 5a) has been chosen to produce the tool of required dimension (60 mm length). Each piece underwent turning and facing operation in a lathe machine, till the required dimensions were obtained. A fresh tool (Fig. 5b) is used for each experiments.

Table 1 Experiment process parameters and their levels

Process parameters	Levels
Pulse current (A)	3, 6, 9, 12, 15, and 18
Pulse ON time (μs)	21, 50, 100, 200, 400, and 600
Pulse OFF/Interval time (μs)	6, 11, 20, 30, 40, and 75
Polarity	Positive (electrode +ve)

Table 2 Fluid properties of dielectrics kerosene and PBD

Properties	Kerosene [6]	Polanga bio-dielectric (PBD)
Density	0.80 g/ml	0.71 g/ml
Viscosity 40 °C	1.2199 cSt	14.5 cSt
Flash point	100–185 °F	421.8 °F
Thermal conductivity	0.145 W/mk	
Specific heat capacity	2.01 J/gk	
Break down voltage	60 kV/2.5 mm	80 kV/2.5 mm
Acid value	0.049 mg KOH	4.0–12 mg KOH
Dielectric strength	60 kV/2.5 mm	80 kV/2.5 mm
Color	Colorless	Greenish yellow

Table 3 Mechanical properties of P20 plastic mold steel [6]

Properties	Hardness (Brinell)	Hardness (Rockwell)	Tensile strength (MPa)	Yield strength (MPa)	Compressive strength (MPa)
Value	300	30	965–1030	827–862	862

Table 4 Chemical composition (wt%) of P20 plastic mold steel [6]

Carbon	0.28–0.40
Silicon	0.10–0.50
Manganese	0.20–0.80
Phosphorous	0.030 max
Sulfur	0.030 max
Chromium	1.40–2.00
Molybdenum	0.30–0.55

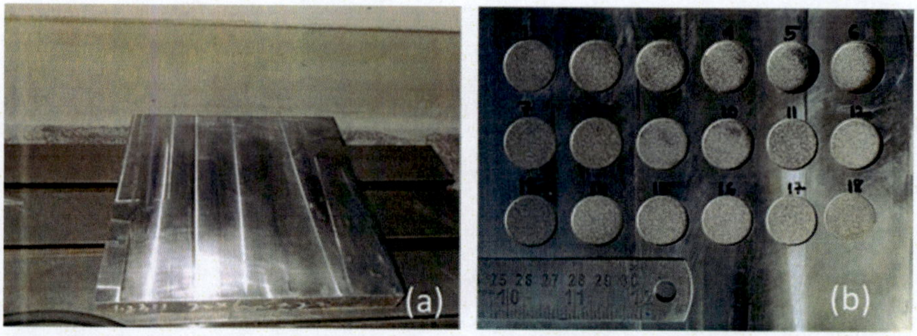

Fig. 4 **a** Workpiece before and **b** workpiece after electro-discharge machining

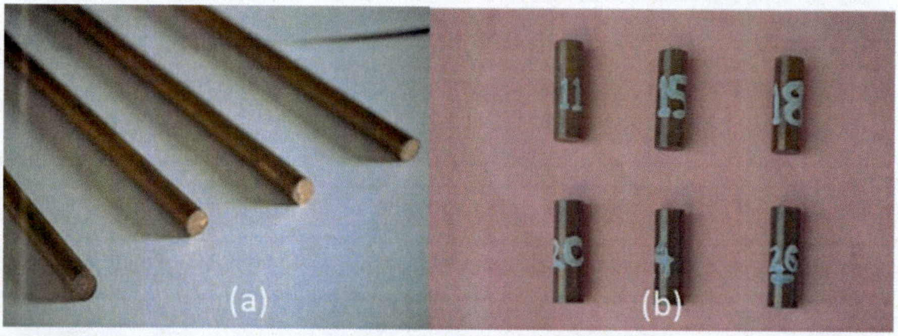

Fig. 5 **a** Raw copper rod and **b** copper electrode after machining

4 Result and Discussion

4.1 Material Removal Rate

The MRR can be defined as the total volume of material machined and divided by the time required for machining. The mathematical expression of MRR is

$$\mathrm{MRR} = \frac{W_{\mathrm{B}} - W_{\mathrm{A}}}{t_{\mathrm{m}}}$$

where W_{B} and W_{A} are the weight of the workpiece before and after machining, respectively, and t_{m} is the machining time.

Figure 6a illustrates the variation of MRR with peak current. For both the dielectrics, the MRR is increasing with increase in the peak current. The percentage of increment in MRR is 59.81–78.45% when PBD is used as the dielectric. This could be due to the higher average temperature generated at machining zone when PBD is used as a dielectric. The reason of higher average temperature in case of

PBD is due to high amount of oxygen content in PBD fluid [10, 11]. This causes vaporization and melting. Another reason for greater MRR in case of PBD is because of efficient flushing of debris from machining zone. The higher flushing ability of PBD is because of its high viscosity [10]. Figure 6b represents the influence of T_{on} time on MRR. Longer discharge energy cycle and higher discharge channel might be the reason of improvement of MRR with increase in T_{on}[12, 13]. Higher MRR is recorded when used PBD as compared with the kerosene. The percentage of improvements varies from 30 to 51%. This could be owing to greater thermal conductivity of PBD, which causes higher heat transformation toward the machining zone. Lower specific heat of PBD could be another reason for better MRR [11]. Figure 6c shows the influence of T_{off} on MRR. With increase in T_{off} time MRR decreases. Higher T_{off} time causes narrow plasma channels, which further minimizes the striking area of positive ions on workpiece surface. The percentage of improvement is varying from 6 to 40% when PBD is used as the dielectric. Lower specific heat and higher thermal conductivity of PBD are the reason behind it [14].

4.2 Surface Roughness

The surface finish reflects the nature of the machined surface. The tribological connection between the mating parts relies upon the surface finish of individual parts. Better surface finish boosts the service life of the mating parts. In this investigation, arithmetic average surface roughness (R_a) was recorded and presented in Fig. 7a–c. Each machined surface is measured twice, and average values have been recorded for analysis. A Taylor Hobson profilometer (0.8 mm cutoff length) has been used to record the arithmetic average roughness parameter. Figure 7a represents the variation of R_a with the current. The trend indicates that R_a increases with increase in the current. The higher current increases the impingement energy, resulting in deeper and uneven crater on the work surface. The R_a value recorded for the machined surface for machining with PBD as a dielectric is comparatively low (in comparison with machining with kerosene). The percentage of decrement varies between 0.04 and 17.79%. The lower R_a when PBD is used might be because of the good thermal property and lower specific heat of the dielectric, which might have limited the energy density. The efficient heat transfer to the environment, i.e., dielectric media produces shallower cavities. Figure 7b illustrates the variation of R_a with T_{on}. More sparking time drags out vaporization and melting, resulting in an increase in R_a as more profound and extensive cavities. The lower R_a obtained with PBD as a dielectric might be because of the greater thermal conductivity, which ensures improved heat distribution on the workpiece surface. In addition, increased dielectric temperature as an effect of lower specific heat of PBD minimizes the viscosity which enhances the flushing effectiveness. Figure 7c indicates the impact of T_{off} time on R_a. It is clearly seen that T_{off} has marginal effect on R_a. Higher T_{off} ensures maximum gap condition for ensuing

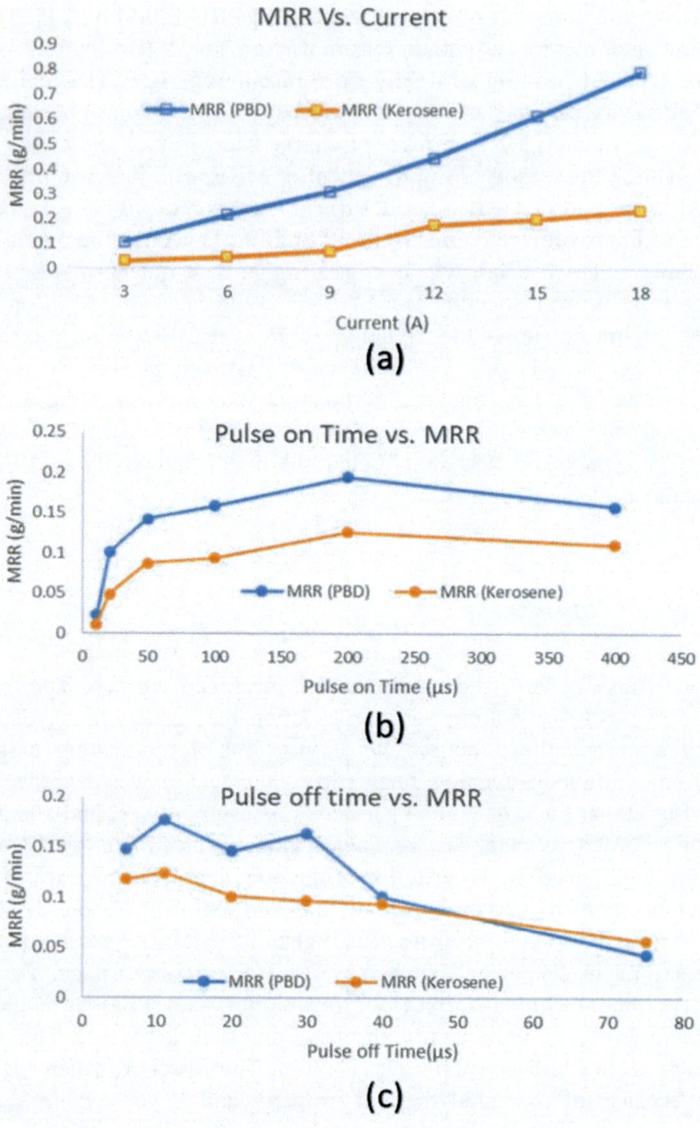

Fig. 6 Variation of material removal rate with **a** current, **b** pulse ON time and **c** pulse OFF time

starting cycles by permitting re-ionization of the ionized gap. Also, smaller and narrower cavities are shaped because of reduced sparking time to diminish R_a. Marginally, lower R_a values recorded with PBD might be because of the good thermal property which could have resulted in well distribution of energy density on the workpiece surface [7].

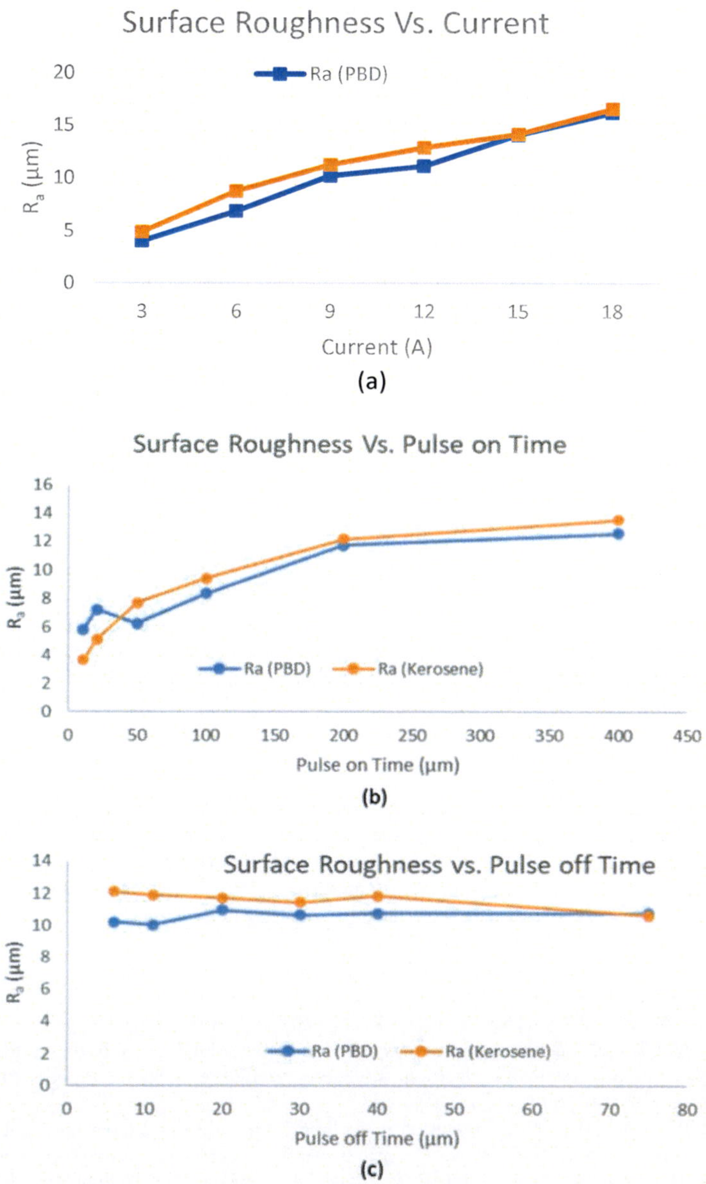

Fig. 7 Variation of surface roughness with **a** current, **b** pulse ON time and **c** pulse OFF time

5 Summary

The above reported experimental study can be outlined as follows:

- Experimentations were conducted, with primary objective of operational and functional feasibility of using transesterification polanga oil as bio-dielectric (PBD) and proved successful.
- Study was conducted, and information is gathered for two predominant output responses such as MRR and R_a.
- Comparative analysis of output responses (MRR and R_a) was done for machining with hydrocarbon-based oil, i.e., kerosene as dielectric and transesterification polanga oil as bio-dielectric.
- 0.08–0.77 times improvement in MRR is observed in case of EDM with PBD.

References

1. Altintas Y, Ber A (2001) Manufacturing automation: metal cutting mechanics, machine tool vibrations, and CNC design. Appl Mech Rev 54:B84–B84. https://doi.org/10.1115/1.1399383
2. Abbas NM, Solomon DG, Bahari MF (2007) A review on current research trends in electrical discharge machining (EDM). Int J Mach Tools Manuf 47:1214–1228. https://doi.org/10.1016/j.ijmachtools.2006.08.026
3. Munoz A, Sheng P (1995) An analytical approach for determining the environmental impact of machining processes. J Mater Process Technol 53:736–758. https://doi.org/10.1016/0924-0136(94)01764-R
4. Zia MK, Pervaiz S, Anwar S, Samad WA (2019) Reviewing sustainability interpretation of electrical discharge machining process using triple bottom line approach. Int J Precis Eng Manuf Green Technol 1–15. https://doi.org/10.1007/s40684-019-00043-2
5. Kou Z, Han F (2018) On sustainable manufacturing titanium alloy by high-speed EDM milling with moving electric arcs while using water-based dielectric. J Clean Prod 189:78–87. https://doi.org/10.1016/j.jclepro.2018.04.072
6. Valaki JB, Rathod PP, Sankhavara C (2016) Investigations on technical feasibility of Jatropha curcas oil based bio dielectric fluid for sustainable electric discharge machining (EDM). J Manuf Process 22:151–160. https://doi.org/10.1016/j.jmapro.2016.03.004
7. Das S, Paul S, Doloi B (2019) An experimental and computational study on the feasibility of bio-dielectrics for sustainable electrical discharge machining. J Manuf Process 41:284–296. https://doi.org/10.1016/j.jmapro.2019.04.005
8. Mishra B, Routra B (2020) Evaluation of technical feasibility and environmental impact of *Calophyllum inophyllum* (Polanga) oil based bio-dielectric fluid for green EDM. Measurement 107744. https://doi.org/10.1016/j.measurement.2020.107744
9. Das S, Paul S, Doloi B (2020) Feasibility investigation of neem oil as a dielectric for electrical discharge machining. Int J Adv Manuf Technol 106:1179–1189. https://doi.org/10.1007/s00170-019-04736-5
10. Kiyak M, Aldemir BE, Altan E (2015) Effects of discharge energy density on wear rate and surface roughness in EDM. Int J Adv Manuf Technol 79:513–518. https://doi.org/10.1007/s00170-015-6840-9

11. Wang X, Liu Z, Xue R, Tian Z, Huang Y (2014) Research on the influence of dielectric characteristics on the EDM of titanium alloy. Int J Adv Manuf Technol 72:979–987. https://doi.org/10.1007/s00170-014-5716-8
12. Daneshmand S, Kahrizi EF, Abedi E, Abdolhosseini MM (2013) Influence of machining parameters on electro discharge machining of NiTi shape memory alloys. Int J Electrochem Sci 8:3095–3104
13. Shabgard M, Seyedzavvar M, Oliaei SNB (2011) Influence of input parameters on the characteristics of the EDM process. J Mech Eng 57:689–696
14. Fonseca J, Marafona JD (2014) The effect of deionisation time on the electrical discharge machining performance. Int J Adv Manuf Technol 71:471–481. https://doi.org/10.1007/s00170-013-5516-6

Analysis of Tensile and Microstructural Properties of Al–Ni Joints by Ultrasonic Spot Welding

Soumyajit Das, Mantra Prasad Satpathy, Bharat Chandra Routara, and Susanta Kumar Sahoo

1 Introduction

In recent years, growing concerns about global warming and energy consumption increase a lot of attention in the automotive and aerospace industries. The automotive manufactures have started accepting a key method of light-weighting so that the amount of fuel being consumed can be utilized economically [1–3]. To manufacture multi-material structures vehicle aluminum (Al) and nickel (Ni) alloy are the most preferred materials due to their high specific strength, low density, easy processing, and recyclability properties [4]. The density of Al and Ni alloys is 2.7 g/cm^3 and 8.9 g/cm^3, respectively. To joint high strength, Al–Ni alloys by using traditional fusion welding (FSW) process may create solidification defects like micro-cracks and gas porosity [5, 6]. Similarly, other solid-state bonding techniques such as resistance spot welding (RSW) were known by their feasibility of bonding dissimilar materials [7]. However, the possibility of brittle intermetallic compounds (IMCs) formation which occurs during the RSW process because of lead to local variations of the alloy functional characteristics [8]. To defeat these problems, solid-state welding technique at shorter weld time without excessive temperature is required.

Form the above viewpoint, ultrasonic metal spot welding (USMW) is a solid-state welding method that uses high-frequency ultrasonic acoustic vibrations (generally 15–40 kHz) to join the workpieces which are placed under the action of clamping force (usually 500–2000 N). The temperature increased between workpieces is close to 30–80% at the melting point of the materials. Also, this process

S. Das (✉) · M. P. Satpathy · B. C. Routara
School of Mechanical Engineering, KIIT, Deemed to be University, Bhubaneswar, Odisha 751024, India

S. K. Sahoo
Department of Mechanical Engineering, National Institute of Technology, Rourkela, Odisha 769008, India

S. Bag et al. (eds.), *Next Generation Materials and Processing Technologies*, Springer Proceedings in Materials 9, https://doi.org/10.1007/978-981-16-0182-8_21

will not be affected by any chemical properties of the workpieces. The main advantages of the USMW technique over other solid-state welding technique were a shorter welding cycle and lower energy input [9, 10]. This method is particularly appropriate for joining thin components, for example wires, metallic foils, packaging materials [11, 12]. A schematic diagram of the USMW setup is elucidated in Fig. 1.

Numerous researchers have worked in the USMW area to demonstrated evidence about the welding of dissimilar metals. Kim et al. [13] studied the USW process by using conductive materials such as Ni-plated Cu and Cu. As a result, a better-quality benchmark was provided by classifying the failures natures in T-peel tests. The tests were performed using a full factorial experimental design according to the varying weld pressures and times. A tensile machine was used to determine the quality of the joint and recognize the weldability of the T-peeled tests. Ni et al. [14] performed the aluminum to nickel joints with and without interlayer to improve the temperature of the weld region. The result concluded that the interlayer improves weld region temperature compare to the non-interlayer specimen. Meanwhile, the maximum tensile shear strength of joint with interlayer achieved the highest value which was 67.8% higher than without interlayer welded specimen. Das et al. [15] analyzed the effect of different weld parameters for joining aluminum (AA1060) with cupronickel (UNS 71500) sheets. The results showed that the tensile shear load increased with increasing weld time and achieved the maximum value of the joint. Several types of weld quality were also noticed based upon the joint formation at different weld parameters. The weld interface discovered the bonding mechanism with the mechanical interlocking feature. Matsuoka [16] presented an explanation of an investigational study of the USW of metals. The investigates were showed to join Al and Cu together. The outcomes summarized that the suitable range of process parameters, like vibration amplitude, welding time, and welding pressure have to be made for every welding condition [17, 18]. Rajathi et al. [19] investigated the mechanical analysis at various process parameters, for example vibration amplitude, weld time, and weld pressure on Al–Cu sheets. The results described that the weld strength achieved high value at low weld pressure, high vibration amplitude, and welding time. After that, it decreased with further increase in weld pressure because of the crack formation occured in the weld

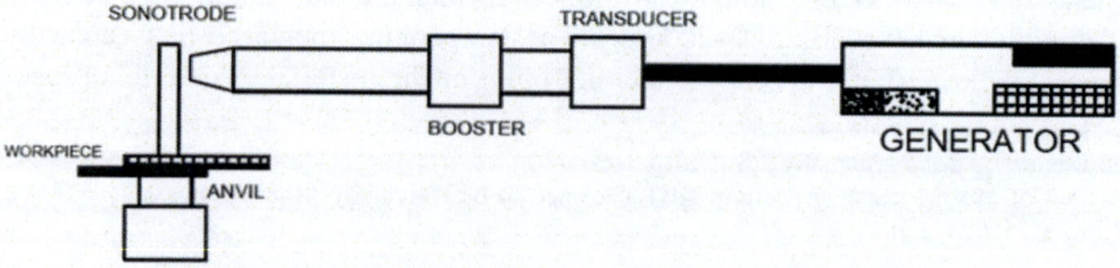

Fig. 1 Schematic diagram of USMW setup

region. In this study, 0.3 mm aluminum (AA1100) sheet was welded to the 0.3 mm pure nickel sheet by using the USMW technique, and the effect of welding parameters on this dissimilar material was investigated. To clarify the weld strength of these Al–Ni joints, mechanical analysis and microstructural characterizations for the weld interface are introduced to improve the weld quality.

2 Experimental Details

In this experiment, 0.3 mm aluminum (AA1100) was lap welded with 0.3 mm pure nickel sheets by using the USMW method. The physical properties and chemical composition of both materials are detailed in Tables 1, 2, and 3. Specimens with a dimension of 75 mm length and 20 mm width were cut from the sheet metals for the preparation of welding samples. During the welding process, Ni was kept at the anvil side, whereas Al was placed on top of the Ni surface. For all the configurations, 20 mm overlap of Al and Ni specimens was ensured. The schematic diagram of the lap configured welding specimen is shown in Fig. 2a.

A lateral drive ultrasonic spot welder (Telsonic® M4000) was used to perform experiments with the maximum output of 3 kW and the frequency of 20 kHz. The complete diagram of the USMW setup was represented in Fig. 2b. The rectangular shape of the horn tip (9 mm × 11 mm) with knurled patterns was also presented in Fig. 2c. Throughout the USMW process, three input parameters such as weld pressure (WP), weld time (WT), and vibrational amplitude (*A*) were varied one at a time whereas other input parameters were kept constant. The weld parameters with

Table 1 Physical properties of aluminum (AA1100) and nickel alloy

Materials	Ultimate tensile strength (MPa)	Yield strength (MPa)	Thermal conductivity ($Wm^{-1}\,°C^{-1}$)	Specific heat ($Jg^{-1}\,°C^{-1}$)
AA1100	89.6	20	222	0.904
Pure nickel	317	59	60.70	0.460

Table 2 Chemical composition of aluminum (AA1100) alloy

Elements	Al	Cu	Fe	Si	Be	Mn	Others
wt. (%)	$\geq$99.00	$\leq$0.20	$\leq$0.95	$\leq$0.95	$\leq$0.08	$\leq$0.05	$\leq$0.15

Table 3 Chemical composition of nickel alloy

Elements	Ni	Cu	Fe	Mn	C	Si	S
wt. (%)	99	$\leq$0.25	$\leq$0.40	$\leq$0.35	$\leq$0.02	$\leq$0.35	$\leq$0.01

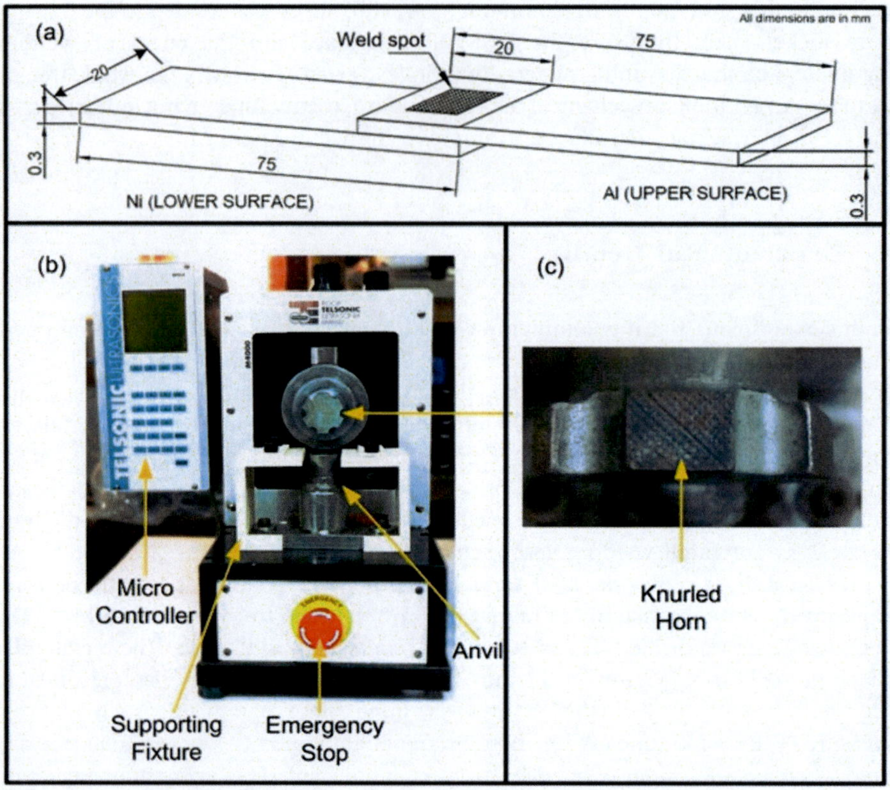

Fig. 2 **a** Schematic diagram of the lap configured welding specimen, **b** lateral drive USMW machine, **c** horn tip with knurls

their various levels are shown in Table 3. Tensile shear strength was conducted to evaluate joint strength by applying 20 kN load in an Instron 3300 machine. The strain rate was constant at 2 mm/min for every welding conditions. To investigate the welding behavior, field emission scanning electron microscopy (FESEM) was used for the analysis of various microstructural conditions (Table 4).

Table 4 Various levels of weld parameters

Factors	Level 1	Level 2	Level 3	Level 4	Level 5	Level 6
Amplitude (μm)	47	68	–	–	–	–
Weld pressure (MPa)	0.22	0.24	0.26	–	–	–
Weld time (s)	0.1	0.2	0.25	0.3	0.35	0.4

3 Results and Discussion

3.1 Tensile Shear Failure Load Analysis

The tensile shear strength results of the welded specimens were determined from the ultimate failure load. A good weld specimen always discloses an acceptable joint strength during the tensile shear test. The current study considers that the weld formation in the USMW method took a very short time with a moderate amount of WP under high-frequency vibration. The effect of WT, WP, and vibration amplitude of 47 μm on the tensile strength was demonstrated in Fig. 3a. The highest tensile shear failure load 453 N of the weld specimen was reached at 0.24 MPa (WP) and 0.3 s (WT). It was noticed that the tensile lap shear load first increased with the increase of WT and afterward it decreased with the further increase of WT for all the input parameters. This was due to the weld time that can cause several plastic deformations at the weld region which finally occurred the formation of cracks around the weld periphery. Figure 3b indicated that when the USMW happened at the vibration amplitude of 68 μm, the maximum tensile shear failure load of 561 N was achieved at 0.3 s (WT) and 0.24 MPa (WP). From the two different graphs, it was concluded that the increase of vibration amplitude caused more relative motion between the weld specimens and it ultimately increased a huge amount of interface temperature. As a result, more amount of plastic deformation and brittle IMCs were formed at the weld region which was directly affecting the weld strength of the Al–Ni joints.

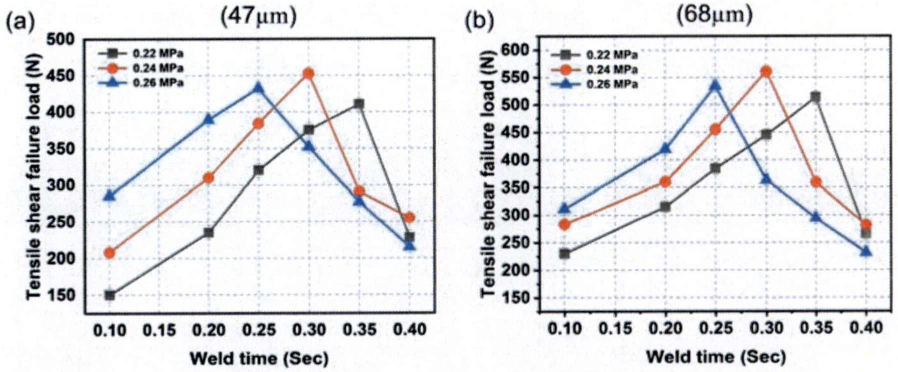

Fig. 3 Tensile shear failure load of Al–Ni welded specimens at **a** 47 μm, **b** 68 μm

3.2 Microstructural Analysis of Weld Interface

To evaluate the effect of various input parameters on the weld quality, the microstructural analysis was an effective way to reveal the complete weld mechanism. Vibration amplitude and weld time played a vital role in Al–Ni joints of the USMW process. Throughout the USMW process, the weld interface temperature was increased, and the horn tip penetrates deeper into the softer Al sheets as compared to Ni because Al is more malleable than Ni alloy. The FESEM analysis along the bond-line of Al–Ni joints at different WTs and vibration amplitudes is demonstrated in Fig. 4a–c. In this microstructure analysis, intermetallic diffusion at the weld region, mechanical interlocking, metallurgical adhesion, and local melting of weld materials were the most observed. The interfacial gaps around the bond-line are shown in Fig. 4a. The Al–Ni sheet was not completely welded when the WT of 0.1 s and vibration amplitude 47 μm. When the WT was increased from 0.1 to 0.3 s and the vibration amplitude of 68 μm, these gaps were minimized (Fig. 4b). However, the wavy-like interface pattern reveals the weld interface was affected by WT in Fig. 4c. The swirling-type trend exhibits the effective diffusion of atoms at the bond-line. Also, a small crack formation was present at the Al surface due to the penetration of horn increased with the increase in weld time across the weld spot. As a result, it was exposed that the intermetallic bond formed between the welded specimens affected the weld quality of the joint as well as weld strength.

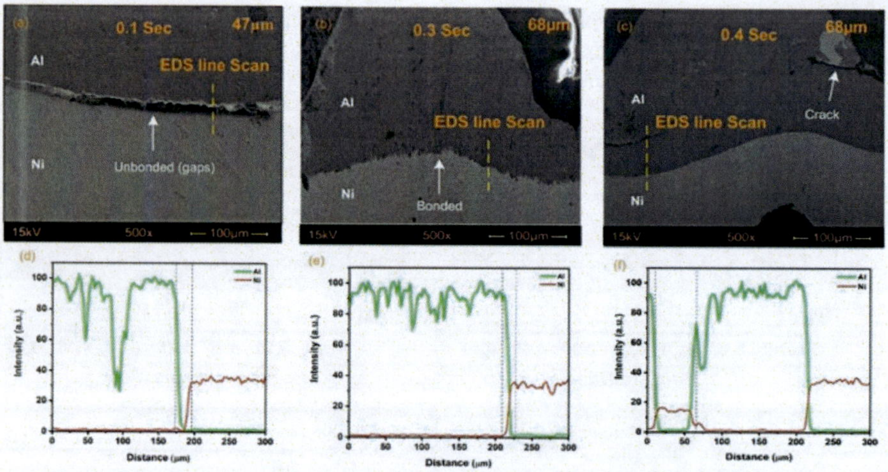

Fig. 4 FESEM images showing different weld qualities at **a** 0.1 s **b** 0.3 s **c** 0.4 s, EDS line scan results for different weld times at **d** 0.1 s **e** 0.3 s **f** 0.4 s

4 Conclusions

The mechanical behavior and microstructural analysis of Al (AA1100) and pure nickel joints at different weld parametric conditions were investigated in this present study. The following conclusions can be obtained from the current investigation:

- I The tensile shear failure load disclosed that the maximum weld strength was reached 561 N at 0.68 μm of vibration amplitude and 0.3 s of WT. After that, this failure load decreased with an increase in WT, WP, and vibration amplitude due to excessive surface crack formation around the weld spot.
- II FESEM analysis revealed the different levels of weld strength quality depending upon the weld input parameters. During the low weld time, the weld pattern was improved heterogeneously at areas below the horn tip edges. For that, the weld interfaces were straight, as the welding time increased plastic deformation area also expanded, and the weld zone changed from typical shape to a convoluted wavy pattern.
- III The EDS line scan analysis discloses the diffusion process at the weld region, and the thickness of IMC increases with the weld time. At the maximum weld time, there is an excessive reduction of the weld strength of the materials, and it was leading to the cracks at the weld interface.

References

1. Huang Y, Wang J, Wan L, Meng X, Liu H, Li H (2016) Self-riveting friction stir lap welding of aluminum alloy to steel. Mater Lett 185:181–184
2. Padhy GK, Wu CS, Gao S (2016) Subgrain formation in ultrasonic enhanced friction stir welding of aluminium alloy. Mater Lett 183:34–39
3. Lu Y, Song H, Taber GA, Foster DR, Daehn GS, Zhang W (2016) In-situ measurement of relative motion during ultrasonic spot welding of aluminum alloy using Photonic Doppler Velocimetry. J Mater Process Technol 231:431–440
4. Matsuoka S, Imai H (2009) Direct welding of different metals used ultrasonic vibration. J Mater Process Technol 209:954–960
5. Oliveira JP, Fernandes FMB, Schell N, Miranda RM (2016) Martensite stabilization during superelastic cycling of laser welded NiTi plates. Mater Lett 171:273–276
6. Song YG, Li WS, Li L, Zheng YF (2008) The influence of laser welding parameters on the microstructure and mechanical property of the as-jointed NiTi alloy wires. Mater Lett 62:2325–2328
7. Sam S, Kundu S, Chatterjee S (2012) Diffusion bonding of titanium alloy to micro-duplex stainless steel using a nickel alloy interlayer: interface microstructure and strength properties. Mater Des 40:237–244
8. Oliveira JP, Barbosa D, Fernandes FMB, Miranda RM (2016) Tungsten inert gas (TIG) welding of Ni-rich NiTi plates: functional behavior. Smart Mater Struct 25:03LT01
9. Fujii HT, Goto Y, Sato YS, Kokawa H (2016) Microstructure and lap shear strength of the weld interface in ultrasonic welding of Al alloy to stainless steel. Scr Mater 116:135–138

10. Kim J (2014) Weldability of $Cu_{54}Zr_{22}Ti_{18}Ni_6$ bulk metallic glass by ultrasonic welding processing. Mater Lett 130:160–163
11. Fujii HT, Endo H, Sato YS, Kokawa H (2018) Interfacial microstructure evolution and weld formation during ultrasonic welding of Al alloy to Cu. Mater Charact 139:233–240
12. Ao S, Li C, Zhang W, Wu M, Dai Y, Chen Y, Luo Z (2019) Microstructure evolution and mechanical properties of Al/Cu ultrasonic spot welded joints during thermal processing. J Manuf Process 41:307–314
13. Kim TH, Yum J, Hu SJ, Spicer JP, Abell JA (2011) Process robustness of single lap ultrasonic welding of thin, dissimilar materials. CIRP Ann 60:17–20
14. Ni ZL, Ye FX (2016) Weldability and mechanical properties of ultrasonic welded aluminum to nickel joints. Mater Lett 185:204–207
15. Das S, Satpathy MP, Pattanaik A, Routara BC (2019) Experimental investigation on ultrasonic spot welding of aluminum-cupronickel sheets under different parametric conditions. Mater Manuf Process 34:1689–1700
16. Matsuoka S (1994) Ultrasonic welding of ceramic/metal. J Mater Process Technol 47:185–196
17. Ishikuro T, Matsuoka S-I (2005) Ultrasonic welding of thin alumina and aluminum using inserts. JSME Int J Ser A Solid Mech Mater Eng 48:317–321
18. Imai H, Matsuoka S (2006) Direct welding of metals and ceramics by ultrasonic vibration. JSME Int J Ser A Solid Mech Mater Eng 49:444–450
19. Ganesh M, Praba Rajathi R (2013) Experimental study on ultrasonic welding of aluminum sheet to copper sheet. Int J Res Eng Technol 2:161–166

Cutting Force Analysis in Micro-milling of Al6061-SiCp Composite

Satyendra Kumar Patel, Ashwani Pratap, Priyabrata Sahoo, Binayaka Nahak, and Tej Pratap

1 Introduction

The micro-featured products of aluminum metal matrix composites (Al6061-SiCp MMC's) have attracted interest of industrial community due to their enhanced surface functionality based on different applications and its extraordinary material properties such as extremely good material properties, lightweight, dimensional stability, wear resistance and directional properties over other used materials [1–5]. The increasing application of such products in different areas like automobile, aerospace, marine, structural, electronics, etc., leads to the increase in industrial wealth and growth. However, the economical and mass scale production of such micro-features on Al6061-SiCp MMCs is very much challenging with any of the manufacturing processes due to hard SiCp abrasive in the soft matrix and the limitations of different material processing methods [2, 4–6]. Pratap and Patra reviewed the applicability of different micro-fabrication processes, their process capabilities and limitations and identified that mechanical micro-end milling has great capability to produce 3D micro-features on wide range of materials. The flexibility and extremely high accuracy make it capable of producing 3D complex micro-features on variety of engineering materials, freeform, cost effectiveness and repeatability of the features which makes it as a leading technology in the present time [5]. Despite several advantages of the micro-end milling process, the micro-scale machinability of Al6061-SiCp MMCs is very much challenging due to the

S. K. Patel · B. Nahak · T. Pratap (✉)
Department of Mechanical Engineering, Motilal Nehru National Institute of Technology Allahabad, Prayagraj 211004, India
e-mail: tpratap@mnnit.ac.in

A. Pratap · P. Sahoo
Micro-Fabrication Laboratory, Department of Mechanical Engineering, Indian Institute of Technology, Patna, Bihar 801103, India

S. Bag et al. (eds.), *Next Generation Materials and Processing Technologies*,
Springer Proceedings in Materials 9, https://doi.org/10.1007/978-981-16-0182-8_22

non-homogeneous distribution and anisotropic behavior of brittle SiCp abrasive as a reinforcement in the soft metal matrix and the formation of surface defects due to interaction of tool with individual grains [2–6]. Even the complexity of the process gets increased due to the scaling phenomenon of the tool, controlling of burr formation and minimization of machined surface defects [4–8]. The miniaturization of cutting tools increases the aspect ratio while the fluctuation of cutting forces caused by varying chip thickness with tool rotation and cutting of different grains of the particulate composite causes uneven force on the tool leading to tool failure [4–10].

Further, it is a very well-known fact that matrix and reinforced particles have different hardness and material properties and hence both show completely different machining characteristics [4–6, 11]. Even, the properties of Al-MMC's workpiece are very much prone to the use of its manufacturing methods. Stir casting is most preferable methods for the fabrication of MMCs due to its advantageous characteristics like simple and low cost, uniform reinforcement, good bonding with metal matrix and its reinforced particles, precise control of the matrix structure and applicability to large volume production [2, 12].

From the literature, it was observed that the extensive analysis of cutting force in micro-milling of Al6061-SiCp MMCs is limited, and it is very much essential to avoid the shortcomings of the process for any kind of surface defect caused by machining of different grains and premature tool breakage due to downscaling of the machining process. Therefore, initially the Al-6061-SiCp MMCs with different range of reinforcement (0, 5, 10 and 15% of SiCp by weight) have been fabricated using stir casting process. Further, an extensive study of micro-milling cutting force analysis is performed on the fabricated sample to analyze the effects of different reinforcement percentage and cutting parameters. The cutting force trend is further correlated with reinforcement percentage and machining parameters to study the cutting mechanics of composites.

2 Experimental Procedure

2.1 *Preparation of Al6061-SiCp MMCs*

Stir casting is used for the fabrication of Al6061-SiCp MMCs with different range of reinforcement (0, 5, 10 and 15% of SiCp by weight). Initially, an ingot of Al6061 alloy is used as soft metal matrix element, and micro-sized SiCp particle is used as reinforcement material. The average particle size (APS) of SiCp is 60–80 μm as per the manufacturer details. First the ingot of Al6061 alloy is completely melted in the furnace, and then the SiCp particle is introduced into the melt (5, 10 and 15% of SiCp by weight) while stirring the solution continuously to achieve the uniform SiCp particles in the matrix. To enhance the bonding between metal matrix and reinforced SiCp particle, a very small amount (1% by weight) of Mg powder is also

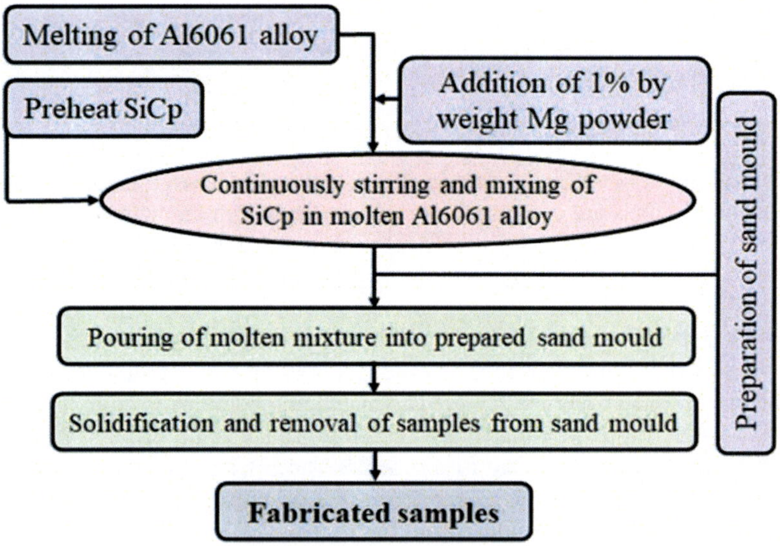

Fig. 1 Fabrication flowchart of the Al6061-SiCp MMC's samples

added. After proper mixing of the solution, it is casted into the sand mold cavity of standard shapes. The fabrication flowchart of aforementioned samples is shown in Fig. 1.

For the experimentation, total four numbers of samples of Al6061 base material (0% SiCp) and Al6061-SiCp MMCs (5, 10 and 15% of SiCp by weight) are fabricated. From the casted sample of Al6061-SiCp MMCs, the workpiece of size 40 mm × 40 mm × 15 mm is prepared for the micro-milling experiments. Before the experiments , the workpiece of aforementioned sized is polished and cleaned in order to avoid any surface contamination and flatness issues.

2.2 *Setup for Micro-milling Experiment*

Micro-milling of the fabricated Al6061-SiCp MMCs is performed on the multi-purpose three-axis machining center (Mikrotools DT-110). All the experiments are done with a low-speed spindle having range of spindle speed 10–3000 rpm and positioning of ±1 μm. The axial movements of the machine tool in three perpendicular directions, i.e., *X*, *Y* and *Z* directions are achieved by the servo controller to control the machining parameters feed and axial depth of cut while spindle speed is controlled by motor. The *Z*-axis of DT-110 moves vertically up and down with respect to workpiece to get desired depth of cut, and *Y*-axis gives feed. All the experiments are performed by a 500 μm diameter fresh two-fluted solid carbide micro-end milling tool (Make: YG-1 Tool Company).

The workpiece is clamped on the stationary Kistler dynamometer which was set on the machine table of the DT-110. The flatness and reference plane is selected by the use of inbuilt contact probe. Further, the cutting force is acquired by the piezoelectric dynamometer from Kistler (Type 9256C2) with sampling frequency of 10,000 Hz and resolution of 0.002 N. The charge signals generated by the piezoelectric sensors are converted into voltage signal by the charge amplifier, and then these voltage signals are converted into force signals with the help of dynamometer sensitivity by the data acquisition (DAQ) system (Kistler, type 5697) of Dynoware software. All experiments are done under dry condition. The experimental setup along with one set of micro-milled channels and cutting force acquisition is shown in Figs. 2 and 3, respectively.

Table 1 shows the experimental cutting conditions by setting the machining parameters and their assigned levels. The machining parameters (spindle speed and feed) have four levels at equal interval, and constant value of axial depth of cut 20 μm is maintained for all the experiments. The selection of parameters is done based on tool edge radius to check the effect of minimum chip thickness phenomenon and low-speed spindle limitations (limited to 3000 rpm) set by the manufacturer. Therefore, the lower feed values are close to the tool edge radius value and spindle speed limited to less than 3000 rpm. By the use of machining parameters combination, total 16 microgrooves have been cut on each specimen (combination of four spindle speed and four feed values at fixed value of axial depth

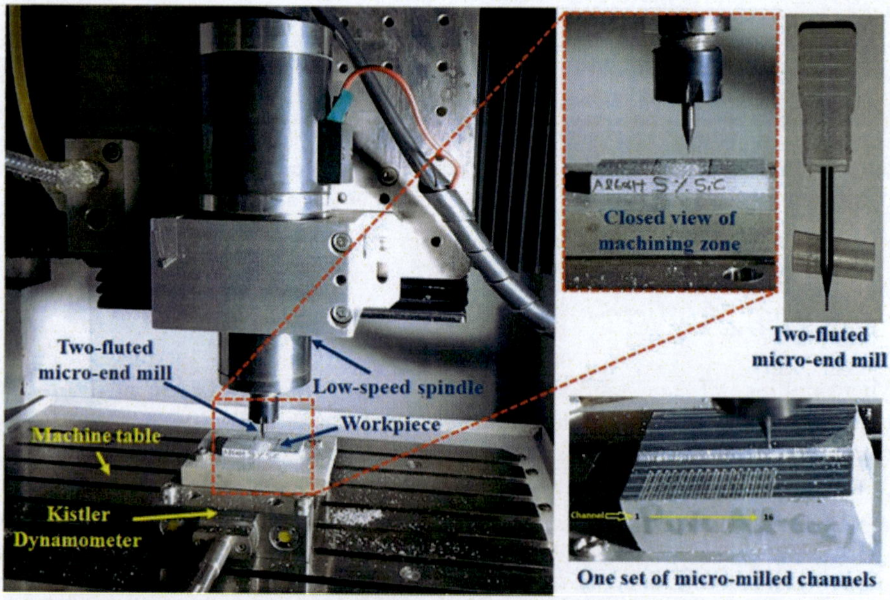

Fig. 2 Experimental setup for micro-milling of Al6061-SiCp MMCs and one set of micro-milled channels

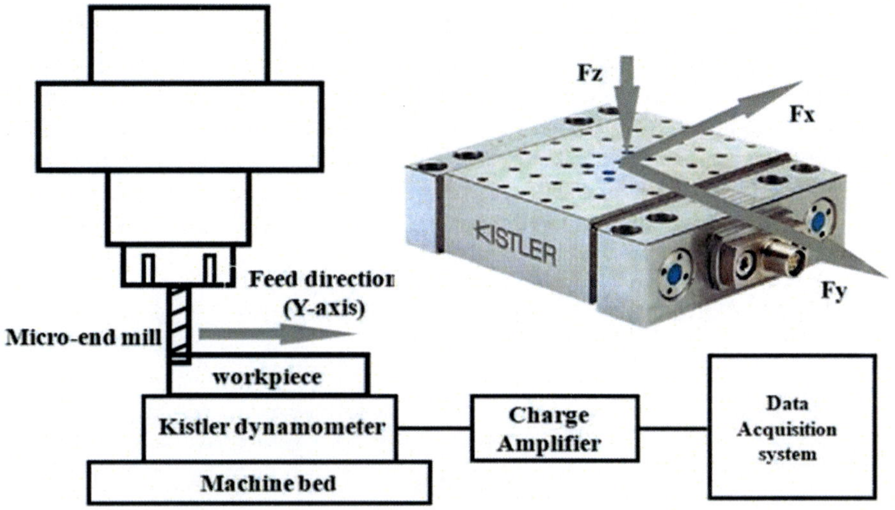

Fig. 3 Cutting force acquisition in micro-milling of Al606-SiCp MMCs using Kistler dynamometer (Type 9256C2)

Table 1 Experimental cutting conditions (machining parameters and their assigned level)

Machining parameters (unit)	Level			
	1	2	3	4
Spindle speed (rpm)	1000	1500	2000	2500
Feed (μm/tooth)	0.5	1.5	2.5	3.5

of cut). For each experimental run, the responses are measured in terms of cutting force components. All the experiments have been repeated thrice on each workpiece.

3 Results and Discussion

The analysis of cutting force is done by considering the root mean square (RMS) values as the normal and tangential forces continuously change their directions with the rotation of two-fluted micro-end milling tool. The average of three RMS values of cutting force in *X*, *Y* and *Z* directions is used for the analysis of cutting forces: F_x (tangential force normal to feed direction), F_y (feed force) and F_z (axial force), respectively. Here, the experimental results have been investigated in three different sections: (i) the effect on feed, (ii) the effect on spindle speed and (iii) the effect on reinforcement (change in SiCp wt%) on the cutting force components. The detailed discussion of each of them is given below.

3.1 Effect of Feed on Cutting Forces

The cutting forces variations with feed and different weight % of SiCp reinforcement (0, 5, 10 and 15) for all the four levels of spindle speed are shown in Figs. 4, 5, 6 and 7, respectively. For each level of spindle speed, the trends of cutting forces are increased with increasing feed. It is also seen that the tangential force (F_x) and feed force (F_y) have similar force signature with feed while the axial force (F_z)

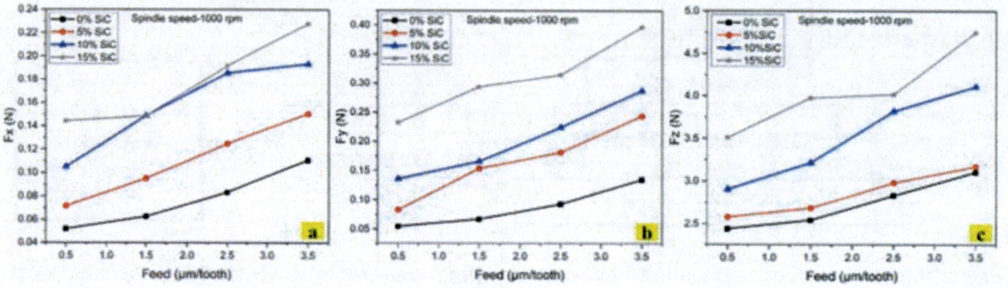

Fig. 4 Variation of cutting forces with feed under fixed values of spindle speed of 1000 rpm **a** F_x **b** F_y **c** F_z

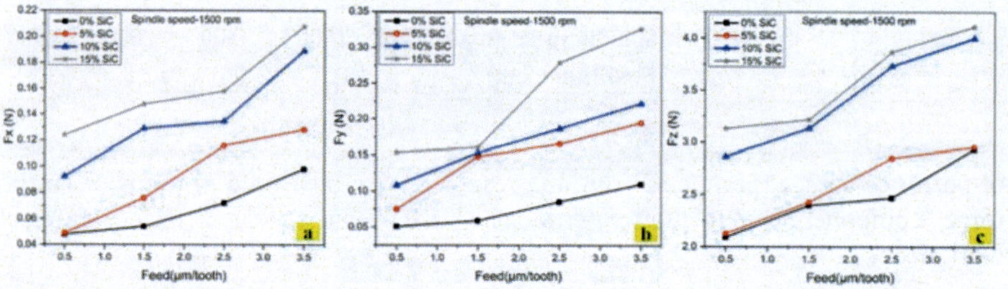

Fig. 5 Variation of cutting forces with feed under fixed values of spindle speed of 1500 rpm **a** F_x **b** F_y **c** F_z

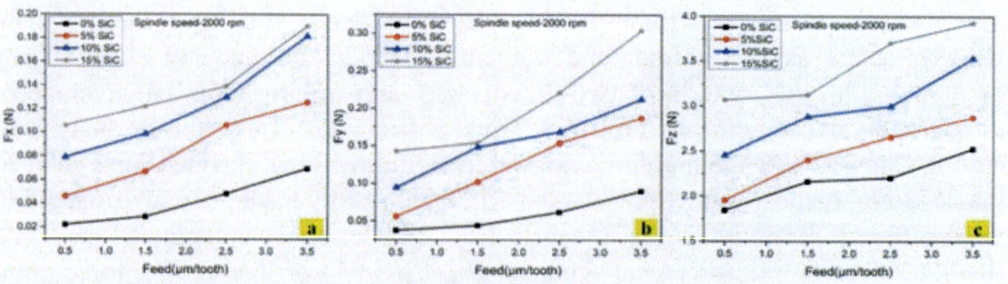

Fig. 6 Variation of cutting forces with feed under fixed values of spindle speed of 2000 rpm **a** F_x **b** F_y **c** F_z

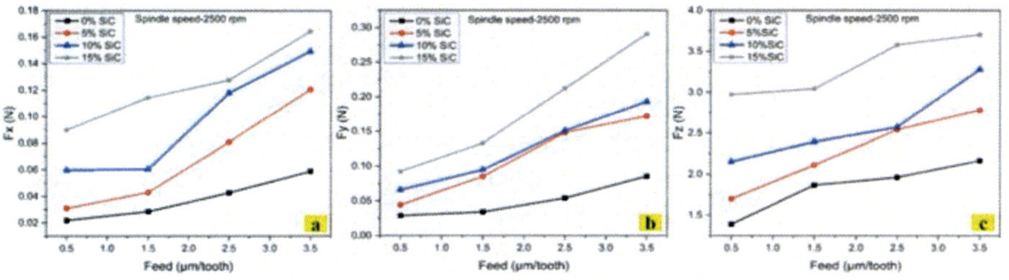

Fig. 7 Variation of cutting forces with feed under fixed values of spindle speed of 2500 rpm **a** F_x **b** F_y **c** F_z

shows different trends. Cutting forces (F_x and F_y) at feed of 0.5 and 1.5 μm/tooth have similar values which shows the influence of size effect. However, a further increase in feed from 1.5 to 3.5 μm/tooth shows significant change in the chip load and shows that cutting forces increase with higher rate. The similar trend of cutting forces is observed for machining of almost all the workpiece (variation in reinforcement). The axial force (Fz) trend shows that the magnitude of force is significantly high due to plunging in the *z*-direction. It is also seen that the axial force (Fz) continuously increases with feed, but the rate of increase is significantly high at feed of 2.5–3.5 μm/tooth for almost all the workpiece. This sudden increase shows the very high chip load and rubbing of the end portion at the tool center due to zero cutting velocity at center of the tool.

3.2 Effect of Spindle Speed on Cutting Forces

The cutting forces variations with spindle speed and different weight % of SiCp reinforcement (0, 5, 10 and 15) for all the four levels of feed are shown in Figs. 8, 9, 10 and 11, respectively. For each level of feed, the trends of all the cutting forces (F_x, F_y and F_z) are decreased with increase in spindle speed. It is due to the

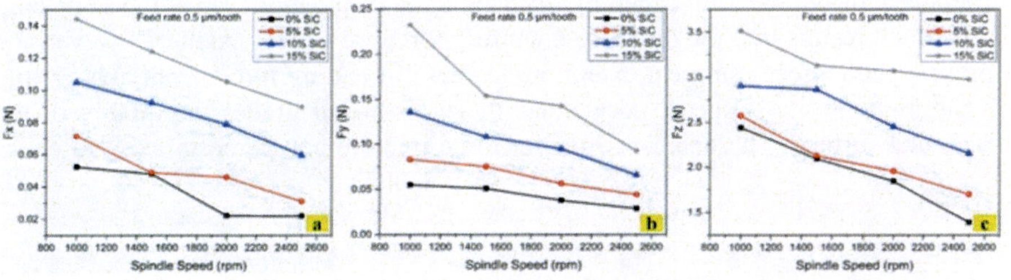

Fig. 8 Variation of cutting forces with spindle speed at feed of 0.5 μm/tooth **a** F_x **b** F_y **c** F_z

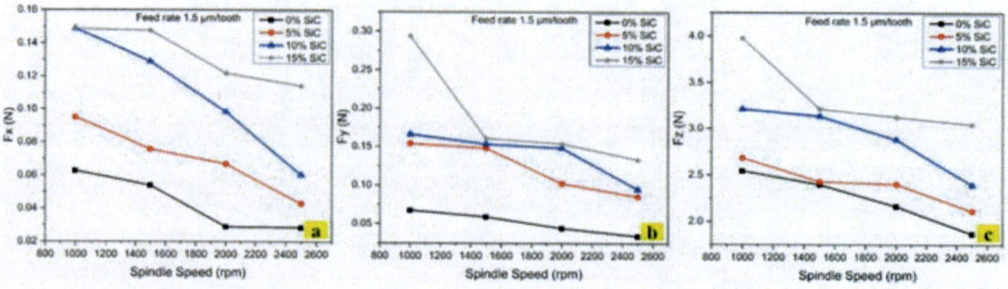

Fig. 9 Variation of cutting forces with spindle speed at feed of 1.5 μm/tooth **a** F_x **b** F_y **c** F_z

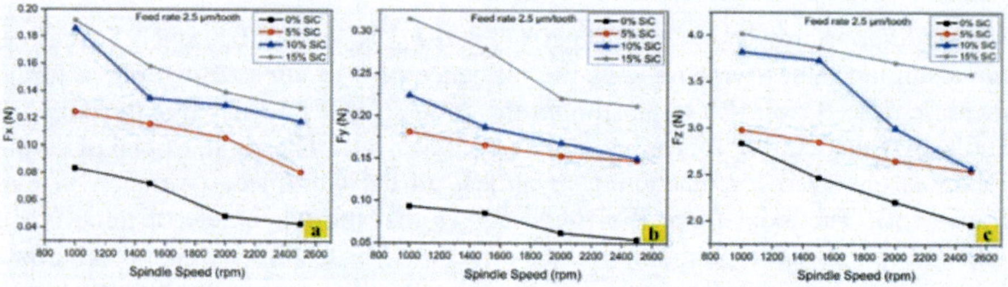

Fig. 10 Variation of cutting forces with spindle speed at feed of 2.5 μm/tooth **a** F_x **b** F_y **c** F_z

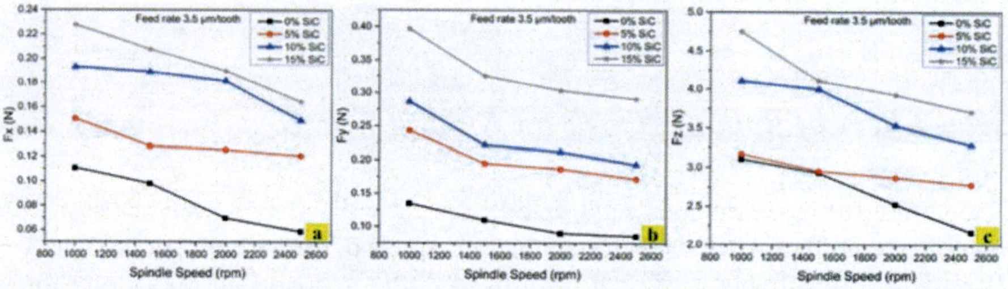

Fig. 11 Variation of cutting forces with spindle speed at feed of 3.5 μm/tooth **a** F_x **b** F_y **c** F_z

dominance of the shearing capability over plowing of the tool even at the same chip load which results into the decrease in cutting forces at a great extent. However, the spindle speed effect at a great extent on decrease in cutting force is not always true as the higher spindle speed, accelerates the wear and degrades the quality of the machined surface. The decrease in axial force is more compared to axial force.

3.3 *Effect of Reinforcement (Change in SiCp wt%) on Cutting Forces*

The cutting forces variations with reinforcement (change in SiCp wt%) and feed for all the four levels of spindle speed are shown in Figs. 12, 13, 14 and 15, respectively. For each level of spindle speed, the trends of cutting forces are increased with increase in reinforcement of SiCp. Results also indicate that the cutting forces for base material, i.e., Al6061 (0% SiCp) are significantly low among others. The

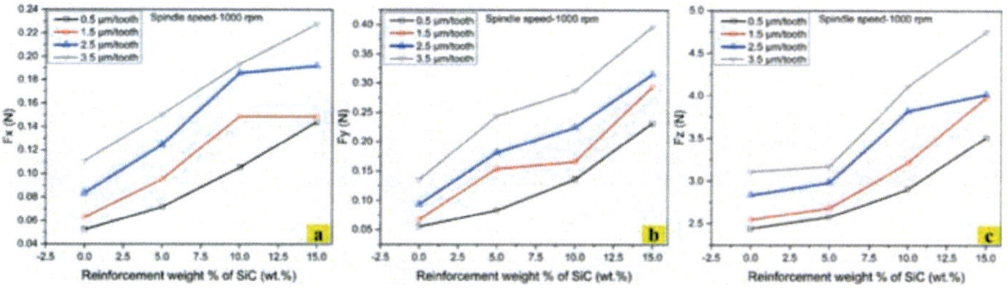

Fig. 12 Variation of cutting forces with reinforcement at spindle speed of 1000 rpm **a** F_x **b** F_y **c** F_z

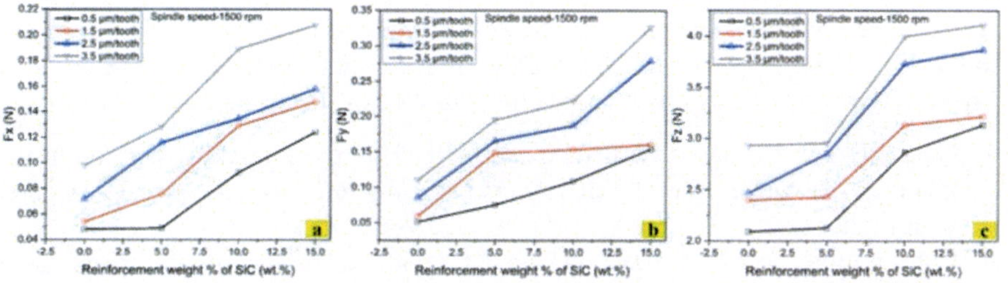

Fig. 13 Variation of cutting forces with reinforcement at spindle speed of 1500 rpm **a** F_x **b** F_y **c** F_z

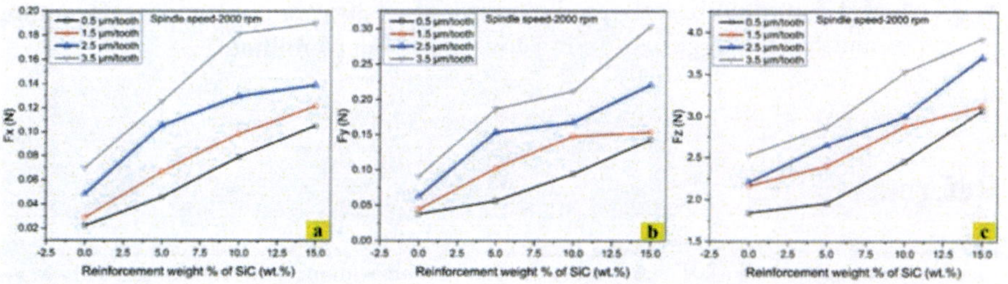

Fig. 14 Variation of cutting forces with reinforcement at spindle speed of 2000 rpm **a** F_x **b** F_y **c** F_z

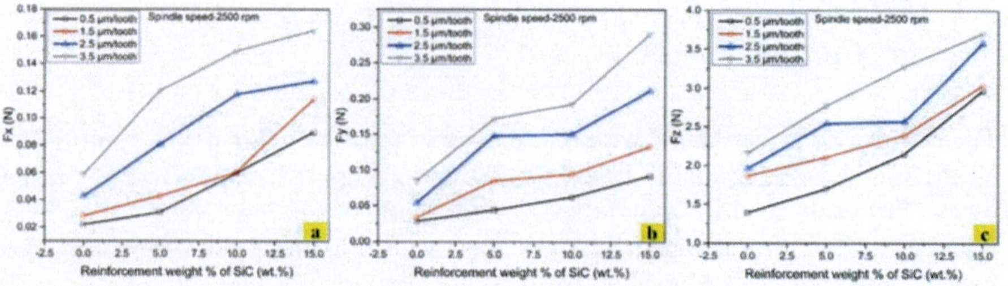

Fig. 15 Variation of cutting forces with reinforcement at spindle speed of 2500 rpm **a** F_x **b** F_y **c** F_z

cutting forces start increasing with increasing reinforcement, and its rate also increases in similar manner due to closed packing of brittle SiCp in the matrix. Higher reinforcement in the matrix allows the frequent contact of cutting edge with hard and brittle SiCp which leads to the significant increase in the cutting force.

4 Conclusions

Present study experimentally demonstrates the effect on machining parameters and SiCp reinforcement on cutting forces generated during micro-milling of Al6061-SiCp MMCs. The analysis shows that the micro-milling process is shown to be a promising approach for fabrication of such micro-features on Al6061-SiCp MMCs. The major observations from the analysis are listed below:

- Cutting force increases with increase in feed as the chip load per tooth also increases. However, the rate of increase at lower feed value is small compared with higher feed.
- Cutting forces in all the directions decrease with increase in spindle speed due to dominance of shearing capabilities.
- Cutting forces increase with increase in reinforcement of SiCp under similar condition due to closed packing of the hard SiCp particles which leads to frequent interaction between tool and SiCp.
- Feed of 1.5 µm/tooth and spindle speed of 2500 rpm gave moderate cutting forces and hence suggested during low-speed micro-milling.

References

1. Cardoso P, Davim JP (2012) A brief review on micromachining of materials. Rev Adv Mater Sci 30:98–102
2. Bains PS, Sidhu SS, Payal HS (2016) Fabrication and machining of metal matrix composite: a review. Mater Manuf Process 31:553–573

3. Reddy PS, Kesavan R, Ramnath BV (2018) Investigation of mechanical properties of aluminium 6061-silicon carbide, boron carbide metal matrix composite. Silicon 10:495–502
4. Liu J, Cheng K, Ding H, Chen S, Zhao L (2016) An investigation of surface defect formation in micro milling the 45% SiCp/Al composite. Proc CIRP 45:211–214
5. Pratap T, Patra K (2018) Micro ball-end milling-an emerging manufacturing technology for micro-feature patterns. Int J Adv Manuf Technol 94:2821–2845
6. Hasan M, Zhao J, Jiang Z (2019) Micromanufacturing of composite materials: a review. Int J Extrem Manuf 1(012004):1–24
7. Pratap T, Patra K (2018) Fabrication of micro-textured surfaces using ball-end micromilling for wettability enhancement of Ti-6Al-4V. J Mater Process Technol 262:168–181
8. Kumar SPL, Jerald J, Kumanan S, Prabakaran R (2014) A review on current research aspect in tool-based micromachining processes. Mater Manuf Process 29:1291–1337
9. Afazov SM, Zdebski D, Ratchev SM, Segal J, Liu S (2013) Effects of micro-milling conditions on the cutting forces and process stability. J Mater Process Technol 213:671–684
10. Sahoo P, Pratap T, Patra K, Dyakonov AA (2018) Size effect in micro-end milling of hardened P-20 steel. Mater Today: Proc 5:23726–23732
11. Joel J, Xavior MA (2018) Aluminium alloy composites and its machinability studies: a review. Mater Today: Proc 5:13556–13562
12. Hashim J, Looney L, Hashmi MSJ (1999) Metal matrix composites: production by the stir casting method. J Mater Process Technol 92–93:1–7

Simulation

A Hybrid Slicing Method to Eliminate the Need of Support Structures in Direct Energy Deposition and Material Extrusion-Based Processes

Ritam Sarma, Sajan Kapil, and S. N. Joshi

1 Introduction

Because of the ability for the realization of geometrical and material complexities, AM has become one of the integral parts of modern manufacturing technologies. It has transformed modern manufacturing processes to a different level by allowing industries to cope up with the competitive market of the present world. The total automation of the process makes it an important component of Industry 4.0. AM process starts with the slicing of the digital model of the component. In the traditional planar slicing strategy, the slicing plane is perpendicular to build direction. Overhanging features of the CAD model require support structures during fabrication in lower kinematics machines. These support structures are required to be removed later on once the fabrication of the entire component is over. The process of removal of the support structure is a very tedious and time-consuming task. Building the support structure also contributes to the increase in the energy consumption of the process. In most cases, human intervention is required resulting in an increase in the lead time of manufacturing. The cohesive force of the feedstock materials triggers damage to the parent component during the removal of the support structure. In such cases, the support structures can be minimized by reorienting the build direction such that the least support is required to build the component [1–3]. Some researchers have proposed the use of different sacrificial material for support mechanisms so that it can be removed later on by dissolving into warm water, acid bath, ultrasonic vibrations, etc. For example, PLA is used as a sacrificial material for a parent part built with ABS in *fused deposition modeling* (*FDM*). The sacrificial material can be later on removed by dipping the entire part into isopropyl alcohol and potassium hydroxide solution [4]. Similarly, Hopkins [5]

R. Sarma (✉) · S. Kapil · S. N. Joshi
Department of Mechanical Engineering, Indian Institute of Technology Guwahati, Guwahati, Assam, India
e-mail: ritam@iitg.ac.in

S. Bag et al. (eds.), *Next Generation Materials and Processing Technologies*, Springer Proceedings in Materials 9, https://doi.org/10.1007/978-981-16-0182-8_23

and Ni et al. [6] also advised the use of different polymer for water-soluble sacrificial materials. While in the case of metal AM, the sacrificial material should have similar weldability as that of the parent material. Hildreth et al. [7] used carbon steel as a sacrificial material for fabricating stainless steel components. Finally, carbon steel was eradicated by electrochemical etching of nitric acid and bubbling oxygen molecules. However, the practice of etching of sacrificial material materials in the case of metal AM may erode the parent material also and are limited to a certain pair of materials only. Some other structures (lattice structure, pin structures, etc.) of similar materials can be seen in the literature to be used as a support mechanism to minimize the effort required in taking it away from the final component during the finishing process. However, it consumes a substantial amount of time during milling and grinding. Sometimes, it becomes difficult to strip off the support from certain positions even by using higher kinematics milling machines also. Moreover, the effect on the surface roughness of the final component due to the support structures cannot be ignored. As it was already discussed, by virtue of higher kinematics, the need for support structures can be shrunk by an ample amount. Few researchers have shown the capability of five-axis kinematics in eliminating support structures. The addition of two rotary axes to the bed together with the three linear axes helps in capturing the materials for overhanging features by tilting the bed in such a way that it won't require any support mechanism. The authors have also classified the different AM processes into two broad categories based on the limitation in kinematics during deposition: (a) 2.5-axis deposition and (b) 5 or higher axis deposition. The classification is based on the fact that there are certain AM processes that are not limited to 2.5-axis deposition and can be incorporated higher kinematics for deposition as shown in Fig. 1. The AM processes such as SLS and SLM are limited to 2.5-axis layer deposition while in case of different DED and material extrusion-based processes, higher axis can be incorporated for realization of layers.

Dutta and Singh [8] developed a machine-independent framework for the decomposition of the CAD model into buildable and non-buildable sub-volumes

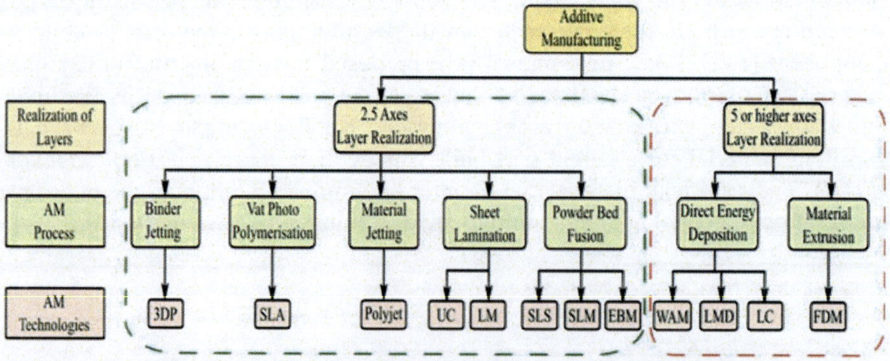

Fig. 1 Classification of AM based on kinematics limitation in layer realization

along a given build direction based on silhouette edge. The build directions for non-buildable volumes were evaluated with the help of Gauss Map [9] or visibility map [10]. The subvolumes were then sequenced in such a way that it would be possible to deposit material either in the substrate or in the pre-built subvolumes without any nozzle collision with the pre-built subvolumes. However, the subvolumes that would require support were not demonstrated in their work. A similar approach was followed by Sundaram and Choi [11] for building a component in the Z-direction using 5-axis laser assisted direct metal deposition process. The unbuildable subvolumes were deposited by reorienting the pre-built volume around Z-axis and Y-axis such that it is possible to capture the material on pre-built volume without a support structure. Ding et al. [12] proposed a decomposition—regrouping strategy for components with holes. The entire component was decomposed into different subvolumes based on curvature criterion followed by generation of topological depth tree for sequenced slicing of the subvolumes. Similar to the approach followed by Dutta and Singh, the build directions for subvolumes were identified by Gauss Map [9]. The initial orientation of the CAD model was taken as a user input, hence making it less suitable for total automation. Moreover, the algorithm was limited to closed concave loops during the decomposition of the volumes. Singh and Dutta [13] demonstrated another method of offset slicing for machines having multiple degrees of freedom for deposition. The non-buildable volumes were identified with the help of angle between the normal to the offset to the base and the component surface at their intersection point. The non-buildable volumes were further subdivided into unprocessed volumes based on practical constraints. Ding et al. [14], in 2017, presented a method of slicing for complex revolved parts. Based on the Silhouette edges, the component was decomposed into different subvolumes. Planar slicing was performed for the core volume while the other subvolumes were sliced using cylindrical coordinate systems, taking the benefit of revolved shape. They manufactured a propeller using 8 axes robotic DED system. However, the algorithm was limited to specific shaped objects and required higher axis kinematics, making it confined to a particular set of machines only. Alkaladi et al. [15] presented an algorithm for depositing material over a freeform substrate. But the method was restrained to thixotropic ink only. The demonstration of components with an overhanging feature that may require support was also missing. Yuan et al. [16] developed an algorithm for multi-directional slicing of STL files using MATLAB following a similar approach. It performed horizontal planar slicing directly for the buildable volumes while they transformed the slicing plane perpendicular to the build directions for each of the non-buildable subvolumes and then planar slicing was performed for the same. The method was not capable of completely eliminating the support for certain complex-shaped objects. It is not necessary to perform only nonplanar slicing for realizing overhanging features without a support mechanism. Kapil et al. [17] have attempted to develop an algorithm for planar slicing of components with undercut without the need for support structures through the use of five-axes kinematics. The researchers were able to fabricate an impeller without support structure by using the proposed algorithm in a WAAM setup. However, the algorithm was not capable of handling abrupt changes in geometry.

From the literature, it can be concluded that through the use of higher kinematics, and it is possible to eliminate the need for the support structure. However, the higher the axis involved in the machine, costlier the process would become limiting its application to sophisticated industries. Moreover, a generic algorithm that has the capability of handling complex-shaped components is also missing. Many researchers have also emphasized five-axis machining for developing non-uniform layers while dealing with five or higher axis based deposition methods. In this work, the authors have proposed a hybrid slicing method for depositing the layers continuously through the five-axis CNC machines or robotic manipulators. The proposed strategy has been demonstrated through CAPP of different illustrative examples. Finally, the process planning of an airplane model is shown by using the proposed hybrid slicing strategy without the requirement of any support structure. This methodology can be implemented to any AM processes that have the capability of five or higher axis deposition of layers, as shown in Fig. 1.

2 Methodology

Slicing may be considered as the process of finding the intersection between the CAD model and the slicing plane. Based on the layer height, it can be classified into two categories, viz. (a) *Uniform slicing* and (b) *Adaptive slicing*. In the case of uniform slicing, the layer height remains constant throughout the process of slicing, while in the later mentioned strategy, the layer height is varied whenever there is a change in the curvature of the component. On the other hand, based on the geometry of the slicing plane, slicing can be grouped into two more categories: (i) *Horizontal planar slicing* in which the slicing plane is horizontal and (ii) *Nonplanar Conformal slicing*. Apart from the aforementioned criteria, slicing can be further categorized based on the order of edge approximation: (A) *Zero-order approximation*, (B) *First-order approximation*, and (C) *Higher-order approximation*. Higher-order edge approximation leads to the elimination of certain inherent errors of the AM process, such as the *Staircase effect*. In most cases, zero-order edge approximation and first-order edge approximations can be seen in practice by the researchers. The detailed classifications of slicing are shown in Fig. 3. Hybridization may be considered as the practice of availing benefits of two or more different processes by combining them together. The different possibilities for hybridization of slicing are shown in Table 1.

The hybridization of uniform layer slicing and adaptive layer slicing is most commonly used by the researchers for realizing the overhanging feature with minimal requirement of support [18–22]. It should be noted that up to a certain

Table 1 Different possibilities of hybrid slicing strategies

Edge approx.	Slicing plane	Uniform layer thickness	Adaptive layer thickness
Zero-order edge approximation	Horizontal	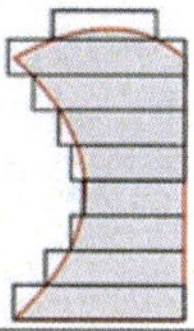	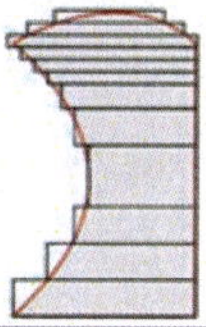
	Conformal	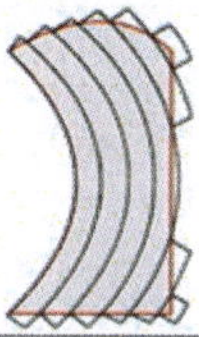	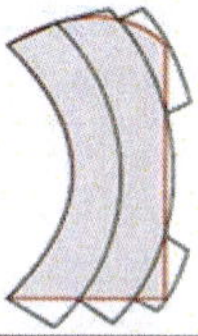
First-order edge approximation	Horizontal	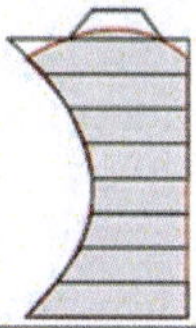	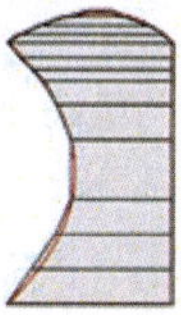
	Conformal	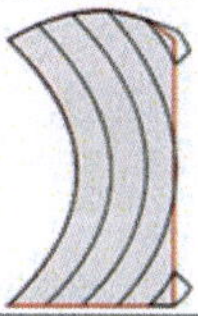	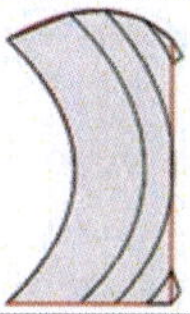
Higher-order edge approximation	Horizontal	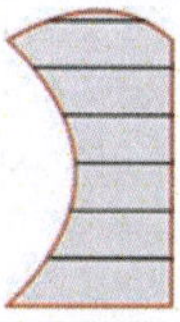	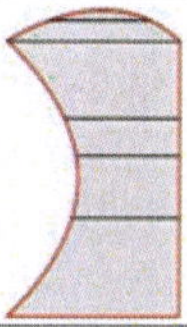
	Conformal	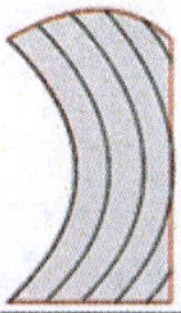	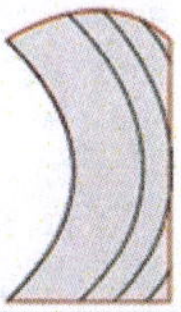

angle of sloping, it possible to realize the overhanging feature without support. This limiting angle is influenced by different factors such as surface tension, cohesive force, and gravitational force and is usually found to be 30° to the vertical axis. In other words, a slanted wall of mild steel can be deposited by virtue of 3-axis kinematics up to an angle of 30° to the vertical axis without support in the WAAM process. Once this limiting angle is exceeded, external supports need to be provided to deposit the subsequent layers. Some researchers have proposed the use of a smaller layer height to avoid excessive support. However, lower layer height results in an increase in the build time [23]. Adaptive slicing also helps in the reduction of support up to a specific limit. The maximum possible layer height can be used for slicing the region without overhang. This maximum possible layer thickness is governed by the machine hardware. For example, the nozzle diameter in the case of FDM is considered to be one of the important parameters influencing the maximum possible layer thickness. The layer height can be adaptively adjusted as the curvature of the component is changed. This will help in capturing the material for depositing overhanging features through cantilever action. The process can be illustrated with the help of slicing a tapered component with an internal duct, as shown in Fig. 2. The entire component can be divided into two regions based on the overhang. In region I, the maximum possible layer thickness can be utilized to deposit material. Once the internal duct is encountered, the layer thickness is adaptively decreased as the curvature of the duct is also varying. With an increase in the slope of the curvature, the layer thickness is required to be decreased in region II. After realizing the duct, the remaining volume can be deposited with maximum possible layer thickness. Hence, this remaining volume can be deposited by utilizing the same parameters as that of region I. In this example, the horizontal planar slicing is performed. This hybrid slicing strategy can be utilized in any additive manufacturing technology as it involves 2.5 axes for the deposition of the layers. The hybridization of uniform–adaptive–conformal nonplanar slicing is also possible in a similar way, as shown in Fig. 5b. The only difference will be it will require 5-axis deposition process.

The proposed hybrid slicing strategy consists of a planar horizontal slicing strategy and nonplanar conformal slicing strategy, as shown in Fig. 3. The former strategy is usually adopted while slicing the volume without overhang. In most of the cases, the overhang features are conformal to some surface. Such features are ideal for performing nonplanar conformal slicing. In such cases, the support can be eliminated by using 5 axes or higher kinematics for deposition. Both the slicing strategies are briefly discussed in the succeeding sections.

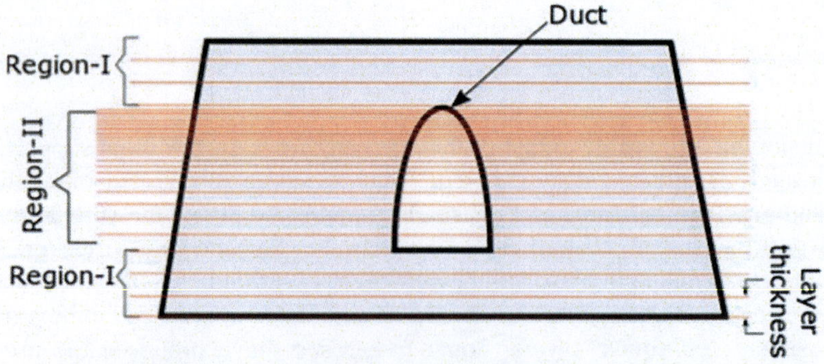

Fig. 2 Uniform—adaptive layer hybrid slicing strategy

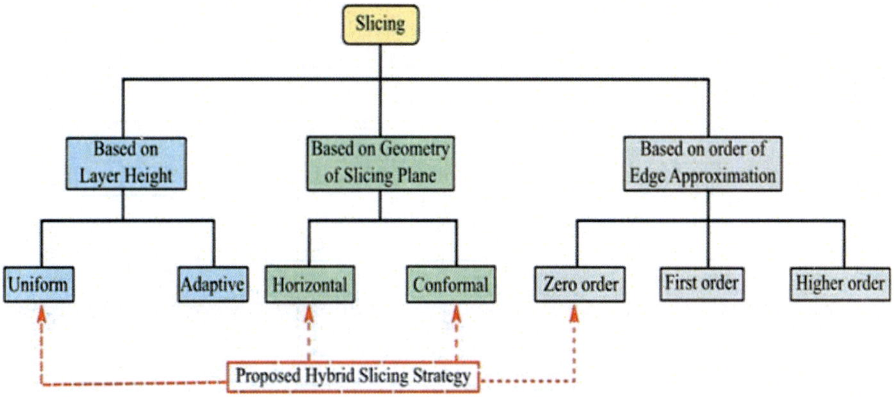

Fig. 3 Classification of slicing along with proposed hybrid slicing strategy

2.1 Horizontal Planar Slicing Strategy

In planar slicing, the slicing plane is perpendicular to the build direction. Hence, each of the obtained slices would have constant z-coordinate if the build direction is aligned along Z-axis. The outer contour of each slice is deposited foremost, followed by the deposition of the inner area of the contour. This process is repeated for the succeeding layers. The volume without overhang can be easily built by using a horizontal planar slicing strategy with 3-axis kinematics. The substrate or pre-built volume for horizontal slicing should be planar. It should be noted that dimensional precision is maintained by performing in situ face milling operation after deposition of each layer in certain DED processes [17]. However, there are some other sophisticated ways of monitoring dimensional accuracy through inspection tools, and by varying process parameters, the errors can be compensated during deposition.

For a given ith slice, $\widehat{n_i}$ is the contact normal vector at a given point. The overhanging feature is required to be identified in the component. This can be done by checking if the angle between the contact normal and the normal to the planar slice is greater than 90°. For a build direction of $+\widehat{k}$ of an overhang feature, the following condition has to be satisfied (Fig. 4).

$$\widehat{n_i} \cdot \widehat{k} < 0 \tag{1}$$

2.2 Nonplanar Conformal Slicing

There are some components in which the overhanging volume may grow conformal to some pre-built volume. In such cases, a conformal slicing strategy is found to be

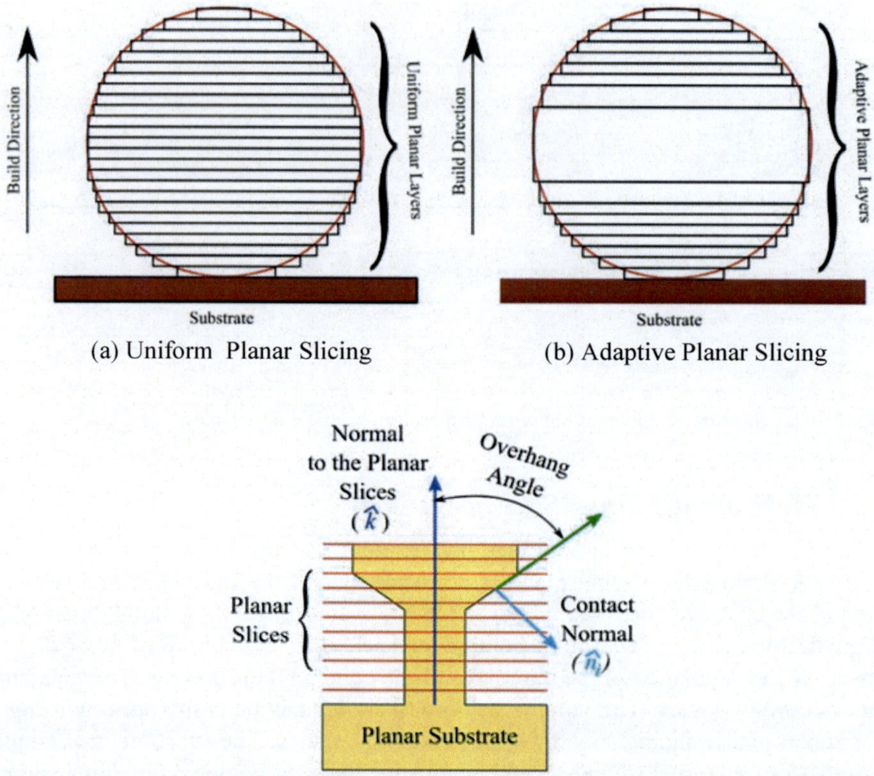

(a) Uniform Planar Slicing (b) Adaptive Planar Slicing

(c) Overhang feature in Planar slicing

Fig. 4 Horizontal planar slicing

more efficient and reliable in realizing the near-net shape of the component. This method can be illustrated with the help of the example, as shown in Fig. 5. In Fig. 5a, the conformal slices of a sphere are shown. The component grows conformal to brown colored pre-built spherical volume or substrate. The layers are deposited on the substrate conformal to it. Hence, the component grows in the radial directions during deposition. It should be noted that the substrate or the pre-built volume are nonplanar in case of conformal slicing. The near-net shape of the deposited sphere would be very close to the final product. In Fig. 5b, the conformal slicing of the yellow feature is shown, which is grown conformal to brown cylindrical volume. The yellow feature has a non-uniform overhanging feature in the middle. Therefore, the layer thickness has to be varied so that it can be deposited without any support. The entire yellow feature can be further divided into two regions. The layer thickness for the region above and below the undercut can be kept maximum. But it has to be adaptively varied while realizing the region with an undercut. The overhang features for conformal slices can be identified in a similar way to that of horizontal slicing as discussed in the previous subsection. For an overhanging feature, the angle between the contact normal and the build direction will be more than 90°. In other words, their dot product will be negative in the case of undercut features. The overhang feature can be identified with the help of the following inequality.

$$\widehat{n_i} \cdot \widehat{M_i} < 0 \tag{2}$$

where $\widehat{n_i}$ is the contact normal vector of the contour at a given point of the ith slice and $+\widehat{M_i}$ is the build direction, normal to the conformal slice at the same point. These vectors are displayed in Fig. 5c. The information of the contact normal can be easily retrieved from the STL files. If the CAD model is in the form of freeform surfaces, the contact normal needs to be evaluated by using the information of the surface of the overhanging feature.

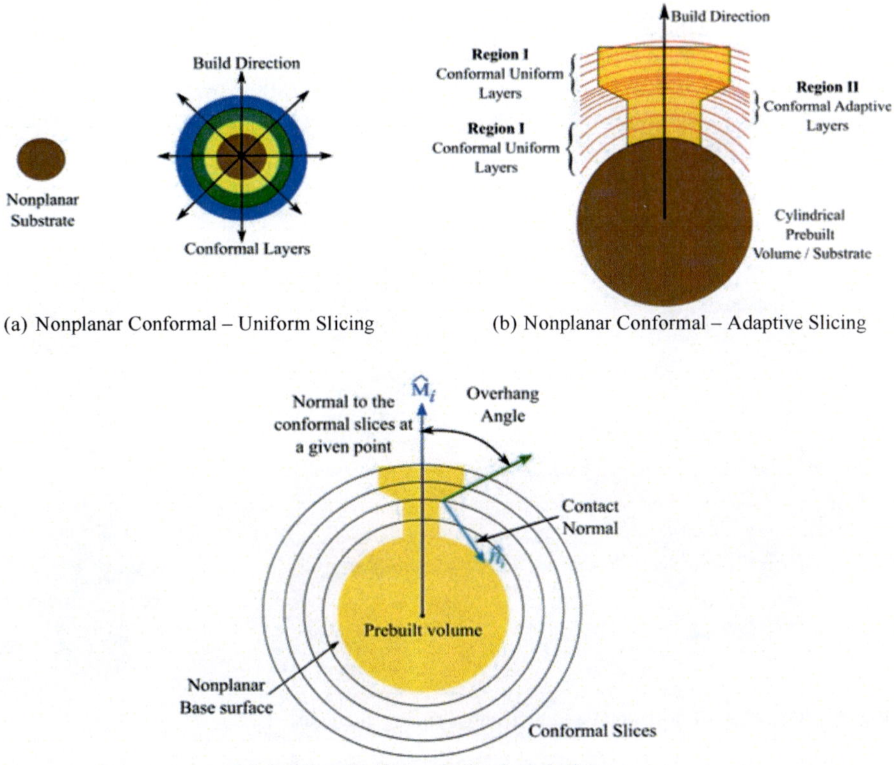

Fig. 5 Nonplanar conformal slicing

2.3 Horizontal—Conformal Hybrid Slicing

The aforementioned horizontal and nonplanar conformal slicing is hybridized in order to avail the advantages of both methods. This can be easily explained with the help of Fig. 6. In this method, the first step is to identify the features according to the feasible build directions. The component can be decomposed into two sub-volumes: green volume and red volume. The green cylindrical core can be easily built-in +Z-direction by using 3-axes kinematics and horizontal planar slicing, as discussed in the previous section. This part will be deposited first by using either uniform layer horizontal slicing. The red-colored subvolume has grown conformal to the green cylindrical volume. Therefore, this subvolume can be realized only after depositing the green part. If horizontal slicing is adopted for this volume too, it would require an enormous amount of support materials from the base. This would have resulted in an increase in the overall built time of the component. Hence, this volume can be sliced by using nonplanar conformal slicing. The slicing planes for this volume will be conformal to the green cylinder and are generated by offsetting the outer surface of the cylinder. The offset distance is governed by the layer thickness. Unlike planar slicing, the contours obtained from conformal slicing have variable x-, y-, and z-coordinates. The contours are deposited initially by using 5-axis kinematics followed by 3-axis infill of the contours. The process is repeated for entire slices. The intermediate slicing planes of horizontal and conformal slicing are shown in Fig. 6.

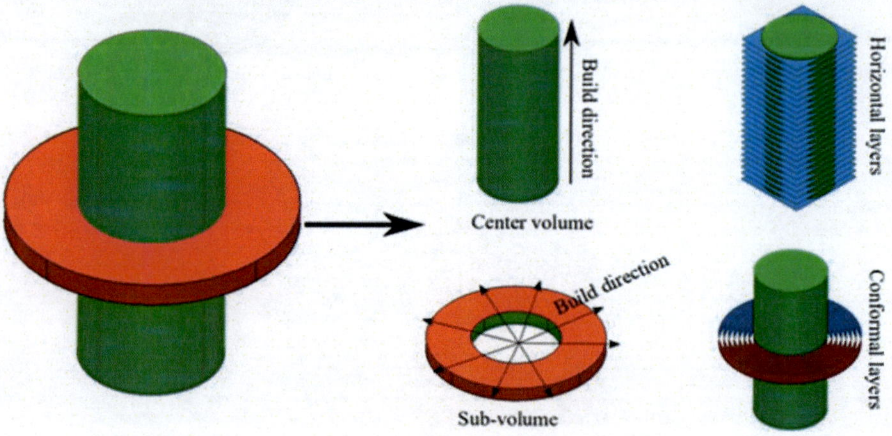

Fig. 6 Illustration of horizontal and conformal slicing strategies

3 Illustration

An airplane like model has been sliced by using the proposed hybrid slicing strategy. The CAD model of the component is shown in Fig. 7a. As already discussed in the previous section, the first step is to decompose the volume and identify feasible build direction of each feature, as shown in Fig. 7b. The entire model can be decomposed into two features: the central body and the wings. The entire procedure of manufacturing is discussed in the subsequent subsections.

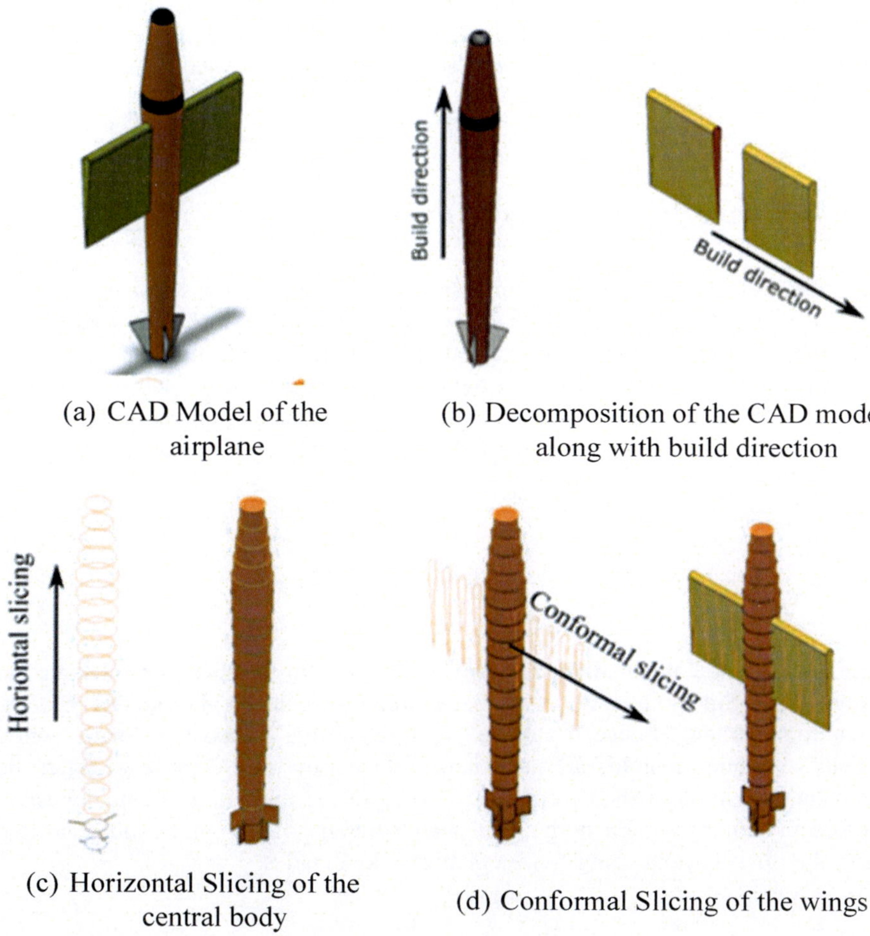

(a) CAD Model of the airplane

(b) Decomposition of the CAD model along with build direction

(c) Horizontal Slicing of the central body

(d) Conformal Slicing of the wings

Fig. 7 Illustration of horizontal planar—conformal nonplanar hybrid slicing strategy

3.1 Slicing

The central compartment does not have any undercut feature as it violates overhang condition, as mentioned in Eq. 1. Consequently, planar slicing can be performed for the central body portion. The contours of the central body obtained after slicing are shown in Fig. 7c. Since wings are identified as overhanging features and can be seen to be grown conformal to the central body, these are sliced using nonplanar conformal strategy so as to avoid the requirement of unnecessary support that would have resulted if horizontal slicing strategy would have been adopted. The slicing plane for nonplanar conformal slicing is obtained by offsetting the outer surface of the central body in the radial direction by layer thickness. Once the slicing planes are obtained, the intersection between the slicing planes and the wings is calculated. The slices obtained after conformal slicing are shown in Fig. 7d.

3.2 Offsetting for Machining Allowance

Due to the relatively high deposition rate of certain AM processes such as in the case of WAAM, poor dimensional accuracies are often reported. Hence, the near-net shape of the component obtained after deposition is required to be finished using the milling process. Therefore, the contours obtained after slicing need to be offset in order to provide a machining allowance for finishing operation. This is an optional step and can be purged in the case of material extrusion-based AM processes.

3.3 Deposition

Once the slices are obtained, the deposition toolpaths for the slices are calculated in the next step. Since the contours obtained from the central body violates the condition for overhang, hence it can be deposited using 2.5-axis kinematics only. Initially, the outer contours are deposited, followed by infill deposition. After the complete deposition of the central body, the layers of the wings can be deposited conformally to the central body. The outer contours of the wings are deposited along the torch vectors using 5 axes or higher kinematics.

3.4 *Finish Milling*

The final finish milling is required to be performed on the near-net shape of the airplane like a model in order to accomplish the desired level of dimensional accuracy and surface finish in case of certain AM processes. The excess material provided for machining is removed in the CNC machine using 5-axis kinematics for milling. Therefore, the toolpaths for the milling process are generated. However, as already discussed in Sect. 3.2, this process is noncompulsory for material extrusion-based AM processes.

4 Conclusion

The need for support structures for realizing overhanging features in the case of metal AM is still recognized as a key issue of the process. However, the use of higher kinematics may abolish the requirement of support mechanisms. In this paper, the authors have presented a hybrid slicing method to eliminate the need for support structures while manufacturing components with overhang. There are different possibilities available for the hybridization of slicing. The hybridization yields advantages of different slicing strategies and improves the quality of the final fabricated component. The hybrid slicing approach comprising of a higher order of edge approximation is found to be one of the least explored areas and has the potential to overcome different inherent issues of AM process. The methodology proposed in this paper consists of horizontal planar slicing and nonplanar conformal slicing strategy. The overhang features are identified using the contact normal vector of the contours and the feasible build direction. The volume without overhang has been sliced using a horizontal planar slicing strategy followed by 2.5-axis deposition of the layers. The overhang features are then sliced using the surface to which it is conformal. The obtained slices are deposited using 5-axis kinematics. The proposed strategy is demonstrated by performing the process planning of an airplane model. A significant reduction in build time, as well as wastage of material, can be observed in comparison with the conventional slicing strategies. The developed slicing method can also be utilized in different AM processes like DED and FDM.

References

1. Luo Z, Yang F, Dong G, Tang Y, Zhao YF (2016) Orientation Optimization in Layer-Based Additive Manufacturing Process. In: Volume 1A: 36th Computers and Information in Engineering Conference, Charlotte, NC, USA, August, p V01AT02A039, https://doi.org/10.1115/detc2016-59969
2. Das P, Mhapsekar K, Chowdhury S, Samant R, Anand S (2017) Selection of build orientation for optimal support structures and minimum part errors in additive manufacturing.

Computer-Aided Design and Applications 14(sup1):1–13. https://doi.org/10.1080/16864360.2017.1308074
3. Cantrell JT et al (2017) Experimental characterization of the mechanical properties of 3D-printed ABS and polycarbonate parts. Rapid Prototyping J 23(4):811–824. https://doi.org/10.1108/RPJ-03-2016-0042
4. Jiang J, Xu X, Stringer J (2018) Support Structures for Additive Manufacturing: A Review. JMMP 2(4):64. https://doi.org/10.3390/jmmp2040064
5. Hopkins PE, Priedeman WR, Bye JF (2012) Manufacturing Systems. US 8,246,888 B2, August 21
6. Ni F, Wang G, Zhao H (2017) Fabrication of water-soluble poly(vinyl alcohol)-based composites with improved thermal behavior for potential three-dimensional printing application: article. J. Appl Polym Sci 134(24), June. https://doi.org/10.1002/app.44966
7. Hildreth OJ, Nassar AR, Chasse KR, Simpson TW (2016) Dissolvable Metal Supports for 3D Direct Metal Printing. 3D Printing and Additive Manufacturing 3(2):90–97, June. https://doi.org/10.1089/3dp.2016.0013
8. Singh P, Dutta D (2001) Multi-Direction Slicing for Layered Manufacturing. J Comput Inf Sci Eng 1(2):129–142. https://doi.org/10.1115/1.1375816
9. Gan JG, Woo TC, Tang K (1994) Spherical Maps: Their Constructions, Properties and Approximation. J Mech Des 116(2):357–363. https://doi.org/10.1115/1.2919386
10. Woo TC (1994) Visibility maps and spherical algorithms. Comput Aided Des 26(1):6–16
11. Sundaram R, Choi J (2004) A Slicing Procedure for 5-Axis LaserAided DMD Process. J Manuf Sci Eng 126(3):632–636. https://doi.org/10.1115/1.1763180
12. Ding D, Pan Z, Cuiuri D, Li H, Larkin N, van Duin S (2016) Automatic multi-direction slicing algorithms for wire based additive manufacturing. Robotics and Computer-Integrated Manufacturing 37:139–150. https://doi.org/10.1016/j.rcim.2015.09.002
13. Singh P, Dutta D (2008) Offset Slices for Multidirection Layered Deposition. J Manuf Sci Eng 130(1):011011. https://doi.org/10.1115/1.2783217
14. Ding Y, Dwivedi R, Kovacevic R (2017) Process planning for 8-axis robotized laser-based direct metal deposition system: A case on building revolved part. Robotics and Computer-Integrated Manufacturing 44:67–76. https://doi.org/10.1016/j.rcim.2016.08.008
15. Alkadi F, Lee K-C, Bashiri AH, Choi J-W (2020) Conformal additive manufacturing using a direct-print process. Additive Manufacturing 32:100975. https://doi.org/10.1016/j.addma.2019.100975
16. Yuan L et al (2020) Application of Multidirectional Robotic Wire Arc Additive Manufacturing Process for the Fabrication of Complex Metallic Parts. IEEE Trans Industry Inf 16(1):454–464. https://doi.org/10.1109/TII.2019.2935233
17. Kapil S, Joshi P, Kulkarni PM, Negi S, Kumar R, Karunakaran KP (2018) Elimination of support mechanism in additive manufacturing through substrate tilting. Rapid Prototyping J 24(7):1155–1165. https://doi.org/10.1108/RPJ-07-2017-0139
18. Tyberg J, Bohn JH (1998) Local Adaptive Slicing. Rapid Prototyping J 4(3):118–127
19. Zhao D, Guo W, Gao F (2020) Research on Curved Layer Fused Deposition Modeling With a Variable Extruded Filament. J Comput Inf Sci Eng 20(4):041014. https://doi.org/10.1115/1.4045637
20. Mani K, Kulkarni P, Dutta D (1999) Region-based adaptive slicing. Comput Aided Des 31 (5):317–333. https://doi.org/10.1016/S0010-4485(99)00033-0
21. Qian X, Dutta D (2001) Feature Based Fabrication in Layered Manufacturing. J Mech Des 123(3):337–345. https://doi.org/10.1115/1.1377282
22. Mao H, Kwok T-H, Chen Y, Wang CCL (2019) Adaptive slicing based on efficient profile analysis. Comput Aided Des 107:89–101. https://doi.org/10.1016/j.cad.2018.09.006
23. Singh P, Dutta D (2003) Multi-Direction Layered Deposition—An Overview of Process Planning Methodologies. In: International Solid Freeform Fabrication Symposium, vol 279, p 10

Discrete Element Analysis of Gravity-Driven Powder Flow in Coaxial Nozzles for Directed Energy Deposition

Ambrish Singh, Sajan Kapil, and Manas Das

1 Introduction

Directed Energy Deposition (DED) is a *Metal Additive Manufacturing (MAM)* technology with applications in part fabrication and repair. It has also found extensive use in laser cladding and surface alloying. The technology uses feedstock, in the form of powder or (and) wire with a heat source, typically a laser, to fabricate parts in a layer-by-layer manner. The heat source first creates a melt pool on the CNC substrate, which is then supplied with feedstock using an appropriate arrangement. Upon solidification of the melt pool, a track is formed with the geometry representing the part contour for a given part cross-section (layer). Generally, due to ease of operation, powder feedstock handling in DED is carried out pneumatically, as illustrated in Fig. 1. Metered powder is conveyed to the melt pool using a stream of inert gas. This results in high particle velocity leading to poor catchment efficiency, decreased surface finish, and improper aspect ratio of the deposited track [1–4]. Pneumatic systems also increase the cost of operation due to increased consumption of inert gas.

The study presented here tries to mitigate some of the challenges associated with a pneumatic system through a gravity-based, non-pneumatic feedstock handling unit. Several aspects of such a gravity-based system are explored in this work, aimed primarily at identifying key design considerations and operating conditions most favorable for its implementation. *Discrete Element Method (DEM)* simulations are used as the method of investigation, providing insight into several critical flow phenomenons and parameters affecting them.

A. Singh (✉) · S. Kapil · M. Das
Department of Mechanical Engineering, Indian Institute of Technology Guwahati, Guwahati, India
e-mail: singh176103107@iitg.ac.in

S. Bag et al. (eds.), *Next Generation Materials and Processing Technologies*, Springer Proceedings in Materials 9, https://doi.org/10.1007/978-981-16-0182-8_24

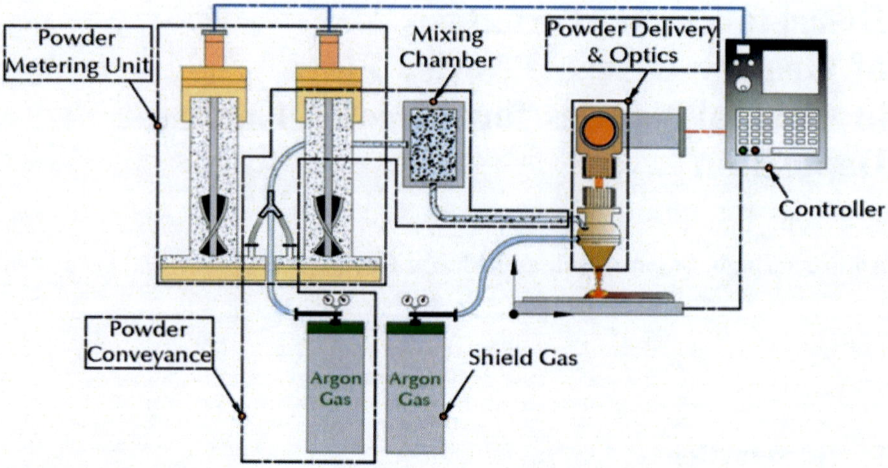

Fig. 1 Schematic illustration of a DED unit with pneumatic powder handling

1.1 Powder Feedstock Handling in DED

The handling of powder feedstock in DED can be categorized into three phases; these are metering, conveyance, and delivery. In the metering phase, a preset value of powder is isolated from the feedstock reservoir. This metered amount of powder is transported to the delivery head (or nozzle) in the second phase of conveyance. In the third and the final phase, powder delivery nozzles create a convergent powder stream with a focal plane matching the location of the melt pool. These three phases of feedstock handling and the methods through which they can be accomplished are illustrated in Fig. 2. This work is aimed exclusively at the final phase of powder delivery.

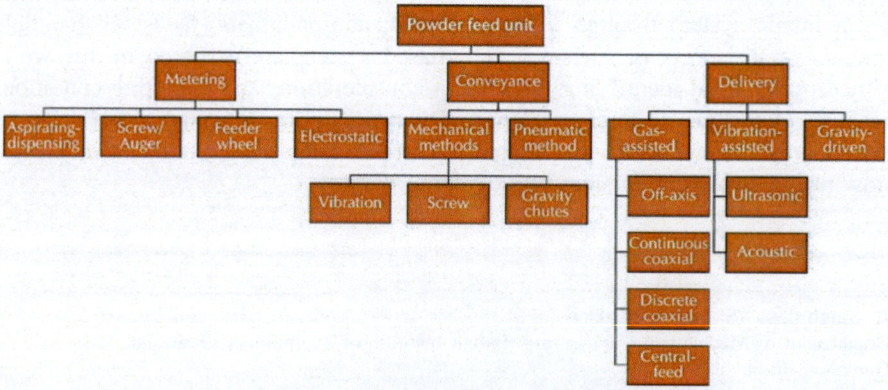

Fig. 2 Classification of methods for powder handling in DED at various phases

1.2 Powder Delivery Nozzles

Several designs of powder delivery nozzles exist in the literature. Some of these design features a central heat source with a coaxial and conical feedstock delivery [5], while others are based on centrally fed feedstock surrounded by a coaxial and conical heat source [6]. Off-axis nozzles (or lateral feed) are yet another variety, popular with laser cladding of axis-symmetric substrates [4]. This study is focused on DEM analyses of coaxial nozzles with a central heat source, and conical envelope of feedstock material. The two popular variants in this category, are discrete coaxial and continuous coaxial nozzles, schematically shown in Fig. 3a and b, respectively. A continuous coaxial nozzle features an annular channel whereas, a discrete coaxial nozzle contains multiple and distinct streams of powder flow. All coaxial powder delivery nozzles (discrete or continuous) should maintain a uniform particle distribution around the heat source. This ensures uniformity in geometric and material properties of the deposited track [7].

2 Methodology

DEM investigations were carried out to establish the feasibility of a gravity-based powder delivery nozzle. The study primarily aims at identifying propitious operating conditions under which, implementation of such a system will yield desirable performance. The study address geometry of the conical nozzle and the effects it has on the powder stream and, consequently, on the deposition characteristics.

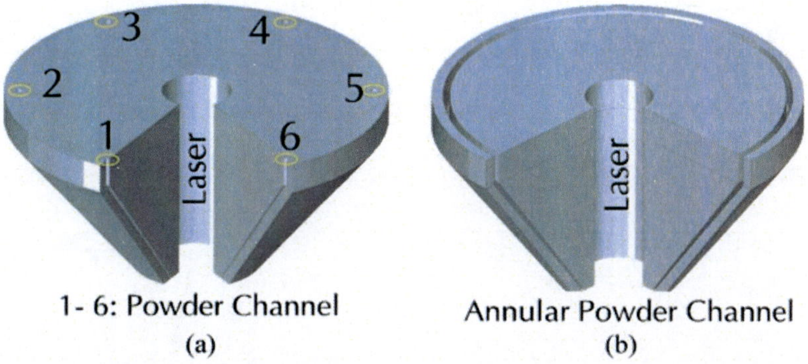

Fig. 3 **a** Discrete and **b** continuous coaxial nozzle with circular and annular powder outlet, respectively

2.1 Effect of Nozzle's Cone Angle

The powder delivery nozzle closely represents a frustum of a cone in geometry, thus allowing the resulting powder stream to be conical and convergent in nature. A cone angle, in the present context, is defined as the angle of inclination of the conical nozzle surface the vertical, shown schematically in Fig. 4. Cone angle significantly affects the powder stream, specifically the plane of convergence (or focal length) and the 'depth of focus.' The 'depth of focus' indicates the distance till which a powder stream remains convergent, having converged at the focal point. Measurements for focal length and 'depth of focus' are made from the nozzle outlet parallel to the laser beam axis. Five values of cone angle, i.e., 55°, 45°, 35°, 25°, and 15°, as illustrated in Fig. 4a–e, were selected to study the effect of cone angle on the powder stream.

2.2 Effect of Head Tilt

As mentioned previously (Sect. 1), the distribution of feedstock around the laser beam is an important factor in nozzle design. Uniformity is this distribution provides accurate builds with better part properties. In order to determine the mass distribution within a conical feedstock envelope, generated by a continuous coaxial nozzle, multiple measurements were made in the 'XY' plane at a fixed location in 'Z.' Consider Fig. 5; virtual bins marked 1–18 records the average powder mass

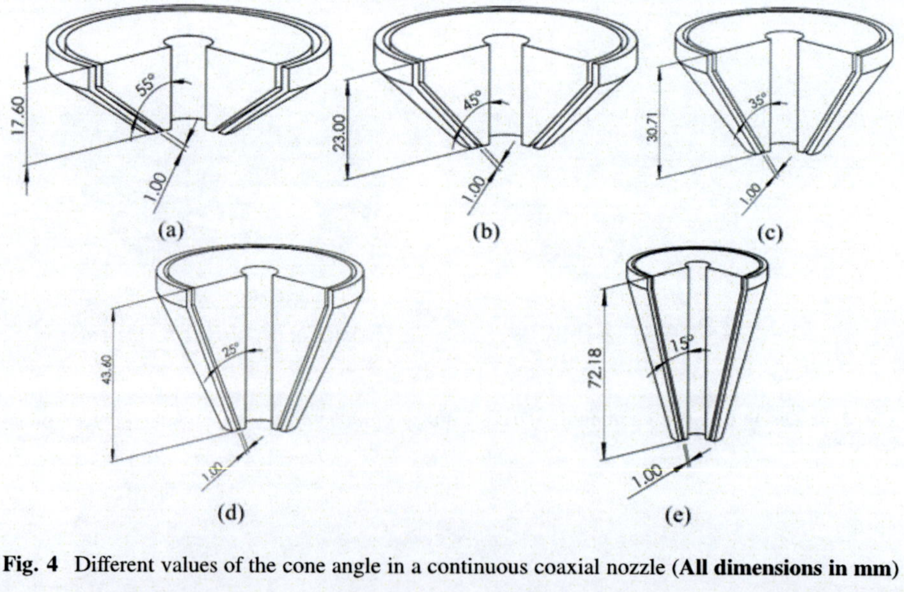

Fig. 4 Different values of the cone angle in a continuous coaxial nozzle (**All dimensions in mm**)

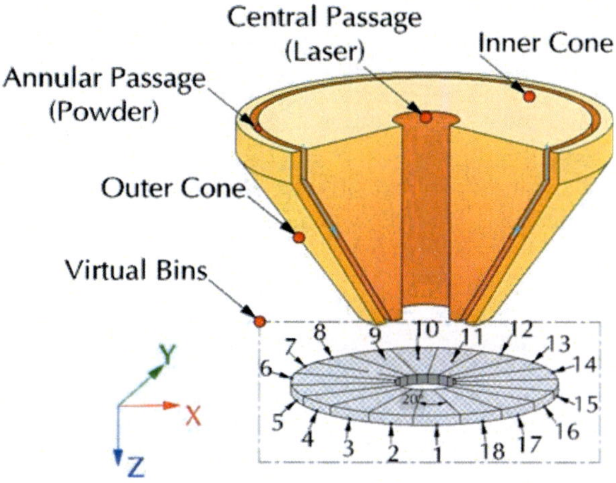

Fig. 5 Virtual bins (mass sensors) placed symmetrically around the central vertical axis

traversing across the bin's domain during the set simulation time. These bins are symmetrically placed, in a circular manner, around the central axis of the nozzle at a distance of 3 mm from the outlet. If the variation amongst each of the bins recorded powder mass is low, then it can be inferred that each of the 18 bins receives (more or less) equal powder quantity. This will signify a uniform powder distribution. If, on the other hand, variation amongst each of the bins recorded powder mass is high, with some of the bins receiving higher powder quantity as compared to others, a non-uniform powder distribution can be inferred. As the head tilt angle increases, a shift from uniform powder distribution to a non-uniform one is expected. The percentage 'coefficient of variation' amongst the average powder mass recorded by each of the bins can indicate the spread of the dataset. In order to implement head tilt in the simulation, the direction of the gravity force vector was manipulated while keeping the magnitude constant. Table 1 shows the head tilt angles along with the orientation of the gravity force vector.

2.3 *Continuous and Discrete Coaxial Nozzle: A Comparative Study*

A distinction is made between a continuous and a discrete coaxial nozzle. The amount of literature on a comparative study between these two is limited, especially

Table 1 Orientation of gravity vector for various values of nozzle inclination

Head tilt angle	0°	2°	4°	6°	8°	10°
g_z (m/s^2)	9.81	9.80	9.78	9.75	9.71	9.66
g_x (m/s^2)	0	0.342	0.68	1.02	1.36	1.70

in the context of a gravity-based design. This part of the study compares discrete and continuous coaxial nozzles in terms of stability of operation. Granular matter, when subjected to flow under gravity in confined channels, is prone to jamming. This makes the selection of channel geometry and dimensions an important consideration in nozzle design. These two variants of nozzles are compared with each other in the context of ease of powder flow in order to determine the minimum dimensions of the flow channel necessary for stable and consistent powder delivery.

3 Discrete Element Method

In contrast to the continuum approach, a discrete approach places emphasis on individual particles, which, when assembled, gives the macroscopic (or overall) outlook. In other words, in a discrete approach, the macroscopic behavior of a system results from microscopic interactions within the constituent particles. The discrete approach is especially helpful in studying phenomena occurring at length scales comparable to particle diameter. Thus, given the advantages of the Discrete Element Method (DEM) in studying granular flow and other bulk material behavior, EDEM™ was used for granular flow simulation. Through multiple simulations with varying geometry of nozzle cones, preliminary results concerning conducive operating conditions for a gravity-based powder delivery nozzle were obtained.

3.1 Theoretical Background

The collision of granular material can be modeled either by using a hard or a soft-sphere approximation. The hard-sphere model, as illustrated in Fig. 6a, depends on the advanced prediction of a collision based on particle trajectory [8]. This model assumes particles to be geometrically rigid and collisions to be instantaneously occurring over a point of contact; thus, resulting in poor accuracy especially, in the simulation of multi-particle interactions. A soft-sphere model, on the other hand, introduces a parameter called 'overlap (δ).' This parameter allows two geometrically rigid particles to overlap during a collision, thus converting an instantaneous collision to a finite duration collision. The finite duration collision allows for multi-particle interaction with an incorporation of dissipative effects such as friction and particle (elastic) deformation. Also, unlike hard-sphere model which assumes point contact, a soft-sphere model features an area contact during collision. Therefore, a soft-sphere model (illustrated in Fig. 6b) is more accurate in simulating bulk matter behavior when compared with hard-sphere model [8–11].

The overlap parameter (δ) can be resolved into normal δ_N and tangential δ_T components. The magnitude and direction of these components govern the forces acting on a particle, which subsequently sets the trajectory of motion. The detection

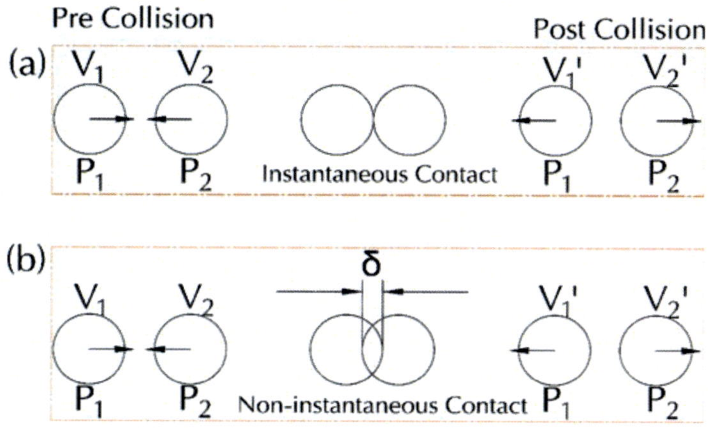

Fig. 6 Particle interaction based on **a** hard-sphere and **b** soft-sphere model

of collision is the most computationally intensive aspect of a DEM simulation. If the center-to-center distance for two spherical particles is less than the summation of their radii, a contact is marked.

Generic DEM Algorithm

A generic soft-sphere DEM model is illustrated in Fig. 7. The forces on particles due to inter-particle or (and) particle-geometry contact (or due to user-specified initial conditions) are computed first. Using Newton's second law of motion, corresponding to the forces, acceleration, velocity, and subsequently, the position of the particles is computed through numerical integration. A contact detection algorithm is used to keep the computation time reasonable while maintaining acceptable accuracy. Each particle within a granular flow has six degrees of freedom, resulting in two types of motion: translation and rotation. Given the force obtained from the contact model, Newton's second law is used to calculate translational and rotational accelerations, which are then numerically integrated over a time step to arrive at new particle states.

For rotation,

$$I\frac{d\omega}{dt} = M \tag{1}$$

For translation,

$$m\frac{dv}{dt} = F_g + F_c + F_{nc} \tag{2}$$

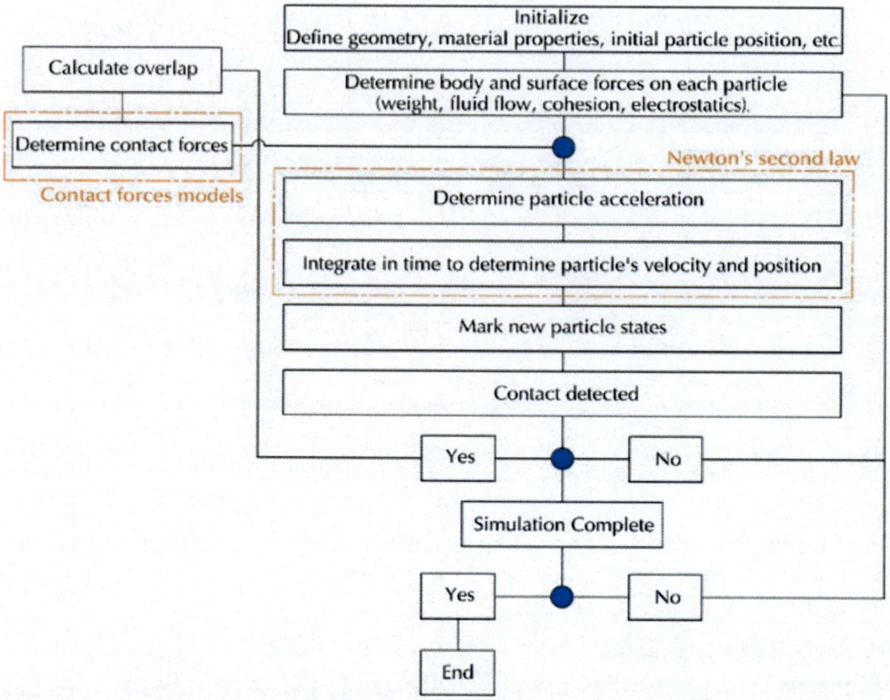

Fig. 7 A generic approach of a soft-sphere DEM model

where,

I = Moment of inertia
v = Translational velocity of a particle
ω = Angular velocity
m = Mass of particle
M = Resultant torque on a particle
F_g = Resultant gravitational force
t = Time
F_c; F_{nc} = Contact and non-contact force

3.2 *Particle Contact Model*

The contact forces can be categorized into two components (i) Normal (F_N) and (ii) Tangential (F_T) [12]. The amount of overlap (δ) can be expressed as

$$\delta = (R_1 + R_2) - d \tag{3}$$

where, R_1 and R_2 are the particle radii, and 'd' is the distance between the particle centers (C_1 and C_2), as illustrated in Fig. 8. The study presented here is based on Non-Linear Elastic Contact and is modeled based on the Hertz-Mindlin No-Slip theory.

According to Hertz theory, as described by Horabik et al. [12], for two contacting spheres over a finite circular area with radius 'a' the normal contact force is expressed as

$$F_N = \frac{4E^*}{3R^*} a^3 \tag{4}$$

where E^* and R^* are effective young's modulus and particle radius, expressed as

$$E^* = \frac{E_i E_j}{E_i\left(1 - v_j^2\right) + E_j(1 - v_i^2)} \tag{5}$$

$$R^* = \frac{R_i R_j}{R_i + R_j} \tag{6}$$

E_i, E_j and v_i, v_j denotes Young's moduli and Poisson's ratio of the two contacting spheres of radii R_i and R_j. The normal contact force is related to the degree of particle overlap (δ) using the relation

$$F_N = k_N \delta_N^{3/2} \tag{7}$$

where the spring stiffness (k_N) is expressed in terms of effective Young's modulus (E^*) and Particle Radius (R^*) of the two contacting particles.

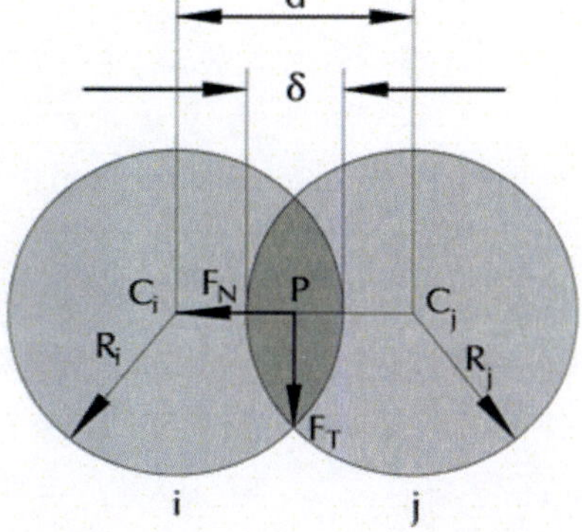

Fig. 8 Normal and tangential force acting on particles during contact

$$k_N = \frac{4}{3} E^* R^{*\,1/2} \tag{8}$$

The radius of the circular contact area is related to overlap (δ_N) by the expression

$$a = \sqrt{R^* \delta_N} \tag{9}$$

The pressure distribution over the circular area of contact is given by the expression

$$p(r) = \frac{3F_N}{2\pi a^2}\left[1 - \left(\frac{r}{a}\right)^2\right]^{1/2} \tag{10}$$

In addition to the normal force, a normal damping force (F_N^d) is also considered, which is expressed as

$$F_N^d = -2\sqrt{\frac{5}{6}}\beta\sqrt{S_N m^*} V_N^{Rel} \tag{11}$$

where $m^* = \left(\frac{1}{m_i} + \frac{1}{m_j}\right)^{-1}$ is the equivalent mass of two contacting particles '*i*' and '*j*'. V_N^{Rel} is the normal component of the relative velocity, β (depends on the coefficient of restitution 'e') and S_N which are expressed as

$$\beta = \frac{\ln e}{\sqrt{ln^2 e + \pi^2}} \tag{12}$$

$$S_N = 2E^* \sqrt{R^* \delta_N} \tag{13}$$

Similar to the above expressions, which considered the normal component of force, tangential force component (F_T) is expressed as

$$F_T = -k_T \delta_T \tag{14}$$

$$k_T = 8G^* \sqrt{R^* \delta_N} \tag{15}$$

where k_T and δ_T are the tangential stiffness and overlap, respectively, and G^* is the equivalent shear modulus. Additionally, tangential damping (F_T^d) is also incorporated in the model which is dependent on the normal component of the relative velocity $\left(V_T^{Rel}\right)$.

$$G^{*} = \left(\frac{2 - \nu_1}{4G_1} + \frac{2 - \nu_2}{4G_2}\right)^{-1} \tag{16}$$

$$F_T^d = -2\sqrt{\frac{5}{6}}\beta\sqrt{S_T m^{*}}V_T^{Rel} \tag{17}$$

$$S_T = 8G^{*}\sqrt{R^{*}\delta_n} \tag{18}$$

3.3 Modeling Setup

A continuous coaxial nozzle was used to carry out the investigation on the effect of the nozzle's cone angle and head tilt angle on granular flow characteristics. Copper was assigned as the nozzle material, which was provided with a uniform influx of stainless steel powder across its periphery. Some of the relevant modeling parameters are provided in Table 2.

Table 2 Relevant setup parameters of DEM study

Bulk material properties		
1.	Material	Stainless Steel
2.	Solid density	8.05 g/cm^3
3.	Shear modulus	7.7 × 10^{10} Pa
4.	Coefficient of restitution	0.6 [13]
5.	Particle diameter	100 μm
6.	Particle injection rate	30 g/min
Equipment (Nozzle) material properties		
1.	Material	Copper
2.	Solid density	8.96 g/cm^3
3.	Shear modulus	4.5 × 10^{10} Pa
4.	Poisson's ratio	0.35
Simulation parameters		
1.	Simulation time	1 s
2.	Integration time-step	5 × 10^{-8} s
3.	Grid size	2R (Twice the radius of a particle)

4 Results and Discussion

4.1 *Effect of the Cone Angle*

The powder stream characteristics for a continuous coaxial nozzle, with a cone angle of 55°, operating under the effect of gravity is illustrated in Fig. 9. Observations are made along the vertical nozzle axis at multiple locations, starting from the nozzle outlet progressing downwards (in the direction of powder flow) in steps of 2 mm. Sectional planes, in the front view (left side), are marked as 1-1 to 9-9, and the adjacent (right side) figure shows the powder distribution on these planes, as observed from the top. These illustrations obtained from DEM analysis confirm the conical and convergent nature of the powder stream. Also, it can be observed form the illustrations that the powder stream converges approximately 10 mm from the nozzle outlet marking its focal point. Observations beyond this plane show the powder stream to diverge again, primarily due to inter-particle collisions. Based on similar qualitative analysis, focal lengths were obtained for a selection of cone angles.

The test on the effect of cone angle on the powder stream shows a negative correlation with the focal length and the 'depth of focus,' as illustrated in Fig. 10. It can be observed for Table 3, as the cone angle decreases, the focal length increases. The 'depth of focus' significantly affects the standoff distance at which deposition should be carried out. A higher value of 'depth of focus' is linked with stable deposition properties [14]. Qualitatively, the DEM results show an increased value of this parameter with a decreased cone angle. Although it is desirable to have a

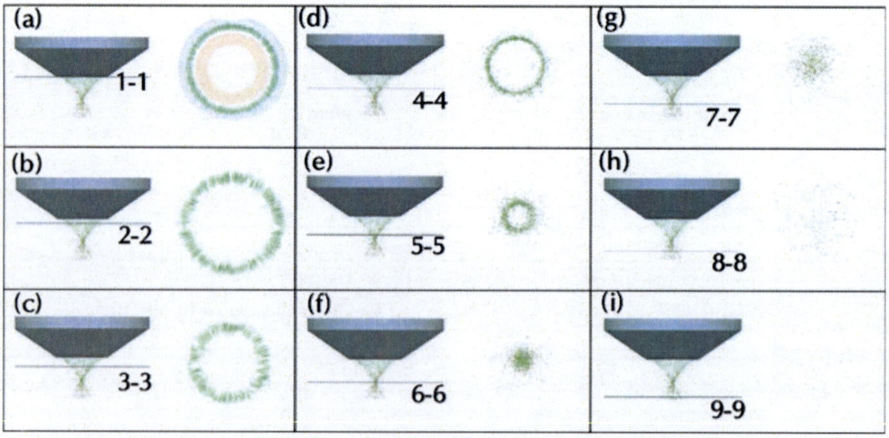

Fig. 9 Powder flow pattern observed at multiple locations measured from the nozzle outlet along the flow stream at **a** 0 mm, **b** 2 mm, **c** 4 mm, **d** 6 mm, **e** 8 mm, **f** 10 mm, **g** 12 mm **h** 14 mm, and **i** 16 mm

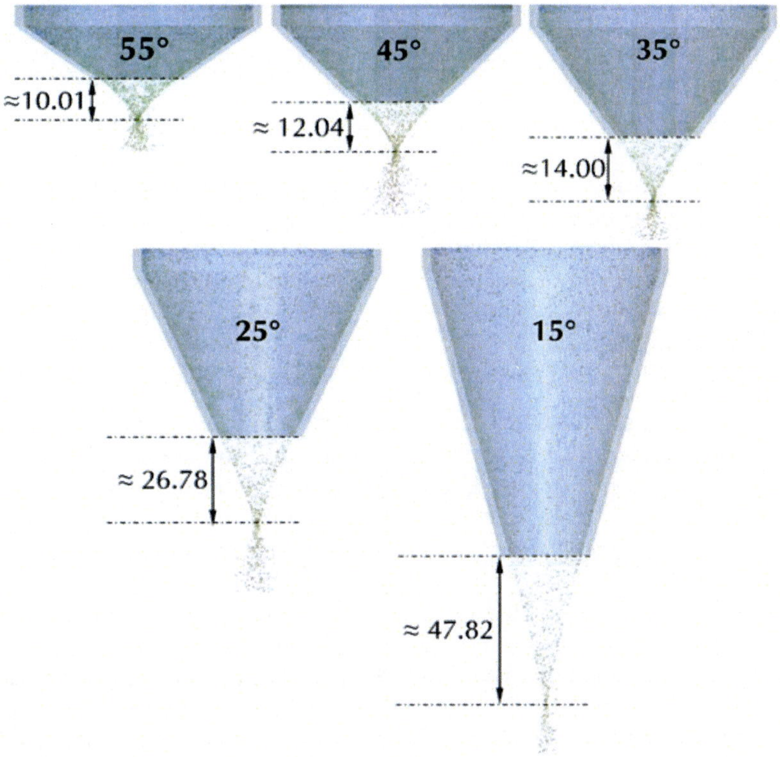

Fig. 10 Increment in the powder stream's focal length with an increase in the cone angle (**All dimensions in mm**)

longer 'depth of focus,' a large focal length often attenuates the laser beam and causes inflight particle vaporization due to excessive radiative heating [15].

Therefore, it is essential to identify the optimal value of the focal length to achieve higher catchment efficiency and a stable range of process parameters.

Powder Particle Distribution at the Focal Length
Through a similar approach of virtual bins, an effort is made to obtain particle distribution at the focal plane of the nozzle with a cone angle of 55°. This analysis is essential to determine the powder stream's focal spot diameter, which plays an important role in determining the catchment efficiency of a DED system. At the focal length, thirty concentric virtual bins were placed, with the radius ranging from 0.1 (innermost bin) to 3 mm (outermost bin) in steps of 0.1 mm, as illustrated in Fig. 11. Each bin records the average powder mass traversing across it within the set simulation time. In order to arrive at the average powder mass passing through an area within two consecutive bins (an annular region), a simple subtraction operation was performed. For instance, subtracting the average mass received by bin 2 (radius of 2.9) from the average mass received by bin 1 (radius of 3.0) will

Table 3 Variations of focal length with the cone angle

Cone angle	55°	45°	35°	25°	15°
Focal length (mm)	≈10.01	≈12.04	≈14.00	≈26.78	≈47.82

provide average mass within the outermost annular region of 0.1 mm. In a similar manner, the average powder mass received by all annular regions, marked as 2, 3… 30 in Fig. 11, was obtained. This data was then plotted against the area of the respective annular regions (obtaining area of the annular regions follows the same subtractive methodology as outlined above), as illustrated in Fig. 12a. This plot of the data, when overlayed with a Gaussian Curve (curve fitting MATLAB R2014a), as illustrated in Fig. 12b, reveals an R-square value of 0.90, confirming a good fit.

Given the Gaussian nature of the curve ($1/e^2$) rule can be applied to determine the focal spot radius. The ($1/e^2$) rule states that; all the areas within which the average powder mass is higher than 13.6% of the maximum can be included in consideration of the focal spot area. Through this rule, the focal radius was determined to be 1.3 mm, and correspondingly, the focal area was calculated to be around 5 mm^2.

As mentioned previously, the spot size of the powder stream at the convergence point governs the catchment efficiency of the deposition. Although a catchment efficiency can be defined in several ways, in this context, comparing the area of the melted substrated with the stream's spot size is most relevant. Any powder particle that falls outside the melted substrate area will not become a part of the track thus, by comparing the areas of the melt pool and the stream spot size, catchment efficiency can be defined as

$$\%\eta = \frac{A_{ms}}{A_P} \times 100$$

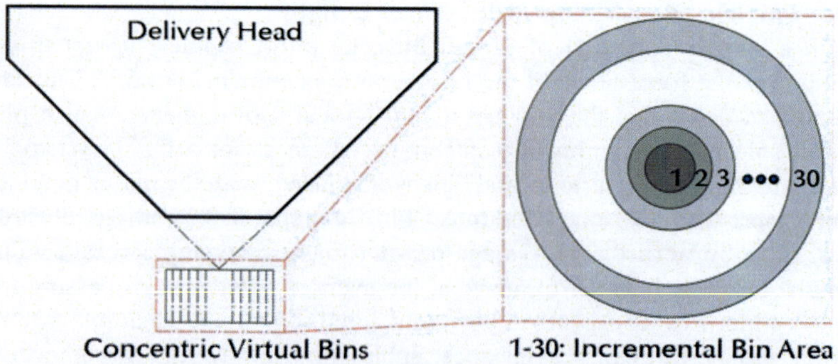

Fig. 11 Virtual bin setup of obtaining particle distribution at the focal plane

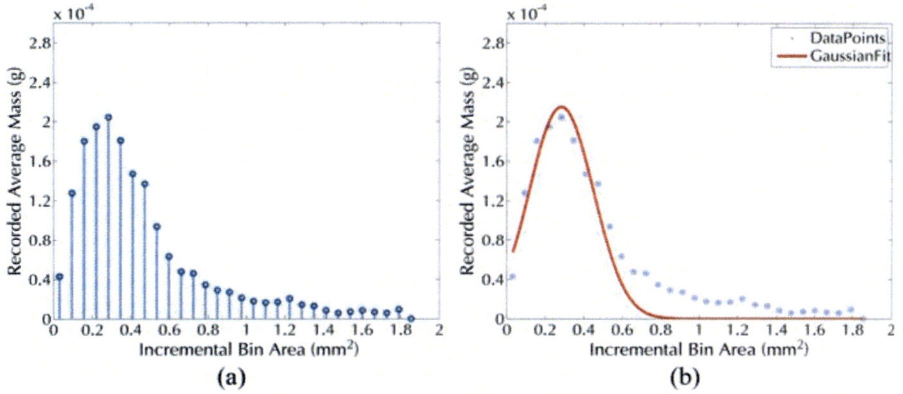

Fig. 12 **a** Average powder mass received in the incremental bin areas and **b** Gaussian fit over the data points

where, A_{ms} is the area of the melted substrate and A_P the spot area of the convergent powder stream. Experiments performed by Goodrazi et al. [4], under a similar mass feed rate of 30 g/min at a laser power of 4500 W and a cladding speed of 400 mm/min, resulted in a melt pool area of 3.17 mm^2. Comparing this substrate melt area with the powder stream spot size results in a theoretical catchment efficiency of 63.4%.

4.2 *Effect of Nozzle Inclination*

In order to evaluate the effect of nozzle inclination on the powder stream behavior, multiple measurements were made around the central nozzle axis. Through the setup highlighted in Sect. 2.2, shown in Fig. 13, a particle distribution was obtained at various values of nozzle tilt angle.

It can be observed from the results that, as the value of tilt increases, the distribution of powder particles becomes increasingly non-uniform. Initially, with zero tilt angle, powder particles are distributed more or less evenly around the central axis. This is inferred from the observation of 'relatively' equal mass received by each of the bins placed symmetrically around the vertical axis, as illustrated in Fig. 14a. As the tilt angle is increased, the particle distribution becomes increasingly non-uniform, with particles favoring one bin over the other. Consequently, on average, some of the bins record more powder particles passing through them as compared with others. Figure 14 shows the increase in the non-uniformity of particle mass with an increase in the tilt angle. The '*coefficient of variation (COV)*,' which signifies the spread of the dataset, also increases with increasing title angle, thus confirming the results.

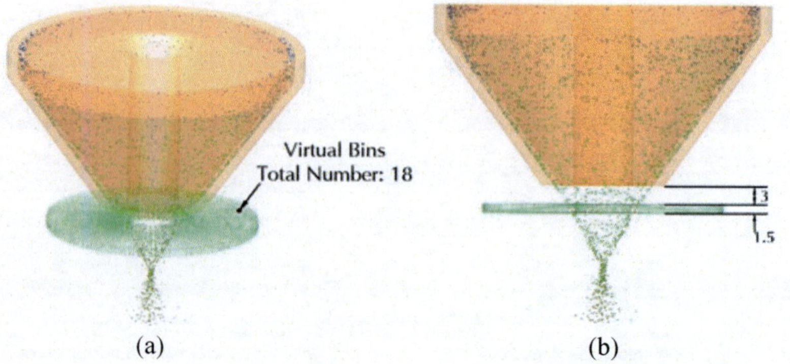

Fig. 13 Implementation of virtual bins (mass-sensors) to measure powder distribution **a** perspective view and **b** location of virtual bins from the nozzle outlet (**All dimensions in mm**)

Furthermore, the results can also be qualitatively verified from Fig. 15, which shows a higher distortion of the powder stream's focal point with an increase in the inclination of the powder delivery head.

Apart from the powder particle distribution, DEM results also suggest an increase in the focal spot as the angle of inclination increases. However, it should be noted that with an increase in the angle of inclination, identifying a distinct focal spot is a challenging task.

4.3 Discrete and Continuous Coaxial Nozzle

As mentioned previously, discrete and continuous coaxial nozzles exhibit a difference in geometry of the powder flow channel. The discrete coaxial nozzle has distinct powder streams, which emerge from circular outlets whereas, a continuous coaxial nozzle has an annular outlet. Figure 16a and c shows the simulation setup for a comparative study of the two nozzle types. Both the nozzles have the same channel dimensions and receive the same mass influx. The discrete coaxial nozzle has six streams, receiving 5 g/min each, whereas continuous coaxial is provided with 30 g/min evenly distributed across its periphery. From the simulation results, it was observed that for an annular flow channel, with the selected material properties, a minimum channel dimension of 'three times the particle diameter (3 × D)' is sufficient to avoid jamming. In the case of cylindrical flow channels with a circular outlet, a minimum dimension of 'five times the particle diameter (5 × D)' is essential to prevent blocking. These results, as shown in Fig. 16, suggest that a continuous coaxial nozzle efficiently avoids particle blocking in gravity-based granular systems as compared to discrete coaxial nozzles.

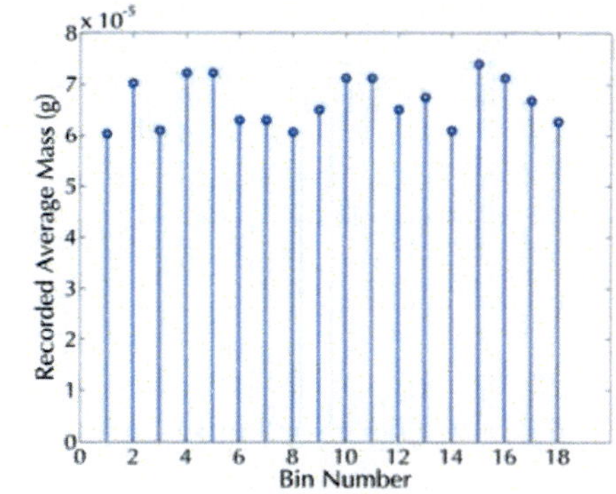

(a) 0° (COV: 6.8%)

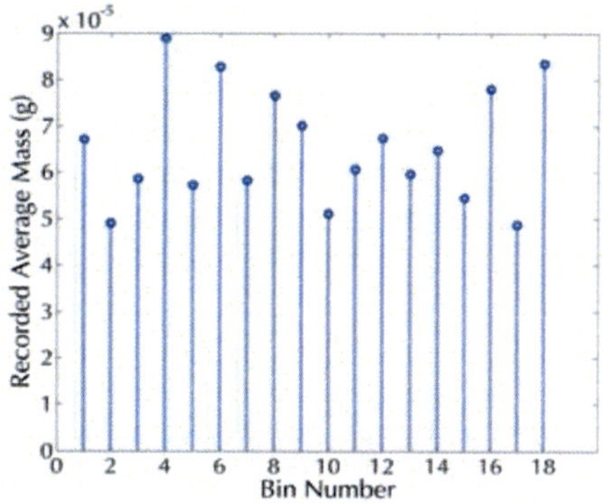

(b) 2° (COV:18.33%)

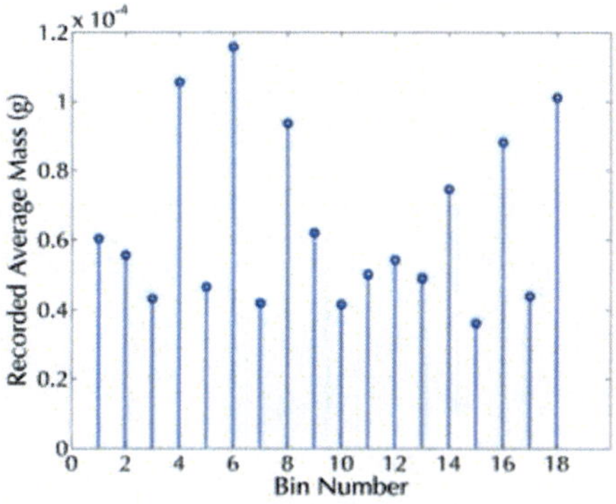

(c) 4° (COV:38.02%)

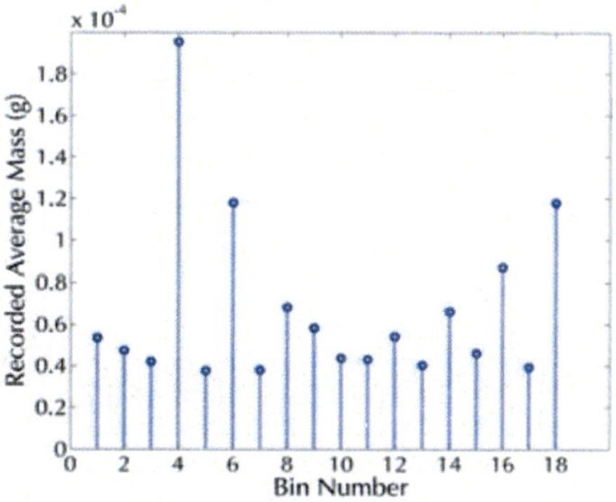

(d) 6° (%COV:59.75%)

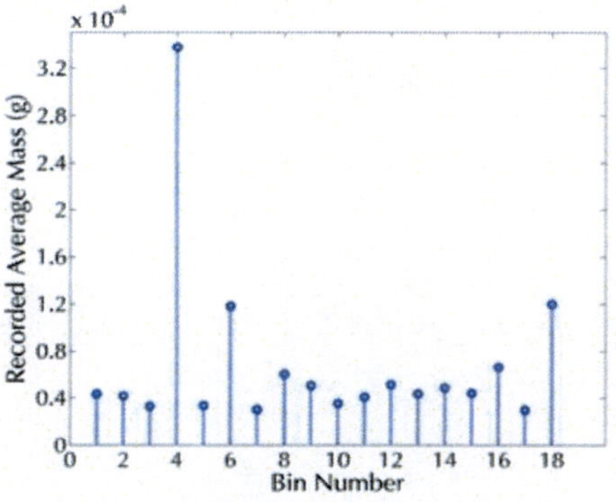

(e) 8° (COV:103.11%)

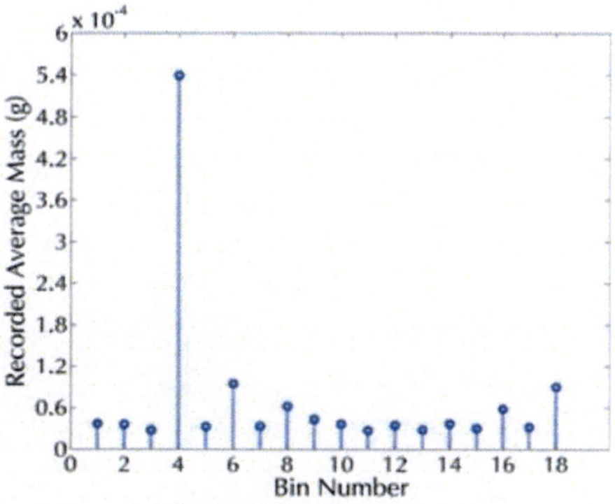

(f) 10° (COV:162.20%)

Fig. 14 Non-uniformity in powder mass distribution with the increased nozzle tilt angle

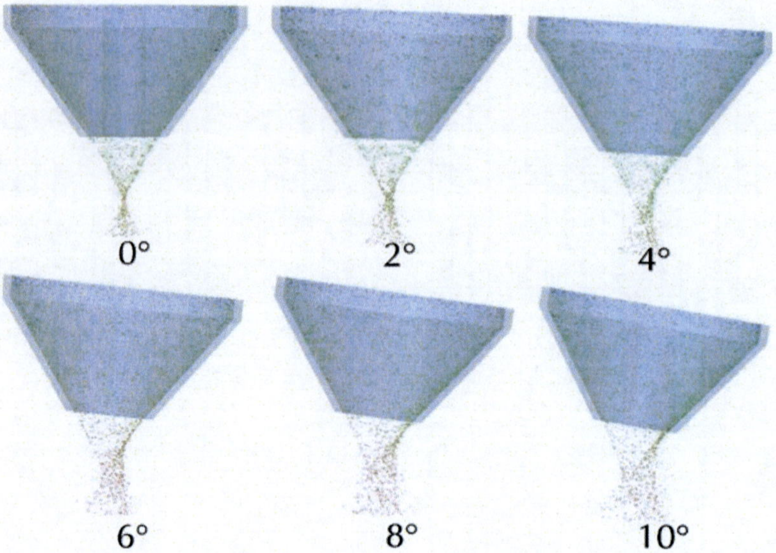

Fig. 15 Distortion of the focal spot and non-uniform powder distribution with increased head tilt

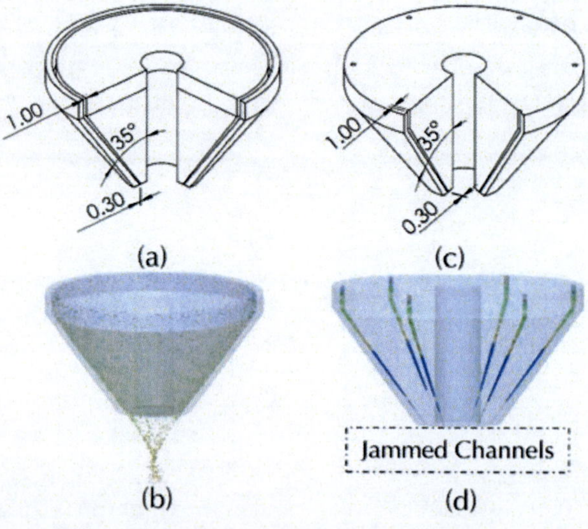

Fig. 16 Jamming observed in **c**, **d** the discrete coaxial nozzle with the same channel dimensions as the **a**, **b** continuous coaxial nozzle

5 Conclusion

Through multiple DEM simulations, the feasibility of the gravity-based powder delivery head for applications in DED is established. Hereafter, the main conclusions reached are summarized below.

- Powder flow under the effect of gravity alone, in the continuous and discrete coaxial nozzle, is of conical and convergent nature. The powder distribution at the focal plane shows a Gaussian behavior.
- For a gravity-based system, with a decrement in nozzle's cone angle, the focal length and the 'depth of focus' of the powder stream increases.
- The nozzle tilt angle, measured from the vertical, has a significant influence on powder particle distribution around the central axis. A tilt angle, as low as 2°, is sufficient to distort the focal point and the particle distribution around the heat source. An implication of this result suggests that a gravity-based powder delivery head should be mounted vertical, and tilt motion should be provided to the substrate, either through a robotic arm or a five-axis CNC, to build complex geometry parts.
- The powder particle distribution becomes increasingly more non-uniform as the nozzle tilt increases. This, without any corrective measures, will negatively impact the deposition process.
- Continuous coaxial nozzles show better stream characteristics than a discrete coaxial nozzle and are less likely to jam. The minimum channel dimensions required for a jamming-free operation in the case of an annular outlet is much lower than a circular outlet

References

1. Pinkerton AJ (2010) Laser direct metal deposition: Theory and applications in manufacturing and maintenance. Adv Laser Mater Process Technol Res Appl, 461–491. https://doi.org/10.1533/9781845699819.6.461
2. Wang W (2011) High-Quality High-Material-Usage Multiple-Layer Laser Deposition of Nickel Alloys Using Sonic or Ultrasonic Vibration Powder Feeding. Proc Inst Mech Eng Part B J Eng Manuf 225:130–139. https://doi.org/10.1177/09544054JEM2128
3. Kovalenko V, Yao J, Zhang Q et al (2016) Development of Multichannel Gas-powder Feeding System Coaxial with Laser Beam. Procedia CIRP 42:96–100. https://doi.org/10.1016/j.procir.2016.02.197
4. Goodarzi DM, Pekkarinen J, Salminen A (2015) Effect of process parameters in laser cladding on substrate melted areas and the substrate melted shape. J Laser Appl 27:S29201. https://doi.org/10.2351/1.4906376
5. Jeromen A, Kuznetsov A, Govekar E et al (2018) Annular laser beam based direct metal deposition. Procedia CIRP 74:222–227. https://doi.org/10.1016/j.procir.2018.08.099
6. de Oliveira U, Ocelík V, De Hosson JTM (2005) Analysis of coaxial laser cladding processing conditions. Surf Coatings Technol 197:127–136. https://doi.org/10.1016/j.surfcoat.2004.06.029

7. Kapil S, Kulkarni P, Joshi P, et al (2018) Retrofitment of a CNC machine for omni-directional tungsten inert gas cladding. Virtual Phys Prototype 2759. https://doi.org/10.1080/17452759.2018.1552484
8. Schwartz SR, Richardson DC, Michel P (2012) An implementation of the soft-sphere discrete element method in a high-performance parallel gravity tree-code. Granul Matter 14:363–380. https://doi.org/10.1007/s10035-012-0346-z
9. Deen NG, Van Sint Annaland M, Van der Hoef MA, Kuipers JAM (2007) Review of discrete particle modeling of fluidized beds. Chem Eng Sci 62:28–44. https://doi.org/10.1016/j.ces.2006.08.014
10. Wassgren C, Curtis JS (2006) The application of computational modeling to pharmaceutical materials science. MRS Bull 31:900–904. https://doi.org/10.1557/mrs2006.210
11. Wassgren C (Purdue University) DEM Modeling : Lecture 06 Introduction to Soft-Particle DEM Normal Contact Force Models. Part I Introduction to Soft-Particle DEM. Lect Notes 1–37
12. Horabik J, Molenda M (2016) Parameters and contact models for DEM simulations of agricultural granular materials: A review. Biosyst Eng 147:206–225. https://doi.org/10.1016/j.biosystemseng.2016.02.017
13. Geer S, Bernhardt-Barry ML, Garboczi EJ et al (2018) A more efficient method for calibrating discrete element method parameters for simulations of metallic powder used in additive manufacturing. Granul Matter 20:1–17. https://doi.org/10.1007/s10035-018-0848-4
14. Haley JC, Zheng B, Bertoli US et al (2019) Working distance passive stability in laser directed energy deposition additive manufacturing. Mater Des 161:86–94. https://doi.org/10.1016/j.matdes.2018.11.021
15. Wolff SJ, Wu H, Parab N et al (2019) In-situ high-speed X-ray imaging of piezo-driven directed energy deposition additive manufacturing. Sci Rep 9:1–14. https://doi.org/10.1038/s41598-018-36678-5

Computational Modelling of In-Process Mitigation Technique to Control Residual Stress in Thick Plate Welding

Saurav Suman and Pankaj Biswas

1 Introduction

The dimensional accuracy of any final welded structures poorly affected from the post-welding distortions and requires a post-weld remedial which increases the overall cost and effort of production. On the other hand, crack propagation and fatigue and catastrophic failure of the structure result from welding induced residual stresses. Therefore, to prevent extra expenses on post-weld treatments, in-process mitigation techniques are more effective in improving product quality and productivity. Masubuchi studied the that compressive residual stresses in thin plates induces buckling distortions [1]. Michaleris and Sun investigated the thermal tensioning as the mitigation approach in controlling welding tempted buckling distortions [2]. They observed that too high temperature shows an unfavourable effect on the buckling distortions while low temperature develops inadequate tensioning [3]. Deo and Michaleris investigated the impact of transient thermal tensioning using experiment and compared the case at 200 and 250 °C of time–temperature–transformation (TTT) for buckling distortion. The tensile stresses in the longitudinal direction along the welding line owing to side heating effectively condensed the compressive residual stresses at the plate edges. Also, mechanical restraints were found effective in reducing the angular distortions [4]. In the trailing heat sink method, pressurized liquid nitrogen was employed as a cooling jet. The parameters namely cooling distance, flow rate of liquid nitrogen, and diameter of cooling jet were selected for an optimal performance, based on various finite element studies. In the in-process rolling technique, a compressive effect in the through thickness

S. Suman
Department of Mechanical Engineering, National Institute of Technology Mizoram, Mizoram, India

S. Suman (✉) · P. Biswas
Department of Mechanical Engineering, Indian Institute of Technology Guwahati, Guwahati, India

S. Bag et al. (eds.), *Next Generation Materials and Processing Technologies*, Springer Proceedings in Materials 9, https://doi.org/10.1007/978-981-16-0182-8_25

direction and a corresponding tensile effect in the welding direction were generated. A rolling force of 5.5 kN was found to be the minimum force that was required to eliminate the buckling distortion completely. Yang and Dong also studied the effect of roller width and roller to torch distance on the buckling distortions [5]. The cryogenic CO_2 cooling was employed to decrease the weld-induced buckling distortions. The technique of cryogenic cooling had a significant effect on reduction of out-of-plane distortions in butt welded plates compared to that of fillet welded plates [6] adopted a dynamically controlled-low stress no distortion (DC-LSND) technique for the control of weld-induced buckling distortion [7]. In their work, they employed CO_2 snow, produced by compressed liquid CO2 through nozzles, as a cooling media. It was seen that a trailing cryogenic cooling device drastically reduced the welding distortions for a gas metal arc welding (GMAW) process. It was reported that a (DC-LSND) technique had a strong impact on the longitudinal welding distortion and reduced it by 81%. The optimum travel speed of cooling device was found to be 600 mm/min. Further it was concluded that, the microstructure as produced by (DC-LSND) technique was almost the same as that obtained by conventional process. Price et al. [8] investigated the mechanical constraints tensioning effect to control the out-of-plane distortions of friction stir welding (FSW) and tungsten inert gas (TIG) welding. The plates were subjected to the various magnitudes of tensile stress along the weld line. It was reported that 25–40% of yield strength as the tensioning effort eliminates the buckling distortions completely. It was also noticed that the optimum value of tensioning depends on the welding parameters. An increase in the tensioning beyond the optimum value led to the formation of buckling distortions again and formed a zone of compressive stress around the welding area [8]. Richards et al. [9] tried to control the generation of residual stress by employing global mechanical tensioning. It was observed that by imposing a mechanical tensioning of the intensity 40% of yield value, residual stresses tends to zero. It was seen that, the values of mechanical tensioning higher than that is required to eliminate the buckling distortions, reverses the stress field in the welded zone. The decrease of residual stresses was very less with the increase of distance from the weld surface for thicker weld plate [9]. Researchers have carried out extensive studies on various methods of weld-induced distortion control. However, it is evident from the literature review that, majority of the methods proposed lack the ability and feasibility of practical implementation as an active distortion control technique in fabrication of large stiffened ship panels. The TMT approach of distortion control had shown some encouraging results in controlling buckling distortions.

A number of researches are found on the mitigation effect of either side heating or heat sink processes as thermal tensioning technique. Inside heating (static or dynamic heat source) case, the thermal gradient during heating and cooling process is managed across the weld centreline in the transverse direction [3, 10, 11]. Whereas, in heat sink (static or dynamic cooling source) case, sudden drop of temperature during and just after the welding process affect the residual stress distribution, i.e. compressive nature stresses in the fusion zone get amplified (thermal tensioning effect) just after the welding suddenly which results in low value of longitudinal residual stresses [12, 13].

Some of the researchers conducted the thermal tensioning mitigation experiments using both side heating and heat sink methods, though all the studies have been performed on thin sheet weld joints. Ilman et al. [14] conducted static thermal tensioning on MIG welded AA 5083 sheet (3 mm) using static heat source bed on both sides of the weld line on top surface and a static heat sink (cooling water) on bottom surface near the root gap [14]. Recently, Li et al. [22] developed a hybrid technology of combined transient heating (both sides on top surface) and trailing cooling source behind welding torch, to weld (5083-H 112) aluminium alloy sheet (3 mm) which effectively reduced distortion and residual stresses by 58% and 65%, respectively. However, no such studies on mitigation techniques are found on thick plate weld joints, i.e. combined thermal tensioning methods on thick weld plates, especially to control residual stresses.

In the current study, a heating source on both the sides of weld line along with a cooling source just beneath the root gap and both of its sides has been proposed to weld a thick plate weld. It was observed that the combined side heating and cooling is effective in controlling residual stress development in the thick plate welded joints.

2 Methodology

A submerged arc welded AISI 304 steel plate of 10 mm thickness was considered for this study. The welding parameters for the SAW welded AISI steel were taken from a previous literature by the Authors [15] and shown in Table 1 along with P91 steel weld parameters.

The whole modelling process was categorized for comparative study of welding induced stresses for four cases, i.e. welding, welding with side heating, welding with heat sinking and welding with combined side heating and heat sink. The schematic of experimental set-up for proposed combined in-process mitigation technique is shown in Fig. 1.

Table 1 Process parameters for welding

Parameters	Welding voltage (V)	Welding current (A)	Welding speed (m/s)	Length of stickout (mm)
Values	26.51	445	5	25

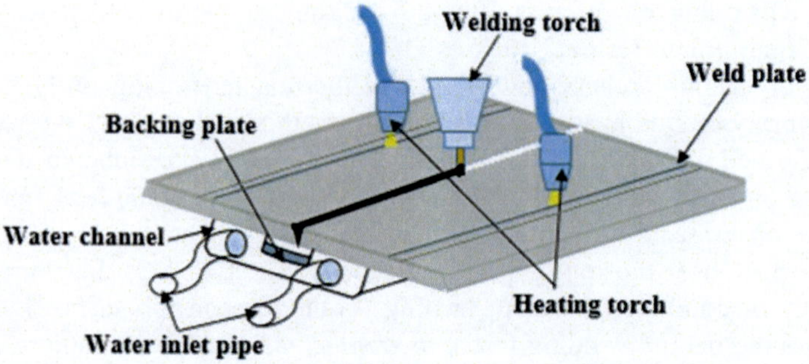

Fig. 1 Schematic of experiment set-up of combined side heating and heat sink

2.1 Mathematical Formulation

The exact amount of heat input similar to experiment was applied using a moving heat source model in Eq. (1). Similarly, the element death and birth technique was employed to model addition of molten metal in the root gap during the welding [16].

$$Q_{arc} = \eta VI \tag{1}$$

where η = Efficiency, V = Source voltage, I = Welding current.

The Gaussian's double ellipsoidal volumetric heat source model was used to model the heat input. The expressions of which are given in Eqs. (2) and (3) for front and rear portion of the heat flux [16–18].

$$Q_{front} = \frac{6\sqrt{3}f_{front}Q_{arc}}{abc_{front}\pi\sqrt{\pi}} e^{\left(\frac{-3x^2}{a^2} - \frac{-3y^2}{b^2} - \frac{-3(z-vt)^2}{c_{front}^2}\right)} \tag{2}$$

$$Q_{rear} = \frac{6\sqrt{3}f_{rear}Q_{arc}}{abc_{rear}\pi\sqrt{\pi}} e^{\left(\frac{-3x^2}{a^2} - \frac{-3y^2}{b^2} - \frac{-3(z-vt)^2}{c_{rear}^2}\right)} \tag{3}$$

where x, y, & z are the coordinates in the imaginary weld joint region and v give the welding speed. 'a', 'b', 'c_{front}', 'c_{rear}' are the constant parameters. 'f_{front}' *and* 'f_{rear}' indicate portion of heat in the front and rear regions of the heat source model.

A surface heat flux model was applied to model the side heating process [19]. It includes heat sources passing through the top surface of the weld plate on both sides of the weld centreline. The rate of heat input is given by

$$Q_{\sup} = \frac{\eta Q}{\pi \bar{r}^2} \exp\left[-2\left(\frac{r}{\bar{r}}\right)^2\right] \tag{4}$$

Where $\overline{r}$ = radius of surface area including 95% of the heat flux, r = distance of the heat torch from the surface.

In finite element analysis, the heat sink or say cooling effect through bottom surface was simulated as the heat transfer between the specimen and the cooling media (water flow parallel to plate length considering laminar flow) is described by forced convection cooling [20] given by Eq. (5).

$$h = (0.664 \times \mathrm{Re}_L^{1/2} \times \mathrm{Pr}^{1/3}) \times k/L \tag{5}$$

where k = Thermal conductivity kJ/s-m-K, L = Length of the plate (m).

$$\mathrm{Re}_L = \frac{LV_\infty \rho}{\mu} \; \& \; \mathrm{Pr} = \frac{\mu C_p}{k}$$

where L = Length of the plate (m), V_∞ = Approach velocity (m/s), ρ = Density of the water (kg/m^3), μ = Viscosity of water (N.s/m^2) and Cp = Heat capacity (kj/kg.K).

2.2 FE Modelling and Meshing

The butt joint model for simulation of various thermal effect processes like welding, side heating and heat sinking was performed in ANSYS Multiphysics software package. The weld bead comprised of actual shape and size of weld bead similar to the experiment. The 3D element **SOLID 70** suggested for thermal modelling with eight nodes and single degree of freedom was selected. While **SOLID 185** element recommended for structural analysis was selected [21].

Fine meshing was preferred in weld zone and side heating zone, i.e. peak temperature zones. The element size was kept coarser for the regions away from these high temperature regions. The complete model along with the meshing is shown in Fig. 2. The element birth–death technique was employed to simulate the real-time addition of weld material during the welding. EKILL and EALIVE commands were carefully applied for deactivating and reactivating the elements before and during the welding process, respectively.

3 Results and Discussion

The time vs temperature graph shown in Fig. 3 below presents the thermal profile of four different cases: welding without in-process mitigation; welding with side heating (region at 25 mm from weld centreline); welding with heat sink (h = 20,000 W/m^2); and welding with combined side heating and heat sink.

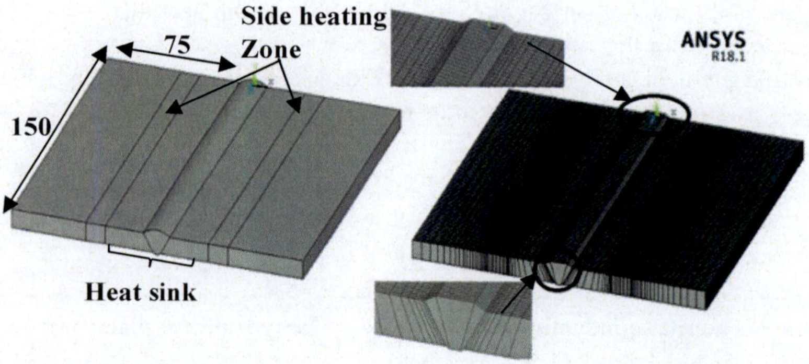

Fig. 2 Butt joint model and meshing

Fig. 3 Time vs temperature graph for combined side heating and heat sink

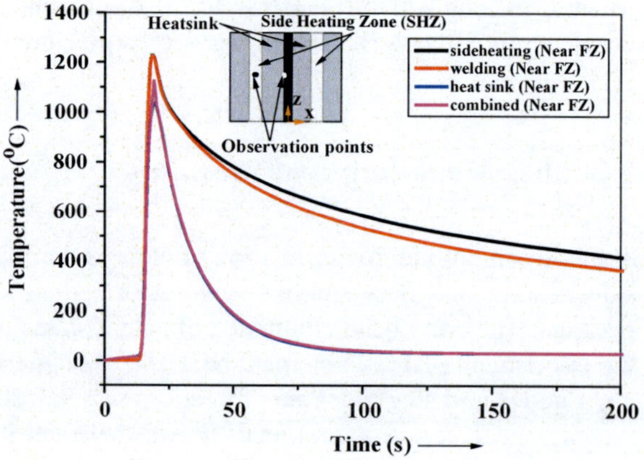

Two points (one near the FZ and another in side heating region) were selected to observe the change in temperature w.r.t time. The effect of heat sink leads over side heating process to control the rate of change in temperature. The effect of heat sink is clearly observed for the point in FZ while, point in the SH region characterizes the change in the cooling rate. The peak temperature for welding and side heating is observed to be same at around 1250 °C, though cooling rate increases for side heating case. On the other hand, the heat sink causes decrease in the peak temperature for the same point from 1250 to 1134 °C, though the cooling rate remains same for combined heat sink and side heating.

Residual stress distribution in longitudinal direction (Sz) is shown in Fig. 4, for all the cases considering in-process thermal tensioning methods. It is clearly observed and also have been explored that both side heating and heat sink efficiently control the residual stress distribution across the weld line. It is also noticeable that, heat sink model is effective in decreasing the intensity of residual stress distribution in the cooling areas, i.e. the stress values reduce to much lower values in the HAZ. From Fig. 4, it is observed that for welding case, the peak residual longitudinal stress is between 200 and 220 MPa, which reduces with minor difference on side heating, but heat sink model effectively reduced it to zero which gradually reached to compressive values up to (−50 to 70) MPa with side heating process. The graph presents the three cases of combined side heating and heat sink thermal tensioning technique for three different locations of heating torch.

The more the distance of the heating torch increases from weld centreline, the narrower the residual stress distribution becomes in the HAZ and FZ. At the same time, the longitudinal residual stress shows sudden peak in the side heating region. The sudden peak in longitudinal residual stress similar peak in the side heating region was observed in thin sheet combined thermal tensioning mitigation [22]. It occurred under the influence of slight compressive plastic strain during heating. In fact, this newly developed tensile stress in the heating region reduces the discrepancy of residual stress distribution in the weld and nearby regions. As a result, it increased the critical buckling load of the plate [23], which successfully controlled the welding distortion.

Hence, for the heating region at 25 mm, the compressive peak converts to tensile nature and provide a proper balance to prevent buckling distortion. Similarly, the combined thermal tensioning effect is visible on transverse stress distribution from the comparative plot shown in Fig. 5. It is observed that side heating slightly increases the transverse residual stress distribution across the weld, whereas heat sink thermal tensioning effect is found to be very effective in mitigation of transverse residual stress by almost 66.7%, though combined side heating and heat sink model is found equally effective for side heating at far away locations. Hence, it can be said that for collective mitigation effect of longitudinal and transverse residual stress distribution combined thermal tensioning shows similar effect on thick plate weld, though it is bit difficult to have same heat sink effect on thick weld plate with same cooling medium.

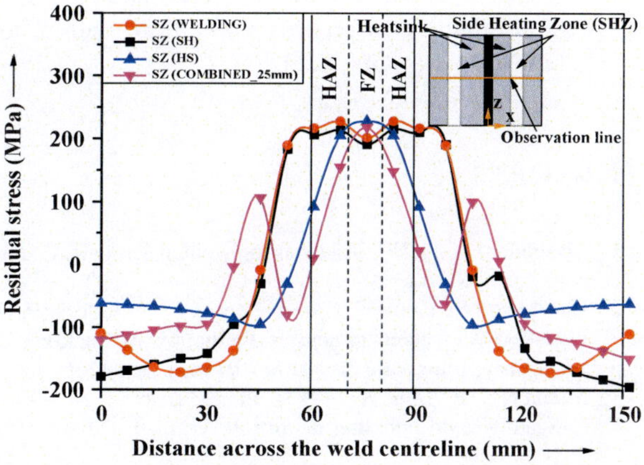

Fig. 4 Longitudinal stress distribution for combined thermal tensioning mitigation

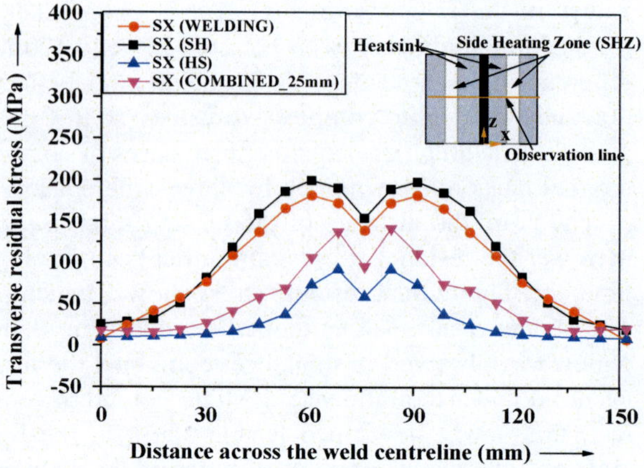

Fig. 5 Transverse stress distribution for combined thermal tensioning mitigation

4 Conclusions

In this study, in-process mitigation methods have been discussed to control welding induced residual stresses in a thick (10 mm) plate weld joint. In-process mitigation methods were thermo-mechanically modelled for a non-ferritic steels like AISI 304 steel (10 mm thick). The conclusions made on the basis of studies performed are as follow-

- Side heating process was found effective on controlling welding induced distortion and residual stresses in thick plate weld by balancing the sudden change residual stress distribution from tensile to compressive across the weld and HAZ region.
- Heat sink thermal tensioning was found appreciably effective, in reducing the intensity of residual stress peak in the weld and nearby region in the thick plate weld.
- Heat sink (using cooling water) was found to reduce the peak residual longitudinal stress by almost 30–35% in thick plate weld.
- Finally, it was discovered that combined side heating and heat sink process can be and effective in-process mitigation method to control residual stress distribution in thick plate weld.

References

1. Masubuchi K (1980) Analysis of welded structures : residual stresses, distortion, and their consequences, p 642
2. Deo MV, Michaleris P, Sun J, Deo MV, Michaleris P, Sun J (2017) Prediction of buckling distortion of welded structures Prediction of buckling distortion of welded structures, vol 1718, July. https://doi.org/10.1179/13621710322500900
3. Michaleris P, Sun X (1997) Finite element analysis of thermal tensioning techniques mitigating weld buckling distortion. Weld. J. (Miami, Fla) 76(12):451

4. Deo MV, Michaleris P, Sun J (2003) Prediction of buckling distortion of welded structures. Sci. Technol Weld Join 8(1):55–61. https://doi.org/10.1179/13621710322500900
5. Yang YP, Dong P (2012) Buckling Distortions and Mitigation Techniques for Thin-Section Structures 21, February, pp 153–160. https://doi.org/10.1007/s11665-011-9928-x
6. Camilleri D, Nash D, Camilleri D, Gray TGF, Nash DH (2008) Mitigation of welding distortion and residual stresses via cryogenic CO2 cooling—a numerical investigation. In : 17th International Conference on Computer Technology in Welding and Manu, June, 2008
7. Guan DL, Guo (2006) Low stress no distortion welding based on thermal tensioning effects for thin materials. Int Sci Prod J, December, 1–52
8. Price DA et al (2007) Distortion control in welding by mechanical tensioning. Sci Technol Weld Join 12(7):620–633. https://doi.org/10.1179/174329307X213864
9. Richards DG, Prangnell PB, Williams SW, Withers PJ (2008) Global mechanical tensioning for the management of residual stresses in welds. Mater Sci Eng, A 489(1–2):351–362. https://doi.org/10.1016/j.msea.2007.12.042
10. Deo MV, Michaleris P (2003) Mitigation of welding induced buckling distortion using transient thermal tensioning, 8(1):49–54. https://doi.org/10.1179/136217103225008919
11. Damale AV, Nandurkar KN (2015) Numerical simulation of side heating for controlling angular distortion in multipass MMAW butt welded plates, 40, April, pp 487–502
12. Li J, Guan Q, Shi Y, Guo D, Du Y, Sun Y (2004) Studies on characteristics of temperature field during GTAW with a trailing heat sink for titanium sheet. J Mater Process Technol 147 (3):328–335. https://doi.org/10.1016/j.jmatprotec.2003.12.012
13. Guo Y, Wu D, Ma G, Guo D (2014) Trailing heat sink effects on residual stress and distortion of pulsed laser welded Hastelloy C-276 thin sheets. J Mater Process Technol 214(12):2891–2899. https://doi.org/10.1016/j.jmatprotec.2014.06.012
14. Ilman MN, Muslih MR, Subeki N, Wibowo H (2016) Mitigating distortion and residual stress by static thermal tensioning to improve fatigue crack growth performance of MIG AA5083 welds. JMADE 99:273–283. https://doi.org/10.1016/j.matdes.2016.03.049
15. Suman S, Sridhar PVSS, Biswas P, Das D (2020) Prediction of welding-induced distortions in large weld structure through improved equivalent load method based on average plastic strains. Weld. World 64(1):179–200. https://doi.org/10.1007/s40194-019-00805-1
16. Suman S, Biswas P (2019) Comparative study on SAW welding induced distortion and residual stresses of CSEF steel considering solid state phase transformation and preheating, vol 51, October, pp 19–30. https://doi.org/10.1016/j.jmapro.2020.01.012
17. Suman S, Biswas P (2020) Thermo-mechanical study of single and multi-pass welding of CSEF steel for residual stresses and deformations considering solid state phase transformation. Mater Today Proc, xxxx (2020). https://doi.org/10.1016/j.matpr.2019.12.299
18. Suman S, Biswas P, (2020) Numerical Study of Welding Distortion in SAW Welded Creep Strength Enhanced Ferrite Steel Joint. In: Biswal B, Sarkar B (ed.) Lecture Notes in Mechanical Engineering. Springer, Singapore, pp 641–648
19. Suman S, Biswas P, Baranwal S, Mekala V (2020) Finite Element Modelling of side heating for mitigation of residual stress and distortion in SAW welded P91 steel weld. Mater. Today Proc. 28:2511–2521. https://doi.org/10.1016/j.matpr.2020.05.004
20. Bengtson HH (2010) Convection Heat Transfer Coefficient Estimation, pp 1–40
21. Suman S, Biswas P (2020) Thermo-mechanical study of single and multi-pass welding of CSEF steel for residual stresses and deformations considering solid state phase transformation. Mater Today Proc 28:789–795. https://doi.org/10.1016/j.matpr.2019.12.299
22. Li M, Ji S, Yan D, Yang Z (2019) Controlling welding residual stress and distortion by a hybrid technology of transient thermal tensioning and trailing intensive cooling. Sci Technol Weld Join 24(6):527–537. https://doi.org/10.1080/13621718.2018.1564473
23. Xu J, Li W (2007) The nonlinear time-varying response of dynamic thermal tensioning for welding-induced distortion control. J Manuf Sci Eng Trans ASME 129(2):333–341. https://doi.org/10.1115/1.2540708

Phenomenological Modelling of Surface Morphology in Friction Stir Welding of Aluminium Alloy

Debtanay Das, Swarup Bag, and Sukhomay Pal

Nomenclature

A_y, M_s, C_s	Material constants for the J-C model
e	Work-hardening exponent
m	Thermal softening coefficient
$\overline{\varepsilon}$	Plastic strain
$\overline{\sigma_s}$	Flow stress
$\dot{\varepsilon}_0$	Reference strain rate at 1 s^{-1}
$\dot{\overline{\varepsilon}}$	Effective strain rate
T_m	Melting point temperature
T_r	Reference temperature
T	Temperature
Δt	Time increment
l_{min}	Minimum element length
c	Dilatational wave speed
λ_L, μ_L	Lame's coefficient
ρ	Density
ω_{max}	Maximum Frequency
ξ	Fraction of critical damping
k	Thermal conductivity
$\dot{Q}$	Rate of heat generation
C	Heat capacity
$\dot{T}$	Rate of change of temperature
q	Convection heat loss
h_{conv}	Coefficient of heat convection
$T_{ambient}$	Ambient temperature
r	Radiative heat loss
ε_e	Emissivity
σ_s	Stefan–Boltzmann constant

D. Das (✉) · S. Bag · S. Pal
Indian Institute of Technology, Guwahati, Assam, India
e-mail: debtanay@iitg.ac.in

S. Bag et al. (eds.), *Next Generation Materials and Processing Technologies*,
Springer Proceedings in Materials 9, https://doi.org/10.1007/978-981-16-0182-8_26

1 Introduction

Nearly three decades have passed since the introduction of friction stir welding (FSW) as a low-temperature welding technique. The core idea behind this method was to weld the materials below the melting point temperature. Initially, this method was applied to weld low strength and low melting point aluminium alloys [1]. After successful application of FSW to low melting point alloys, it is now also being used to weld relatively high strength alloys with a higher melting point than that of aluminium alloys such as different grades of steel and Inconel [2, 3]. In the simplest form, the FSW comprises two components, namely the welding tool and the workpiece. The workpiece can be divided into three zones based on microstructural differences, i.e., the stir zone (SZ), the parent material (PM) and the transitional zone (TZ) between SZ and PM. The TZ comprises the thermo-mechanically affected zone and the heat-affected zone.

The FSW is a multi-physics problem that encompasses several physical phenomena within itself, like the frictional interaction between the workpiece and the tool material, thermal interaction between the outer surface of the workpiece and the surrounding, material mixing and flow within and out of the workpiece [3]. Due to such complicated nature of the process, this method requires thorough investigation for several aspects. It is not always feasible to perform multiple experiments to reach a certain level of confidence concerning an obtained result. Therefore, for such scenarios, computational modelling should be utilized as a support module of the actual experimental process. Numerical models have some inherent benefits making it a widely acceptable choice whenever experimentation is not a viable option. Moreover, numerical models help to study some phenomena, which are otherwise extremely difficult to observe in practical cases. Material flow and its mixing is an integral part of FSW, but its visualization is quite tricky in practical situations as it can be occurring inside of the workpiece. Numerical models not only help in material flow visualization, but also useful in cost-cutting and producing repeatability results. These models can also be used to optimize the tool and fixture designs, process parameter window, improve joint efficiency and predict the inconsistencies in the joint, which leads to defect formation. Multiple numerical methods have been developed and used by the researchers. These methods can be categorized under the umbrella of either the computational solid mechanics (CSM) or computational fluid dynamics (CFD) approaches [4].

A basic CFD model considers an Eulerian domain for the workpiece through which the material flows. The tool is considered to be non-deformable and rigid. A fluid domain is generally used for the region where the significant material flow takes place, i.e., the region surrounding the stir zone. For this region, both heat and momentum equations are solved. The remaining regions in the workpiece do not observe significant material flow, and therefore, only the heat conduction equation is solved for these regions. A notable issue with this model is that neither of these regions follows the combined elasto-viscoplastic behaviour of the material generally observed in the case of FSW process [5]. It can be argued that the FSW is a high strain and high strain rate process [6, 7]. Therefore, it would be a clever approach to ignore the negligible elastic deformation of 0.002 in favour of larger incremental time steps [8]. This method significantly helps to lessen the computational time. The computational domain is divided into multiple sections to represent the complete FSW system. These sections are the backing plate, the workpiece and the tool. The workpiece is further divided into a rotating and stationary mesh. The rotating mesh is the region surrounded by the tool. This region is considered as a fluid domain for CFD analysis. Adjacent to the rotating mesh is the stationary mesh that extends throughout the remaining length of the workpiece. This is used for thermal and flow behaviour modelling of the material escaping out of the stirred zone. The traverse speed is applied to the material as inlet and outlet velocities equal to the welding speed. One of the disadvantages of the CFD model is that it cannot predict residual stresses, which is very important to determine weld quality and life [9].

CSM method comprises the individual Eulerian approach, the Lagrangian approach, Coupled Eulerian–Lagrangian (CEL) approach and arbitrary Lagrangian–Eulerian (ALE) approach. Although these individual approaches are easier to model, they suffer from certain drawbacks, which can be rectified by the combination of both of these classical approaches. The Eulerian approach considers a fixed node which makes modelling free boundaries difficult as the bounding nodes need not necessarily coincide with the element edges [10]. Thus, modelling of the plunge stage during FSW with the Eulerian approach is very difficult. Contrary to the Eulerian approach, the Lagrangian approach considers a moving node. Such an approach is beneficial for low deformation problems, but during FSW, this model can suffer from severe node deformation, which makes it difficult for the model to converge [11]. To compensate for these drawbacks, the best possible approach is the coupled use of both Lagrangian and Eulerian approaches. Currently, the literature reveals only two strategies where coupled use of the Lagrangian and Eulerian approach leads to solution convergence. One of the methods is ALE, where the mesh is fixed to the workpiece and the nodes represent an individual location in the material domain. This allows the material to easily distort during the deformation process. An adaptive remeshing criterion is generally employed by researchers with ALE to ease the severe deformation issue that might result in non-convergence of the solution [12]. The second method is the CEL. Although the mathematical background for the CEL technique was laid as early as 1963 by Noh [13], it was not until recently this technique was implemented by researchers for the modelling of FSW process [14]. In this approach, the workpiece is modelled as an Eulerian domain containing immovable nodes. Inflow and outflow of the material from the domain indicates the movement of the material in the system. A rigid Lagrangian body is used to model the FSW tool [15].

Heat generation is one of the most important phenomena during the FSW process, where heat is generated because of the friction between the workpiece and the tool. Different friction laws are used to model the heat generation in the workpiece.

Coulomb's friction law predicts the temperature profile more accurately compared to other friction laws [12]. The heat generation during FSW depends on not only the friction law but also the size and shape of the welding tool and process parameters used. Therefore to calculate the total heat generated during FSW, a heat generation equation as a function of tool radius and process parameters is suggested in the literature [16, 17]. Although many FSW models only study the heat generation behaviour, it is not the only phenomenon occur during FSW process. Another significant aspect of the process is the mixing of material and the flow of the material. The material flow can be understood with the help of flow stresses. Therefore, it is of paramount interest to use a material model like the Sheppard–Wright model [18] or Johnson–Cook model [19] that can help model the flow stresses in the workpiece. Insufficient flow stresses can lead to insufficient material flow and improper material mixing results in the generation of defects in the weld [20]. Another approach that can study the material flow in FSW is the smoothed particle hydrodynamics (SPH) [21, 22]. This technique is different from the CEL and ALE approaches due to the fact that it is a meshless technique. The SPH technique is based on Lagrangian approach and therefore obtaining data at each node point even with severe deformation is quite simple. This process can be very easily applied to track free surfaces during FSW and therefore study the various defects that might occur during the FSW process. Over the years, the researchers studied the material mixing with the help of tracer particle [23]. It can be used both experimentally and for numerical prediction. In this case, a tracer particle, which is different from the base material but similar in properties, is embedded in the path of tool movement. Studying the final position of the tracers at the end of the weld gives a clear idea about the material flow and movement in the SZ. However, the tracking of the tracer particle is a little bit difficult for material mixing at high rotational speed.

The current work is developed after observing that multiple studies have already been performed on FSW focusing on the parameter window, thermal model and material flow. But much focus is not provided on the surface morphology of the welded sample. It must be recognized that the surface morphology develops because of the material flow on the workpiece while rotating around the tool. Here, an attempt has been made to formulate a numerical model following the CEL approach that predicts the flow of material during FSW process. Further, this study can be utilized as a precursor to predict the defect due to insufficient material flow. Moreover, the emphasis is provided on methods such as mass scaling that can be utilized to speed up the total simulation time. It is a very relevant problem of the CEL method, which although beneficial over the classical Eulerian method or Lagrangian method. It is quite notorious for its requirement of substantially high time to converge for a solution.

2 Theoretical Formulation

A fully coupled thermo-mechanical model is developed following the CEL approach to study the FSW process. The workpiece is an Eulerian domain with fixed mesh, while the tool is a rigid Lagrangian domain. The rotation and linear velocity boundary condition is applied to the rigid tool. The dimension for the workpiece is considered as 72 mm × 75 mm × 6.35 mm for the current study. An ordinary cylindrical tool having a shoulder radius of 9 mm and a featureless pin of radius 2.5 mm is used. Aluminium alloy is considered as workpiece while tool steel is used as the tool material. The temperature-dependent physical properties of Al-6061 aluminium alloy are provided in Table 1. Direct visualization of material movement is not possible in the CEL approach. Therefore, a separate domain is fixed on top of the workpiece, i.e., the Eulerian domain consists of two parts, namely the void domain and the material domain, as shown in Fig. 1. The bottom domain is the actual workpiece. The top void domain is designed to contain the flow of material that leads to chip formation.

2.1 Material Model

Plastic deformation occurring during the FSW is captured with the help of a Johnson–Cook (J-C) material model. It is a rate-dependent viscoplastic material model. The material constants of Al-6061 for the J-C model are provided in Table 2. The mathematical representation of the J-C material model is

$$\overline{\sigma}_s = \left[A_y + M_s\overline{\varepsilon}^{e}\right]\left[1 + C_s \ln\left(\frac{\dot{\overline{\varepsilon}}}{\dot{\varepsilon}_0}\right)\right]\left[1 - \left(\frac{T - T_r}{T_m - T_r}\right)^m\right] \quad (1)$$

Table 1 Physical properties of Al 6061 [14]

Temperature (K)	Thermal conductivity (W m^{-1} K^{-1})	Poisson's ratio	Density (kg m^{-3})	Elastic modulus (GPa)	Specific heat (J kg^{-1} K^{-1})
298	162	0.33	2690	66.94	945
373	177	0.334	2690	63.21	978
422	184	0.335	2670	61.32	1000
477	192	0.336	2660	56.80	1030
533	201	0.338	2660	51.15	1052
592	207	0.36	2630	47.17	1080
644	217	0.4	2630	43.51	1100
700	229	0.41	2600	28.77	1130

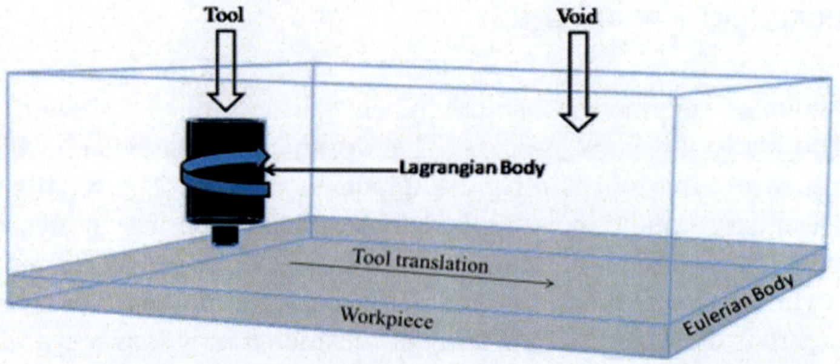

Fig. 1 Arrangement of the workpiece, the tool and the void region for the model

Table 2 Johnson–Cook material constant for Al-6061 [14, 15]

A_y (MPa)	M_s (MPa)	C	m	e	T_r (°C)	T_m (°C)
324	114	0.002	1.34	0.42	24	583

where $\overline{\sigma_s}$ is flow stress, A_y, M_s, C_s are material dependent constants, e is the work-hardening exponent, $\bar{\varepsilon}$ and $\dot{\bar{\varepsilon}}$ are plastic strain and effective strain rate respectively, m is the thermal softening coefficient, $\dot{\varepsilon}_0$ is reference strain rate (1 s^{-1}), T_m and T_r are the melting point and reference temperatures, respectively.

2.2 *Mass Scaling and Stability of the Model*

One of the most significant disadvantages of the CEL technique is high convergence time. This problem can be rectified with the help of mass scaling. The artificial increase in the mass of an element is referred to as mass scaling. This increase in mass leads to an increase in stable time increment. The change in the time increment due to weight manipulation is based on Eqs. 2 and 3, which are expressed as:

$$\Delta t \leq \left[\frac{l_{min}}{c}\right] \tag{2}$$

$$c = \sqrt{\frac{\lambda_L + 2\mu_L}{\rho}} \tag{3}$$

where Δt is the time increment, c is the dilatational wave speed, l_{min} is the minimum length of the element, λ_L and μ_L are Lame's constant. On the application of the mass scaling, i.e., increasing the mass of the model can lead to the fluctuation in the inertia effects in the model. Therefore, a universal method to ensure that mass scaling does not have an adverse effect on the model is to compare the internal and kinetic energy values of the system. The kinetic energy (KE) to the internal energy (IE) ratio of the system must remain below 5–10% [24].

Implicit integration is a common approach to solve inertia dominated problems. But, due to the convergence difficulties that might occur in highly nonlinear contact problems, explicit integration is often preferred [25]. The time increment, in this case, is only conditionally stable. Thus, the time increment in an undamped case must satisfy Eq. 4 to achieve stability. The Courant, Friedrichs and Lewy (CFL) criterion for wave propagation decide the stability condition. It states the required time for an individual step should be less than the time needed by dilatational stress waves to pass through the smallest element [25]. The stable time step for the undamped condition is expressed as:

$$\Delta \mathrm{t} \leq \frac{2}{\omega_{\max}} \tag{4}$$

while with damping the Eq. 4 is modified to:

$$\Delta \mathrm{t} \leq \frac{2}{\omega_{\max}} \times \left(\sqrt{1+\xi_{max}^2} - \xi_{\max}\right) \tag{5}$$

where ω_m is the maximum eigenfrequency of the assembled mesh and ξ_{max} is the critical damping fraction with the highest frequency.

2.3 Boundary Condition and Contact Interaction

During FSW, two different boundary conditions act on the model, i.e., the thermal and the mechanical boundary conditions. The external surface of the Eulerian domain is constrained so that material flowing out of the domain can be restricted. The bottom surface is restricted in the Z-direction to model the effect of the backing plate during welding. The rotation and translation velocities are both applied to the rigid tool. The whole process can be divided into three steps, viz. the plunging step, dwell step and the linear step.

The transient heat equation is solved for the thermal model. The convection and radiation heat loss from the workpiece are accounted for boundary condition. The mathematical form of the thermal governing equation and boundary conditions is given as:

$$\nabla[k(T)\nabla T] + \dot{Q} = \rho C\dot{T} \tag{6}$$

$$q = h_{\mathrm{conv}}(T - T_{\mathrm{ambient}}) \tag{7}$$

$$r = \varepsilon_e \times \sigma_s \left(T^4 - T^4_{ambient}\right) \tag{8}$$

where *C*, *k(T)* and ρ are temperature-dependent heat capacity, thermal conductivity and material density, respectively. $\dot{Q}$ is the rate of heat generation, $\dot{T}$ is the rate of changing temperature, *q* is the heat loss due to convection, *r* is the radiative heat loss, h_{conv} is coefficient of heat convection,ε_e is emissivity, σ_s is Stefan–Boltzmann constant, $T_{ambient}$ is ambient temperature and *T* is the varying temperature.

The heat generation during FSW occurs due to tool–workpiece frictional interaction. The widely accepted friction law proposed by Coulomb is used to model the friction contact between the workpiece and the tool. A constant value of 0.4 is used as the coefficient of friction [1].

3 Results and Discussion

The developed numerical model is validated by comparing the predicted temperature profile with experimental data [26]. The comparison of the experimental and numerical data is provided in Fig. 2. The numerical data is obtained for the nodes at half the thickness of the workpiece. The process parameters used for the welding are a linear speed of 50 mm/min, a rotational speed of 600 rpm and a depth of plunge as 0.05 mm. The highest temperature predicted on the workpiece on the top surface is about 520 °C. It must be noted that the temperature rise in FSW is an indirect phenomenon; i.e., the heat generation is because of the frictional contact between the tool and the workpiece. Thus, the accurate prediction of temperature depends significantly on the frictional law and the coefficient of friction used. Various literature shows the use of different coefficients of friction, generally ranging between 0.3 and 0.5 [27] because the frictional coefficient and slip rate cannot be derived by any straightforward method [7]. This can lead to variations in numerically predicted and experimentally obtained temperature values. The temperature distribution and flow behaviour of FSW process is non-symmetrical as compared to other conventional welding methods because either side of the weld centreline observes a slightly different temperature. As shown in Fig. 3, the temperature disparity observed between the advancing and the retreating side is about 5–10 °C. The rotation of the tool results in the deposition of cooler material behind the tool in the retreating side, while the comparatively hot material is deposited in the advancing side [28]. Therefore, this temperature difference appears and is more prevalent near the weld centreline. This temperature difference becomes almost negligible, moving away from the weld centreline.

Since the FSW process is subjected to different temperatures on the advancing and retreating sides, this ensures that numerical simulation must be performed for the complete model rather than one-half of the workpiece as generally done in the case of arc welding techniques [29, 30]. In addition, the whole solution domain for

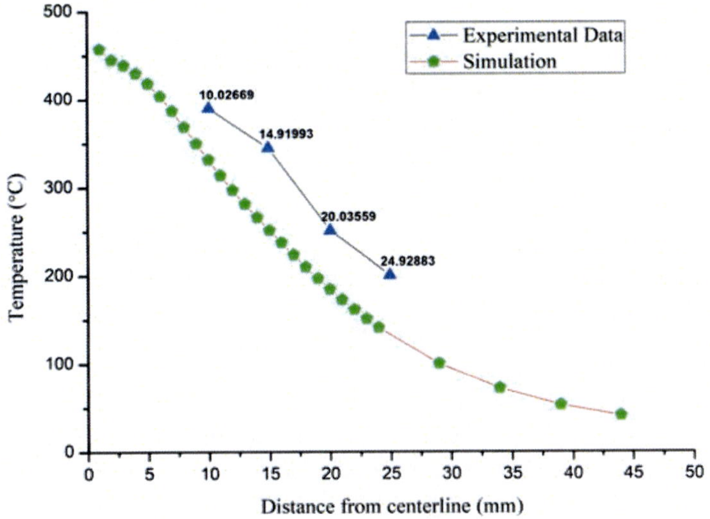

Fig. 2 Comparison between simulated and experimentally measured temperature [26]

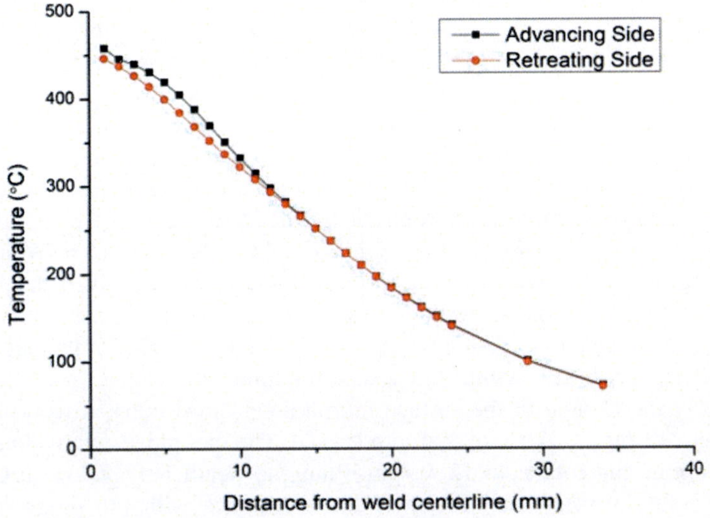

Fig. 3 Predicted temperature difference between advancing and retreating sides of the workpiece

FSW process is computationally expensive. Mass scaling has proven an effective tool to counteract this problem. The time increment ignoring damping is given by [31].

$$\Delta t \leq \min\left(l_{min} \times \sqrt{\frac{\rho}{\lambda_L + 2\mu_L}}\right) \tag{9}$$

$$n_{\text{inc}} = t_{\text{tot}}/\Delta t \tag{10}$$

where n_{inc} is the number of increments required and t_{tot} is the total time. From Eq. 10, it is understood that either reducing the total time or natural time of the system can reduce the number of increments or it can be reduced by increasing the time increment. The first approach is applied by increasing the loading speed. In case of a quasi-static method, this approach does affect the numerical simulation significantly. In the current work, we used the second approach of increasing the increment time. This method has already proved to provide acceptable results [24, 31]. For the second approach, the time increment is controlled by controlling the material properties. The material property manipulated here is the density of the workpiece. From Eq. 9, Δt increases by a factor of x if ρ is increased by x^2. Thus, effectively reducing n_{inc} by a factor of 1/x. This method of reducing the number of increments or increasing the stable time is called mass scaling. Mass scaling is used with rate-dependent problems. The only point of concern is that it should not be so high that inertia forces start dominating. To ensure that the assumed mass scaling is not adversely affecting the obtained result, the ratio of kinetic to internal energy must be below 10% [24]. We have adopted a uniform mass scaling with a factor of 10^7. The ratio of the kinetic energy to internal energy is shown for the first two stages of FSW in Fig. 4. The highest value achieved is little above 2%, which is well within the acceptable range. This shows that the internal energy of the system remains much higher compared to the kinetic energy, meaning that the FSW remains a quasi-static process even when we are using a mass scaling factor of 10^7, which can also be corroborated from the independent work of Zhang and Zhang [24]. This opens the possibility of using mass scaling approach for FSW simulation considering it to be a quasi-static process. Moreover, it is observed that a sudden spike in the ratio value occurs as the tool shoulder ultimately touches the workpiece resulting in a sudden temperature rise. Although there is a spike at the beginning of the dwell stage, it is still within the acceptable limit.

To study the change in the surface morphology of Al-6061, a workpiece with dimension 240 mm × 140 mm × 4 mm is used. The tool moves with a linear speed of 90 mm/min and rotates at 1200 rpm. Plunging depth is kept low at 0.05 mm. Change in the surface morphology of the workpiece with the progress of tool traverse can be observed in Fig. 5. We can follow a small bulging out of the workpiece at the back of the tool when the tool is inserted into the workpiece. This bulge appears due to the pushing of the material behind the tool. From Figs. 5 and 6, it is observed that the temperature of the chip is almost equal to the base material. Chip formation is a surface defect appearing during the FSW process due to excessive heat generation. Generally, the chip formation is more pronounced when the tool rotating speed is comparatively higher than the tool traverse speed. In such a condition, the workpiece gets into a hot metal processing condition aiding in

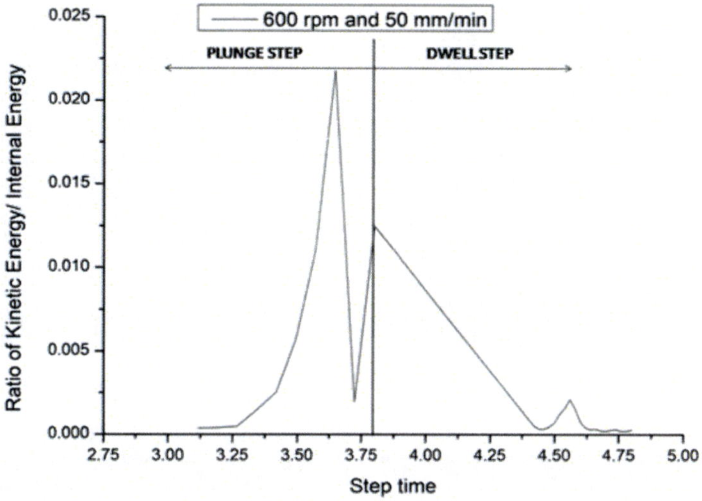

Fig. 4 Kinetic energy and internal energy ratio distribution with step time during the plunge and dwell step with a mass scaling factor of 10^7

severe plastic deformation. The rise in temperature also reduces the flow stress of the material, especially in the stirred zone. These factors, viz. high temperature, high plastic deformation and lower flow stresses, lead to the expulsion of the material out the stir zone in the form of a chip. Figure 6 shows the image of the chip formation at the end of the weld. It is evident from the image that the chip formation is more continuous on the advancing side than the retreating side. Low plunge depth in the present work can be a reason for the discontinuous chip on the retreating side. It is evident from Fig. 3 that the advancing side observes a higher temperature than the retreating side. This difference in temperature may arise from the disparity of the plastic work done between the advancing and the retreating sides. This can be observed from Figs. 7 and 8. Figures 7a and 8a show the cross-sectional view of plastic strain and equivalent plastic strain at the beginning of the traverse stage. It can be observed from the figure that the plastic strain distribution is almost equally spaced in the advancing and retreating sides at the initiation of the welding step. This changes by the end of the welding, when the plastic strain and equivalent plastic strain are more widely spread and have a higher value on the advancing side than the retreating side. It is observed in Figs. 7b and 8b. Imam et al. [32] made a similar observation for Al 5083 alloy. Moreover, the width of the plastic distribution is broader towards the top of the workpiece, i.e. where the tool shoulder contacts the workpiece. In contrast, the distribution becomes sufficiently narrow with the increase in workpiece depth. This observation concurs with the general FSW process, where severe plastic deformation is observed near the tool–workpiece interaction.

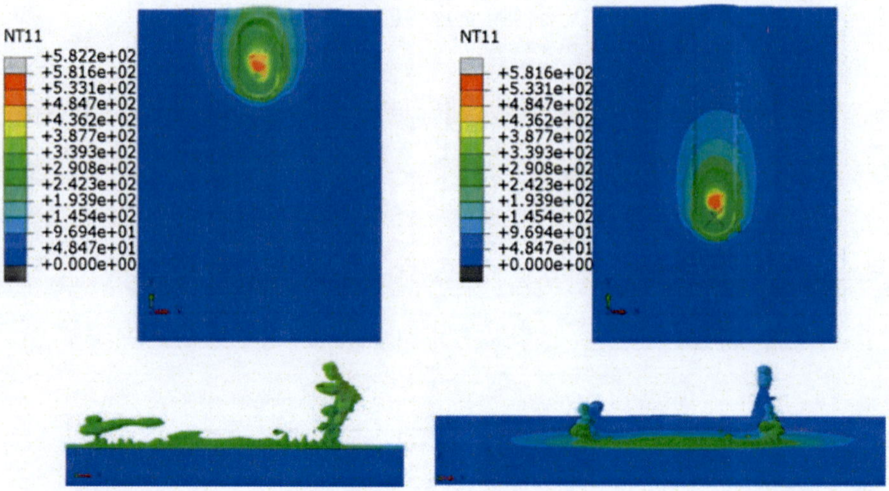

Fig. 5 Evolution of chip morphology and temperature contours (°C) during the FSW process at **a** 25 s and **b** 80 s

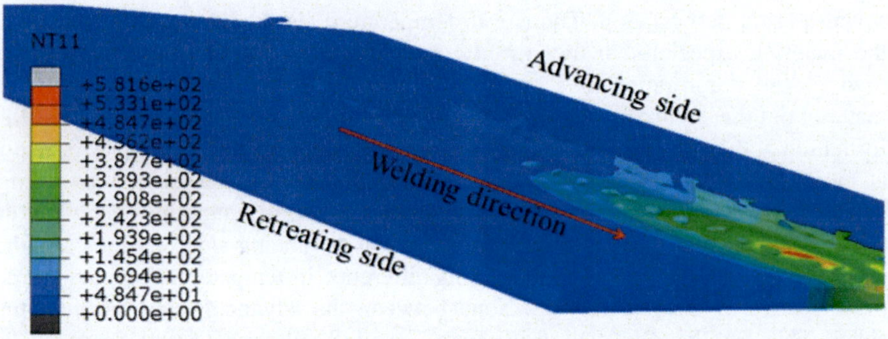

Fig. 6 Surface morphology and temperature contours (°C) during FSW on both advancing and retreating sides

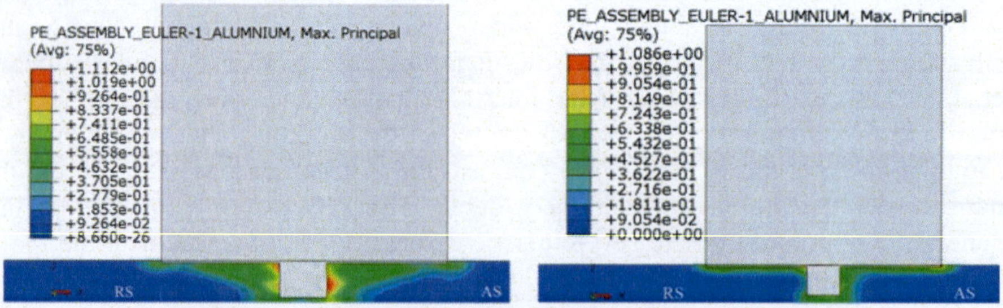

Fig. 7 Plastic strain distribution at **a** beginning of traverse stage and **b** end of the traverse stage

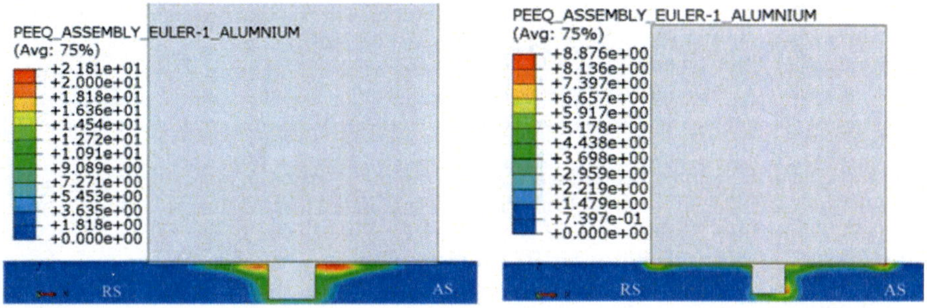

Fig. 8 Equivalent plastic strain distribution at the **a** beginning of traverse stage—at 0 s and **b** end of traverse stage—at 110 s

4 Conclusions

A coupled Eulerian and Lagrangian model is successfully developed to study the various features of the FSW process. The developed model is capable of predicting both temperature distribution and material flow in the weld zone. This work describes the change in surface morphology observed during the FSW process. In addition, the model predicts the stress and strain distributions appearing in the solution domain. Chip formation, a vital aspect of the FSW process, is analysed using this method. With less plunge depth, the chip formation is less and discontinuous. Thermal analysis predicts a temperature difference of about 10 °C between the advancing and the retreating sides. In CEL approach, the computational cost is reduced by using mass scaling method while not affecting the results adversely such that the FSW process can be considered as a quasi-static process. An apparent variation in plastic strain distribution is observed on the advancing and retreating sides of the workpiece. Similar non-uniform distribution of strain is also observed along with the depth of the workpiece.

Acknowledgements The authors gratefully acknowledge the computational facility provided by the Department of Mechanical Engineering, IIT Guwahati, India.

References

1. Mishra RS, De PS, Kumar N (2014) Friction stir welding and processing: Science and engineering, 9783319070
2. Lienert TJ, Stellwag WL, Grimmet BB, Warke RW (2003) Friction Stir Welding Studies on Mild Steel. AWS Weld J, January, pp 1–9
3. Shi L, Wu CS (2017) Transient model of heat transfer and material flow at different stages of friction stir welding process. J Manuf Process 25:323–339. https://doi.org/10.1016/j.jmapro.2016.11.008

4. Chen GQ et al (2013) Computational fluid dynamics studies on heat generation during friction stir welding of aluminum alloy. Comput Mater Sci 79:540–546. https://doi.org/10.1016/j.commatsci.2013.07.004
5. Colegrove PA, Shercliff HR (2005) 3-Dimensional CFD modelling of flow round a threaded friction stir welding tool profile, vol 169, pp 320–327. https://doi.org/10.1016/j.jmatprotec.2005.03.015
6. Kuykendall K, Nelson T, Sorensen C (2013) On the selection of constitutive laws used in modeling friction stir welding. Int J Mach Tools Manuf 74:74–85. https://doi.org/10.1016/j.ijmachtools.2013.07.004
7. Nandan R, DebRoy T, Bhadeshia HKDH (2008) Recent advances in friction-stir welding - Process, weldment structure and properties. Prog Mater Sci 53(6):980–1023. https://doi.org/10.1016/j.pmatsci.2008.05.001
8. Jain R, Pal SK, Singh SB (2017) Numerical modeling methodologies for friction stir welding process. In Computational Methods and Production Engineering, Elsevier, pp 125–169
9. Neto DM, Neto P (2013) Numerical modeling of friction stir welding process: A literature review. Int J Adv Manuf Technol 65(1–4):115–126. https://doi.org/10.1007/s00170-012-4154-8
10. Priyadarshini A, Pal SK, Samantaray AK (2012) Finite element modeling of chip formation in orthogonal machining. Stat Comput Tech Manuf 9783642258:101–144. https://doi.org/10.1007/978-3-642-25859-6_3
11. Dialami N, Chiumenti M, Cervera M, Agelet de Saracibar C (2017) Challenges in Thermo-mechanical Analysis of Friction Stir Welding Processes. Arch Comput Methods Eng 24(1):189–225. https://doi.org/10.1007/s11831-015-9163-y
12. Assidi M, Fourment L, Guerdoux S, Nelson T (2010) Friction model for friction stir welding process simulation: Calibrations from welding experiments. Int J Mach Tools Manuf 50 (2):143–155. https://doi.org/10.1016/j.ijmachtools.2009.11.008
13. Noh WF (1963) CEL: A time-dependent, two-space-dimensional, coupled Eulerian-Lagrange code
14. Al-Badour F, Merah N, Shuaib A, Bazoune A (2013) Coupled Eulerian Lagrangian finite element modeling of friction stir welding processes. J Mater Process Technol. https://doi.org/10.1016/j.jmatprotec.2013.02.014
15. Al-Badour F, Merah N, Shuaib A, Bazoune A (2014) Thermo-mechanical finite element model of friction stir welding of dissimilar alloys. Int J Adv Manuf Technol 72(5–8):607–617
16. Arora A, Mehta M, De A, Debroy T (2012) Load bearing capacity of tool pin during friction stir welding. Int J Adv Manuf Technol 61(9–12):911–920. https://doi.org/10.1007/s00170-011-3759-7
17. Mehta M, Arora A, De A, Debroy T (2011) Tool geometry for friction stir welding - Optimum shoulder diameter. Metall Mater Trans A Phys Metall Mater Sci 42(9):2716–2722. https://doi.org/10.1007/s11661-011-0672-5
18. Dialami N, Cervera M, Chiumenti M, De Saracibar CA (2017) Local – global strategy for the prediction of residual stresses in FSW processes. Internatioanl J Adv Manuf Technol 88:3099–3111. https://doi.org/10.1007/s00170-016-9016-3
19. Meyghani B, Awang M, Wu CS (2020) Finite element modeling of friction stir welding (FSW) on a complex curved plate. J. Adv. Join. Process 1, November, p 100007. https://doi.org/10.1016/j.jajp.2020.100007
20. Chauhan P, Jain R, Pal SK, Singh SB (2018) Modeling of defects in friction stir welding using coupled Eulerian and Lagrangian method. J. Manuf. Process 34, November, 158–166, https://doi.org/10.1016/j.jmapro.2018.05.022
21. Fraser KA, Fraser K, St-Georges L, Kiss LI (2013) Prediction of Defects in a Friction Stir Welded Joint using the Smoothed Particle Hydrodynamics Method. Proc. 7th Asia Pacific IIW Int. Congr., July, pp 474–479 [Online]. Available: https://www.researchgate.net/publication/275271401

22. Fraser K, Kiss LI, St-Georges L, Drolet D (2018) Optimization of friction stir weld joint quality using a meshfree fully-coupled thermo-mechanics approach. Metals (Basel) 8(2). https://doi.org/10.3390/met8020101
23. Dialami N, Chiumenti M, Cervera M, Agelet De Saracibar C (2013) An apropos kinematic framework for the numerical modeling of friction stir welding. Comput Struct 117:48–57. https://doi.org/10.1016/j.compstruc.2012.12.006
24. Zhang Z, Zhang HW (2009) Numerical studies on controlling of process parameters in friction stir welding. J Mater Process Technol 209(1):241–270. https://doi.org/10.1016/j.jmatprotec.2008.01.044
25. Cocchetti G, Pagani M, Perego U (2013) Selective mass scaling and critical time-step estimate for explicit dynamics analyses with solid-shell elements. Comput Struct 127:39–52
26. Zhu Z, Wang M, Zhang H, Zhang X, Yu T, Wu Z (2017) A finite element model to simulate defect formation during friction stir welding. Metals (Basel) 7(7):256
27. Su H, Wu CS, Pittner A, Rethmeier M (2014) Thermal energy generation and distribution in friction stir welding of aluminum alloys. Energy 77:720–731. https://doi.org/10.1016/j.energy.2014.09.045
28. Hamilton C, Kopyściański M, Senkov O, Dymek S (2013) A coupled thermal/material flow model of friction stir welding applied to Sc-modified aluminum alloys. Metall Mater Trans A 44(4):1730–1740
29. Baruah M, Bag S (2016) Influence of heat input in microwelding of titanium alloy by micro plasma arc. J Mater Process Technol 231:100–112. https://doi.org/10.1016/j.jmatprotec.2015.12.014
30. Kumar B, Kebede D, Bag S (2018) Microstructure evolution in thin sheet laser welding of titanium alloy. Int J Mechatronics Manuf Syst 11(2–3):203–229. https://doi.org/10.1504/IJMMS.2018.092875
31. Riahi M, Nazari H (2011) Analysis of transient temperature and residual thermal stresses in friction stir welding of aluminum alloy 6061-T6 via numerical simulation. Int J Adv Manuf Technol 55(1–4):143–152. https://doi.org/10.1007/s00170-010-3038-z
32. Imam M et al (2018) Deformation characteristics and microstructural evolution in friction stir welding of thick 5083 aluminum alloy. Int J Adv Manuf Technol 99:663–681

A Numerical Model for Piezoelectric Beam with Graded Properties

Pankaj Sharma

1 Introduction

Functionally graded materials (FGMs) are the efficient materials in comparison with other conventional materials. In previous decades, the smart structures are used piezoelectric materials [1]. But there were certain restrictions of piezoelectric materials. Due to mitigate the certain restrictions, the functionally graded piezoelectric materials (FGPMs) have been introduced [2–4]. FGPMs are the combination of functionally graded materials and piezoelectric materials. In the present scenario, the FGPMs are used in several engineering applications. The functionally graded piezoelectric materials (FGPMs) are the advanced materials which have been used in design of smart structures, sensors, atomic force microscope, actuators medical imaging and ultrasound [5, 6].

2 Literature Survey

The different mathematical theories are used to analyze the vibration characteristics of piezoelectric beam with graded properties. The Euler–Bernoulli theory is used for thin beams, and Timoshenko beam theory is employed for study of thick beams with consideration of shear correction factor. Yang et al. [7] used Timoshenko beam theory to predict the modal behavior of FGPM beam. Li et al. [8] adopted three-dimensional theory to calculate the natural frequencies of FGPM beam. The Euler–Bernoulli beam theory was used by the Armin et al. [9] to evaluate the resonant frequencies of FGM beam with graded properties. The first shear defor-

P. Sharma (✉)
Department of Mechanical Engineering, Rajasthan Technical University,
Kota, Rajasthan, India
e-mail: psharma@rtu.ac.in

S. Bag et al. (eds.), *Next Generation Materials and Processing Technologies*,
Springer Proceedings in Materials 9, https://doi.org/10.1007/978-981-16-0182-8_27

mation theory is employed by Doroushi et al. [10] to analyze the free vibration behavior of FGPM beam. The same authors solved the same problem using higher-order shear deformation theory [11]. Yao et al. [12] studied force vibration analysis of FGPM beam using modified Timoshenko beam theory. The static and dynamic analysis of FGPM beam using Timoshenko beam theory was given by Lezgy-Nazargah et al. [13]. Using three-dimensional theory, Pandey et al. [14] studied of FGPM beam under electromechanical loads. Li et al. [15] derived the formulation for FGPM beam using strain gradient theory. Parashar et al. [16] studied free vibration behavior of axially polarized FGPM using modified Timoshenko beam theory. Using the same theory, Pankaj et al. [17] gave the exact solution of flexural modal behavior of axially polarized FGPM beam. Sharma et al. [18] used first-order shear deformation theory to predict the natural frequencies of FGPM annular plate. The Timoshenko beam theory was used by the same author to study the vibration analysis of thickness polarized FGPM beam using Harmonic differential quadrature method [19]. The detail study of FGPM beam is available with the literature [20]. As review from the literature, it revealed that the different theories are employed to evaluate the natural frequencies of FGPM beam. The Euler–Bernoulli theory is used for thin beams, and Timoshenko beam theory is employed for study of thick beams. A new theory is introduced called refined trigonometric shear deformation theory which does not required the shear correction factor to analyze the modal study of axially polarized FGPM beam.

To the author's best knowledge, the vibration analysis of axially polarized FGPM beam using trigonometric shear deformation theory is not available in the literature. The material properties are assumed to be varied continuously in the thickness direction according to power law distribution. The energy principle is used to derive the equations of motion using refined trigonometric shear deformation theory. The approximate efficient numerical technique called generalized differential quadrature (GDQ) is adopted to convert the partial derivative equations to linear algebraic equations. The vibration behavior of axially polarized FGPM beam is analyzed under clamped-clamped boundary condition. Finally, the linear algebraic equations along with available boundary conditions are solved using MATLAB tool. It is observed that the refined trigonometric shear deformation theory is given accurate results for thin and thick beams.

3 Modeling of FGPM Beam

In this study, the geometry of axially polarized FGPM beam model is shown in Fig. 1. The length, width, and thickness of the beam are represented by l, b, and h, respectively. The beam is polarized in the direction of length. The extreme left and extreme right faces are clamped. When a single electric alternating current pulse is given to this beam and then short-circuited; i.e., the external voltage magnitude is maintained at zero value, the thickness mode vibrations are observed.

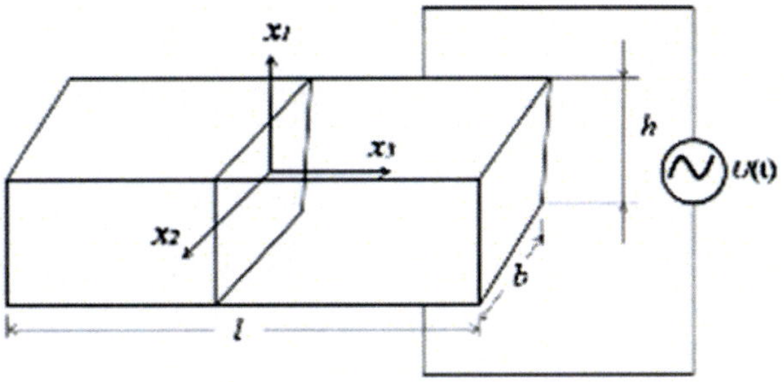

Fig. 1 Geometry of FGPM beam

As per the rule of power law, the material properties at any location in the thickness direction can be obtained using [20].

$$\xi(x_1) = \xi_{ul}\left(\frac{2x_1 + h}{2h}\right)^n + \xi_l \tag{1}$$

where $\xi_{ul} = \xi_u - \xi_l$ and ξ_u, ξ_l are the constants at the top and the bottom surfaces, respectively. The power exponent is represented by n. The displacement components are written according to refined trigonometric shear deformation theory as follows

$$u = u(x_3, t) \tag{2}$$

$$w = -x_1 w_{,3} + \frac{h}{\pi} sin \frac{\pi x_1}{h} \psi(x_3, t) \tag{3}$$

Here w is axial displacement, whereas the transverse displacement is denoted by u. The rotation is represented by ψ.

Now the normal strain (λ_3) is written as

$$\lambda_3 = w_{,3} \tag{4}$$

and the shear strain (λ_5) can be obtained as

$$\lambda_5 = u_{,3} + w_{,1} \tag{5}$$

The electric potential (Φ) can be written as [20].

$$\Phi(x_1, x_3, t) = -\left(\sin\frac{2\pi x_1}{h}\right)\varphi(x_3, t) + \frac{V}{h} x_1 cos(Ut) \tag{6}$$

The electric field induced in x_1- and x_3-directions is expressed as E_1 and E_3, respectively, which are given as

$$E_1 = -\Phi_{,1} \tag{7}$$

$$E_3 = -\Phi_{,3} \tag{8}$$

Now the normal stress induced in the beam is T_3 which is written as [16].

$$T_3 = c_{33}\lambda_3 - e_{33}E_3 \tag{9}$$

The shear stress (T_5) could be written as [16].

$$T_5 = c_{55}\lambda_5 - e_{15}E_1 \tag{10}$$

Here c_{33}, c_{55} are the stiffness coefficients and e_{33}, e_{15} are the piezoelectric coupling coefficients. The dielectric displacements (D_1 and D_3) are described as [16].

$$D_1 = e_{15}\lambda_5 + \varepsilon_{11}E_1 \tag{11}$$

$$D_3 = e_{33}\lambda_3 + \varepsilon_{33}E_3 \tag{12}$$

Here ε_{11} and ε_{33} are the dielectric constants for piezoelectric material.
The energy term enthalpy (H) can be expressed as

$$T_a = \frac{\partial H}{\partial \lambda_a} \tag{13}$$

Here a = 3, 5
The dielectric displacement can be written as

$$D_i = -\frac{\partial H}{\partial E_i} \tag{14}$$

Here i = 1, 3.
The energy term called kinetic energy (κ) can be expressed as

$$\kappa = \frac{1}{2}\int_v \rho\left(\dot{u}^2 + \dot{w}^2\right)dv \tag{15}$$

The Hamilton energy principle is used to derive the equations of motion as written as

$$\delta\int_{t_0}^{t_1} L dt + \int_{t_0}^{t_1} \delta W dt = 0 \tag{16}$$

The relation between Lagrangian (L) function and energy terms is as follows

$$L = \int_{v} (\kappa - H)dv \tag{17}$$

In this study, the term virtual work done (W) is taken as zero. Substitute Eqs. (13–15) in Eq. (16), the equations of motion are obtained as

$$-F_{11}u_{,3333} + F_{12}\psi_{,333} + F_{13}\varphi_{,33} = -I_0\ddot{u}_{,33} - I_2\ddot{\psi}_{,3} - I_3\ddot{u} \tag{18}$$

$$-F_{12}u_{,333} + F_{55}\psi_{,33} - (G_{33} - G_{11})\varphi_{,3} - F_{15}\psi = I_2\ddot{u}_{,3} + I_1\ddot{\psi} \tag{19}$$

$$F_{13}u_{,33} + (G_{11} - G_{33})\psi_{,33} - G_{16}\varphi_{,33} + G_{66}\varphi = 0 \tag{20}$$

The constants available in Eqs. (18–20) are listed here

$$F_{11} = \int_{\frac{-h}{2}}^{\frac{h}{2}} c_{11}x_1^2 dx_1$$

$$F_{15} = \int_{\frac{-h}{2}}^{\frac{h}{2}} c_{55}\left(cos\,cos\frac{\pi x_1}{h}\right)^2 dx_1$$

$$F_{12} = \frac{h}{\pi}\int_{\frac{-h}{2}}^{\frac{h}{2}} c_{11}x_1\left(sin\,sin\frac{\pi x_1}{h}\right)dx_1$$

$$G_{11} = 2\int_{\frac{-h}{2}}^{\frac{h}{2}} e_{31}\left(sin\,sin\frac{\pi x_1}{h}\right)\left(cos\,cos\frac{2\pi x_1}{h}\right)dx_1$$

$$F_{55} = \frac{h^2}{\pi^2}\int_{\frac{-h}{2}}^{\frac{h}{2}} c_{11}\left(sin\,sin\frac{\pi x_1}{h}\right)^2 dx_1$$

$$G_{16} = \int_{\frac{-h}{2}}^{\frac{h}{2}} \in_{11}^{s}\left(sin\,sin\frac{2\pi x_1}{h}\right)dx_1$$

$$G_{66} = \frac{4\pi^2}{h^2}\int_{\frac{-h}{2}}^{\frac{h}{2}} \in_{33}\left(cos\,cos\frac{2\pi x_1}{h}\right)^2 dx_1$$

$$I_2 = -\frac{h}{\pi}\int_{\frac{-h}{2}}^{\frac{h}{2}} \rho\left(sin\,sin\frac{\pi x_1}{h}\right)x_1 dx_1$$

$$I_0 = \int_{\frac{-h}{2}}^{\frac{h}{2}} \rho x_1^2 dx_1$$

$$I_1 = \frac{h^2}{\pi^2} \int_{\frac{-h}{2}}^{\frac{h}{2}} \rho \left(sin\, sin \frac{\pi x_1}{h} \right)^2 dx_1$$

$$F_{13} = -\frac{2\pi}{h} \int_{\frac{-h}{2}}^{\frac{h}{2}} e_{31} x_1 \left(cos\, cos \frac{2\pi x_1}{h} \right) x_1 dx_1$$

$$G_{33} = \int_{\frac{-h}{2}}^{\frac{h}{2}} e_{15} \left(sin\, sin \frac{2\pi x_1}{h} \right) \left(cos\, cos \frac{\pi x_1}{h} \right) dx_1$$

The clamped-clamped boundary condition can be written as $u = 0$, $\psi = 0$ at the extreme left and the extreme right faces.

4 Numerical Technique

The efficient numerical technique called "generalized differential quadrature (GDQ)" method is adopted in this work. The partial derivative equations are converted into linear algebraic equations by using this method. Finally, the MATLAB tool is used to calculate the natural frequencies of the beam.

Bellman et al. [21] introduced the concept of differential quadrature method in 1971. As per this method, the *rth* order partial derivative of a function $f(\alpha)$ at a grid point i is approximated as written as

$$f_{,\alpha}^{(r)}(\alpha_i) = \sum\nolimits_{j=1}^{N} \mu_{ij}^{(r)} f(\alpha_j) \quad (21)$$

where i represents the internal grid points and $\mu_{ij}^{(r)}$ is the weighting coefficients for rth order derivative. N is the total number of grid points. The weighting coefficients for first order derivative $\mu_{ij}^{(1)}$ are calculated as

$$\mu_{ij}^{(1)} = \frac{\Theta^{(1)}(\alpha_i)}{(\alpha_i - \alpha_j)\Theta^{(1)}(\alpha_j)} \quad (22)$$

All above-weighting coefficients are the non-diagonal elements of a matrix. But the diagonal elements $\mu_{ii}^{(1)}$ are evaluated as [22].

$$\mu_{ii}^{(1)} = -\sum\nolimits_{j=1, j\neq i}^{N} \mu_{ij}^{(1)} \quad (23)$$

The primary function $\Theta(\alpha)$ according to GDQ method can be expressed as

$$\Theta(\alpha) = \prod_{j=1}^{N} (\alpha - \alpha_j) \tag{24}$$

The first order derivative of primary function $\Theta^{(1)}$ could be obtained as

$$\Theta^{(1)} = \prod_{j=1, j\neq k}^{N} (\alpha_k - \alpha_j) \tag{25}$$

In general, the weighting coefficients for r[th] order derivate are obtained as

$$\mu_{ij}^{(r)} = r\left[\mu_{ii}^{(r-1)}\mu_{ij}^{(1)} - \frac{\mu_{ij}^{(r-1)}}{(\alpha_i - \alpha_j)}\right] \tag{26}$$

The diagonal terms can be calculated as

$$\mu_{ii}^{(r)} = -\sum_{j=1, j\neq i}^{N} \mu_{ij}^{(r)} \tag{27}$$

The grid points can be written as [22].

$$\alpha_i = -\frac{1}{2}\left[\cos\cos\frac{(i-1)\pi}{(N-1)}\right] \tag{28}$$

i = 1, 2, 3….. N.

Now Eqs. (18–20) can be written as per GDQ method with substituting $u = u$ eiωt, $\psi = \psi e^{i\omega t}$ and $\varphi = \varphi e^{i\omega t}$

$$\begin{aligned} &-F_{11}\sum_{j=1}^{N}\mu_{ij}^{(4)}u_j + F_{12}\sum_{j=1}^{N}\mu_{ij}^{(3)}\psi_j + F_{13}\sum_{j=1}^{N}\mu_{ij}^{(2)}\varphi_j \\ &= -I_0\omega^2\sum_{j=1}^{N}\mu_{ij}^{(2)}u_j - I_2\omega^2\sum_{j=1}^{N}\mu_{ij}^{(1)}\psi_j - I_3\omega^2 u_i \end{aligned} \tag{29}$$

$$\begin{aligned} &-F_{12}\sum_{j=1}^{N}\mu_{ij}^{(3)}u_j + F_{55}\sum_{j=1}^{N}\mu_{ij}^{(2)}\psi_j - (G_{33} - G_{11})\sum_{j=1}^{N}\mu_{ij}^{(1)}\varphi_j - F_{15}\psi_i \\ &= I_2\omega^2\sum_{j=1}^{N}\mu_{ij}^{(1)}u_j + I_1\omega_j^2\psi_i \end{aligned} \tag{30}$$

$$F_{13}\sum_{j=1}^{N}\mu_{ij}^{(2)}u_j + (G_{11} - G_{33})\sum_{j=1}^{N}\mu_{ij}^{(2)}\psi_j - G_{16}\sum_{j=1}^{N}\mu_{ij}^{(2)}\varphi_j + G_{66}\varphi_i = 0 \tag{31}$$

Here i = 2, 3, …, N−1.

The boundary conditions are descretized as

$$u_i = 0 \text{ for } i - 1, \text{N and } \psi_i = 0 \text{ for } i = 1, \text{N} \quad (32)$$

Now substituting Eq. (32) in Eqs. (29–31), the equations of motion are available in the following form

$$\Omega - 4\pi^2 f^2 M = 0 \quad (33)$$

Here Ω is the stiffness matrix, M represents the inertia matrix, and f is the natural frequency.

5 Numerical Results and Discussions

5.1 Convergence Study

The natural frequencies (f) of axially polarized FGPM beam using refined trigonometric shear deformation theory are reported here. The material properties and beam composition are taken from [7]. The clamped-clamped boundary condition is used in this study. The non-dimension frequency ($\check{f}$) is calculated by ($\check{f}$) $= 2\pi f l \sqrt{\frac{\rho}{c_{11}}}$.

Here f is the natural frequency in Hz. Figure 2 shows the effect of number of sampling points on non-dimensional frequency of FGPM beam (l = 20 mm and h = 1 mm). It can be observed that the twelve grid points are sufficient to achieve well convergence. Therefore, twelve grid points are used to compute the results in further study.

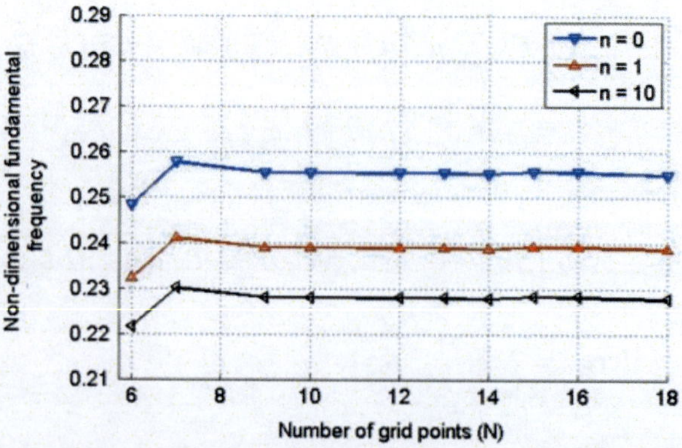

Fig. 2 Variation in non-dimensional frequency of axially polarized FGPM beam with number of sampling points (l = 20 mm and h = 1 mm)

5.2 Validation Study

The refine trigonometric shear deformation theory is used in this work to evaluate the modal characteristics of FGPM beam. So results are compared with previous published work in which modified Timoshenko theory was employed [16] which are displayed in Table 1. The non-dimensional frequency ($\check{f}$) is calculated by using $\check{f} = 2\pi f h\sqrt{\frac{\rho}{c_{55}}}$. It is revealed that the results are very close to the results obtained in the literature [16].

5.3 Parametric Study

Now the parametric study is carried out. The first two non-dimensional frequencies are given for different lengths of the beam for different power exponent values displayed in Table 2. It is revealed that the natural frequency decreases with increase in the value of length. It can be observed that the frequencies decrease with increase in volume fraction exponents.

Table 1 Comparison of natural frequencies of FGPM beam using two different theories Timoshenko beam theory (TBT) and trigonometric shear deformation theory (TSBT)

Frequency ($\check{f}$)	$n = 0$		$n = 1$		$n = 2$	
$l = 7$ mm, $h = 1$ mm						
	TBT [16]	TSDT	TBT [16]	TSDT	TBT [16]	TSDT
I	0.1902	0.1900	0.1802	0.1800	0.1782	0.1789
II	0.4611	0.4598	0.4391	0.4396	0.4343	0.4401
$l = 15$ mm, $h = 1$ mm						
I	0.0449	0.0501	0.0422	0.0431	0.0417	0.0421
II	0.1194	0.1201	0.1125	0.1123	0.1111	0.1121
$l = 25$ mm, $h = 1$ mm						
I	0.0164	0.0166	0.0154	0.0159	0.0152	0.0157
II	0.0446	0.0449	0.0419	0.0422	0.0414	0.0412

Table 2 First two non-dimensional frequencies of FGPM beam under clamped-clamped boundary condition with different power exponents $n = 0.5, 1.5, 9$

Frequency ($\check{f}$)	$n = 0.5$	$n = 1.5$	$n = 9$
$L = 26\ mm,\ h = 1\ mm$			
I	0.2548	0.2385	0.2276
II	0.6927	0.6491	0.6198
$L = 7\ mm,\ h = 1\ mm$			
I	0.8909	0.7999	0.7909
II	2.1005	2.0011	1.9331
$L = 6\ mm,\ h = 1\ mm$			
I	0.9001	0.8512	0.8101
II	2.111	2.0299	1.959

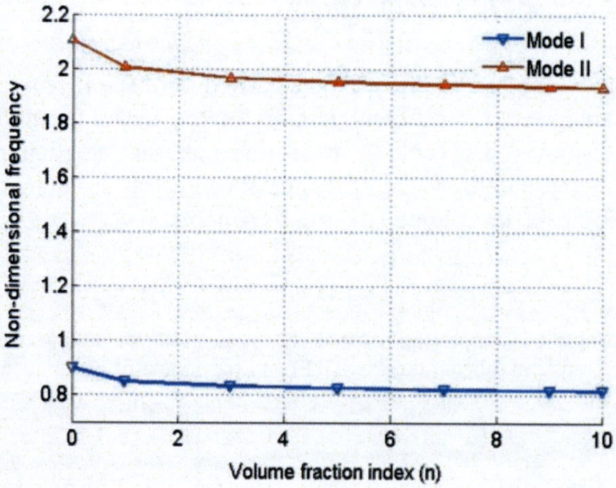

Fig. 3 Effect of power exponent on first two lower order non-dimensional frequency of an FGPM beam ($l = 6$ mm, $h = 1$ mm)

Figure 3 displays the effect of power exponent on first two lower order natural frequencies for length $l = 6$ mm and $h = 1$ mm. It can be seen that the natural frequencies are decreased with increase in power exponents.

6 Summary and Conclusion

The vibration analysis of axially polarized FGPM beam is studied. The refined trigonometric shear deformation theory is used in this work. For this theory, no shear correction factor is needed. The material properties are assumed to be varied continuously in the thickness direction. The energy principle is used to derive the equations of motion. The approximate efficient numerical technique called generalized differential quadrature (GDQ) is used to convert the partial derivative equations of motion into linear algebraic equations. The vibration behavior of axially polarized FGPM beam is analyzed under clamped-clamped boundary condition. Finally, the linear algebraic equations along with available boundary conditions are solved with the help of through MATLAB tool. It is observed that the refined trigonometric shear deformation theory is given accurate and comparable results without consideration of shear correction factor. It is revealed that the natural frequency decreases with increase in the value of length. It can be observed that the frequencies decrease with increase in volume fraction exponents. This theory could be used to study of modal behavior of FGPM thin and FGPM thick beams without consideration of shear correction factor.

References

1. Yang J (2005) An introduction to the theory of piezoelectricity. Springer, New York, NY. https://doi.org/10.1007/978-3-030-03137-4
2. Chi S-H, Chung Y-L (2006) Mechanical behavior of functionally graded material plates under transverse load—Part I: Analysis. Int J Solids Struct 43(13):3657–3674
3. Uchino K (1996) Piezoelectric actuators and ultrasonic motors. Springer, ISBN: 978-1-4612-8638-7
4. Zhu X, Meng Z (1995) Operational principle, fabrication and displacement characteristics of a functionally gradient piezoelectric ceramic actuator. Sens Actuators, A 48(3):169–176
5. Wu CC, Kahn M, Moy W (1996) Piezoelectric ceramics with functional gradients: a new application in material design. J Am Ceram Soc 79(3):809–812
6. Rubio W, Montealegre S, Vatanabe L, Paulino GH, Silva ECN (2011) Functionally graded piezoelectric material systems–a multiphysics perspective. Computational Chemistry of Solid State Materials
7. Yang J, Xiang HJ (2007) Thermo-electro-mechanical characteristics of functionally graded piezoelectric actuators. Smart Mater Struct 16(3):784
8. Li Y, Shi Z (2009) Free vibration of a functionally graded piezoelectric beam via state-space based differential quadrature. Compos Struct 87(3):257–264
9. Armin A, Behjat B, Abbasi M, Eslami MR (2010) Finite element analysis of functionally graded piezoelectric beams, 45–72
10. Doroushi A, Akbarzadeh AH, Eslami MR (2010) Dynamic analysis of functionally graded piezoelectric material beam using the hybrid Fourier-Laplace transform method. Engineering Systems Design and Analysis 49156:475–483
11. Doroushi A, Eslami MR, Komeili A (2011) Vibration analysis and transient response of an FGPM beam under thermo-electro-mechanical loads using higher-order shear deformation theory. J Intell Mater Syst Struct 22(3):231–243
12. Yao RX, Shi ZF (2011) Steady-state forced vibration of functionally graded piezoelectric beams. J Intell Mater Syst Struct 22(8):769–779
13. Lezgy-Nazargah M, Vidal P, Polit O (2013) An efficient finite element model for static and dynamic analyses of functionally graded piezoelectric beams. Compos Struct 104:71–84
14. Pandey VB, Parashar SK (2016) Static bending and dynamic analysis of functionally graded piezoelectric beam subjected to electromechanical loads. Proceedings of the Institution of Mechanical Engineers, Part C: Journal of Mechanical Engineering Science 230(19):3457–3469
15. Li YS, Feng WJ, Cai ZY (2014) Bending and free vibration of functionally graded piezoelectric beam based on modified strain gradient theory. Compos Struct 115:41–50
16. Parashar SK, Sharma P (2016) Modal analysis of shear-induced flexural vibration of FGPM beam using Generalized Differential Quadrature method. Compos Struct 139:222–232. https://doi.org/10.1016/j.compstruct.2015.12.012

17. Sharma P, Parashar SP (2016) Exact analytical solution of shear-induced flexural vibration of functionally graded piezoelectric beam. In: AIP conference proceedings, vol. 1728, no. 1, p. 020167. AIP Publishing LLC, https://doi.org/10.1063/1.494621817
18. Sharma P, Parashar SK (2016) Free vibration analysis of shear-induced flexural vibration of FGPM annular plate using generalized differential quadrature method. Compos Struct 155:213–222. https://doi.org/10.1016/j.compstruct.2016.07.077
19. Sharma P (2018) Efficacy of Harmonic Differential Quadrature method to vibration analysis of FGPM beam. Compos Struct 189:107–116. https://doi.org/10.1016/j.compstruct.2018.01.059
20. Sharma P (2019) Vibration analysis of functionally graded piezoelectric actuators. Springer, New York, NY. https://doi.org/10.1007/978-981-13-3717-8
21. Shu C (2012) Differential quadrature and its application in engineering. Springer. https://doi.org/10.1007/978-1-4471-0407-0
22. Bellman R, Kashef BG, Casti J (1972) Differential quadrature: a technique for the rapid solution of nonlinear partial differential equations. J Comput Phys 10(1):40–52

Design and Analysis of Microgripper Using COMSOL for Drug Delivery Applications

Murugappan Elango and Adithyan Annamalai

1 Introduction

Gripping and manipulating micro-objects are required for a wide range of important applications such as the assembly of micro-parts to obtain miniature systems or component assembly in electronics packages [1]. A good micromanipulator should grasp objects of different orientations in a sturdy way to achieve high positioning accuracy. The manipulators must control grasping forces in order to prevent any damage to micro-sized objects.

The model discussed in our paper is a piezoelectrically actuated microgripper with mechanical contact. The microgripper contains a piezoelectric actuator that operates in the longitudinal mode. Elongation in the longitudinal direction creates a lifting movement to the structure. Microgrippers have effective applications in manufacturing industry, drug delivery and materials research.

Boudaoud et al. [2] described a modelling approach of a MEMS-based micro-gripper with integrated force sensor while handling micro-glass balls of 80 μm diameter. A state-space representation was developed to couple both the dynamics of the actuation and sensing subsystems of the gripper through the stiffness of the manipulated object. Bank [3] presented the design (mechanical and electrical structure) of a novel interaction force sensor, in which a microgripper can be easily integrated. The mechanical and electrical study was carried out for the development of the interaction force sensor according to typical requirements in micro-handling. Dechev et al. [4] developed a system to create joints between micro-parts that can be in-plane or out-of-plane with respect to the substrate. This system could accommodate micro-parts of different shapes, micro-parts fabricated from different materials and could also combine micro-parts from multiple chips to assemble a single microstructure.

M. Elango · A. Annamalai (✉)
Thiagarajar College of Engineering, Madurai 625015, Tamil Nadu, India

S. Bag et al. (eds.), *Next Generation Materials and Processing Technologies*,
Springer Proceedings in Materials 9, https://doi.org/10.1007/978-981-16-0182-8_28

Tamadazte et al. [5] developed a gripper having 4-DOF thereby allowing open–close motions as well as up–down motions. It was based on piezoelectric actuators which consisted two parallel piezoceramic PZT PIC 151 bimorphs. Feng et al. [6] delineated a linear micro-actuator having positive properties like high response and displacement. They can be used in microsatellite remote sensing applications. The prototype was developed using a non-silicon process. The authors could finally realize a bi-stable mechanism having a response time of 0.96 ms [7] designed a microgripper that can effectively grip objects having a diameter of 80 μm. In fact, a thermal force sensor was embedded on the arms of the microgripper. The thermal force sensor besides preventing the gripped object from getting damaged also has high resolution and large measurement range.

Martinez and Panepucci [8] modelled polymer-based grippers having parallel displacement geometry meant to operate objects having a diametrical range of 5–50 μm. Piezoelectric and mechanical actuation systems were employed and simulated. Researchers focused on different types of microgripper, namely electrostatic microgripper, piezoelectric microgripper, thermal microgripper and shape memory alloy microgripper. Among all these types, piezoelectric microgripper has advantageous properties like high power density, high speed, extremely high efficiency and low temperature sensitivity. So, this paper focused on the performance improvement of piezoelectric microgripper. Varona et al. [9] designed a microgripper revolving around the concept of parallel plate electrostatic actuation. This type of actuator, owing to its compact nature can effectively be used in microrobotics and biomedical applications.

The key parameters affecting the performance of microgrippers are delineated in [10]. They are: material used, gripping range, base material, piezo material, contact pressure analysis and stress distribution. Based on the above parameters, our microgripper is designed and simulated using COMSOL Multiphysics package.

2 Gripper Modelling

A microgripper is one of the key elements in drug delivery applications for handling micro-objects (tablets, capsules, soft gelatin capsules, suspensions, elixirs) safely. An essential component of all microgrippers is the actuator, which offers the desired applied force thereby making it operate as a gripper. In this project, the piezoelectric actuator is used. 2D and 3D elements are used to model the microgripper mechanism and predict the amplification ratio, stress concentration and displacements of the microgripper mechanism.

The existing design as depicted in [4] is modelled using COMSOL. Polysilicon is chosen as the base material. Figure 1 shows the existing model of the microgripper. This model emulates the standards of a piezoelectrically actuated microgripper with mechanical contact. This microgripper contains a piezoelectric actuator that operates in longitudinal mode. Elongation in the longitudinal direction creates a lifting movement to the structure. There are two arms in the microgripper and its dimensions are delineated in Fig. 1 The centre actuation piezo material is used to close and open the arms of the microgripper.

In Fig. 2, the geometry of the existing microgripper arm has been modified by employing a V-shaped inclusion on both arms. The material and other properties are the same as that of model 1, discussed in Fig. 1. The dimensions of the arms of the microgripper are reduced compared to model 1 as shown in Fig. 2.

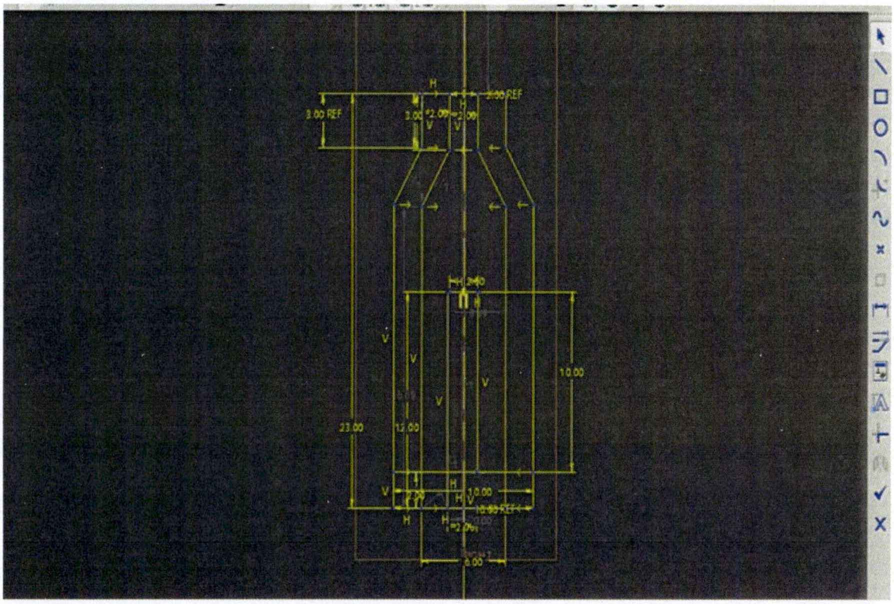

Fig. 1 Microgripper model 1 with straight arm

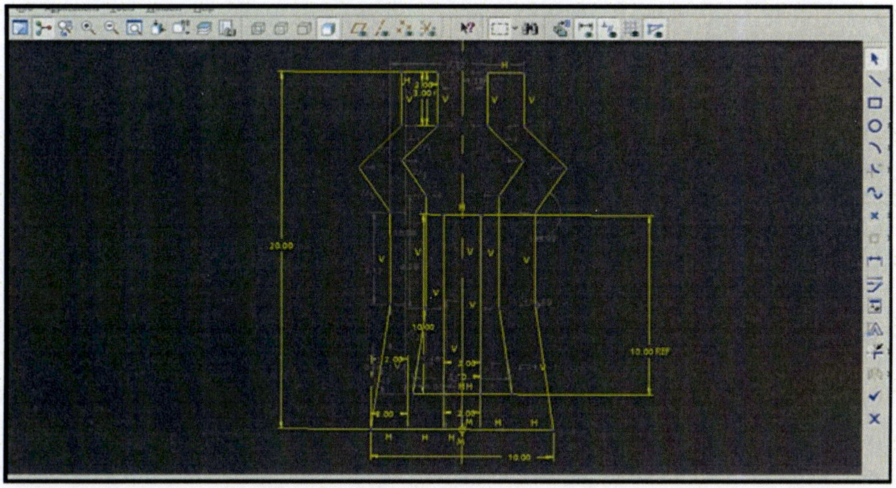

Fig. 2 Microgripper model 2 with arm modification

3 Simulation

The effectiveness of the gripper depends on nature various materialistic and design parameters. Three-dimensional simulations are the key to determine the rigidity of the proposed designs and thereby finally enlisting the structural and layout parameters. Also, the accidental off-plane actuation, an ensuing result of non-planarity of the motion is addressed in the design. This would enforce a pushing action once the gripper is open. The design is simulated using COMSOL Multiphysics software. For both the designs, the base material is chosen as polysilicon and the piezo material is chosen as lead zirconate. The boundary conditions considered are fixed and ground. Triangular meshing is used to run the simulation.

3.1 Base Material Analysis

Four plausible base materials were chosen for analysis, namely polysilicon, silicon, silicon oxide and silicon nitride. They were analysed and the material showing highest displacement properties was concluded to silicon nitride as detailed in Table 1. Figures 3, 4, 5 and 6 show the modified gripper design modelled using various base materials. Also, a plot showing the displacement values pertaining to the gripper with the four base materials is shown in Fig. 7.

3.2 Piezoelectric Material Analysis

Three piezo materials that are commonly used in the previously modelled micro-grippers were considered for analysis. The materials considered are zinc oxide, aluminium nitride and lead zirconate. Analysis was done to find the displacement values of the microgripper when different piezo materials were used as shown in Figs. 8, 9 and 10. Lead zirconate exhibits the highest displacement value when used as piezo material in microgrippers. Table 2 shows the displacement values of microgripper when zinc oxide, aluminium nitride and lead zirconate are used as the piezo materials.

Table 1 Displacement values of gripper modelled using different materials

S. No.	Base material	Displacement (x e^{-5}) (μm)
1	Polysilicon	2.9731
2	Silicon	2.9721
3	Silicon oxide	2.996
4	Silicon nitride	3.012

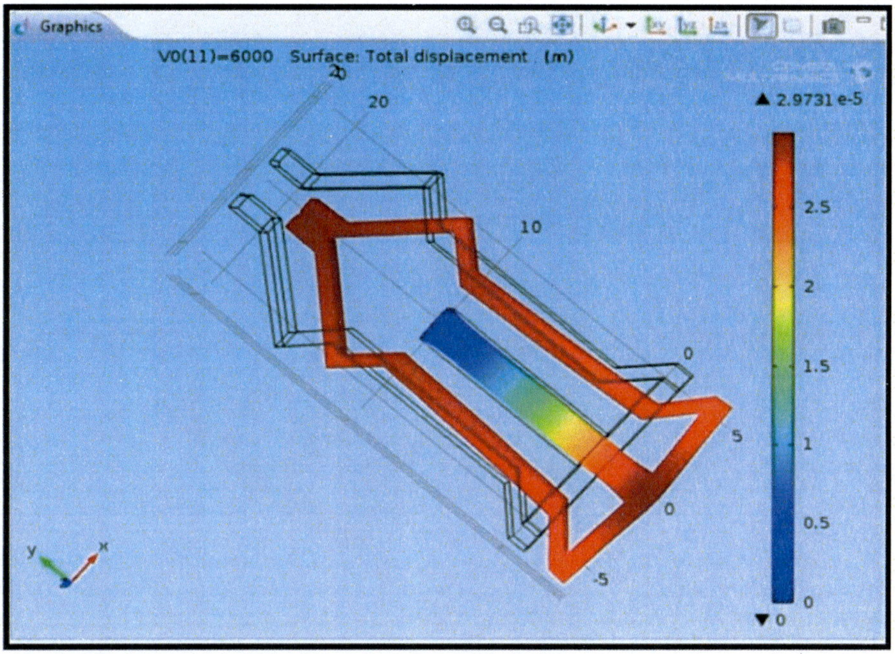

Fig. 3 Displacement analysis for base material polysilicon

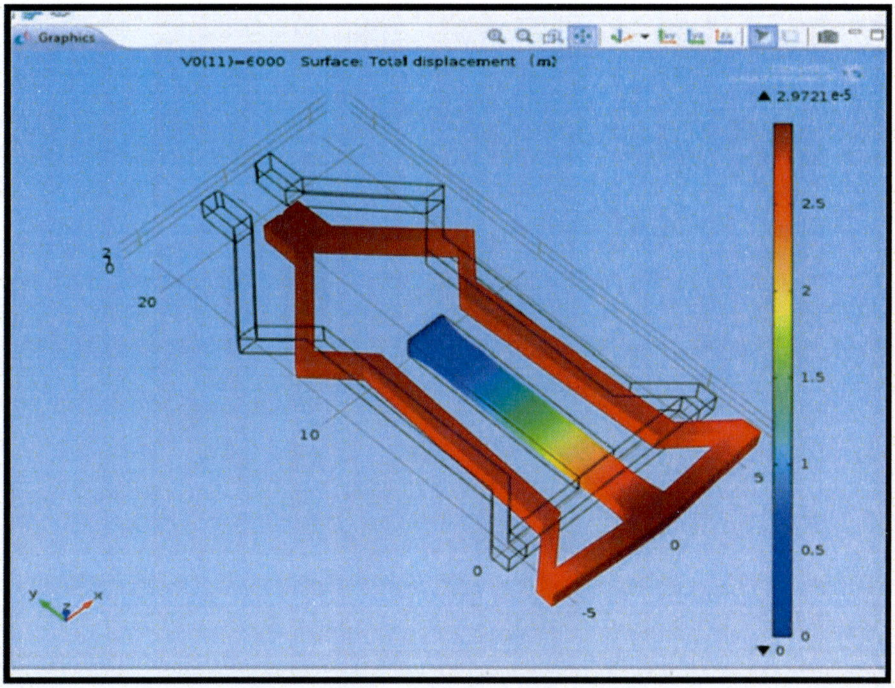

Fig. 4 Displacement analysis for base material silicon

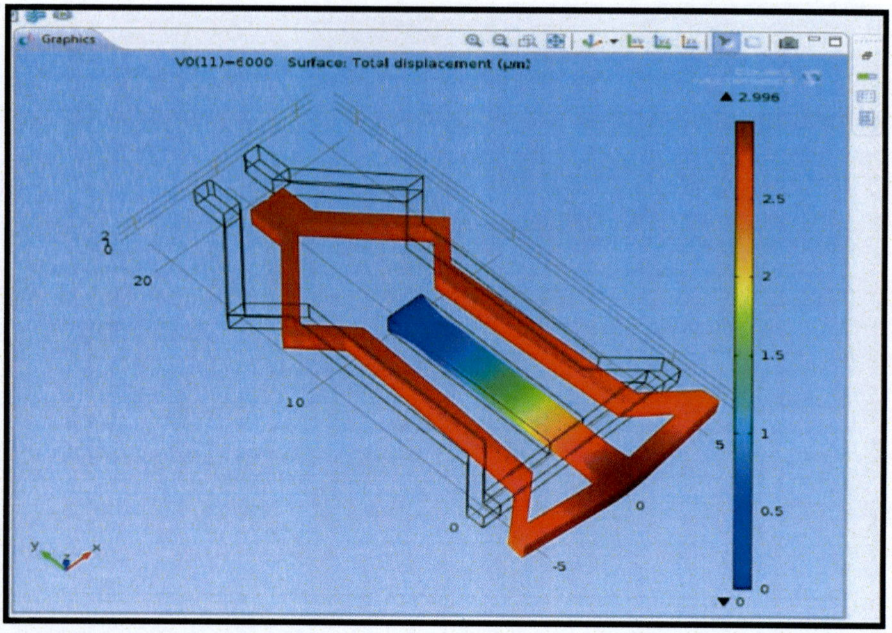

Fig. 5 Displacement analysis for base material silicon oxide

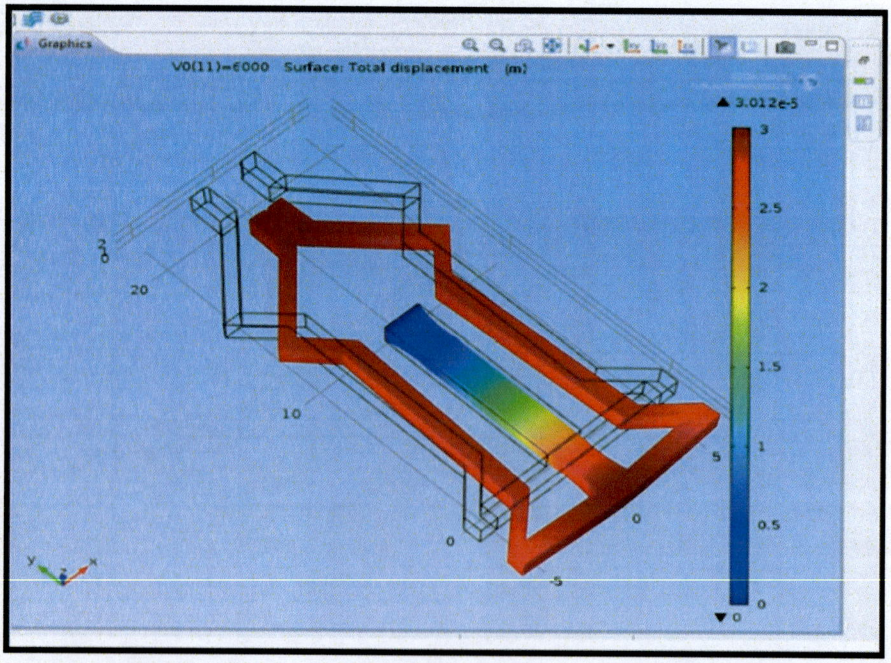

Fig. 6 Displacement analysis for base material silicon nitride

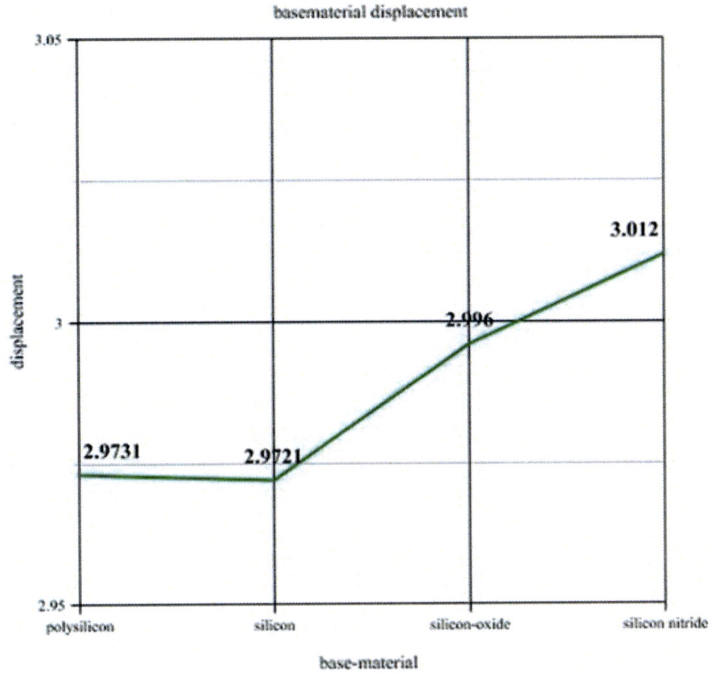

Fig. 7 Displacement values for various materials

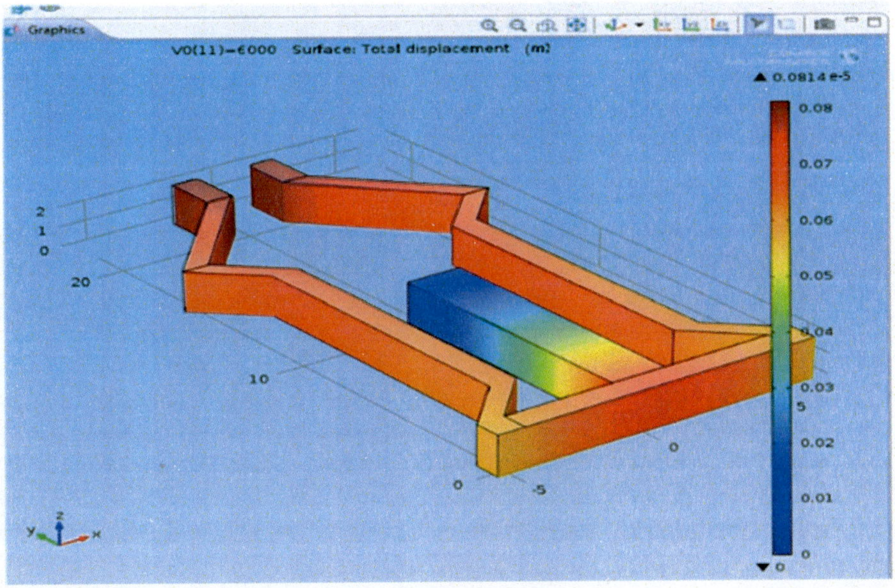

Fig. 8 Displacement analysis for piezo material zinc oxide

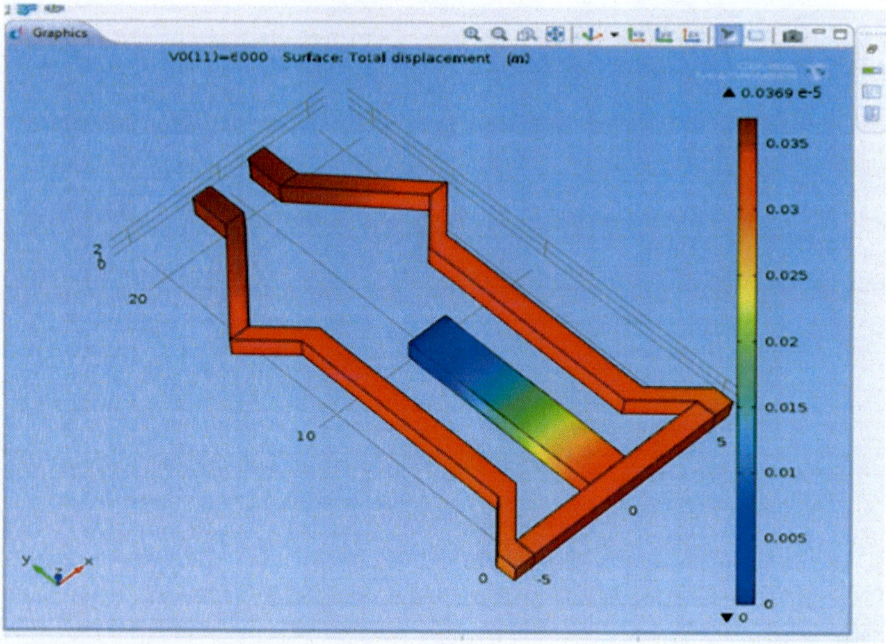

Fig. 9 Displacement analysis for piezo material aluminium nitride

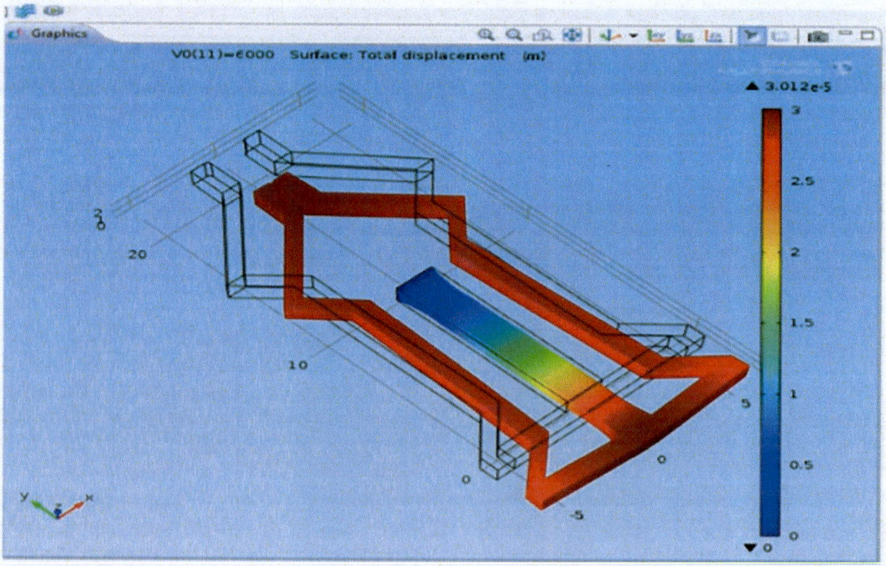

Fig. 10 Displacement analysis for piezo material lead zirconate

Table 2 Displacement values of gripper modelled using different piezo materials

S. No.	Piezo material	Displacement (μm)
1	Zinc oxide	0.0814
2	Aluminium nitride	0.0369
3	Lead zirconate	3.012

3.3 Contact Pressure Analysis

The contact pressure of the new design was compared with that of the existing design. The contact pressure of the new design increased by 28% from 2.8723×10^6 N/m^2 to 3.7004×10^6 N/m^2. Thus, the new design could hold components more effectively than the existing design owing to the increased contact pressure. The contact pressure of the new design and the existing design is shown in Figs. 11 and 12.

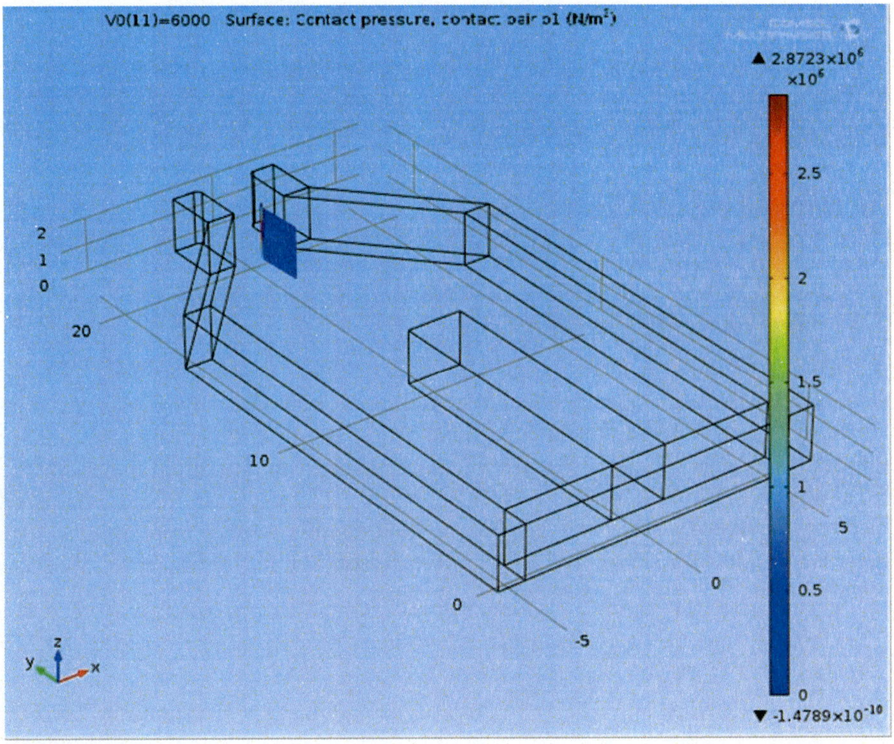

Fig. 11 Contact pressure of the existing design

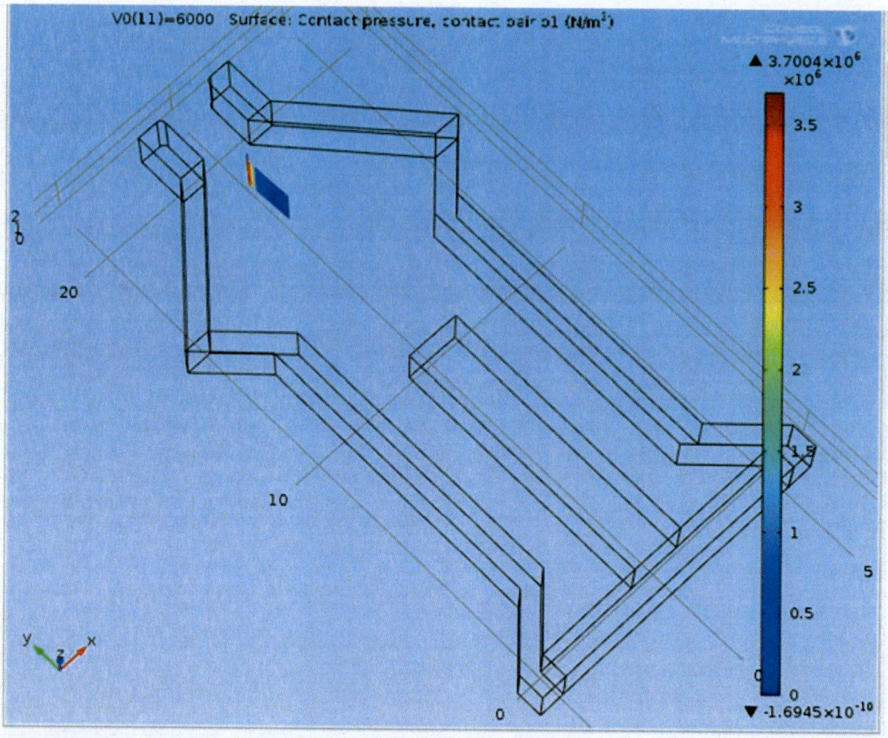

Fig. 12 Contact pressure of the new design

3.4 Gap Distance Analysis

Gap distance is the distance between the holding protrusions. An analysis was done aimed to measure the gap distance in the existing and modified design. It could be inferred from Figs. 13 and 14 that the gap distance in the modified experienced a 31% decrease from 0.1468 μm to 0.1114 μm. This decrease could effectively increase the efficiency of the gripper in holding minuscule sized objects. This decrease in gap also serves as the primitive reason behind the increase in contact pressure.

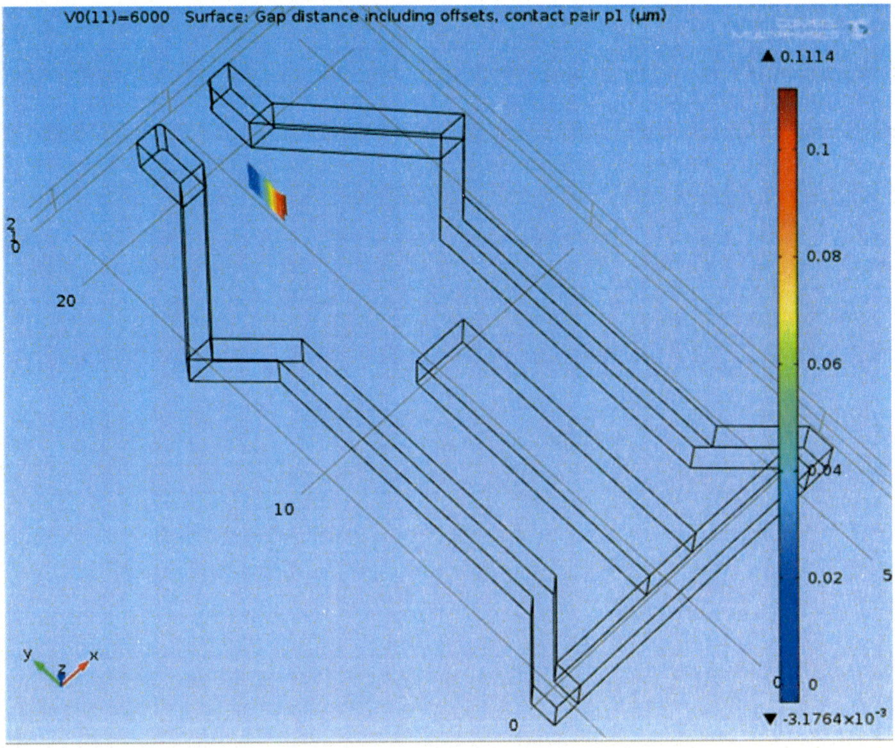

Fig. 13 Gap distance including offsets in the current design

3.5 Stress Distribution Analysis

Stress distribution analysis is performed on the modified and the existing design. It is evident from Figs. 15 and 16 that there was no significant difference in the stress values in the modified design.

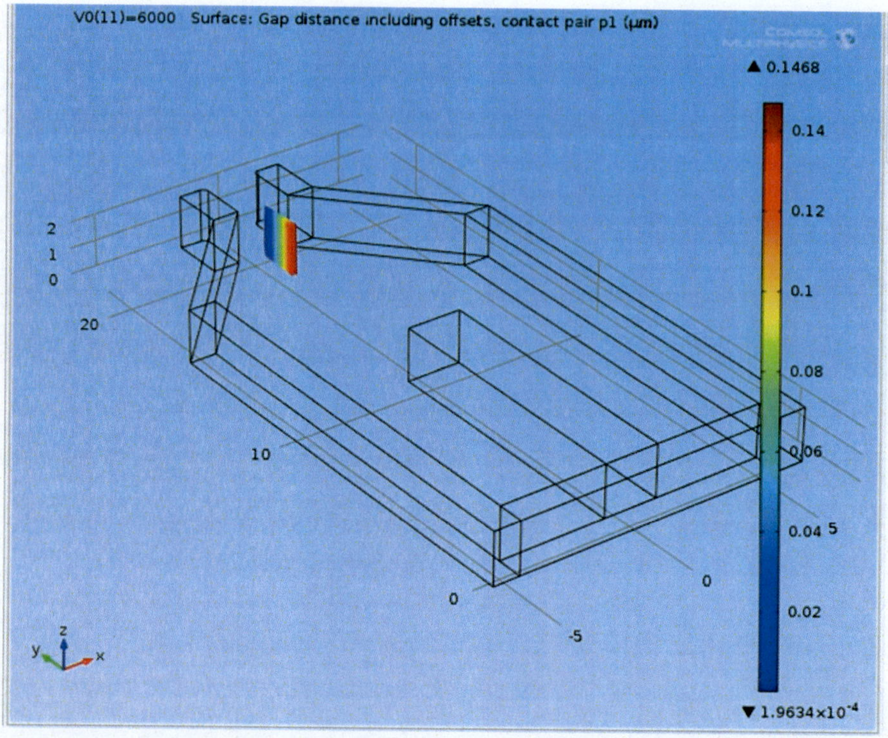

Fig. 14 Gap distance including offsets in the current design

4 Conclusion

Design and analysis of a microgripper for micro-electromechanical systems (MEMS) applications were performed using COMSOL 4.2. From the simulation results, it could be found that a maximum displacement value of 30 μm, contact pressure value of, 3.7004e^{6} N/m^{2} can be achieved by utilizing silicon nitride as the base material and lead zirconate as the piezoelectric material. The contact pressure increase was attributed to the decrease in gap distance between the holding

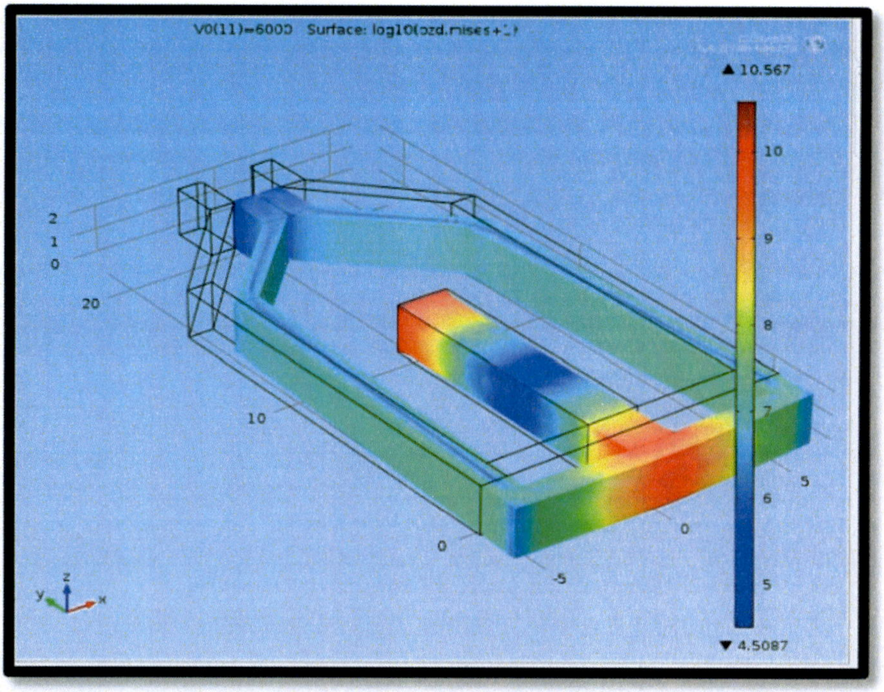

Fig. 15 Stress distribution in the current design

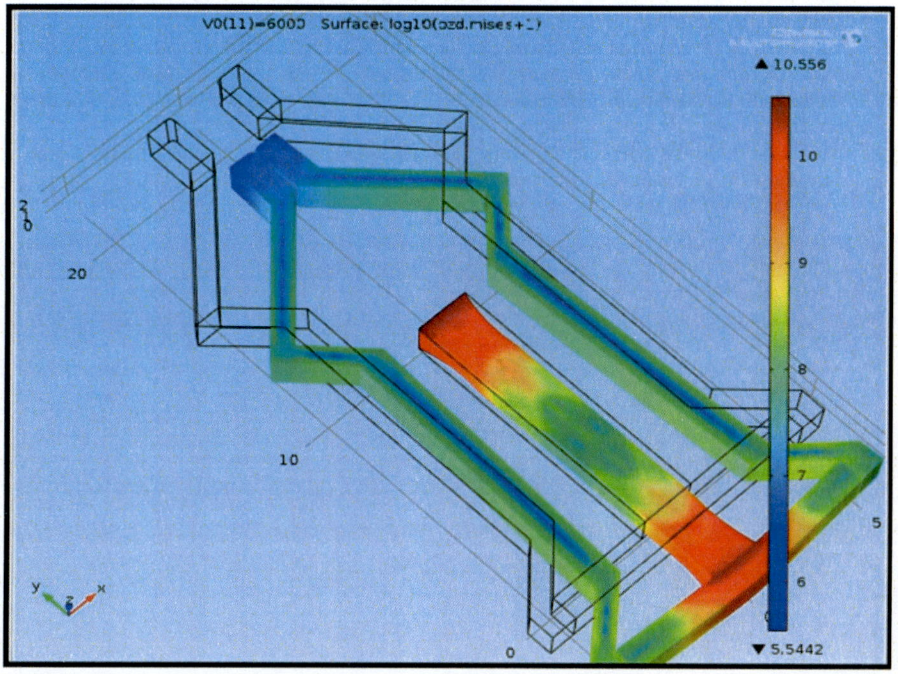

Fig. 16 Stress distribution in the modified design

protrusions. Also, there was minimal change in stress distribution throughout the microgripper system of the modified design. This work could also be extended to other microgrippers used in varied applications.

References

1. Yang S, Xu Q (2017) A review on actuation and sensing techniques for MEMS-based microgrippers. J Micro-Bio Robot 13:1–14. https://doi.org/10.1007/s12213-017-0098-2
2. Boudaoud M, Haddab Y, Le Gorrec Y (2010) Modelling of a MEMS-based microgripper: application to dexterous micromanipulation. IEEE International Conference on Intelligent Robots and Systems, Taipei, Taiwan
3. Bank JM, Development of novel 6 DOF interaction force sensor for microgripper applications, Department of Precision and Microsystems Engineering, TU Delft
4. Dechev N, Cleghorn WL, Mills JK (2004) Microassembly of 3-D microstructures using a compliant passive microgripper. J Microelectromech Syst 13(2):176–189. https://doi.org/10.1109/JMEMS.2004.825311
5. Tamadazte B, Paindavoine M, Agnus J, Petrini V, Le-Fort Piat N (2012) Four DOF Piezoelectric Microgripper Equipped with a Smart CMOS Camera. J Microelectromechanical Syst 21(2):256–258. https://doi.org/10.1109/JMEMS.2011.2180363
6. Feng H, Miao X, Yang Z (2018) Design, Simulation and Experimental Study of the Linear Magnetic Microactuator. Micromachines 9:454. https://doi.org/10.3390/mi9090454
7. Yang S, Xu Q (2016) Design and simulation a MEMS microgripper with integrated electrothermal actuator and force sensor. International Conference on Advanced Robotics and Mechatronics (ICARM), Macau, pp 271–276, https://doi.org/10.1109/icarm.2016.7606931
8. Martinez JA, Panepucci RR (2007) Design, Fabrication, and Characterization of a Microgripper Device. Florida Conference on Recent Advances in Robotics, FCRAR 2007, Florida
9. Varona J, Saenz E, Fiscal-Woodhouse S, Hamoui AA (2009) Design and fabrication of a novel microgripper based on electrostatic actuation, 52nd IEEE International Midwest Symposium on Circuits and Systems, Cancun, pp 827–832, https://doi.org/10.1109/mwscas.2009.5235896
10. Nikoobin A, Niaki MH (2012) Deriving and analyzing the effective parameters in microgrippers performance. Scientia Iranica 19(6):1554–1563. https://doi.org/10.1016/j.scient.2012.10.020

Automated Modeling of Swaged Reissner–Mindlin Plates Using FEniCS

G. Verma, S. Sengupta, S. Mammen, and S. Bhattacharya

1 Introduction

Swaging process is frequently used to connect thin plates to comparatively thicker orthogonal plates in parallel plate-type fuel assemblies. This is done as welding and other such processes cannot be performed either due to limitation of space or because of functionality. This type of joining technique helps in maintaining a compact core in order to obtain high neutron flux. In this process, the thin fuel plates are swaged to the comparably thick orthogonal plates by inserting them into the corresponding slots and roll swaging it. A CFCF-type edge condition is established for the plates where all the translational movements are constrained and except one (along the axis parallel to the swage joint), and all other rotational movements are restricted [1–5]. Caresta and Wassink [3, 4] suggested that for small rotations, a perfect swage joint can be modeled using torsional springs whose stiffness is depended on the quality of swage joint.

In the present work, FEniCS package is used to model this swaged joint of thin Reissner–Mindlin plates using its automated computational modeling approach (Fig. 1). This modeling technique is benchmarked by comparing its modal characteristics to that of a finite element model created in a commercial code for different boundary conditions and further validating and comparing with the works of Caresta and Wassink [3, 4]. Mesh sensitivity analysis is performed by varying the number of discretized elements and the order of the interpolating polynomial.

G. Verma (✉) · S. Sengupta · S. Mammen · S. Bhattacharya
Research Reactor Design and Projects Division, Bhabha Atomic Research Centre, Mumbai, India

S. Bag et al. (eds.), *Next Generation Materials and Processing Technologies*, Springer Proceedings in Materials 9, https://doi.org/10.1007/978-981-16-0182-8_29

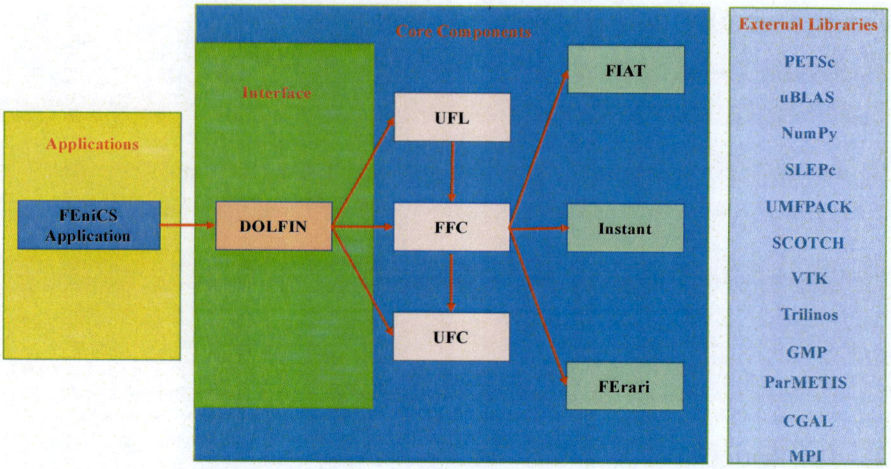

Fig. 1 Flow diagram representing FEniCS architecture [6]

2 Formulation for Thin Swaged Reissner–Mindlin Plate

2.1 Finite Element Formulation of the Reissner–Mindlin Plate [5, 6]

For an isotropic Reissner–Mindlin thin plate of thickness h and edge lengths a and b under bending, the neutral surface belongs to domain $\Omega \subset R^2$ as shown in Fig. 2. The transverse deflection w and the rotation of the neutral surface along y-axis and x-axis are given by $\underline{\beta} = \begin{bmatrix} \beta_x & \beta_y \end{bmatrix}^T$. The vector unknown in the neutral axis is given by $\underline{u}$.

$$\underline{u} = \begin{bmatrix} w \\ \beta_x \\ \beta_y \end{bmatrix} \tag{1}$$

The deflected plate curvature $\underline{\kappa}$ and the shear strain $\underline{\gamma}$ are given as,

$$\underline{\kappa} = \underline{\underline{L_d}}\,\underline{\beta} \tag{2}$$

$$\underline{\gamma} = \nabla \underline{w} + \underline{\beta} \tag{3}$$

where $\underline{\underline{L_d}}$ is a differential operator matrix given as (Fig. 3).

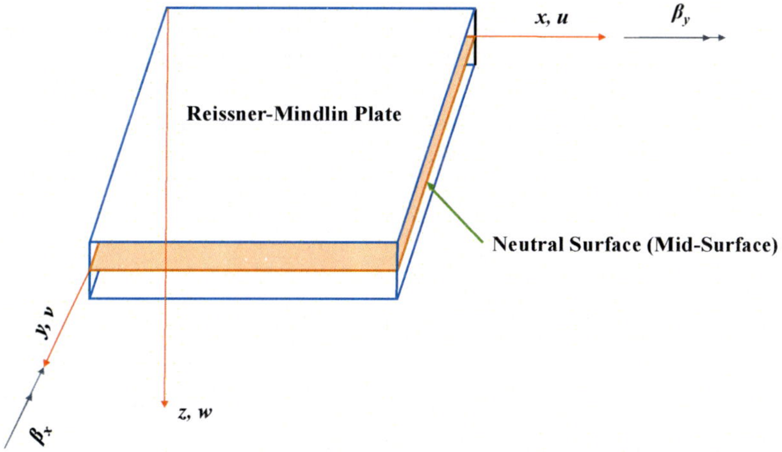

Fig. 2 Schematic of a Reissner–Mindlin plate with positive vertical displacement w and two rotations β_x, β_y [6]

Fig. 3 Schematic of a swage joint [4]

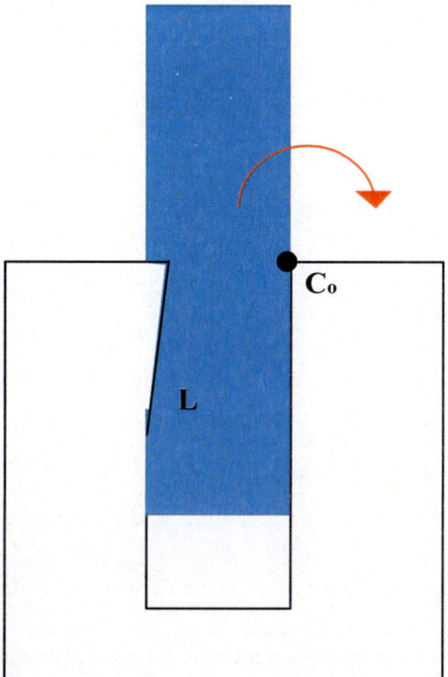

$$\underline{\underline{L_d}} = \begin{bmatrix} \partial/\partial x & 0 \\ 0 & \partial/\partial y \\ \partial/\partial y & \partial/\partial x \end{bmatrix} \tag{4}$$

and $\nabla = [\partial/\partial x \quad \partial/\partial y]^T$. For the static Reissner–Mindlin plates, the governing differential equations are given as,

$$\nabla \cdot \underline{\underline{D_b}\kappa(\beta)} + Gkh\underline{\gamma} = 0 \text{ in } \Omega \tag{5}$$

$$Gkh\nabla \cdot \underline{\gamma} + p = 0 \text{ in } \Omega \tag{6}$$

$$w = \overline{w},\ \underline{\beta} = \underline{\overline{\beta}} \quad \text{on } \Gamma = \partial\Omega \tag{7}$$

The Galerkin weak form for the dynamic equilibrium equation of the Reissner–Mindlin plates is given in Eq. (8).

$$\int_\Omega \delta\underline{\kappa}^T \underline{\underline{D_b}}\underline{\kappa} d\Omega + \int_\Omega \delta\underline{\gamma}^T \underline{\underline{D_s}}\underline{\gamma} d\Omega + \int_\Omega \delta\underline{u}^T \underline{\underline{m}}\ddot{\underline{u}} d\Omega = 0 \tag{8}$$

where is the Dirac-Delta function, $\underline{\underline{D_b}}$ is the flexural rigidity given as,

$$\underline{\underline{D_b}} = \frac{Eh^3}{12(1-\nu^2)} \begin{bmatrix} 1 & \nu & 0 \\ \nu & 1 & 0 \\ 0 & 0 & (1-\nu)/2 \end{bmatrix} \tag{9}$$

with E being the Young's modulus, ν as the Poisson's ratio and G is the modulus of rigidity. $\underline{\underline{D_s}}$ is the shear rigidity which is given as per Eq. (10),

$$\underline{\underline{D_s}} = khG \begin{bmatrix} 1 & 0 \\ 0 & 1 \end{bmatrix} \tag{10}$$

where the shear correction factor k depends upon the boundary condition. $\underline{\underline{m}}$ is the mass matrix as per Eq. (11) with density ρ and thickness h.

$$\underline{\underline{m}} = \begin{bmatrix} \rho h & 0 & 0 \\ 0 & \rho h^3/12 & 0 \\ 0 & 0 & \rho h^3/12 \end{bmatrix} \tag{11}$$

The eigenvalues of the Reissner–Mindlin plates are obtained using the form,

$$\left(\underline{\underline{K}} - \omega^2 \underline{\underline{M}}\right) \underline{d} = 0 \tag{12}$$

where global stiffness matrix is $\underline{\underline{K}}$ and global mass matrix is $\underline{\underline{M}}$. $\underline{\underline{K_e}}$ and $\underline{\underline{M_e}}$ is the elemental stiffness and mass matrices given as in Eqs. (13) and (14).

$$\underline{\underline{K_e}} = \int_{\Omega_e} \underline{\underline{B_b}}^T \underline{\underline{D_b B_b}} d\Omega_e + \int_{\Omega_e} \underline{\underline{B_s}}^T \underline{\underline{D_s B_s}} d\Omega_e \tag{13}$$

$$\underline{\underline{M_e}} = \int_{\Omega_e} \underline{\underline{Q}}^T \underline{\underline{mQ}} d\Omega_e \tag{14}$$

where Q_1, Q_2, Q_3 *and* Q_4 correspond to the basis function for each node in an element.

$$\underline{\underline{B_b}} = \begin{bmatrix} 0 & \partial Q_1/\partial x & 0 & 0 & \partial Q_2/\partial x & 0 & 0 & \partial Q_3/\partial x & 0 & 0 & \partial Q_4/\partial x & 0 \\ 0 & 0 & \partial Q_1/\partial y & 0 & 0 & \partial Q_2/\partial y & 0 & 0 & \partial Q_3/\partial y & 0 & 0 & \partial Q_4/\partial y \\ 0 & \partial Q_1/\partial y & \partial Q_1/\partial x & 0 & \partial Q_2/\partial y & \partial Q_2/\partial x & 0 & \partial Q_3/\partial y & \partial Q_3/\partial x & 0 & \partial Q_4/\partial y & \partial Q_4/\partial x \end{bmatrix} \tag{15}$$

$$\underline{\underline{B_s}} = \begin{bmatrix} \partial Q_1/\partial x & Q_1 & 0 & \partial Q_2/\partial x & Q_2 & 0 & \partial Q_3/\partial x & Q_3 & 0 & \partial Q_4/\partial x & Q_4 & 0 \\ \partial Q_1/\partial y & 0 & Q_1 & \partial Q_2/\partial y & 0 & Q_2 & \partial Q_3/\partial y & 0 & Q_3 & \partial Q_4/\partial y & 0 & Q_4 \end{bmatrix} \tag{16}$$

$$\underline{\underline{Q}} = \begin{bmatrix} Q_1 & 0 & 0 & Q_2 & 0 & 0 & Q_3 & 0 & 0 & Q_4 & 0 & 0 \\ 0 & Q_1 & 0 & 0 & Q_2 & 0 & 0 & Q_3 & 0 & 0 & Q_4 & 0 \\ 0 & 0 & Q_1 & 0 & 0 & Q_2 & 0 & 0 & Q_3 & 0 & 0 & Q_4 \end{bmatrix} \tag{17}$$

2.2 Boundary Condition for the Swaged Reissner–Mindlin Plate

According to Caresta and Wassink [3, 4], for a small rotation θ around C_0, the displacement component normal to the line of contact leads to compression in the material. The localized reaction force (per unit rotation) is given as (Figs. 4 and 5).

$$df = \varepsilon E \cdot dA \cdot r cos\delta \tag{18}$$

where $dA = w.d\xi$, ε is the strain, A is the area and w width of the contact respectively. r is the local radius and δ is the angle and are given as,

$$r^2 = \xi^2 + r_0^2 - 2\xi r_0 \cdot cos\gamma \tag{19}$$

$$\gamma = \frac{\pi}{2} + \varepsilon \tag{20}$$

$$\varepsilon = arctan\left(\frac{h_s}{d_s}\right) \tag{21}$$

$$\delta = arccos\left(\frac{(r_N^2 - r_0^2 + \xi^2)}{2r_N\xi}\right) \tag{22}$$

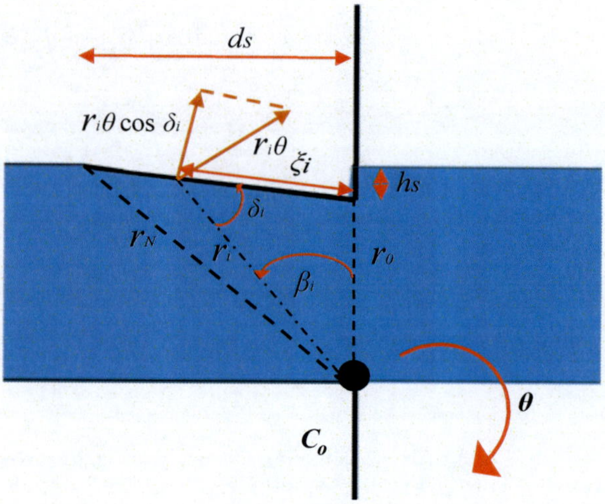

Fig. 4 Schematic of the swage joint kinematics [4]

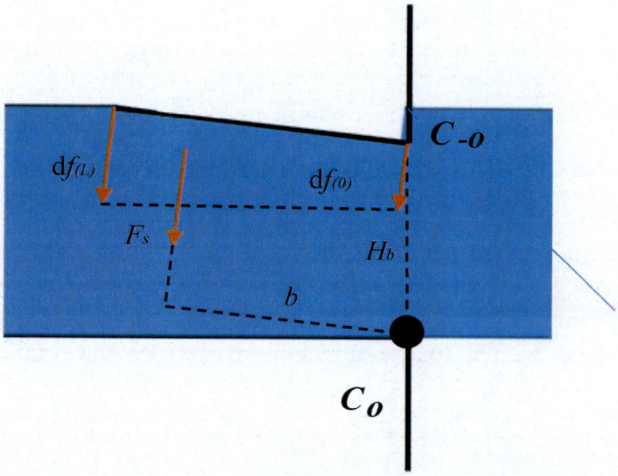

Fig. 5 Schematic representation of force resulting from small rotation of the plate [4]

Normal to the contact line, the resultant force can be represented as,

$$F_s = \int df = \varepsilon Ew \int_0^L rcos\delta.d\xi \tag{23}$$

The lever arm about C_0 is given as,

$$b = H_b sin\varepsilon + \frac{2}{3L} \tag{24}$$

H_b is the distance between C_0 and C_{-0}. This results in a moment about C_0 which is given by,

$$M_s = F_s\theta \cdot b = K_s\theta \tag{25}$$

$$K_s = F_s b \tag{26}$$

where K_s is the equivalent swage joint torsional stiffness. For a very high value of K_s, a perfectly clamped condition is simulated whereas for $K_s = 0$, it corresponds to a simply supported condition. The torsional stiffness is represented in the form of a distributed value in terms of N/rad.

3 Introduction FEniCS Implementation

FEniCS is a Python/C++ based open-source package which uses finite element technique to solve partial differential equations (PDEs). The FEniCS platform was created in 2003 under FEniCS research and software project. The FEniCS package uses a number of C++/Python open-source libraries such as DOLFIN (Dynamic Object-oriented Library for FINite element computation) as its primary user interface, FFC (FEniCS Form Compiler) for automated code generation of matrices (tensors) and vectors assembly from variational forms, FIAT (Finite Element Automatic Tabulator) for numerical construction of finite element basis functions, UFL (Unified Form Language) for declaration of finite element discretization for variational forms and functional, UFC (Unified Form-assembly Code) as interface for the finite element code generation, etc. Figure 1 represents the flow diagram representing the FEniCS architecture.

FEniCS enables simple mathematical construction of the variational form and permits rapid implementation of its high performance solvers for discretization and problem solving. FEniCS features automation for basis function generation, automation for the variational form evaluation, automation for finite element assembly and automation in the adaptive error control. FEniCS uses unique scripting method to form function spaces, select type of elements, setting of solvers, etc. based on the variational formulation of the PDEs.

In the present work, MITC plate elements along with Selective Reduced Integration (SRI) approach prevent shear locking due to strong symmetry. Appropriate forms of PETSc matrices are defined for system stiffness and mass matrices from Galerkin formulation. Symmetry is preserved while applying boundary conditions to stiffness matrix. Finally, modal analysis is performed using SLEPcEigenSolver (Fig. 6) [6–8].

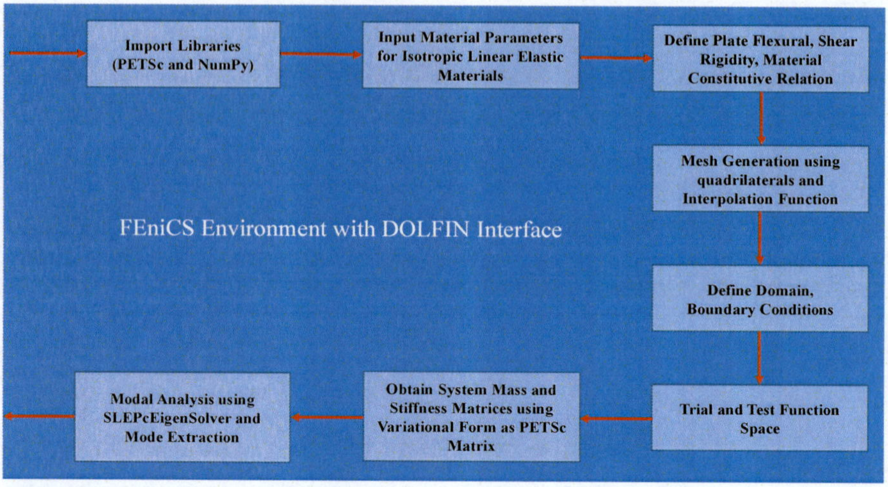

Fig. 6 Flow diagram of the algorithm [6]

4 Results and Discussion

In the present work, modal analysis for C-F-C-F and SS-F-SS-F boundary (edge) conditions are performed in order to validate the FEniCS-automated MITC model of the thin Reissner–Mindlin plate. For validation, the results were compared to analytical results and to the results obtained from a commercial finite element code. Table 1 shows the eigenfrequencies and mode shapes obtained for C-F-C-F and SS-F-SS-F boundary conditions for which the results are in good agreement (error <1%) for the given aspect ratio of the plate.

Table 1 Eigenvalues and modal shapes for Reissner–Mindlin plate with C-F-C-F and SS-F-SS-F boundary conditions through FEniCS and commercial FE

Mode no.	FEniCS (Hz)	Commercial FE (Hz)	Analytical (Hz) [9]	Error (%)	Mode (n, m)
Clamped-Free-Clamped-Free (C-F-C-F)					
1	1143.74	1144.92	1146.02	0.09	(0, 0)
2	1145.17	1146.44	1146.02	0.03	(0, 1)
3	1155.59	1156.13	1156.29	0.01	(0, 2)
4	1171.57	1172.21	1166.57	0.48	(0, 3)
Simply Supported-Free-Simply Supported-Free (SS-F-SS-F)					
1	504.18	504.33	504.65	0.06	(0, 0)
2	507.74	507.97	508.77	0.16	(0, 1)
3	525.03	525.46	526.76	0.25	(0, 2)
4	552.74	553.28	555.02	0.31	(0, 3)

Subsequently, in an effort to establish and validate the FEniCS model for swaged Reissner–Mindlin plates, the variation of normalized frequency parameter with the distributed rotational stiffness for a swaged plate with specific aspect ratio is analyzed as shown in Fig. 7 and the results are compared to Caresta and Wassink experimental results [3, 4]. From Fig. 7, it is observed that for higher rotation stiffness values, the frequency parameter λ_n (Eq. 27) approaches unity, toward fundamental eigenfrequency of the clamped fuel plate edge.

$$\lambda_n = \omega_n a^2 (\rho h/J)^{1/2} \tag{27}$$

Here, ω_n is the circular eigenfrequency, h is the plate thickness and ρ is the material density. J is the flexural rigidity given by Eq. (28),

$$J = Eh^3/12(1 - \nu^2) \tag{28}$$

The normalized λ_n parameter is the ratio of λ_n for a given stiffness to perfectly clamped edge stiffness. Figure 8 displays the first four modes of the swaged plate. An important point to note is that for both C-F-C-F and SS-F-SS-F plates, the mode shapes are equivalent to that of the swaged plate as per Table 1. Mesh sensitivity analysis is performed for three different mesh sizes and three different degrees of interpolation. The results presented in Fig. 8 use 18,000 elements and degree of interpolation as 2.

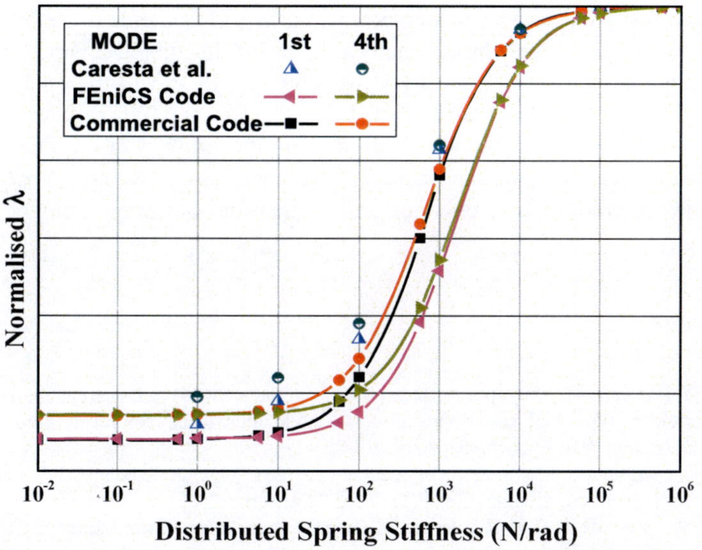

Fig. 7 Variation of frequency parameter with distributed spring stiffness

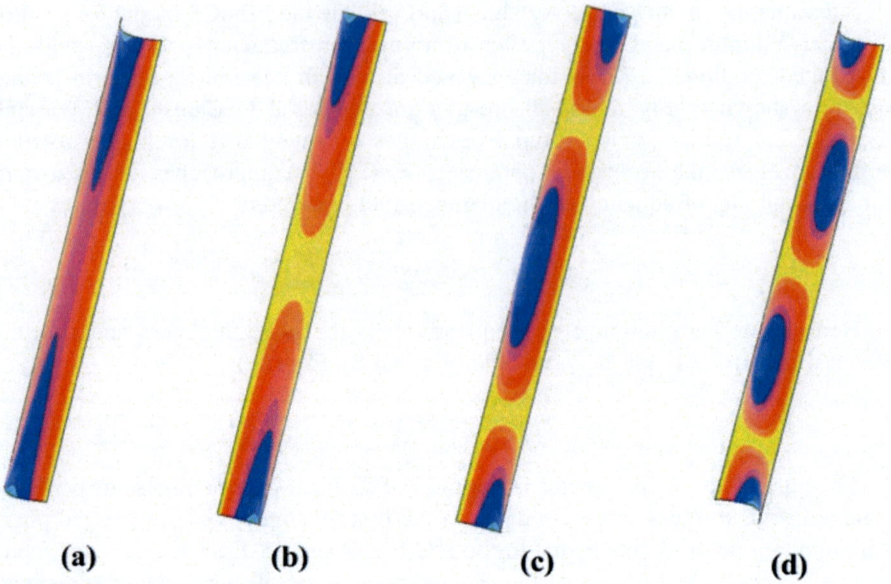

Fig. 8 Mode shapes of swaged Reissner–Mindlin plate **a** 1st mode; (n, m) = (0, 0) **b** 2nd mode; (n, m) = (0, 1) **c** 3rd mode; (n, m) = (0, 2) **d** 4th mode; (n, m) = (0, 3) [2]

5 Conclusion

In the present work, FEniCS-based automated computational model of swaged variable stiffness Reissner–Mindlin plates using MITC elements is established. This modeling technique is benchmarked by comparing its modal characteristics to a finite element model developed in a commercial code and to the analytical solutions; and further validated with the works of Caresta and Wassink. It is observed that eigenvalues and mode shapes obtained from the analysis of the FEniCS model were found to be in good agreement. Sensitivity analyses are carried out by studying the variation in discretization and in the order of interpolating polynomial.

References

1. Verma G, Eswaran M, Sengupta S, Reddy GR, Mammen S (2017) Dynamic characteristics of immersed plate-type fuel assemblies under seismic excitation. Nucl Eng Des 314:11–28. https://doi.org/10.1016/j.nucengdes.2017.01.005
2. Verma G, Sengupta S, Mammen S, Bhattacharya S (2017) Dynamic behavior of swaged plates in water immersed conditions. 13th ICOVP, IIT Guwahati
3. Caresta M, Wassink D (2010) Vibrational characteristics of roll swage jointed plates. Acoust Aust 38(2):82–86

4. Caresta M, Wassink D (2012) Dynamic characterization and longitudinal strength of swaged joints. Appl Acoust 73(5):484–490. https://doi.org/10.1016/j.apacoust.2011.12.002
5. Caresta M, Wassink D (2013) Structural analysis of plate-type fuel assemblies and development of a non-destructive method to assess their integrity. Nucl Eng Des 262:209–218. https://doi.org/10.1016/j.nucengdes.2013.05.003
6. Verma G, Sengupta S, Mammen S, Bhattacharya S (2020) Free Vibration Analysis of Reissner-Mindlin Plates using FEniCS. NCRASE-2020, NIT Jamshedpur
7. Langtengen HP, Logg A (2016) Solving PDEs in Python: The FEniCS Tutorial I, Simula Springer Briefs on Computing 3. https://doi.org/10.1007/978-3-319-52462-7
8. Bleyer J (2018) Numerical Tours of Computational Mechanics with FEniCS, Zenodo, https://doi.org/10.5281/zenodo.1287832
9. Logg A, Mardal KA, Wells GN (2012) Automated Solution of Differential Equations by the Finite Element Method: The FEniCS Book, Lecture Notes in Computational Science and Engineering, 1st Edition, Volume 84, Springer-Verlag Berlin Heidelberg. https://doi.org/10.1007/978-3-642-23099-8

Theoretical Analysis of Composite Blade in Static Condition to Find Material Uncertainty and Validation with COMSOL Software

Avinash K. Parkhe and Sandeep S. Wangikar

1 Introduction

The importance use of composite materials has been increasing consistently in different industries like civil engineering, mechanical engineering, aerospace engineering due to their advantageous characteristics. Sometimes, due to faulty manufacturing process or delamination of beams, it gives variation in results during working and it creates errors during measuring unknown parameters. Few researchers have conducted some static and dynamic analysis on composites. Ronge et al. presented dynamic analysis of rotating blade in undamaged and damaged using health monitoring concept [1]. Kachare et al. represented experimental setup for measuring acceleration in dynamic condition using developed sensor [2]. Hake et al. presented the static and dynamic setup for delamination checking in blade for different loads and acceleration measurement for different rpm. For analyzing the material uncertainty, the deflection is measured. The composite blade is considered like a cantilever beam. A non-contact device working on the magnetic field, i.e., the Hall Effect Sensor is established. A magnet field is created if a magnet has come in front of sensor and further the change in voltage or field is calibrated in terms of deflection of blade. For checking all the blades uncertainty, the same process is followed [3–8]. Here, in this study, theoretical and numerical analysis has carried out on composite blade for checking the material uncertainty. During the manufacturing process of blade, some internal cracks or cavities are produced which will create the delamination in blades and due to this delamination stiffness of blade gets loosed which creates the errors during measurement or during further analysis. So, here, we are going to check such cracks

A. K. Parkhe (✉) · S. S. Wangikar
SVERI's College of Engineering, Pandharpur 413304, Maharashtra, India
e-mail: akparkhe@coe.sveri.ac.in

S. S. Wangikar
e-mail: sswangikar@coe.sveri.ac.in

S. Bag et al. (eds.), *Next Generation Materials and Processing Technologies*,
Springer Proceedings in Materials 9, https://doi.org/10.1007/978-981-16-0182-8_30

presents by measuring deflection of blade to check the uncertainties present in material. The blade is fixed like a cantilever, where one end is fixed and another end is free to apply the load. Due to applied load at free end, deflection will take place at free end. In this study, theoretical analysis has carried out on composite blade to checks the material uncertainty presents it. To validate those theoretical results, numerical analysis has carried out on same blade using COMSOL Multiphysics software for measuring the same deflection parameter. Then, theoretical and numerical results are compared with each other to find % error between them. After this comparison, it is observed that there is no material uncertainty or delamination in the blade during its manufacturing.

2 Theoretical Analysis of Composite Blade

In theoretical analysis, blade is fixed like cantilever type where one is fixed and another end is free for applying point load as shown in Fig. 1. Due to applied load at free end, deflection is taking place, and there are standard formulas for calculating defined parameters. So, the changes are carried out to determine M.I. for hollow composite box beam. Then, further calculations are found out.

Fig. 1 Cantilever type composite blade

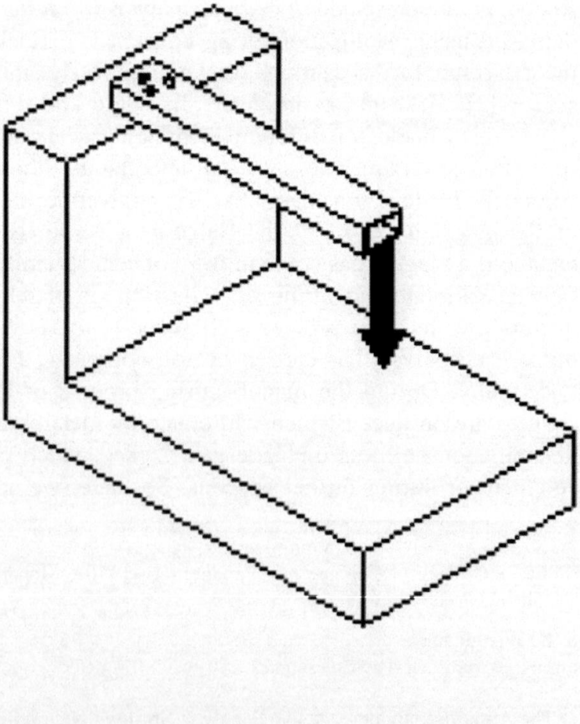

2.1 Moment of Inertia for Composite Box Beam

The standard formula for calculating the deflection at free end is given below:

$$\delta = \frac{Pl^3}{3EI} \quad (1)$$

M.I. is calculated for this by using formula given below.

$$I = \frac{bd^3}{12} \quad (2)$$

$$I = \frac{\pi}{64}\left(D^4 - d^4\right) \quad (3)$$

Total M.I. of beam = 38975.75 mm^4

2.2 Free End Deflection Calculation

By using M.I. calculated above, deflection at free end by varying point load at free end from 10 N to 80 N is calculated and presented in Table 1. Sample calculation for deflection is given below.

Deflection for 10 N:

$$\delta = \frac{10 \times 800^3}{3 \times 135 \times 10^3 \times 38975.75}$$

$$\delta = 0.32\,mm$$

Table 1 Deflection at free end of beam (analytical results)

S. no.	Load (N)	Analytical deflection (mm)
1.	10	0.32
2.	20	0.64
3.	30	0.97
4.	40	1.29
5.	50	1.62
6.	60	1.94
7.	70	2.27
8.	80	2.59

3 Numerical Analysis of Composite Blade

To validate the theoretical results, the numerical analysis of composite blade has carried out using COMSOL Multiphysics software for measuring the same deflection parameter. The composite blade has a ply lay-up [0°/90°]s on all four sides the composite blade contains 8 layers. With the help of solid mechanics, module in the COMSOL Multiphysics deflection is calculated. Figure 2 shows geometry of composite blade in COMSOL software.

Meshing method is used for meshing the geometry in COMSOL Multiphysics. The geometry of composite blade with extremely coarse mesh is shown in Fig. 3.

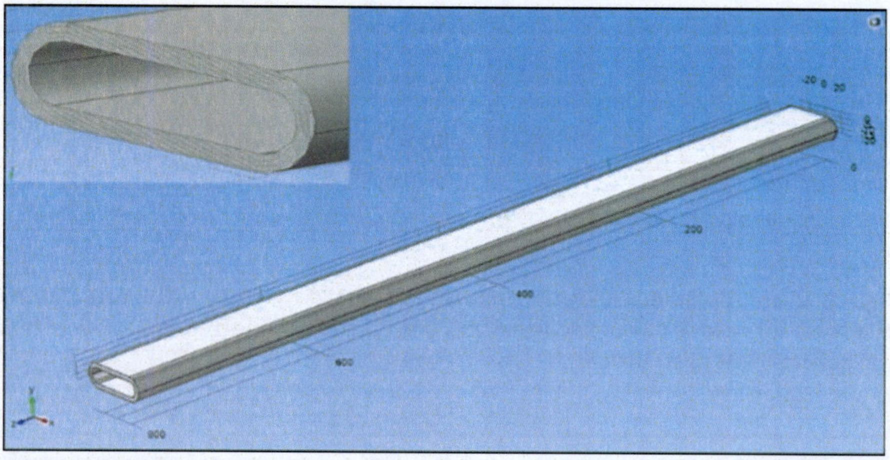

Fig. 2 Geometry of composite blade in COMSOL

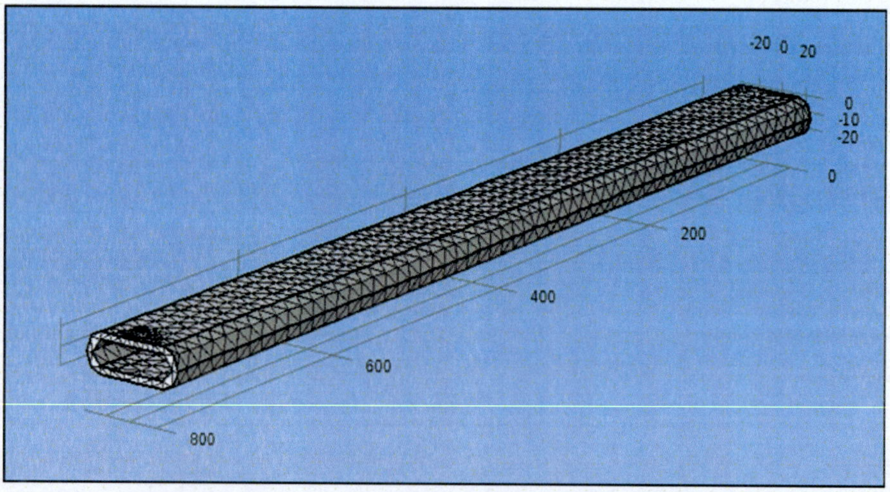

Fig. 3 Meshing of composite blade

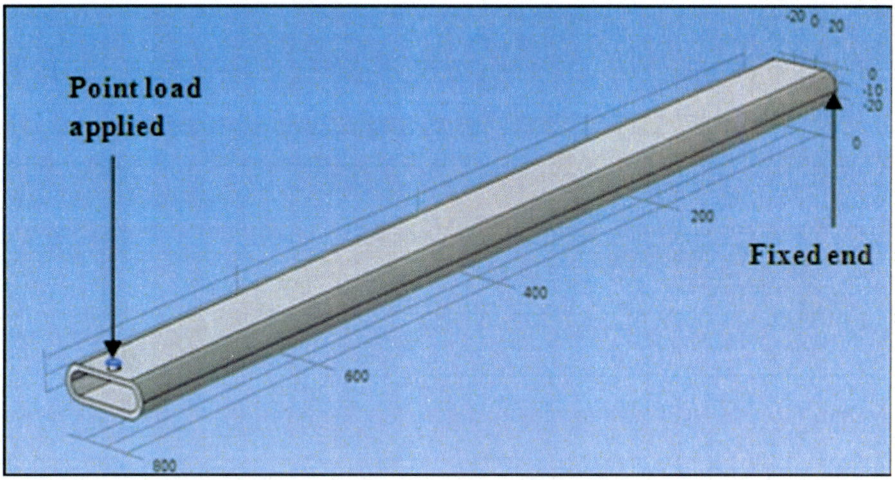

Fig. 4 Boundary conditions applied to blade

During analysis the composite blade will be considered as cantilever type. So, one end is fixed and other end is free to apply the load as shown in Fig. 4.

3.1 Numerical or Simulation Result

The numerical analysis has carried out on same blade to check its deflection at free end after applying load. As there is no any support at free end, the maximum deflection will take place and it will get increases with increase in load at free end. Some of the numerical results obtained from COMSOL software are represented by Figs. 5 and 6.

Figure 5 represents the numerical result or deflection at free end for 10 N load. The red color in Fig. 5 represents the maximum deflection taking place at free end which is equal to 0.3 mm.

Similarly, Fig. 6 represents the numerical result or deflection at free end for 20 N load. The red color in Fig. 6 represents the maximum deflection taking place at free end which is equal to 0.59 mm and it will continuously increases with increase load at free end which is represented by Table 2.

Finally, theoretical and numerical results (COMSOL results) are compared with each other to check % error between them. After the analysis and comparison of both results, it has been observed that there is no much difference in theoretical and numerical results or deflection. So, we can say that there is no any material uncertainty presents in the blade. The comparisons of both results are also represented by the following graph (Fig. 7).

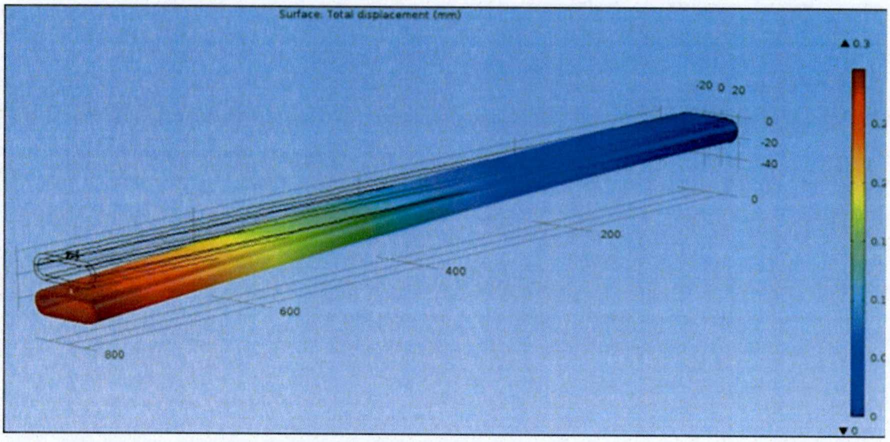

Fig. 5 Deflection at free end for 10 N

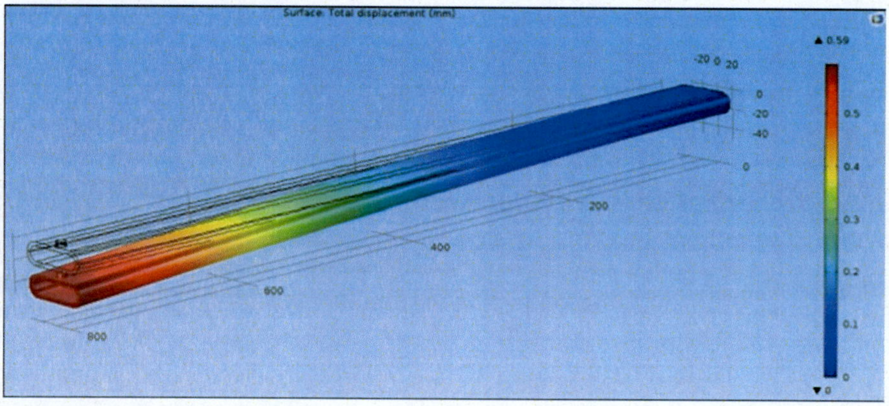

Fig. 6 Deflection at free end for 20 N

Table 2 Deflection at free end of blade (numerical results)

S. no.	Load (N)	Numerical deflection (mm)
1.	10	0.3
2.	20	0.59
3.	30	0.89
4.	40	1.19
5.	50	1.48
6.	60	1.78
7.	70	2.08
8.	80	2.37

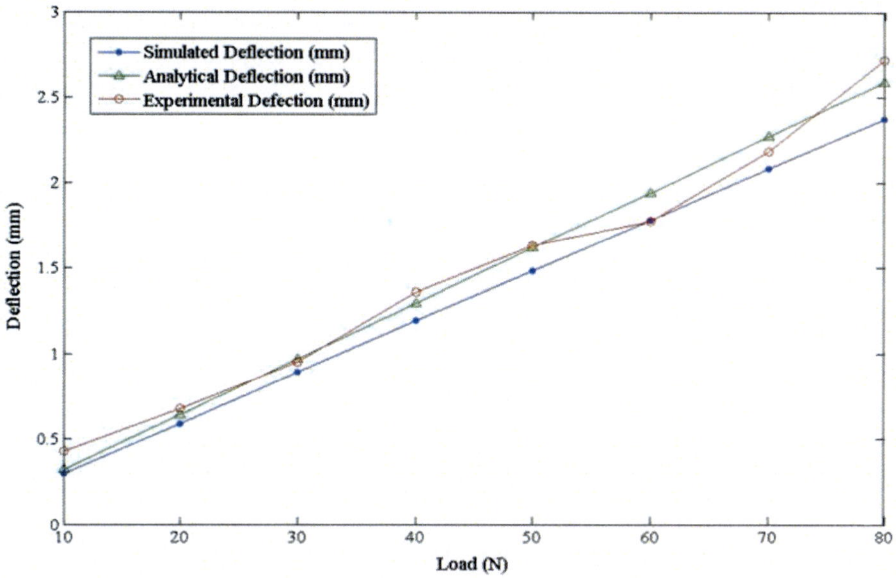

Fig. 7 Graph of theoretical versus numerical results

4 Conclusion

As the composite blades are manufacturing by different processes like using vacuum bagging process or by hand lamination process. Due to some faulty machines or processes some delamination, uncertainty, defects will be observed in such composite products which will create some errors while conducting tests on it. To find such errors, various parameters are measured like stress, deflection, natural frequency, strain, etc. In this study, the theoretical analysis has carried on composite blade for measuring the deflection at free end and same results are compared with numerical results obtained from COMSOL software.

From the above theoretical and numerical study on composite blade and through its comparison, the following conclusions are obtained:

1. The theoretical analysis has carried out on cantilever type composite blade for measuring the deflection at free end. During analysis, load at free end is varied from 10 N to 80 N to find the deflection. The deflection will increases with increase in load at free end.
2. Also, the numerical analysis is carried out using COMSOL software for measuring the same deflection at free end of the blade and accordingly the results are represented by figures.
3. Through this comparison, it has been observed that the deflection values of composite blade obtained from theoretical and numerical analysis are very near to each other. There is no much difference between them which shown by above

graph. So, we can say that there is no any material uncertainty or delamination presents in the blade or there is no any error in manufacturing process of composites.

References

1. Ronge B, Pawar P, Parkhe P (2018) Experimental analysis of composite rotor blade models for damage identification. Advances in Science and Engineering Technology International Conferences (ASET), IEEE
2. Kachare PS, Parkhe AK, Utpat AA (2019) Free Vibration Analysis of Rotating Composite Box Beam using GY-521 Accelerometer. Int J Sci Res Publ 9(2)
3. Hake AA, Ronge NS, Bhingare VA, Parkhe AK, Bhuse PK, More SN (2019) Static & Dynamic Research of Composite Blade using Condition Monitoring Method. IJRTE 8(2):S11
4. Parkhe AK, Shinde AB, Sawant NS, Pawar PM, Haridas P (2018) Experimental Estimation of Material Uncertainty of Composite Beam Using Hall Effect Sensor. Techno-Societal
5. Kachare PS, Parkhe AK, Utpat AA, Salunkhe SY (2019) Health Monitoring of Static Composite Beam for Material Uncertainty and its Numerical Validation. IJNTR 5(3):79–83
6. Parkhe AK, Wangikar SS, Patil PK, Vhare CK, Kashid DT, Pawar MP (2020) Analytical and Numerical Stress Analysis of Composite Box Beam in Dynamic condition and Validation with COMSOL Multiphysics software. AEGAEUM J
7. Parkhe AK, Haridas PD, Kapurkar RB, Pawar PM (2015) Experimental Study on Natural Frequency of Composite Box Beam for Multiple. Int J Res Aeronaut Mech Eng 3(12), December
8. Jafari-Talookolaei Ramazan-Ali, Della Christian (2015) Dynamic behavior of a rotating delaminated composite beam including rotary inertia and shear deformation effects. Ain Shams Engineering Journal 6:1031–1044

Free Vibration Analysis of FGM Structures Using FEM Technique

Pankaj Sharma, Ashish Khinchi, and Ankit Jain

1 Introduction

Benefits of functionally graded materials (FGMs) over composite materials are that in FGMs material properties vary continuously; however, material properties are discontinuous across the layer in composite material. Furthermore, material properties in FGMs are graded in all three direction x, y and z, respectively, but this is not possible in composite materials. In the FGMs, volume fraction varies from zero to a hundred per cent due to this smoother stress distribution, lower stress concentration and higher bonding strength obtained [1]. A more desirable benefit of using functionally graded materials in a dental implant is that it provides a more favourable mechanical environment for attending the better osseointegration and remodelling. FGMs implant produces low von-Mises stress as compared to the titanium implant. FGMs implant improves bone apposition efficiency and biocompatibility in terms of both mechanical and chemical affinities [2, 3]. The ceramic phase of FGMs is able to withstand high-temperature conditions due to their better thermal resistance properties, whereas the metal phase of FGMs provides better mechanical properties and minimizes the possibilities of catastrophic failure. Thus, FGMs possess several advantages over composite material. So, this is appropriate material for various kinds of applications [4, 5].

Garg et al. [6] proposed a model based on FEM method which works on HSDT theory. In this analysis, orthotropic, isotropic, composite and sandwich laminate skew plates are used by the author for harmonic behaviour. Srinivasa et al. [7] dealt with the modal analysis of isotropic and composite skew plate. In this analysis,

P. Sharma · A. Jain
Department of Mechanical Engineering, Rajasthan Technical University, Kota, Rajasthan, India

A. Khinchi (✉)
Department of Mechanical Engineering, Vedant College of Engineering and Technology, Kota, Rajasthan, India

S. Bag et al. (eds.), *Next Generation Materials and Processing Technologies*, Springer Proceedings in Materials 9, https://doi.org/10.1007/978-981-16-0182-8_31

experimental value is compared with results obtained by FEM software. Two opposing sides are clamped, and the remaining side is completely free used by the author as a boundary condition. The effect of two parameters, i.e. aspect ratio and skew angle, was studied by the author.

Zhang and Lei [8] discussed the modal characteristics of FG nano-composite triangular plates based on FSDT theory. Jadhav and Deshmukh [9] discussed combined experimental and numerical approach on modal behaviour of glass fibre epoxy composite plate, having rectangular, circular, triangular and elliptical cut-out with same cross-sectional area with the help of FEM approach. Gormman [10] carried out the modal study of right-angle triangular plate. A basic mathematical approach was applied to a right-angle triangular plate to obtain natural frequencies and corresponding mode shapes. In this analysis, simply supported end condition was used by the author. Sakiyama et al. [11] presented harmonic behaviour of right-angle triangular plate with variable thickness. An estimated procedure was used by the author for obtained natural frequencies and related mode shapes. Belalia [12] presented a nonlinear and linear modal behaviour of thick triangular FGPs having two equal sides. Power law gradation adopted by the author for material properties variation. Effect of several factor, i.e. functionally graded indices, thickness ratio, apex angle, was reported in this analysis. Modal analysis of taper triangular plate under thermal environment was analysed by Kaur [13]. Rayleigh–Ritz technique was used by the author to determine the frequency parameter and respective mode shape. Taper parameter and aspect ratio for a different type of triangular plate under clamped end condition were reported by the author. All the results obtained in this analysis are presented in tabular form only. Free vibration behaviour of symmetrical trapezoidal plates is presented by the Saliba [14]. In this research, superposition technique was adopted by the author to obtain a natural frequency. This superposition technique was invented by Gorman. In this analysis, simply supported boundary conditions were applied to the trapezoidal plate. Results are presented in graphical as well as tabular form under the effect of aspect ratio. Computer-generated mode shape was also presented by the author. Zamani et al. [15] carried out the harmonic analysis on thick trapezoidal plates with different combinations of boundary conditions. FSDT and GDQ approaches were adopted by the author for validation study and other analyses. Effect of several parameters such as boundary condition, thickness and layup set-up on the resonant frequency of skew and trapezoidal plate was presented in this research article. Gupta et al. [16] dealt with the vibration behaviour of non-homogeneous trapezoidal plate under thermal effect. In this analysis, density and thickness varies as parabolic and linearly, respectively. Effect of many factors, i.e. tapering constant, aspect ratio and non-homogenous constant, is studied in detail by the author. The detailed study of FGM structures with and without piezoelectric effect with the help of COMSOL is available in the literature [17–20].

The engineering application of skew plate are such as wings, tails, fins of aircraft, missiles, panels in skew bridges and highways. The engineering application of trapezoidal plates are marine, aerospace engineering fields and in construction of simple and complex structures, etc. The engineering application of triangular plates is aerospace industries and automotive industries. Free vibration analysis of FGM structures, i.e. triangular plate and trapezoidal plate, using COMSOL is not available in the literature before.

2 Modelling of FGM Structures

In the present study, various FGM structures such as skew plate, triangular plate and trapezoidal plates are used in free vibration analysis. The parametric study is presented for FG-triangular plate and FG-trapezoidal plate. The effects of the functionally graded index and width-to-thickness ratio on natural frequencies are reported. Geometries of FGM structures, i.e. skew plate, trapezoidal plate and triangular plate, are shown in Fig. 1.

Power law gradation (P-FGM) also known as rule of mixture is used on FGM structures (triangular plate and trapezoidal plates) for smooth and continuous variation of material constant in the thickness direction. FGM is composition of metallic and ceramic materials. The top surface is hundred per cent ceramic. The bottom surface is hundred per cent metallic phase. For ceramic phase (alumina) E, ρ and v are 380 GPa, 3800 kg m^{-3} and 0.3, respectively, and for metallic phase (aluminium) E, ρ and v are 70 GPa, 2700 kg $m^{-3}$and 0.3, respectively [3] where E, ρ and v are the Young modulus, density and Poisson ratio. The Poisson ratio (v) is

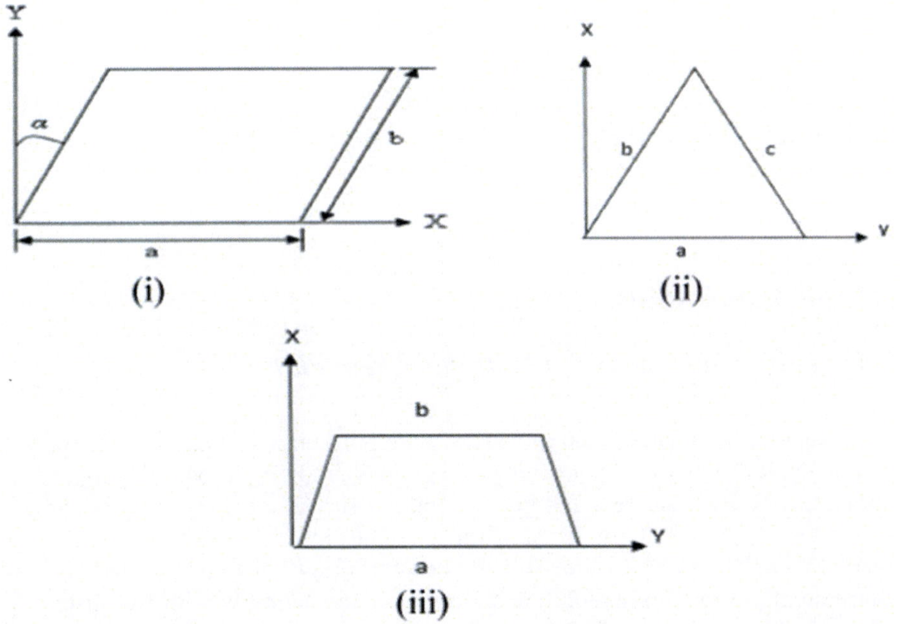

Fig. 1 Geometries of FGM structures **i** skew plate **ii** triangular plate **iii** trapezoidal plate

kept constant. FGM (Al/Al_2O_3) has several good properties such as low weight high strength, good wear resistance and temperature resistance properties. So, in this analysis FGM (Al/Al_2O_3) is used. Material constants are varied according to these equations [3].

$$E_z = (E_c - E_m)V_f^p + E_m \tag{1}$$

$$\rho_z = (\rho_c - \rho_m)V_f^p + \rho_m \tag{2}$$

$$V_f = \left(\frac{z}{h} + \frac{1}{2}\right) \tag{3}$$

Here, V_f and g are volume fraction and functionally graded index (also known as power law index), respectively.

2.1 Boundary Conditions

1. Triangular plate with different boundary conditions are consider as
   - All sides are clamped (CCC).
   - Two sides clamped and the other one is simply supported (CCS).
   - All sides are simply supported (SSS).
2. Trapezoidal plate under end support are consider as
   - All sides are clamped(CCCC).
   - All edges are simply supported (SSSS).
   - Unequal opposite edges are simply supported, and other edges are clamped (CSCS).

3 Validation Study

3.1 Validation Study for Isotropic Skew Plate

Free vibration analysis of isotropic skew plate is taken into consideration for validation study. Two types of boundary conditions are being used, i.e. simply supported (SSSS) and clamped. The geometrical parameters are a = 1 m, b = 1 m and h = 0.001 m. The results are listed in Table 1. The dimensionless form is defined as $w' = w\,(b^2/\pi^2)\,(\rho h/D)^{0.5}$. Here, D is the flexural rigidity of isotropic plate. For the isotropic plate, the present solution shows good compatibility with Ref. [6].

Table 1 Comparison of first dimensionless frequencies of isotropic skew plate under CCCC and SSSS end condition ($a = b = 1$ m and $h = 0.001$ m)

Boundary conditions		Skew angles		
		0	30	45
SSSS	Garg et al. [6]	2.000	2.539	3.629
	Present	1.999	2.531	3.6207
	Error (%)	0.04	0.31	0.24
CCCC	Garg et al. [6]	3.648	4.680	6.699
	Present	3.644	4.677	6.638
	Error (%)	0.10	0.06	0.91

Table 2 Comparison of first dimensionless frequencies of composite laminate plate under different end conditions ($a = b = 1$ m, $h = 0.1$ m)

Boundary Conditions		Skew angles		
		0	30	45
SSSS	Garg et al. [6]	1.5699	2.0884	2.8032
	Present	1.5497	2.0003	2.7006
	Error (%)	1.70	4.21	3.6
CCCC	Garg et al. [6]	2.3820	2.7922	3.4739
	Present	2.3081	2.7351	3.4168
	Error (%)	3.10	2.04	1.64

3.2 Validation Study for Laminate Composite Skew Plate

Now, the free vibration analysis of laminate composite skew plate is taken into consideration for validation study. Two types of boundary conditions are being used, i.e. simply supported (SSSS) and clamped (CCCC). The geometrical parameters are $a = 1$ m, $b = 1$ m and $h = 0.1$ m. For the laminate composite skew plate, the present solution shows good compatibility with Ref. [6]. The material properties are $E_1 = 400$ GPa, $E_2 = E_3 = 10$ GPa, $G_{13} = G_{23} = 5$ GPa $G_{12} = 6$ GPa. The results are shown in Table 2. The dimensionless form is define as $w' = w\,(b^2/\pi^2 h)\,(\rho h/D)^{0.5}$ (Fig. 2).

4 Parametric Study

The modal characteristics of the functionally graded triangular plate and functionally graded trapezoidal plate are considered for parametric study.

	Mode Shapes	
	SSSS	CCCC
0	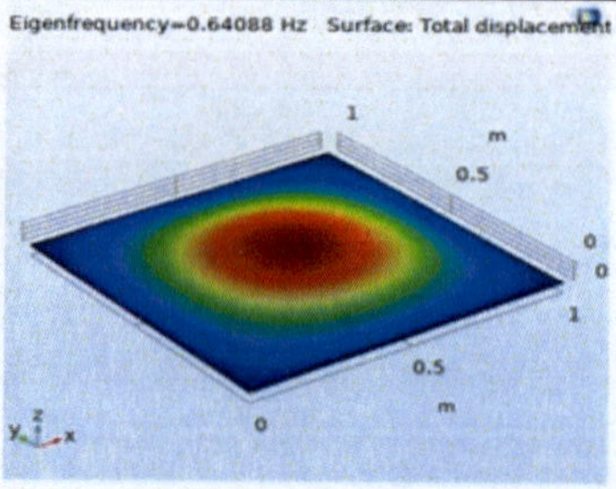	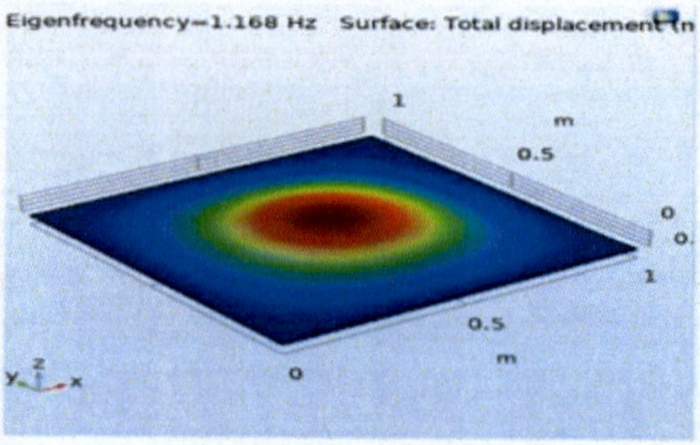
30	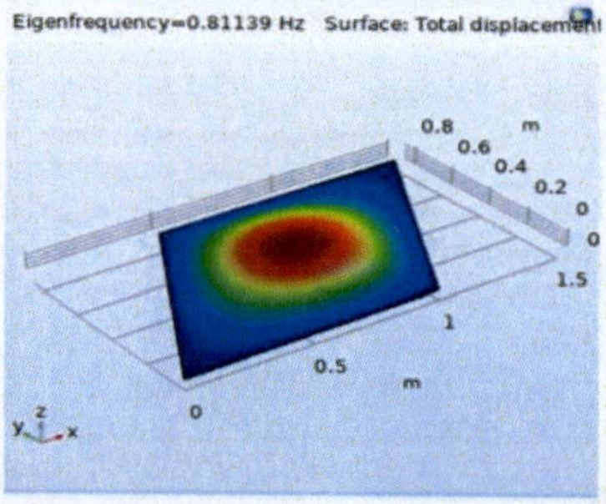	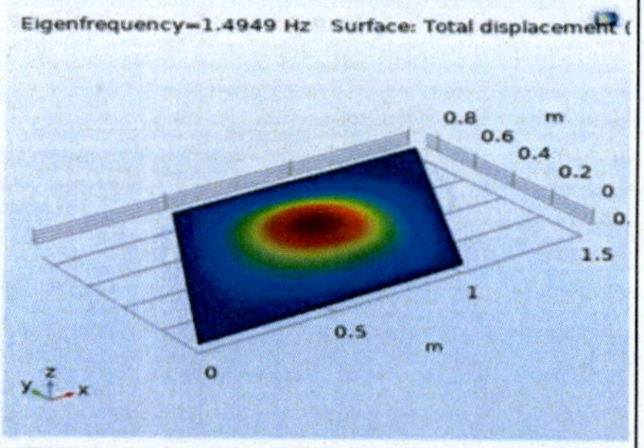
45	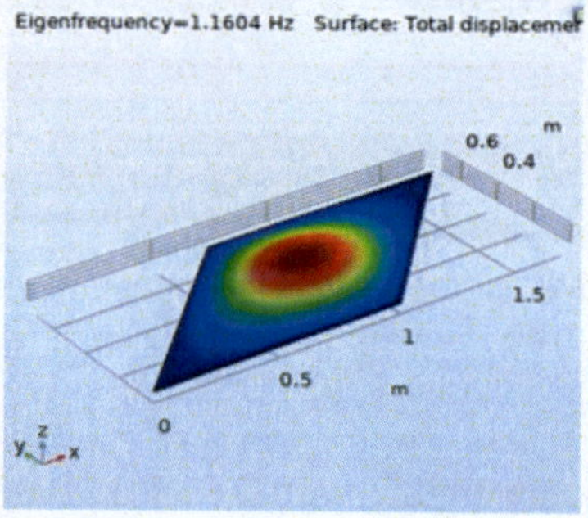	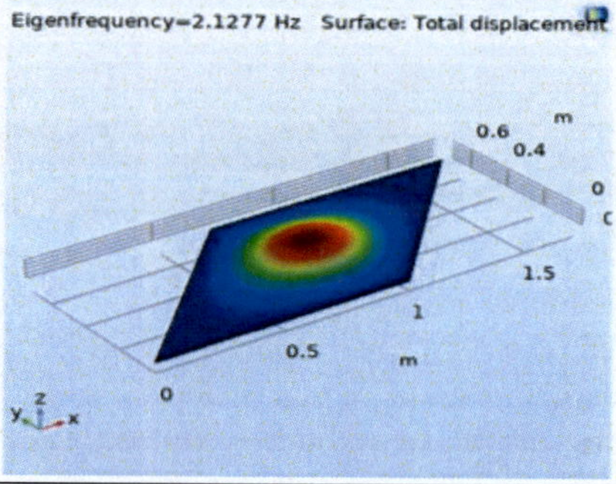

Fig. 2 Mode shapes of skew plates under different end conditions ($a = b = 1$ m and $h = 0.001$ m)

4.1 *Parametric Study for Triangular Plate*

The modal study of FG-triangular plate is considered for parametric study. The geometric parameters are $a = 1$ m, $b = 1$ m, $c = 1$ m and thickness (h) = 0.01. The effect of functionally graded indices and width-to-thickness ratio (k) on natural frequencies is studied in detail.

4.1 (I) Effect of Functionally Graded Index (*g*) on Natural Frequencies
In this study, functionally graded index is varied from range of 1 to 6. Width-to-thickness ratio (k = 100) is kept constant. The results are obtained for different boundary conditions using finite element software COMSOL-5.4 and are listed in Table 3 (Figs. 3 and 4).

4.1 (II) Effect of Width-to-Thickness Ratio (*k*) on Natural Frequencies
The modal analysis of triangular plate is carried out for width-to-thickness ratio k = 100, 20, 10 and 5 with functionally graded index g = 1. The results are obtained for different boundary conditions using finite element software COMSOL-5.4 and are listed in Table 4 (Figs. 5 and 6).

4.2 *Parametric Study for Trapezoidal Plate*

The modal study of FG-trapezoidal plate is considered for parametric study. The geometrical parameter is a = 1 m, b = 0.6 m and thickness (h) = 0.01 m. Effect of functionally graded indices and width-to-thickness ratio (k) is studied.

4.2 (I) Effect of Functionally Graded Indices on Natural Frequencies
In this study, functionally graded index is varied from range 1 to 6. Three types of end conditions are being used, i.e. CCCC, SSSS and CSCS. Width-to-thickness ratio (k = 100) is kept constant. The results are obtained for different end conditions using finite element software COMSOL-5.4 and are listed in Table 5.

Table 3 Natural frequencies (Hz) of FGM triangular plate under various support conditions with different functionally graded index (a = b = c = 1 m, h = 0.01 m and k = 100)

	Mode	g					
		g = 1	g = 2	g = 3	g = 4	g = 5	g = 6
CCC	Mode 1	394.93	334.21	293.85	269.67	256.83	249.11
	Mode 2	751.4	635.88	559.08	513.09	487.52	473.96
	Mode 3	1169.9	990.00	870.42	792.02	759.01	737.91
	Mode 4	1247.9	1056.00	928..47	852.07	809.63	787.12
CCS	Mode 1	325.87	257.77	242.66	222.51	211.42	205.55
	Mode 2	658.42	557.19	489.89	449.59	427.19	415.31
	Mode 3	1052.7	890.83	783.23	718.8	682.98	663.99
	Mode 4	1127.6	954.24	839.99	769.97	731.60	711.26
SSS	Mode 1	210.66	178.27	156.74	143.84	136.68	132.88
	Mode 2	490.66	415.23	365.07	335.04	318.35	309.49
	Mode 3	839.29	710.25	624.47	573.10	544.54	529.39
	Mode 4	908.83	769.11	676.21	620.58	589.66	573.26

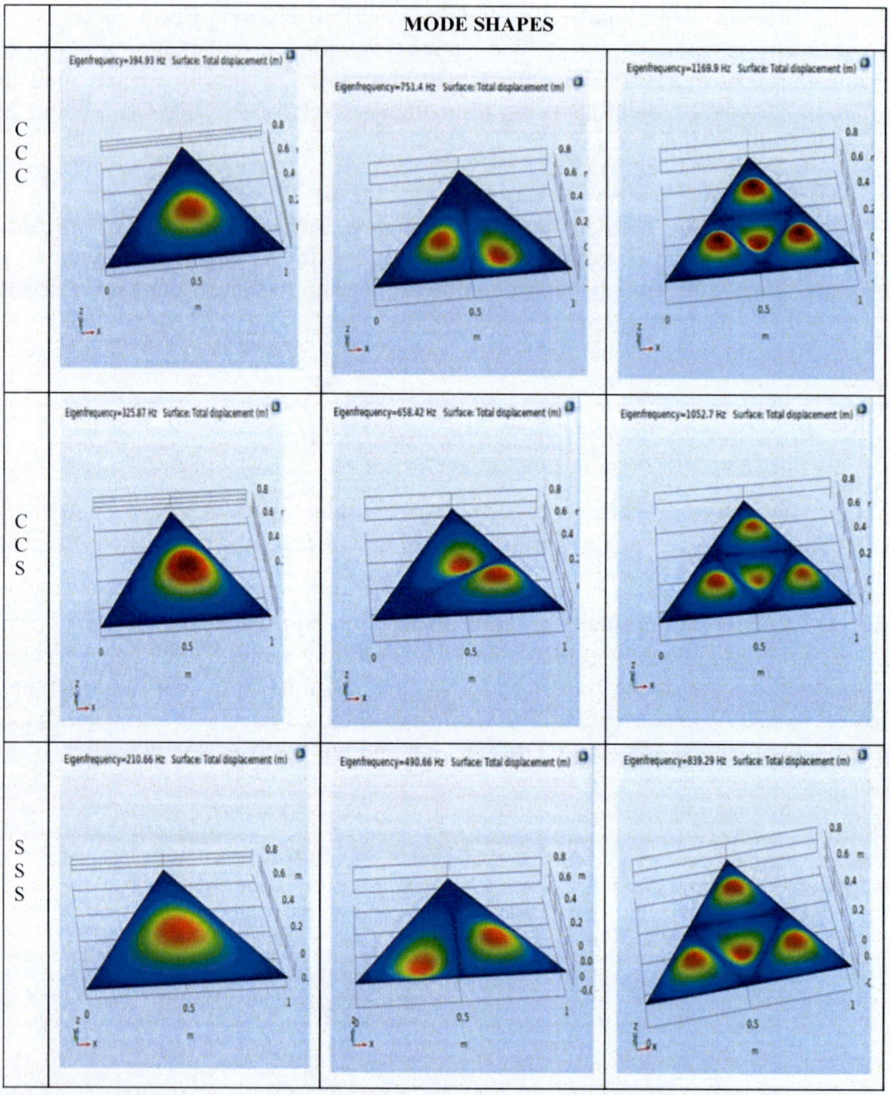

Fig. 3 Mode shapes of triangular plates under different end supports ($a = b = c = 1$ m, $h = 0.01$ and g)

4.2 (II) Effect of Width-to-Thickness Ratio (k) on Natural Frequencies

The modal analysis of trapezoidal plate is carried out for width-to-thickness ratio $k = 100$, 20, 10 and 5 with functionally graded index $g = 1$. The results are obtained for different boundary conditions using finite element software COMSOL-5.4, and results are listed in Table 6.

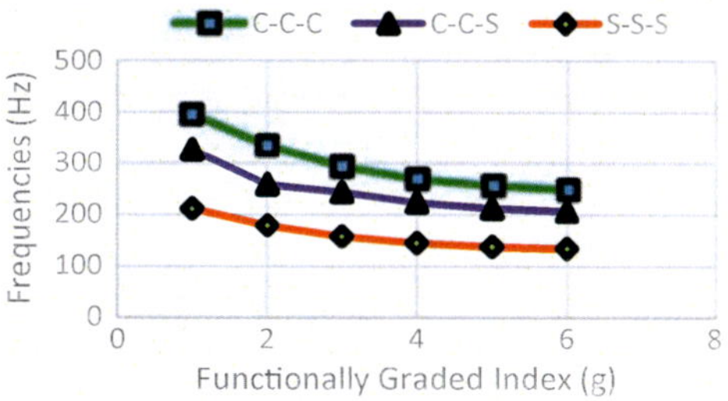

Fig. 4 Effect of functionally graded indices on natural frequencies of FG-triangular plate under various support conditions for mode 1

Table 4 Natural frequencies for various width-to-thickness ratio (k) of FGM triangular plate under several end conditions ($a = b = c = 1$ m, $h = 0.01$ m and $g = 1$)

	k	Mode			
		Mode 1	Mode 2	Mode 3	Mode 4
CCC	100	394.93	751.4	1169.9	1247.9
	20	1839.6	3353.7	5018.3	5304.6
	10	3120	5312	7568.2	7907.2
	5	4358.6	6858.1	9397.8	9714.5
CCS	100	325.87	658.42	1052.7	1127.6
	20	1544	3003.3	4622.9	4912.2
	10	2703.5	4909.1	7179.4	7550.5
	5	3948	6559.7	9158.1	9553.6
SSS	100	210.66	490.66	839.29	908.83
	20	1029.8	2332.56	3864.2	4159.2
	10	1934.1	4115.9	6435.8	6860
	5	3220.2	6112.7	8834.9	9309.1

5 Summary and Conclusion

This modal characteristics of FGM structures such as skewed, triangular and trapezoidal plate are presented in this paper. The accuracy of the present method is validated by the result obtained with previously published work. The effect of functionally graded index, width-to-thickness ratio and boundary condition of the FGM structures on natural frequencies is reported. This study is useful in the design of FGM structure for engineering applications. Some of the significant results are summarized as

- It can be observed that natural frequencies decrease with increase in volume fraction index for power law distribution under different boundary conditions for FG-triangular plate and FG-trapezoidal plate as discussed earlier with help of plots.
- It can also be observed that natural frequencies increase with decrease in width-to-thickness ratio for power law distribution under different boundary conditions.

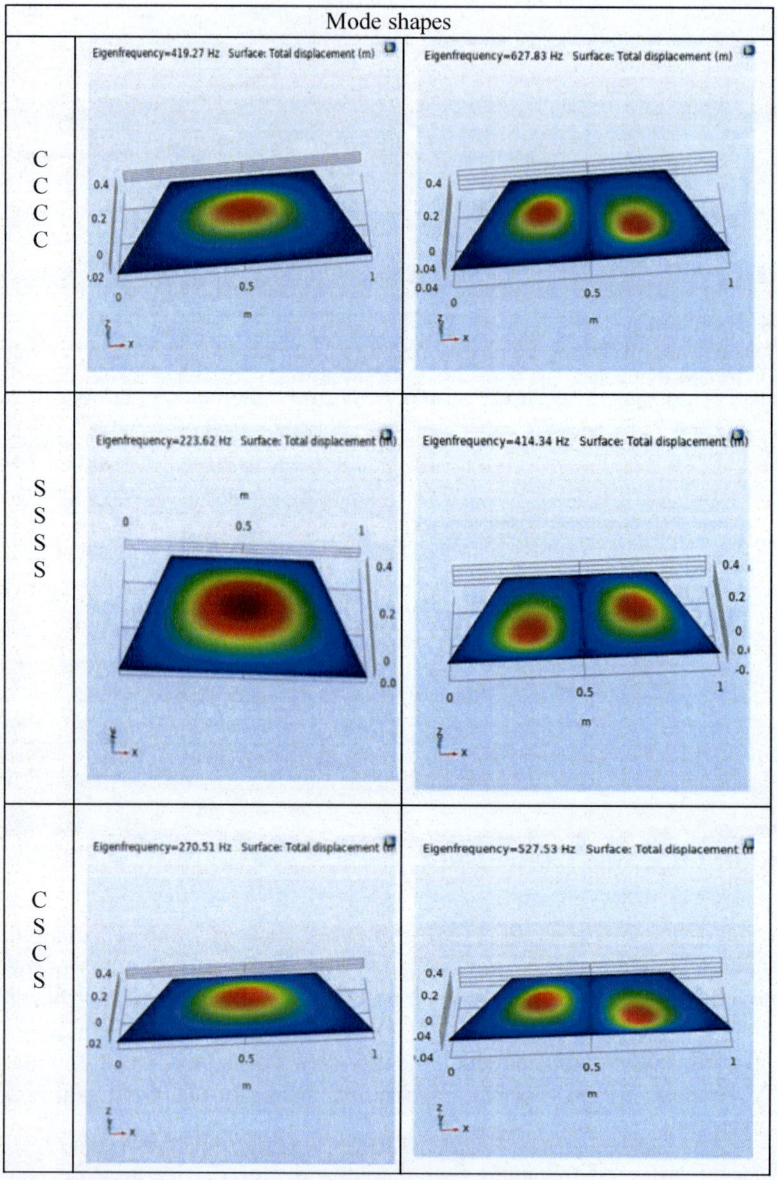

Fig. 5 Mode shapes of trapezoidal plate with different end conditions (a = 1 m, b = 0.6 m and h = 0.01, g = 1

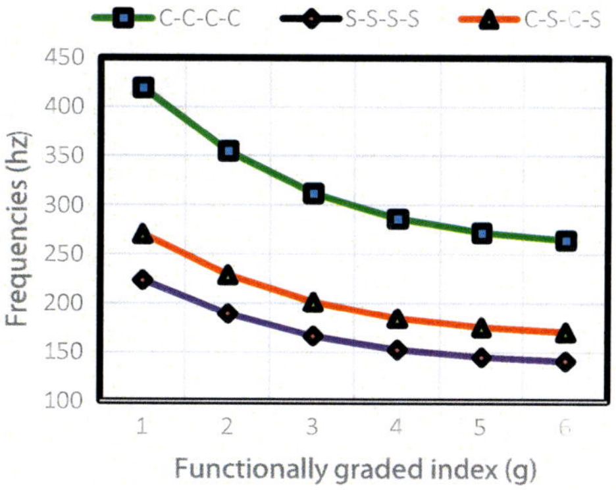

Fig. 6 Effect of functionally graded indices on natural frequencies of FGM trapezoidal plate under various support conditions for mode 1

Table 5 Natural frequencies (Hz) of FGM trapezoidal plate under various support condition with different functionally graded indices (a = 1 m, b = 0.6 m, h = 0.01 m and k = 100)

	Mode	g					
		g = 1	g = 2	g = 3	g = 4	g = 5	g = 6
CCCC	Mode 1	419.27	354.81	311.95	286.29	272.03	264.46
	Mode 2	627.83	531.30	467.13	428.70	407.34	396.01
	Mode 3	949.82	803.79	706.7	648.57	616.25	599.11
	Mode 4	1034.7	875.66	769.89	706.56	671.35	652.38
SSSS	Mode 1	223.62	189.24	166.38	152.69	145.08	141.05
	Mode 2	414.34	350.64	308.29	282.39	268.83	261.35
	Mode 3	668.94	556.10	497.72	456.78	434.02	421.95
	Mode 4	738.57	625.02	549.52	504.32	479.19	465.86
CSCS	Mode 1	270.51	228.92	201.27	184.71	175.51	170.63
	Mode 2	527.53	446.43	392.51	360.22	342.27	332.75
	Mode 3	706.24	597.67	525.48	482.25	458.22	445.8
	Mode 4	892.54	755.32	664.09	609.46	579.02	562.98

Table 6 Natural frequencies for various width-to-thickness ratio (k) of FGM trapezoidal plate under various end condition (a = 1 m, b = 0.6 m, h = 0.01 m and g = 1)

	k	Mode			
		Mode 1	Mode 2	Mode 3	Mode 4
CCCC	100	419.27	627.83	949.82	1034.70
	20	1942	2846	4164.3	4479.30
	10	3279	4648	6472.8	6858.3
	5	4547.1	6208.4	7411.4	8172.4
SSSS	100	223.62	414.34	668.94	738.57
	20	1088.9	1979.5	3123.7	3424.00
	10	2083.30	3548.80	5347.60	5798.50
	5	3372.4	5409.2	7411.4	7606.9

(continued)

Table 6 (continued)

	k	Mode			
		Mode 1	Mode 2	Mode 3	Mode 4
CSCS	100	270.51	527.73	706.24	892.54
	20	1296.2	2493.7	3267.3	3985.8
	10	2340.4	4192.8	5514.4	6362.40
	5	3639.20	5785.4	7411.40	7707.40

- It is also noted that in the study of clamped end conditions, natural frequencies are found to be greater than the simply supported end conditions and other combinations of boundary conditions.

References

1. Silva ECN, Walters MC, Paulino GH (2008) Modeling bamboo as a functionally graded material. In: AIP conference proceedings, vol. 973, no. 1, pp. 754–759. American Institute of Physics https://doi.org/10.1063/1.2896876
2. Sharma P (2019) Vibration analysis of functionally graded piezoelectric actuators. Springer, New York, NY. https://doi.org/10.1007/978-981-13-3717-8
3. Khinchi A, Sharma P (2020) Free vibration analysis of isotropic spherical cap and FG-spherical cap with cut-out using COMSOL. In: AIP Conference Proceedings, vol. 2220, no. 1, p. 130074. AIP Publishing LLC. https://doi.org/10.1063/5.0001299
4. Sharma P, Parashar SK (2016) Exact analytical solution of shear-induced flexural vibration of functionally graded piezoelectric beam. In: AIP conference proceedings, vol. 1728, no. 1, p. 020167. AIP Publishing LLC. https://doi.org/10.1063/1.4946218
5. Sharma P, Singh R (2019) Investigation on modal behaviour of FGM annular plate under hygrothermal effect. In: IOP Conference Series: Materials Science and Engineering, vol. 624, no. 1, p. 012001. IOP Publishing. https://doi.org/10.1088/1757-899x/624/1/012001
6. Garg AK, Khare RK, Kant T (2006) Free vibration of skew fiber-reinforced composite and sandwich laminates using a shear deformable finite element model. J Sandw Struct & Mater 8 (1):33–53. https://doi.org/10.1177/1099636206056457
7. Srinivasa CV, Suresh YJ, Prema Kumar WP (2014) Experimental and finite element studies on free vibration of skew plates. IJASE 6(1):48. https://doi.org/10.1007/s40091-014-0048-3
8. Zhang LW, Lei ZX, Liew KM (2015) free vibration analysis of functionally graded carbon nanotube-reinforced composite triangular plates using the FSDT and element-free IMLS-Ritz method. Compos Struct 120:189–199. https://doi.org/10.1016/j.compstruct.2014.10.009
9. Jadhav YG, Deshmukh PV (2016) Effect of cut-out shape on free vibration of composite plates. IJCESR 3:48–53
10. Gorman DJ (1983) A highly accurate analytical solution for free vibration analysis of simply supported right triangular plates. J Sound Vib 89(1):107–118. https://doi.org/10.1016/0022-460X(83)90914-8
11. Sakiyama T, Huang M (2000) Free-vibration analysis of right triangular plates with variable thickness. J Sound Vib 234(5):841–858. https://doi.org/10.1006/jsvi.2000.2903
12. Belalia SA (2017) Linear and non-linear vibration analysis of moderately thick isosceles triangular FGPs using a triangular finite p-element. Mech Adv Mater Mod Process 3(1): 4. https://doi.org/10.1186/s40759-017-0018-0

13. Kaur N (2020) Vibrational Behavior of Tapered Triangular Plate with Clamped Ends under Thermal Condition. J Inst Eng (India): Ser C, 1–9. https://doi.org/10.1007/s40032-019-00551-9
14. Saliba HT (1986) (1986)"Free vibration analysis of simply supported symmetrical trapezoidal plates.". J Sound Vib 110(1):87–97. https://doi.org/10.1016/S0022-460X(86)80076-1
15. Zamani M, Fallah A, Aghdam MM (2012) Free vibration analysis of moderately thick trapezoidal symmetrically laminated plates with various combinations of boundary conditions. Eur J Mech-A/Solids 36:204–212. https://doi.org/10.1016/j.euromechsol.2012.03.004
16. Gupta AK, Sharma P (2016) Vibration study of non-homogeneous trapezoidal plates of variable thickness under thermal gradient. J Vib Control 22(5):1369–1379. https://doi.org/10.1177/1077546314535280
17. Sharma P (2018) Efficacy of Harmonic Differential Quadrature method to vibration analysis of FGPM beam. Composite Structures. 189:107–116, https://doi.org/10.1016/j.compstruct.2018.01.059
18. Sharma P, Parashar SK (2016) Free vibration analysis of shear-induced flexural vibration of FGPM annular plate using generalized differential quadrature method. Compos Struct 155:213–222. https://doi.org/10.1016/j.compstruct.2016.07.077
19. Parashar SK, Sharma P (2016) Modal analysis of shear-induced flexural vibration of FGPM beam using Generalized Differential Quadrature method. Composite Structures 139:222–232, https://doi.org/10.1016/j.compstruct.2015.12.012
20. Sharma P (2020) Vibration analysis of FGP actuator due to longitudinal piezoelectric coupling coefficient. In: AIP Conference Proceedings, vol. 2220, no. 1, p 130072. AIP PublishingLLC. https://doi.org/10.1063/5.0001180

Numerical Behaviour of Al 6063 Under Varying Striker Velocity and Length of Split Hopkinson Pressure Bar at Large Strain Rates

Manish Kumar Gupta

1 Introduction

The study of aluminium alloy Al 6063 under high strain rate conditions has always been important due to its wider use in aerospace industry, structural materials and space vehicles due to their great specific strength and low density [1, 2]. To study the impact characteristics of wires made up of iron, John Hopkinson [3] used the Hopkinson bar experimental technique in 1872. Later on, some modifications were done to the Hopkinson bar by several researchers to get the exact pressure–time curves, error-free wave dispersion and the strains present in the pressure bar [4–6]. Different techniques such as pulse shaping system [7, 8] and conical striker method [9–11] were also introduced to see the changes in rise and fall of waves. Lee and Lin [12] examined the impact behaviour of stainless steel 304L under dynamic conditions using SHPB. Chen et al. [13] investigated the behaviour of ductile alloys under SHPB system. Heating source equipped with SHPB is adopted by Kajberg and Sundin [14] to observe the mechanical response of steels at high temperatures. Flow behaviour of Al alloy 6061 is investigated by Hall and Guden [15] under the effect of different lubricants used in SHPB system. The characterization of stainless steel 316 L with a high strain rate is analysed by Tasdemirci et al. [16] using SHPB experimental system and finite element simulation of SHPB compression system done for cylindrical specimen [17]. Radial inertia and friction effects had been examined for pure aluminium 1100 under high strain rate using the finite element software [18]. Effects of pulse shaping technique are studied numerically in the SHPB apparatus [19].

There are some literatures available for dynamic behaviour of Al 6063 based on simulation. However, it is hard to find the publications that examine the behaviour based on SHPB parameters variation, especially for striker bar. This paper presents

M. K. Gupta (✉)
Department of Mechanical Engineering, NIT Patna, Patna, India
e-mail: manishkumargupta@springernature.com

S. Bag et al. (eds.), *Next Generation Materials and Processing Technologies*,
Springer Proceedings in Materials 9, https://doi.org/10.1007/978-981-16-0182-8_32

a numerical simulation of Al 6063 in SHPB under different conditions of striker bar to understand the effect of striker bar. In addition, the effect of aspect ratio of specimen is also being studied to see the changes in incident and reflected waves due to change in specimen dimension.

2 Methodology

This paper presents the numerical simulation of Al 6063 under dynamic conditions with the variation of SHPB parameters and specimen diameter. The striker bar length and velocity have been changed 200–400 mm and 20–40 m/s, respectively, for SHPB parameter variation. The diameter of specimen Al 6063 is also changed from 5 to 15 mm by keeping the length constant to see the impact of aspect ratio on behaviour of the material. During the adjustment in velocity of striker bar and specimen diameters across, all simulations are performed under the length of 200 mm striker bar.

2.1 Split Hopkinson Pressure Bar

A compression SHPB contains striker, input bar and output bar. Input bar and output bars are of equal length and diameter as can be seen from Fig. 1. A damper is also devoted at the end part of output bar to absorb the energy transmitted by the output bar. The sample to be tested is inserted between input and output bars. It works on the principle of one-dimensional wave propagation. When striker impacts with certain velocity to the interface of input bar, it creates compression stress/strain wave (incident wave,ε_i) in the input bar. Wave created in the input bar propagates along the input bar and loads the specimen under compression. On reaching the wave to the sample, some part of the incident wave (ε_i) gets reflected back to the input bar as reflected wave (ε_r); the remaining wave gets transmitted to the output bar as transmitted wave (ε_t). Figure 1 represents the working and wave propagation theory of signals obtained by strain gauges attached with the bars which are used to find the strain rate ($\dot{\varepsilon}_s$), strain (ε_s) and stress (σ_s) of test sample as follows [20]:

$$\dot{\varepsilon}_s(t) = -\frac{2C_0}{L_s}\varepsilon_r(t) \quad (1)$$

$$\varepsilon_s(t) = -\frac{2C_0}{L_s}\int_0^t \varepsilon_r(t)dt \quad (2)$$

$$\sigma_s(t) = E_b\frac{A_b}{A_s}\varepsilon_t(t) \quad (3)$$

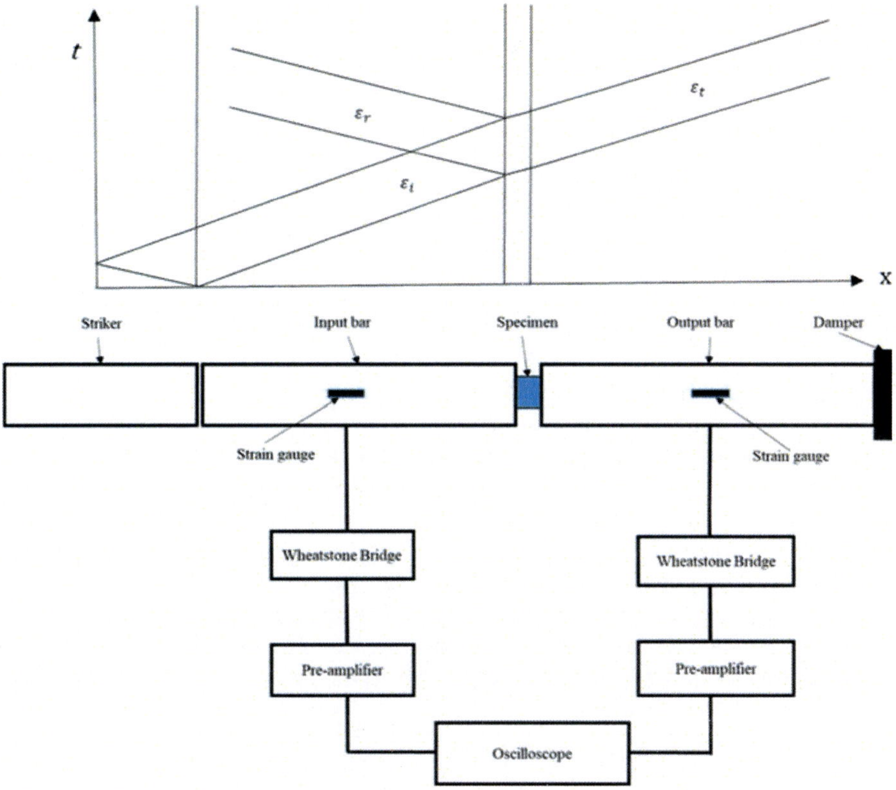

Fig. 1 Illustration of SHPB system and working principle

where C_0 is the wave speed in the bar; L_s is the specimen length; E_b is the modulus of elasticity of bar; A_b and A_s are the cross-sectional area of bar and sample, respectively; t is the time. C_0 is the under-root ratio of elastic modulus to the density of bar material.

2.2 *Finite Element Analysis*

The finite element simulations are investigated using FE software Abaqus/Explicit 6.14 [21]. A 3D model of SHPB system is modelled in ABAQUS to understand the effect of parameters. Each components of SHPB system are created separately in ABAQUS/CAE and subsequently assembled as can be seen from Fig. 2. Table 1 represents the dimensions of bars and specimen. Mild steel is selected as bar material having density, elastic modulus and Poisson's ratio; 7800 kg/m^3, 200 GPa and 0.3, respectively. For specimen Al 6063, the Johnson–Cook (JC) material model [22] is selected. It is widely used to show the behaviour of metals under high strain rate and is often used in SHPB simulation for constitutive behaviour of specimen presented as Eq. 4.

$$\sigma = [A + B\varepsilon^n][1 + C \ln \dot{\varepsilon}^*][1 - (T^*)^m] \tag{4}$$

where σ is true stress; ε is true strain; A, B, C, n, m are material parameters; $\dot{\varepsilon}^*$ is calculated by the division of equivalent strain rate to reference strain rate and T^* is temperature (homologous). It is widely used to show the behaviour of metals under high strain rate and is often used in SHPB simulation for constitutive behaviour of specimen. Constitutive model constants used for Al 6063 are shown in Table 2. Density, modulus of elasticity and Poisson's ratio are considered as 2700 kg/m^3, 69 GPa and 0.25, respectively, for Al 6063.

Meshing of each component of SHPB is done with elements C3D8R. The bars and specimen are meshed using global element size of 0.28 and 0.5, respectively. Total number of elements and nodes are presented in Table 3 with corresponding condition of the simulation.

A general contact algorithm is applied between input bar and specimen and specimen and output bar. Boundary conditions and meshing of 3D models are shown in Fig. 2. To generate the high strain rate, the striker bar is impacted to the input bar with the velocity of 20, 30 and 40 m/s with different lengths. The last end of output bar is kept fixed so that it can absorb all the forces transmitted through the output bar at the end portion.

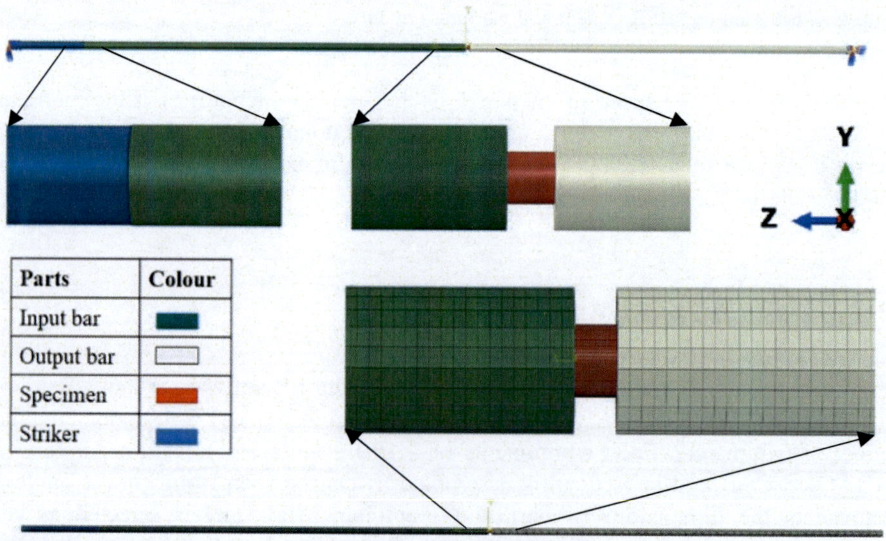

Fig. 2 3D SHPB assembly with boundary conditions and meshed model with zoom at specimen position

Table 1 Geometric properties of SHPB

Parameter	Striker bar	Input bar	Output bar	Specimen
Length (mm)	200–400	1000	1000	10
Diameter (mm)	20	20	20	5–15

Table 2 Constants used to define a constitutive model of Al 6063 [23]

A	B	N	C	m
111.82 MPa	241.4 MPa	0.415	0.012	1.003

Table 3 Total number of elements in assembly with variations in simulation

Parameters		Total number of elements(C3D8R)
Striker velocity	20 m/s	45,755
	30 m/s	45,755
	40 m/s	45,755
Striker length	200 mm	45,755
	300 mm	47,447
	400 mm	49,139
Specimen diameter	5 mm	39,255
	10 mm	45,755
	15 mm	56,515

2.3 *Validation of Numerical Approach*

To validate the numerical methods used in this paper, the result reported by [24] for Al 6063 in dynamic condition is compared here. In their numerical simulation, length of striker bar, input and output bars were 200 mm and 1500 mm, respectively. Each bar's diameter was of 20 mm. Al 6063 specimen taken in their simulation was 10 mm diameter and 10 mm length. The same conditions of SHPB are considered here for validation purpose. Figure 3 shows the comparison between the results obtained by the present approach and the result obtained by Afdal et al. [24]. It can be seen that both results are in close agreement, which validated this numerical approach.

3 Results and Discussion

The present investigation focuses on the effect of varying parameters of SHPB components and specimen aspect ratio on the wave propagation and stress–strain behaviour of Al 6063. In order to see the effect of parameters variation of SHPB

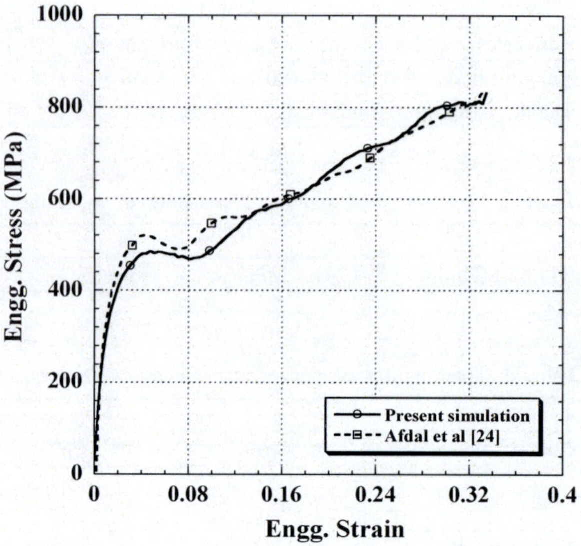

Fig. 3 Comparison of present FE simulation with Afdal et al.'s result for Al 6063 at striker velocity of 20 m/s

components, striker bar length and velocity have been varied from 200–400 mm and 20–40 m/s, respectively, whereas the specimen diameter is varying from 5 to 15 mm to observe the aspect ratio of Al 6063 at high strain rate.

3.1 *Effect of Striker Bar Velocity*

Figure 4 represents the correlation among the waves and true stress–strain curve with various velocities in the striker bar of SHPB framework. From Fig. 4a, it is seen that the peak values and duration, i.e. span of waves, are reasonably relative to the speed of striker bar, and it is least for 20 m/s and most extreme for 40 m/s. Figure 4b shows the typical true stress and true strain curves for Al 6063 at various speeds. As observed, the impacted speed has critical impact on the state of stress–strain curves. The maximum and minimum true ultimate stress is obtained for striker's velocity 40 m/s and 30 m/s, respectively. Therefore, striker's velocity governs both wave's variation and flow curve of Al 6063 under large strain rate.

3.2 *Effect of Length of Striker Bar*

Figure 5 shows the effect of striker bar length on the flow curve of Al 6063 and wave propagations in bars at impact velocity of 20 m/s. The results state that the wave duration of incident, reflected and transmitted signals for each striker bar length is almost same; however, it shifted the location with respect to time as the

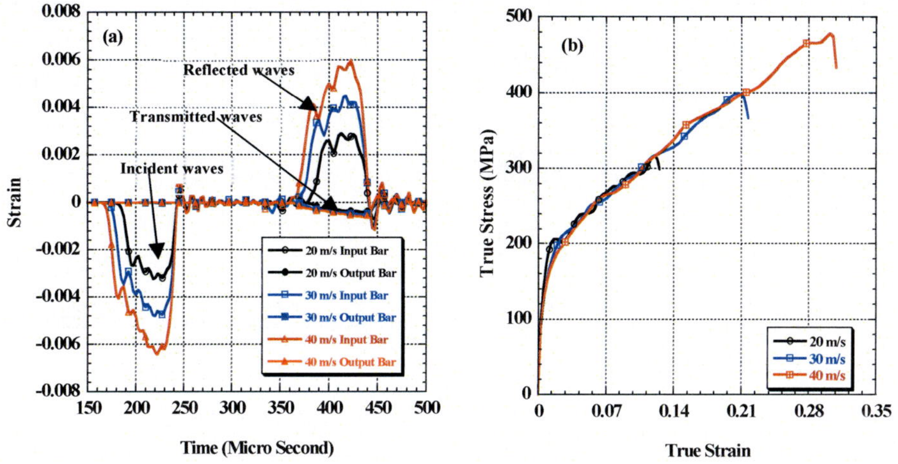

Fig. 4 Effect of striker bar velocity on **a** input and output bar waves **b** true stress–strain curves of Al 6063

striker bar length varies from 200 to 400 mm as shown in Fig. 5a. It is also observed that the peak values of each wave are dependent on the length of the striker bar and are minimum for 200 mm and maximum for 400 mm. Figure 5b presents the true-stress-true-strain curve for Al 6063 with the variation of striker bar length. It is concluded that the ultimate true stress values of specimen increase as the length of the bar increased. Hence, it can be concluded that the striker bar having length of 200 mm and velocity of 20 m/s having the lowest ultimate true stress and peak wave value. The striker bar having length of 400 mm and 300 mm with velocity of 20 m/s having almost constant ultimate true stress and peak wave value higher than that of the 200 mm length with 20 m/s of striker bar.

3.3 *Effect of Diameter of Specimen*

Figure 6 depicts the effect of specimen's diameter from 5 to 15 mm by keeping the length constant as 10 mm on the flow curve of Al 6063 and wave propagations in bars at an impact velocity of 20 m/s. It can be seen in Fig. 6a that nothing is happening with the variation of diameter of the specimen on the incident wave propagated through the bar, whereas the reflected and the transmitted waves are affected as the diameter of specimen changes from 5 to 15 mm. Maximum and minimum peak values of reflected wave are found for diameter 5 and 15 mm, respectively. For transmitted wave, the maximum and minimum peak values are obtained for specimen having diameter of 15 and 5 mm, respectively. The comparison of the true stress vs true strain curve of Al 6063 for different diameters is shown in Fig. 6b. It is observed that the maximum and almost same values of

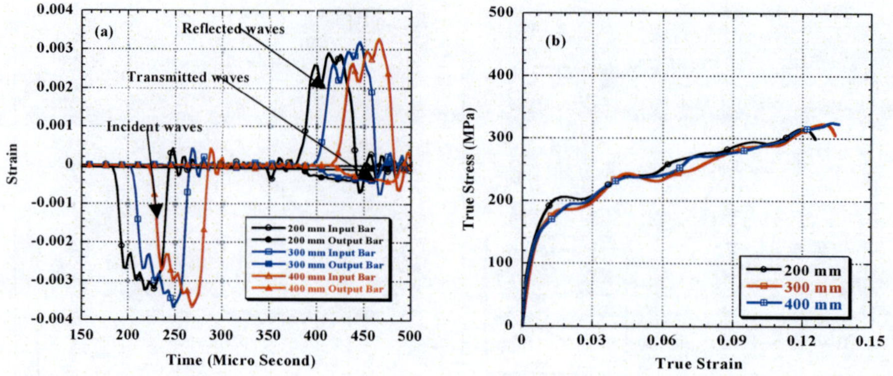

Fig. 5 Effect of striker bar length on **a** input and output bar waves **b** true stress–strain curves of Al 6063

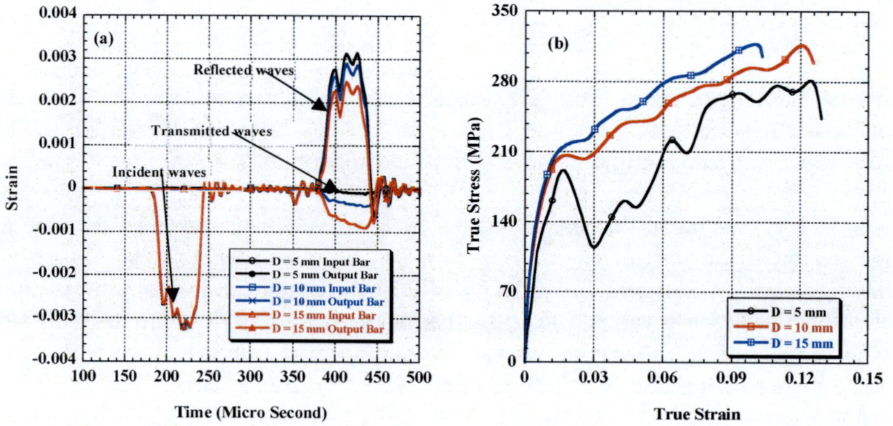

Fig. 6 Effect of specimen diameters on **a** input and output bar waves **b** true stress–strain curves of Al 6063

ultimate true stress value for Al 6063 have diameter 10 and 15 mm and minimum for 5 mm diameter of specimen at impact velocity of 20 m/s with the striker bar length of 200 mm. Similar results are found for yield stress, i.e. higher and almost same for 10 and 15 mm specimen diameter and minimum for 5 mm diameter of specimen.

Table 4 presents the true yield stress and true ultimate compressive stress of each simulation with elongation percentage. It can be understood from Table 4 that the yield stress value for each simulation is almost constant, but the true ultimate compressive stress changes with change in condition of SHPB simulation. The results state that the ultimate compressive stress of Al 6063 is positive strain rate sensitive, and ductility also increases with velocity increment in striker bar. Table 4 also illustrates that there is only marginal increment in ultimate compressive stress

Table 4 Numerical results of compression tests under different conditions

Parameters		True stress–strain curve		Total elongation (%)
		Yield stress (MPa)	Ultimate compressive stress (MPa)	
Striker velocity	20 m/s	160 ± 1	318 ± 4	12 ± 0.1
	30 m/s	162 ± 2	400 ± 2	20 ± 0.12
	40 m/s	160 ± 1	480 ± 3	31 ± 0.2
Striker length	200 mm	160 ± 1	318 ± 4	12 ± 0.1
	300 mm	162 ± 1	321 ± 3	13 ± 0.5
	400 mm	160 ± 2	324 ± 3	13.5 ± 0.2
Specimen diameter	5 mm	162 ± 3	281 ± 2	12.5 ± 0.3
	10 mm	160 ± 1	318 ± 4	12 ± 0.1
	15 mm	163 ± 3	317 ± 2	10 ± 0.2

and ductility with increase in striker bar length. For the maximum value of the ultimate compressive stress is found for specimen having 10 mm diameter, maximum ductility is obtained for specimen having 5 mm diameter.

4 Conclusions

In this paper, numerical simulation of Al 6063 under high strain conditions has been done by varying the striker length, striker velocity and specimen diameter to recognize the result of these parameters on the wave propagation and flow stress behaviour. For SHPB parameters variation, the simulations have been performed under the different length and velocity of the striker bar. To see the effect of aspect ratio, specimen diameters have been changed by keeping the length constant. The following outcomes are found from the simulation:

(1) As the striker velocity increases, both the peak values and duration of the incident, reflected and transmitted waves increased. Maximum and minimum peak values and duration of waves are found for 40 and 20 m/s, respectively.
(2) Al 6063 has the positive strain rate sensitivity for ultimate compressive strength, whereas yield strength remains unaffected from changes in velocity.
(3) Ductility of Al 6063 increased with increasing strain rate.
(4) Wave's location shifted as length of striker bar changes, but the duration of these waves remains constant. Peak values of each wave are dependent on the length of the striker bar and maximum for 400 mm and minimum for 200 mm. However, these values are very nominal and can be ignored.
(5) Effect of striker bar length on true stress–strain can be neglected.

(6) Specimen diameter change does not affect the incident wave of bar. The reflected wave peak values increased and the transmitted wave peak values decreased with the increase in diameter of specimen.
(7) Maximum and almost same values of ultimate true stress value for Al 6063 have diameter 10 and 15 mm and minimum for 5 mm diameter of specimen.

References

1. Nishida M, Hayashi K, Nakagawa J, Ito Y (2012) Influence of temperature on crater and ejecta size following hypervelocity impact of aluminum alloy spheres on thick aluminum alloy targets. Int J Impact Eng 42:37–47
2. Liang XP, Li HZ, Huang L, Hong T, Ma B, Liu Y (2012) Microstructural evolution of 2519-T87 aluminum alloy obliquely impacted by projectile with velocity of 816 m/s. Trans Nonferrous Metals Soc China 22:1270–1279
3. Hopkinson J On the rupture of iron wire by a blow. Proc Manchester Literary Philosophical Society 11:40–45
4. Hopkinson B (1914) A method of measuring the pressure produced in the detonation of high explosives or by the impact of bullets. Philos Trans R Soc Lond Ser A: Math, Phys Eng Sci 213(497–508):437–456
5. Bancroft D (1941) The velocity of longitudinal Waves in cylindrical Bars. Phy Rev 59: 588–593
6. Davies RM (1948) A critical study of the Hopkinson pressure bar. Phil. Trans. R. Soc. A240:375–457 (1948)
7. Chen W, Luo H (2004) Dynamic compressive responses of intact and damaged ceramics from a single split Hopkinson pressure bar experiment. Exp Mech 44:295–299
8. Lu Y, Li Q (2010) Appraisal of Pulse-Shaping Technique in Split Hopkinson Pressure Bar Tests for Brittle Materials. Int J Prot Struct 1:363–390
9. Cloete T, Van Der Westhuizen A, Kok S, Nurick GN (2009) A tapered striker pulse shaping technique for uniform strain rate dynamic compression of bovine bone. In: DYMAT—International Conference on the Mechanical and Physical Behaviour of Materials under Dynamic Loading, Brussels, Belgium, 7–11 September
10. Song B, Casem D, Kimberley J (2009) Springer Science Business Media. Heidelberg, Germany, pp 901–907
11. Bekker A, Cloete T, Chinsamy-Turan A, Nurick G, Kok S (2015) Constant strain rate compression of bovine cortical bone on the Split-Hopkinson Pressure Bar. Mater Sci Eng, C 46:443–449
12. Lee WS, Lin CF (2001) Impact properties and microstructure evolution of 304L stainless steel. Mater Sci Eng, A 308:124–135
13. Chen W, Song B, Frew DJ, Forrestal MJ (2003) Dynamic small strain measurements of a metal specimen with a split Hopkinson pressure bar. Exp Mech 43:20–23
14. Kajberg J, Sundin KG (2013) Material characterisation using high-temperature Split Hopkinson pressure bar. J Mater Process Technol 213:522–531
15. Hall IW, Guden M (2003) Split Hopkinson pressure bar compression testing of an aluminum alloy: E_ect of lubricant type. J Mater Sci Lett 22:1533–1535
16. Tasdemirci A, Ergönenç C, Güden M (2010) Split Hopkinson pressure bar multiple reloading and modeling of a 316 L stainless steel metallic hollow sphere structure. Int J Impact Eng 37, 250–259
17. Zencker U, Clos R (1999) Limiting conditions for compression testing of flat specimens in the split Hopkinson pressure bar. Exp Mech 39:343–348

18. Meng H, Li QM (2003) Correlation between the accuracy of a SHPB test and the stress uniformity based on numerical experiments. Int J Impact Eng 28:537–555
19. Baranowski P, Małachowski J, Gieleta R, Damaziak K, Mazurkiewicz Ł, Kolodziejczyk D (2013) Numerical study for determination of pulse shaping design variables in SHPB apparatus. Bull Pol Acad Sci Tech Sci 61:459–466
20. Kolsky H (1948) An investigation of the mechanical properties of materials at very high rates of loading. In: Proceedings Physics Society. Institute of Physics (Journal) Ltd., London, pp 676–700
21. Abaqus (2014) Abaqus user's. Manual. "Version 6.14." Dassault Systemes Simulia Corp., Providence, RI, USA
22. Johnson GR, Cook WH (1983) A constitutive model for metals subjected to large strains, high strain rates and high temperatures. In: Proceedings of the Seventh International Symposium on Ballistics, Hague, The Netherlands, 19–21 April, Vol 54, pp 1–7
23. Kruszka L, Anaszewicz J, Janiszewski J, Grazka M (2012) Experimental and numerical analysis of Al 6063 duralumin using Taylor impact test. Epj Web of Conferences 26:1062 EDP Sciences. https://doi.org/10.1051/epjconf/20122601062
24. Annisa J, Muhammad AK, Leonardo G (2016) Development of a numerical model for simulations of split hopkinson pressure bar. ARPN J Eng Appl Sci, 11(10)

Mathematical Modelling-Based Solar PV Module and Its Simulation in Comparison with Datasheet of JAPG-72-320/4BB Solar Module

Sriparna Das, Akshit Samadhiya, and Kumari Namrata

1 Introduction

The advancement of country is determined by the growth of industry and it running efficiently by taking less input or power and giving desirable amount of output without polluting the environment. So, electric power is the basic need of all the countries. Extraction of power is the ideal source for running of industries. Using of coal, petroleum and other non-renewable energy resources to run the industry is not the solution as they will take millions of years to form, by the process known as carbonization which mainly deals with carbon. The dead vegetation is buried under the soil with the presence of high temperature and high pressure, it gets changed to coal which takes a lot of years to reform itself, and petroleum is formed by two step process first by producing diagenesis from kerogen, and in second step, catagenesis forms kerogen to petroleum and natural gas. Diagenesis is when under less pressure and temperature the compacting process is done. Under mild pressure and temperature, the mixture of proteins, lipids, water and minerals form into kerogen with the help of microbes, which is costly as well as time consuming. So, one should use renewable energy sources to run industry which includes solar energy, hydro energy, wind energy, nuclear energy, diesel energy, natural gas and also energy from tides. But solar is the most suitable and available form of energy as huge portion of total power production can be generated by placing solar panels and solar cells without polluting the environment as it does not require coal or diesel to burn and produce energy. The design of solar PV cell and comparing its performance analysis with the practical existing cell is explained subsequently. As India stands third in producing as well as consuming of electrical energy, so this evolvement of model will help to examine the performance of the various solar cell that is coming

S. Das (✉) · A. Samadhiya · K. Namrata
Department of Electrical Engineering, NIT Jamshedpur, Jamshedpur 831014, Jharkhand, India
e-mail: namrata.ee@nitjsr.in

S. Bag et al. (eds.), *Next Generation Materials and Processing Technologies*, Springer Proceedings in Materials 9, https://doi.org/10.1007/978-981-16-0182-8_33

into market for production of solar module and giving much more stress for using of solar energy for power production.

2 India's Sources of Power Production

The installed capacity of India according to 2019 is 368.79 GW according to the record taken from 31 December 2019. The generation of power consists of 55.4% that is 205,134.5 MW from coal, 12.3% that is 45,699.22 MW from large hydro, 1.3% that is 4683.16 MW from small hydro, 10.2% that is 37,693.75 MW from wind power, 9.4% that is 34,627.82 MW from solar panel, 2.7% that is 10,022.95 MW from biomass, 1.8% that is 6780 MW from nuclear power, 6.7% that is 24,955.36 MW from gas and 0.1% that is 509.71 MW from diesel sources [1]. One should keep in mind conservation of the energy so that it will not get wasted, it should be transferred into other forms as per requirements and should have provision for storing the extra energy for future purpose. Here, the focus has been given how to represent a solar P.V cell by a single diode model equivalent circuit and its modelling before implementing into practical use so as to have the idea of its power generation capacity and overall efficiency.

3 Methodology

In this paper, a photo cell is represented by a block diagram model using a single diode equivalent circuit diagram analysing from the numerical analysis of the photo cell. In photocell, light energy falls on the junction formed by p-n layer and excites the hole and electron to move within the junction so that the current flowing is taking place [2]. Analytical methods are used basically to make this model. After the modelling is done, the parameters like irradiance, voltage and temperature, needed as input in the model are taken from the datasheet of the solar PV practical module, and the output is seen as approximately close to the output given in the datasheet of the module, followed by the error calculation for analysing the performance of the model created.

4 Mathematical Model Form of Photovoltaic Cell

Solar cell is one of the upcoming technologies of conversion of photons or the rays of the sun to electrical energy. A number of photovoltaic cells connected in either series or parallel to form a PV module. When we connect a number of PV module either in series or in parallel, then it forms PV array. This is one of the good efficient technologies for the generation of electricity from the sun's rays that is available in nature. The solar cell can be represented by a single p-n junction diode. The light or the sun's rays falling on the semiconductor material are absorbed by silicon. The two layers of the PV cell are specially doped with one excess of electrons and another deficit of electron, and this causes occurrence of electric field. The electric field will cause the electron which is not tightly bound to move to one direction which is known as the junction. When the energy of the sun falls, then it allows the electrons to flow which will cause a current to flow. Here, in the representation of single diode equivalent model, there are two resistances attached, one connected in series and other in parallel, the former to reduce the fill factor and later one to decrease losses and for protection purpose [3]. Ideally parallel resistance should be infinite, and series resistance should be shorted.

5 Modelling of an Ideal PV Cell

By applying Kirchhoff's current law from Fig. 1, we get [4]:

$$I_1 = I_{\mathrm{ph1}} - I_{\mathrm{d1}} \tag{1}$$

From the general equation of the diode, we can say

$$I_{\mathrm{d1}} = I_{\mathrm{s1}}\left[\exp\left(\frac{q_1 V_{\mathrm{oc1}}}{N_{\mathrm{s1}} K_1 A_1 T_0}\right) - 1\right] \tag{2}$$

Substituting the value of I_{d1} from Eq. (2) to Eq. (1), we get

$$I_1 = I_{\mathrm{ph1}} - I_{\mathrm{s1}}\left[\exp\left(\frac{q_1 V_{oc1}}{N_{s1} K_1 A_1 T_0}\right) - 1\right] \tag{3}$$

But consideration of the actual ideal circuit is not possible as there is addition of series and parallel resistance in the circuit [5]. The series resistance is due to losses

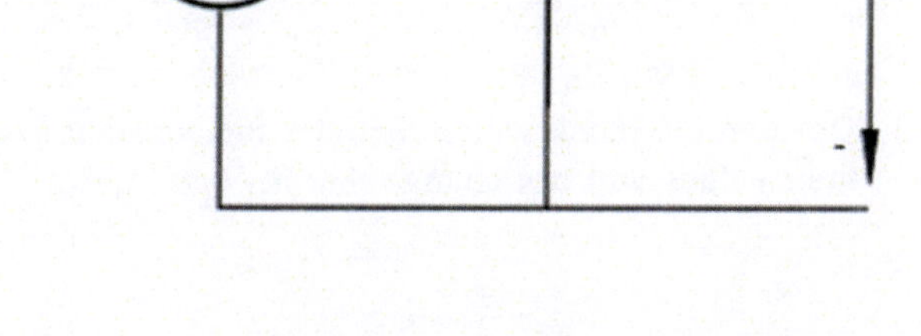

Fig. 1 Ideal structure of a PV cell [6]

for Joules' effect that is losses due to flowing of current in metal grid, semiconductor material, collecting bus and terminal. The parallel resistance is due to seepage of current due to cell thickness and surface effect. So, the above equations are modified into following equations:

$$I_{d1} = I_{s1}\left[\exp\left(\frac{q_1(V_1 + I_1 R_{s1})}{N_{s1} K_1 A_1 T_0}\right) - 1\right] \tag{4}$$

$$I_1 = I_{ph1} - I_{s1}\left[\exp\left(\frac{q_1(V_1 + I_1 R_{s1})}{N_{s1} K_1 A_1 T_0}\right) - 1\right] \tag{5}$$

If we consider the solar cell is acting in series parallel manner, then the above Eq. (5) is transformed to a new equation shown as

$$I_1 = N_{p1} * I_{ph1} - N_{p1} * I_{s1}\left[\exp\left(\frac{q_1(V_1 + I_1 R_{s1})}{N_{s1} K_1 A_1 T_0}\right) - 1\right] \tag{6}$$

There are various parameters that we need to describe as they are not described in the above equation I_{ph1} that is photocurrent of PV cell due to change in irradiance, the reverse saturation current I_{rs1} and the saturation current I_{s1} which is given by the below equations:

$$I_{ph1} = [I_{sc1} + K_{i1}(T_{01} - T_{r1})] * \frac{G_1}{G_{ref1}} \tag{7}$$

$$I_{rs1} = \frac{I_{sc1}}{\left[\exp\left(\frac{q_1 V_{oc1}}{N_{s1} K_1 A_1 T_0}\right) - 1\right]} \tag{8}$$

$$I_{s1} = I_{rs1}\left[\frac{T_{01}}{T_{r1}}\right]^3 \exp\left[\left(\frac{q_1 E_{g1}}{A_1 K_1}\right)\left(\frac{1}{T_{r1}} - \frac{1}{T_{01}}\right)\right] \tag{9}$$

These above equations are used to design the model of solar PV cell in MATLAB, and inputs are given from the datasheet, that is shown in Table 1, and output is obtained, and comparative analysis is done between the datasheet parameters and the generated model (Fig. 2; Table 2).

6 Steps Associated for Making Simulation

1. First develop the temp conversion model for converting the temperature from Celsius to Kelvin.
2. Develop the block which includes the constant that is used in the formulas and their values will not change that is $N_{s1} K_1 A_1 T_{01}$.

Table 1 Given parameters of solar PV used module [8]

S. no.	Parameters	Denoted Name	Values used
1	Max. Power	P_{m1}	320 W
2	Max. Voltage	V_{m1}	37.38 V
3	Max Power Current	I_{m1}	8.56A
4	O.C Voltage	V_{oc1}	46.22 V
5	S.C Current	I_{sc1}	9.06A
6	No. of Series cells	N_{s1}	72
7	No. of Parallel cell	N_{p1}	1
8	Ideality factor of diode	A_1	1.3
9	Temp. coefficient of cell	K_{i1}	0.058/ °C
10	Ref. Temp.	T_{ref1}	25C
11	Solar irradiance	G_{ref1}	1000 at STC

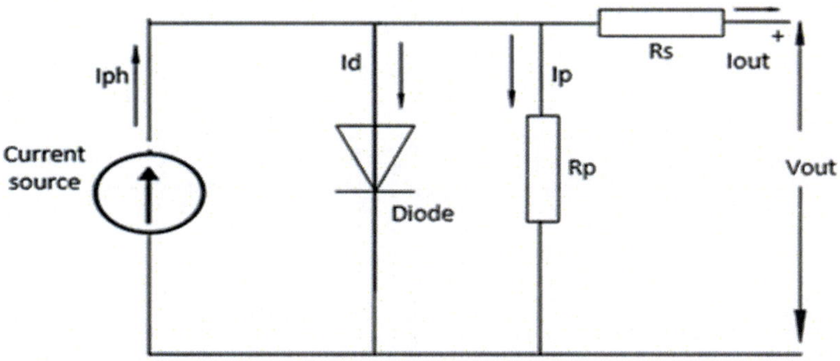

Fig. 2 Circuit diagram of a practical PV cell [7]

Table 2 Comparing the parameter of datasheet of JAP6-72/320/4BB with the simulated model

S. no.	Parameters	Datasheet values	Simulated model values	Rel. error (%)
1	P_{m1}	320 W	316.4 W	1.125
2	I_{sc1}	9.06 A	9.05 A	0.11
3	V_{m1}	37.38 V	41.91 V	12.1
4	I_{m1}	8.56 A	7.551A	11.7
5	V_{oc1}	46.22 V	46.20 V	0.04

3. Develop the photocurrent model.
4. Develop reverse saturation current mode and then saturation current mode.
5. Develop output current mode.

7 Simulated Model

The simulated model is shown in below (Figs. 3, 4, 5) [9]:

8 Results and Discussion

From the above-generated simulation and graph, comparison between the performances of the simulated model of the PV cell with the datasheet values of the practical solar cell can be made easily [11]. Here, single diode representation is basically used to represent the PV model. This can also be replaced by double diode mechanism, but due to cost effective, this policy is considered. Values of power, voltage and current are measured by keeping once irradiance constant and other time temperature constant. In this simulation model, the input required is irradiance, voltage and temperature taken from datasheet. We have set the value of irradiance

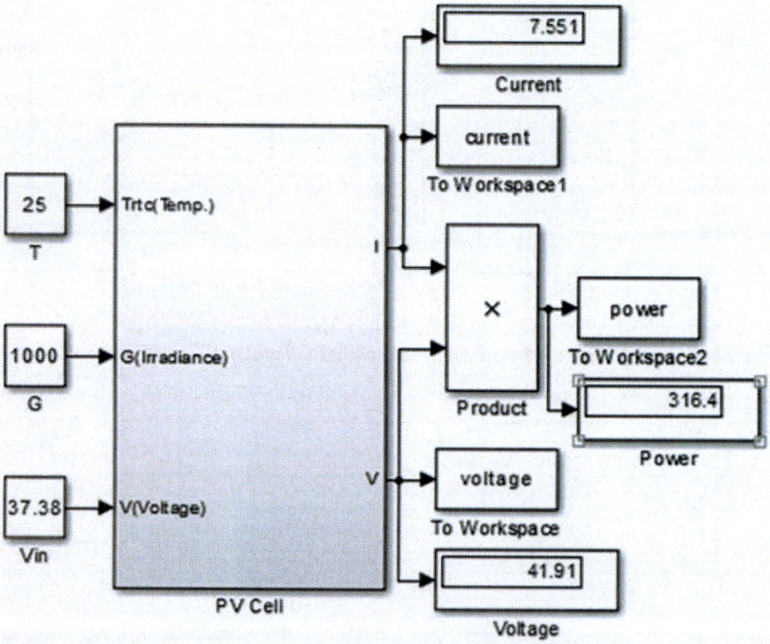

Fig. 3 PV model

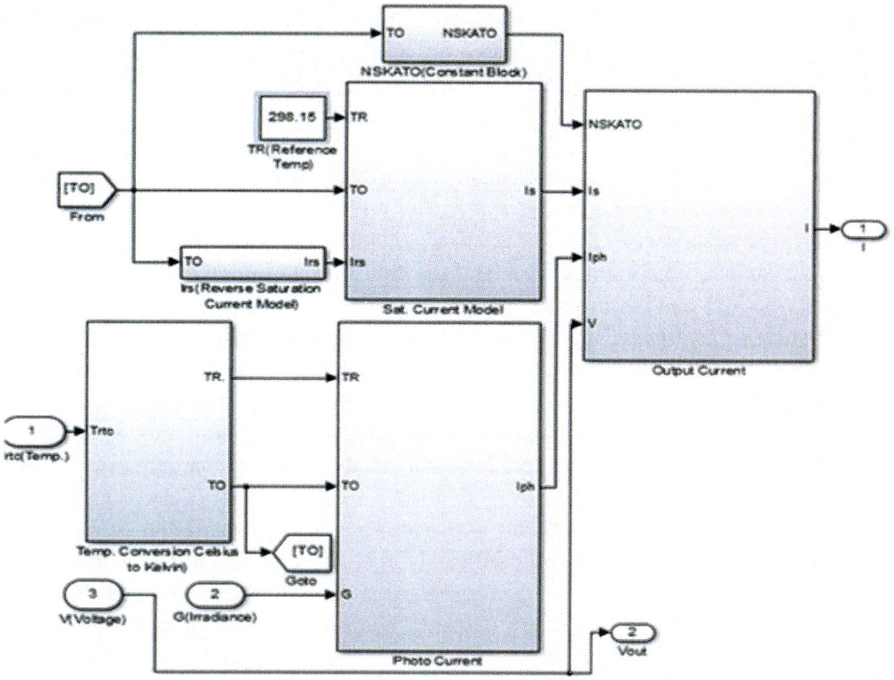

Fig. 4 Inner structure model of PV cell

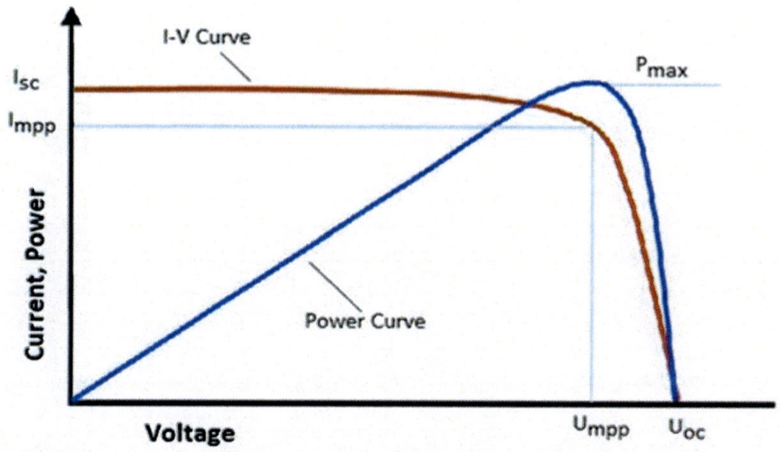

Fig. 5 I-V and P-V graph of a PV cell [10]

as 1000 W/m^2, voltage as 37.38 V and temperature as 25 °C, and we measured the power as the output, as for this case, only the system will have the maximum efficiency. By substituting different values by hit and trial process, this state can easily be obtained. From the datasheet of the practical solar cell, the output comes to 320 W, but the simulated output comes to be 316.4 having an error of 1.125%. So, the power that can be extracted is well enough stating the efficiency of the model is quite high and this model can easily replace the existing one at lower cost. Further, it can be observed that the voltage generated at the output is a bit more than input, so it steps up the voltage a bit, and the simulated current value differs than the practical one and has the error of 11.7%.

The values of various parameters like current, voltage and power vary same as that of ideal exixting solar cell while keeping either temperature or irradiance constant. When keeping the temp at 25 °C and changing the irradiance, we can see the changes in parameters in Table 3 [12], and when keeping the irradiance at 1000 W/m^2, we change the temperature then the change in parameters which is shown in Table 4 [13].

When the temp remains constant and with decreasing the irradiance that is the solar radiation is becoming less, then it can be realized that the current value and the voltage value both decrease, so power will also decrease with decrease in irradiance. When keeping the irradiance constant and changing the temperature that is when we increase warmth of the surrounding, it can be seen there is decrease in voltage very slowly with the decrease in current. So, this is exactly similar to the characteristics of an ideal PV cell.

Table 3 Values of change in power and current with change in irradiance keeping temperature 25 °C

S. no.	Irradiance (W/m^2)	Current (A)	Voltage (V)	Power (W)
1	1000	7.551	41.91	316.4
2	800	6.177	41.09	253.8
3	600	4.696	40.20	188.7
4	400	3.124	39.25	122.6
5	200	1.480	38.27	56.64

Table 4 Values of change in power and current with change in temperature keeping irradiance 1000 W/m^2

S. no.	Temperature (°C)	Current (A)	Voltage (V)	Power (W)
1	25	7.551	41.91	316.4
2	30	6.766	41.44	280.4
3	35	5.766	40.84	235.6
4	40	4.581	40.13	183.8
5	45	3.234	39.32	127.2

9 Conclusion

The single diode detailed modelling [14] from its construction till the simulation performed for taking the results is explained in a stepwise manner. The model has been simulated by using MATLAB software [15]. The maximum error while making this model is 12.1%. This model can be used to test also all the different practical PV cell by basically changing the input parameters. This simulated model can also help to calculate the amount of power that can be generated throughout the year getting the value of irradiance and temperature of that month. From the practical model, values are taken and given as an input to the simulated model and how the model behaves from the original existing module is depicted in this paper with simulation and proper explanation.

References

1. Location wise regional summary of all India installed capacity of utility power stations. Available in www.cea.nic.in/reports/monthly/installedcapacity/2020/installed_capacity-03.pdf
2. Peng C, Wu M, Yue D, (2018) Working Region and Stability Analysis of PV Cells Under the Peak Current-mode Control. In: 2018 IEEE Transactions on Control System Technology, pp 1–8
3. Shannan NM, Yahaya NZ, Singha B, Salam Z, Ahmed KY (2017) Theoretical Analysis of the Series Parasitic Resistance in Photovoltaic Cell. In: 2017 IEEE 8th Control and System Graduate Research Colloquium (ICSGRC 2017), pp 73–76
4. Morsy KA, Abouelatta MA, El-Banna MM, Elsaid MK (2018) Modelling and Simulation of Photovoltaic Module for Space Applications using Simulink/MATLAB. In: 2018 First International Workshop on Deep and Representation Learning (IWDRL), 29 March, pp 55–61
5. Nascimento AJ, Cavalcanti MC, Bradaschia F, Silva EA, Michels L, Pietta LP (2018) Parameter Estimation Technique for Double Diode Model of Photovoltaic Module. In: 2017 Brazilian Power Electronics Conference (COBEP), 5 January 2018, pp 978–984
6. Selmi T, Abdul-Niby M, Devis L, Devis A (2014) P&O MPPT Implementation Using MATLAB/Simulink. In: 2014 9th International Conference on Ecological Vehicles and Renewable Energies (EVER), 26 June 2014, pp 1–8
7. Ahmad T, Sobhan S, Nayan F (2016) Comparative Analysis between Single Diode and Double Diode Method of PV cell: Concentrate Different Parameters Effect on its Efficiency, J Power and Energy Engineering, January 2016, pp 31–46
8. Vinod RK, Singh SK Solar Photovoltaic Modelling and Simulation: As a renewable energy solution, In: Energy Report, Research Paper in Elsevier, Available in ScienceDirect, pp 710–712
9. Bana S, Saini RP A mathematical modelling framework to evaluate the performance of Single Diode and Double Diode based SPV Systems. In: Energy Report, Research Paper in Elsevier. Available in ScienceDirect, pp171–187
10. Rodrigues MG, Godina R, Pouresmaeil E, Porto JPS (2017) Simulation Study of a Photovoltaic Cell with Increasing level of Model Complexity. In 2017 IEEE International Conference on Environmental and Electrical Engineering and 2017 IEEE Industrial and Commercial Power Systems Europe (EEEIC/I&CPS Europe), 6–9 June 2017, pp 1–4

11. Louzazni M, Khouya A, Amechnoue K, Gandelli A, Mussetta M, Craciunescu A (2018) Metaheuristic Algorithm for Photovoltaic Parameters: Comparative Study and Prediction with a Firefly Algorithm. In: MDPI Applied Science. pp 1–22
12. Ibharim MM, Jiya JJ, Idakwo HO Characterization of PV-Cell Using Real Data. Int J Eng Develop Res 4(2):1696–1703
13. Chandel TA, Yasin MY, Mallick MA (2019) Modelling and Simulation of Photovoltaic Cell Using Single Diode Solar Cell and Double Diode Solar Cell Model. Int J Inno Technol Exp Eng 8(10):558–565
14. Selmi T, Niby MA, Devis L, Davis A (2014) P&O MPPT Implementation Using MATLAB/Simulink. In: 2014 Ninth International Conference on Ecological Vehicles and Renewable Energy (EVER), pp 1–4
15. Ali RA, Yasin NM (2019) Simulation of Solar Cell and sinusoidal pulse width modulation inverter Using MATLAB and Proteus. In: 4th Scientific International Conference-Najaf-IRAQ-2019, pp 86–91

Computational Analysis of Efficiency for Non-toxic Perovskite Solar Cell with Inorganic HTMs

Arnab Das and A. B. Deoghare

1 Introduction

In 1896, Swedish scientist, Svante Arrhenius stated that the use of fossil fuel could contribute to global warming [1]. Over the last few decades, this issue has become a hot-buttoned topic. A number of alternative sources of energy have been investigated since then, which also resulted in lower emissions, lesser fuel price and less pollution.

Solar energy is a vital source of renewable energy within which light and heat from the Sun are harnessed employing a variety of ever-evolving technologies like photovoltaics, solar heating, solar architecture, solar thermal energy, molten salt power plants and artificial photosynthesis. In the present scenario, solar cells made of silicons are widely used for commercial purpose. For the manufacturing and production of silicon-based solar cells, sophisticated fabrication techniques are required which makes the solar panel costlier. But recent researches have shown that perovskite materials can be used as a replacement of silicon-based cells as these materials are simple and cheap to produce. Perovskites act as active absorber layer in perovskite solar cells (PSC) to produce electricity from the light and heat of the Sun [2].

However, there are few challenges that need to be addressed before perovskites displace silicon from its dominant position in photovoltaic industry. These are substitution of lead (Pb) by a non-toxic element and resisting the degradation by replacing the expensive hole transport material (HTM) made from 2,20,7,70-tetrakis (N,N-di-p-meth oxyphenylamine)-90, 9-spirobifluorene (spiro-OMETAD). Persistent efforts are being utilized around the globe to handle the above issues. The replacement of lead may be done by replacing the toxic lead atom by selected metals either having similar ionic radius and/or outer electron shell

A. Das · A. B. Deoghare (✉)
NIT Silchar, Silchar 788010, Assam, India

S. Bag et al. (eds.), *Next Generation Materials and Processing Technologies*, Springer Proceedings in Materials 9, https://doi.org/10.1007/978-981-16-0182-8_34

structures. Replacing lead (Pb) by tin (Sn) has shown the foremost effective results. However, it is found that simply replacing lead (Pb) by other metals will not lead to higher photo conversion efficiencies PCEs [3].

On the other hand, spiro-OMETAD is used in perovskite-based solar cells very frequently as HTM. It is found that this material is moisture-sensitive. It is also proved that ultraviolet light and moisture cause degradation within the perovskite-based cells. Hence, it can be said that spiro-OMETAD is liable for the degradation of the solar cells [4, 5]. A reasonable amount of literature can be found dedicated to the effect of HTM layer on the stability of the perovskite-based cells [6–8]. There are also some published works based on the degradation of the performance of dye-sensitized cells (DSC) because of the presence of spiro-OMETAD [9, 10]. The problem is related to the oxidizing effects [7, 8] and thermal stress effects [6, 9, 10]. From this, it may be concluded that the main disadvantage is related to the material itself. To resolve these issues, some inorganic materials are suggested to be used as HTM layer, like nickel oxide (NiO) [11], copper thiocyanate (CuSCN) [12], cuprous oxide (Cu_2O) and copper iodide (CuI) [13].

In this study, SCAPS-1D, developed by the University of Gent, was used to calculate the key parameters of tin iodide perovskite photovoltaic cell with the subsequent configuration: Glass/FTO/TiO_2/$CH_3NH_3SnI_3$/HTM/Au, HTM being NiO, CuSCN, CuI, Cu_2O and spiro-OMETAD. This work focuses on searching for the effects of the absorber thickness layer, HTM thickness, absorber layer defect density on the cell to predict the performance of the device. It should be beneficial to device design and understand the operation mechanism, which may be helpful for innovative experimental studies and fabrication processes.

2 Analysis of the Device

2.1 Method

The construction of perovskite solar cells (PSCs) is almost like most of the inorganic thin-film solar cells. The excitation is Wannier-type in perovskite materials. Within the case of inorganic thin-film solar cells, the excitation type is also similar. Simulation software is often used as a tool for understanding of the device mechanism and determining the optimum design. Solar cell capacitance simulator (SCAPS) is such kind of simulation software which was widely applied for inorganic semiconductor photovoltaic cell devices like silicon, CdTe, $Cu(In,Ga)(S,Se)_2$ (CIGS) and $Cu_2ZnSn(S,Se)_4$, and it is often accustomed for simulating PSCs, based on the two similarities in excitation type and device configuration as discussed earlier. SCAPS is a one-dimensional solar cell simulation program which uses three differential coupled equations that are Poission's equation, continuity equation for electrons and holes.

The equations are

$$\frac{d}{dx}\left(-\varepsilon(x)\frac{d\psi}{dx}\right) = q\left[p(x) - n(x) + N_d^+(x) - N_a^-(x) + p_t(x) - n_t(x)\right] \quad (1)$$

$$\frac{dn_p}{dt} = G_n - \frac{n_p - n_{po}}{\tau_n} + n_p\mu_n\frac{d\xi}{dx} + \mu_n\xi\frac{dn_p}{dx} + D_n\frac{d^2n_p}{dx^2} \quad (2)$$

$$\frac{dp_n}{dx} = G_p - \frac{p_n - p_{no}}{\tau_p} - p_n\mu_p\frac{d\xi}{dx} - \mu_p\xi\frac{dp_n}{dx} + D_p\frac{d^2p_n}{dx^2} \quad (3)$$

where x is the direction along the thickness, q is electron charge, G is generation rate, ε is permittivity, ψ is electrostatic potential, ξ is electric field, and D is diffusion coefficient. $n(x)$, $p(x)$, $n_t(x)$, $p_t(x)$ are free electrons, free holes, trapped electrons and trapped holes, respectively. $N_a^-(x)$ refers to ionized acceptor-like doping concentration, and $N_d^+(x)$ stands for ionized donor-like doping concentration.

2.2 Modeling of the Device

Figure 1 shows the configuration of the photovoltaic cell, used for simulation, where different colors indicate different layers and their thickness. For the HTM layer, different HTMs have been used. These are Cu_2O, CuI, NiO, CuSCN and the traditional spiro-OMETAD. The thickness of each layer is varied with different configurations to get the best results. For the thickness of the perovskite layer, the thickness is varied from 50 to 800 nm, the HTM thickness is varied from 50 to 500 nm, ETM (TiO_2) layer thickness is chosen to be within the range of 50–200 nm, and finally, FTO thickness is varied from 50 to 500 nm for each simulation with different HTMs. For the left contact end, gold (Au) is chosen, and glass is used as the transparent right contact material. The input parameters for different layers are shown in Table 1. These parameters are collected from various experimental data [14–16].

2.3 Simulation Procedure

The thicknesses of the various layers of the cell were optimized by following a certain algorithm as shown in Fig. 2. For the primary iteration, the thicknesses of the ETM layer and the perovskite layer were set to 100 and 500 nm, respectively. For perovskite solar cells having spiro-OMETAD as HTM, these values are found as optimum values [17, 18]. Using these thicknesses for ETM and perovskite layer,

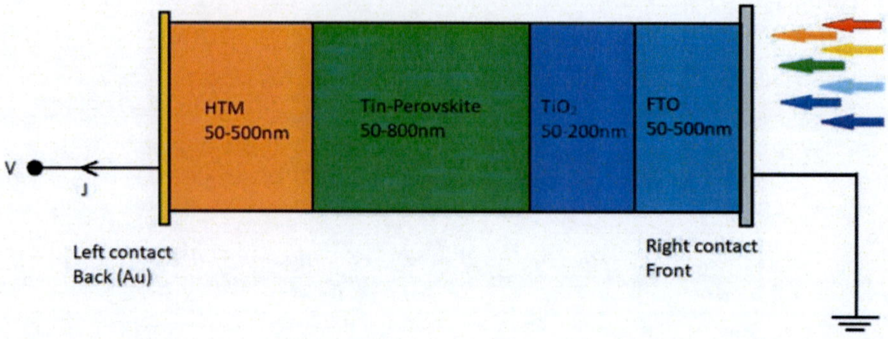

Fig. 1 Configuration of the solar cell structure

Table 1 Input parameters of all the materials for SCAPS simulation

	TiO_2	$CH_3NH_3SnI_3$	Spiro-OMETAD	$p\text{-}Cu_2O$	p-CuI	p-CuSCN	p-NiO
Die-electric permittivity	10	8.2	3	7.11	6.5	10	10.7
Electron mobility (cm^2/Vs)	100	1.6	1e-4	2e2	100	100	12
Hole mobility (cm^2/Vs)	25	1.6	2e-4	8e1	43.9	25	2.8
Acceptor concentration ($1/cm^3$)	0	1e18	1e18	1e18	1e18	1e18	1e18
Donor concentration ($1/cm^3$)	1e17	1e18	0	0	0	0	0
Bandgap (eV)	3.26	1.30	3.06	2.17	3.1	3.6	3.8
CB DOS ($1/cm^3$)	2e17	2.2e18	2.8e19	2.02e17	2.8e19	2.2e19	2.8e19
VB DOS ($1/cm^3$)	6e17	1.8e19	1e19	1.1e19	1e19	1.8e18	1.0e19
Affinity (eV)	4.2	4.17	2.05	3.20	2.1	1.7	1.46
Band to band recombination rate (cm^3/s)	2.3e-9		2.3e-9				2.3e-9

the HTM thickness is optimized by an iterative process. Thus, the optimum thickness of the HTM was identified. Using this value for HTM thickness, iterations were applied again to get the new optimized thickness value for the ETM layer. This method was done again and again to search out the set of optimum values for ETM and HTM layer thicknesses for various combinations of the considered cell structures. Finally, the optimum thickness for the absorber layer was computed using the set of optimized ETM and HTM thickness values. However, it

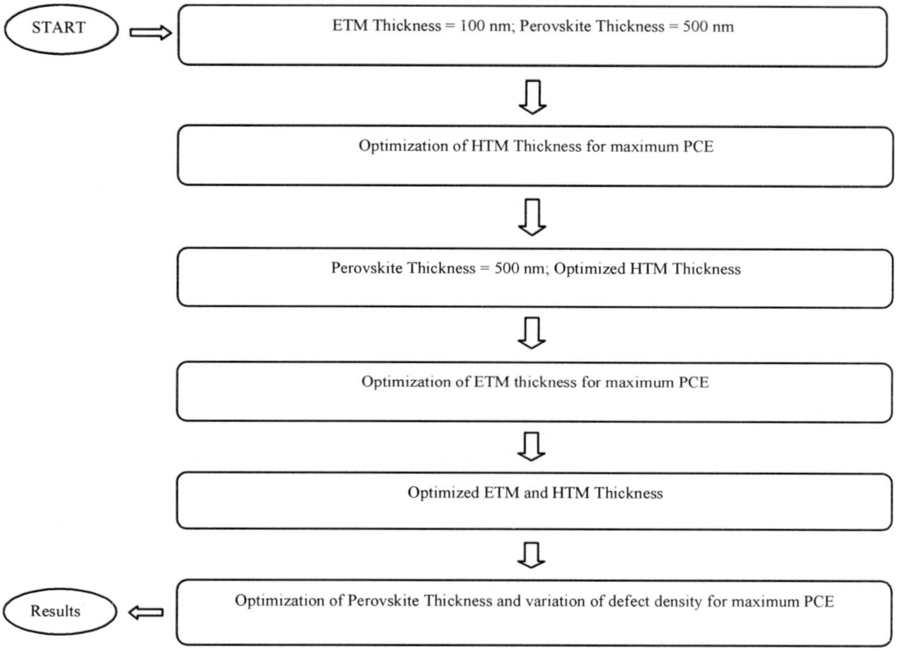

Fig. 2 Flowchart of the simulation procedure

should be noted that these optimum values may be varied by varying the carrier mobilities. After finding out these values, the defect density of the absorber layer is varied to find out the best suitable configuration for maximum efficiency. Finally, these values are compared, and the possible outcome is tried to predict for all the configurations, which is discussed in the result and discussion section. A flowchart of the simulation procedure is shown in Fig. 2 for easy understanding of the whole process.

3 Results and Discussion

The modeling of $TiO_2/CH_3NH_3SnI_3$/HTM/FTO perovskite solar cell structure is presented in this work, while in case of HTM material, various inorganic materials like Cu_2O, CuI, CuSCN, NiO have been tried, and the results are compared with the traditionally used spiro-OMETAD HTM material. Methyl ammonium tin iodide ($CH_3NH_3SnI_3$) is a non-toxic material and can be used for commercial purpose. The analysis was performed in SCAPS-1D software, and various parameters like the J-V characteristics, FF characteristics, PCE were found out. It was found that, Cu_2O and CuI, when used as HTM have outperformed the performance of the solar cell using

Table 2 Optimized thickness for ETM, HTM and absorber layer

Thickness	Spiro-OMETAD as HTM	Cu_2O as HTM	CuI as HTM	CuSCN as HTM	NiO as HTM
HTM	50 nm	100	100	250	200
Perovskite	700 nm	750	650	400	400
ETM (TiO_2)	100 nm	100	100	100	100
FTO (nm)	500	500	500	500	500
PCE (%)	22.33	22.50	22.51	18.58	21.21

spiro-OMETAD as HTM by achieving PCE more than 22.5%. All the simulations are conducted under AM 1.5G illumination, and the input parameters are listed in Table 1. The optimized cell configurations for all the materials as HTM are discussed in the Sect. 3.1, and the final thickness values are listed in Table 2.

3.1 *Best Suitable Configuration*

Simulations were performed by varying the thickness of the materials used for each configuration. From every simulations, the data are collected and finally analyzed, so that best suitable configuration for each of the materials can be predicted. For the analysis, a certain algorithm is used, which is depicted in the methodology section. The optimized cell configuration is chosen on the basis of maximum PCE obtained for each case. The details of the thickness of each material for all the configuration are listed in Table 2.

It should be noted that, all the configuration details shown in the table are for maximum PCE obtained for each case and these results were found when the defect density of the absorber layer was minimum. The effect of defect density is discussed later. But, from Tables 3, 4, 5 it is noticeable that, the configurations of spiro-OMEATD HTM cell, Cu_2O HTM cell and CuI HTM cell are almost similar. While the optimized thickness of the perovskite absorber layer lies in between 650 and 750 nm for all the three configurations, and HTM layer thickness lies in between 50 and 100 nm. The ETM layer thickness was found to be 100 nm for best possible outcome for all the three configurations. Again, it was also found that, the cells having NiO and CuSCN as HTM are having almost similar configuration, as the perovskite layer thickness and the ETM layer thickness for these two cells were found to be 400 and 100 nm, respectively, for both the configuration. The HTM layer thickness was found to be lying in between 200 and 250 nm for NiO and CuSCN HTM cells. The maximum achievable PCE was found in the case of Cu_2O and CuI as HTM, as both of the cell configuration have achieved efficiency more than 22.5%. This is even found to be more than the efficiency achieved by using spiro-OMETAD as HTM. But, the cells, having CuSCN and NiO as HTM, were lagging behind in case of PCE obtained by these two cells, as the maximum achieved PCE by these two cells was 18.58% and 21.21%, respectively.

3.2 *J-V Characteristics*

The J-V curve created by simulation using SCAPS-1D software is shown in Figs. 3 and 4 for Cu_2O HTM cell and CuI HTM cell, respectively, under illumination. This curve is generated, due to carrier generation and separation because of incident of photons into the absorber layer. The J-V curve helps to find out the short circuit current, open circuit voltage, fill factor and the PCE of a certain cell configuration. The curve shown in Figs. 3 and 4 is automatically generated by the SCAPS-1D software, which shows the various parameters at the maximum PCE condition for Cu_2O and CuI HTM cells, respectively.

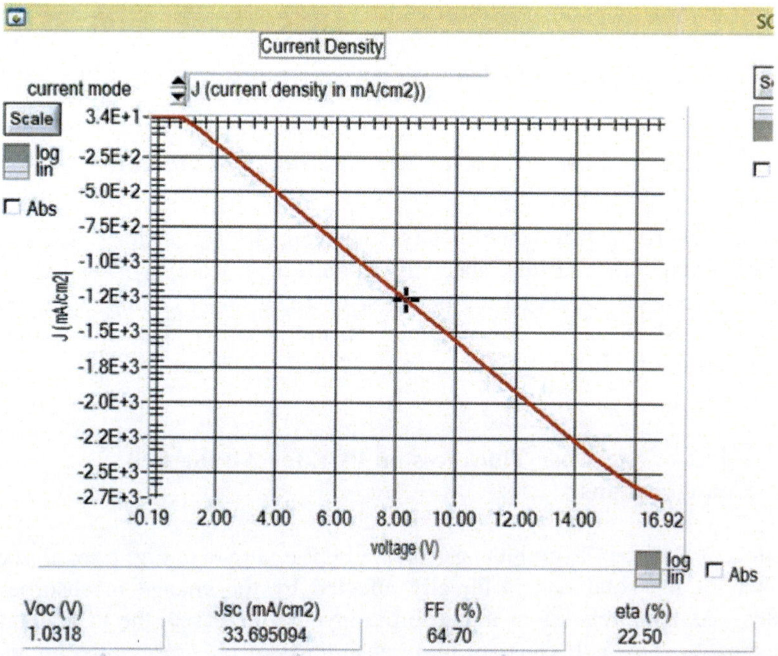

Fig. 3 J-V characteristics curve for Cu_2O HTM cell

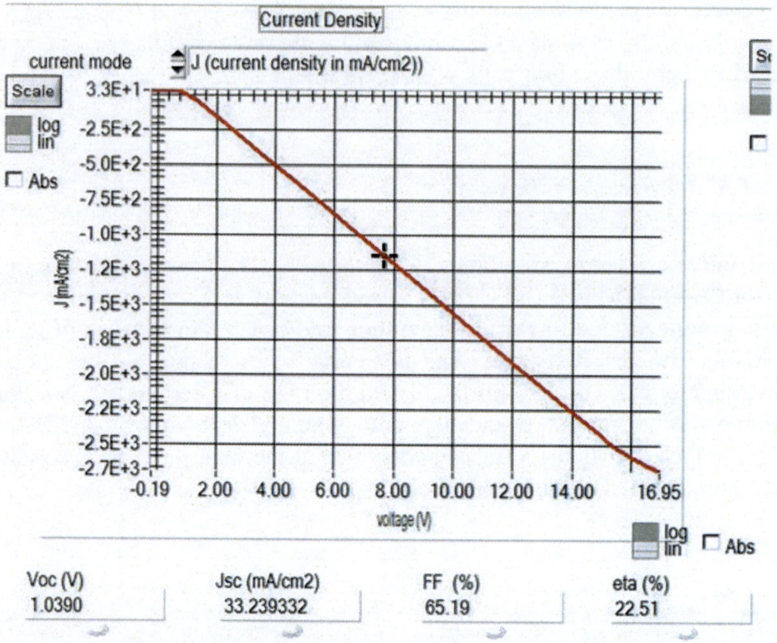

Fig. 4 J-V characteristics curve for CuI HTM cell

3.3 Effect of Absorber Layer on Various Parameters

To understand the performance of any solar cell, the J-V characteristics are not enough. Various other factors, specially, the effect of absorber layer plays a vital role, when it comes to predict the behavior of a cell under certain conditions.

The effect of the defect concentration on the absorber layer and its thickness were carried out to find out the maximum PCE for each configuration.

3.3.1 Effect of Absorber Thickness on PCE for All the Cell Configurations

The effect of absorber layer thickness on the PCE is shown in Fig. 5 for all the cells. The PCE of the solar cell is directly affected by the change in absorber layer thickness. As the thickness of the absorber layer is increased, the PCE obtained is also increased. This is because of more absorption of photons, as the thickness of the absorber layer is more. But, after reaching a certain limit, because of more recombination centers, the PCE starts to decrease. The same trend can be seen for all the cell configurations, as shown in Fig. 5. It can be noticed that the PCE vs perovskite thickness curve for the cells having spiro-OMETAD, Cu_2O and CuI is almost similar and all of these cell configurations have achieved cell efficiency more than 22%, while cells having Cu_2O and CuI have shown the best possible outcome. The cell having CuSCN has the lowest PCE achieved, whereas cell having NiO has reached the PCE of more than 21%, but this value of PCE lies for very short range of absorber thickness, as it falls drastically when the absorber layer thickness is further increased.

3.3.2 Effect of Absorber Defect Density on the PCE

The effect of absorber layer defect density is carried out in this step. For this analysis, the defect density of 10^{15}, 10^{14} and 10^{13} is chosen. Generally, when the defect density is increased, the minority carrier lifetime is reduced. With the reduction of lifetime, the diffusion length of the electron and hole also reduces, and thus, the recombination loss in the absorber layer also increases. Hence, the best value of PCE can be obtained by reducing the defect density in the absorber layer. Figure 6 shows the effect of defect density in the PCE for all the cell configurations. In Fig. 6, the PCE vs the perovskite absorber layer thickness curve is plotted for various defect density of the absorber layer. Figure 6a–e shows the PCE vs perovskite thickness curves for spiro-OMETAD, Cu_2O, CuI, NiO and CuSCN as HTM, respectively. The black line shows the curve for defect density 10^{13}, while the red and blue line show the curves for defect density 10^{14} and 10^{15}, respectively, for all the cell configurations. As expected, the best result could be seen for the configuration, having less defect density in the absorber layer. The parameters obtained by the defect density study are listed in Tables 3, 4, 5, 6, and 7.

From the analysis done by changing the defect density of the absorber layer, the final parameters that were found are listed in Tables 3, 4, 5, 6 and 7 for the cells having spiro-OMETAD, Cu_2O, CuI, NiO and CuSCN as HTM, respectively. From

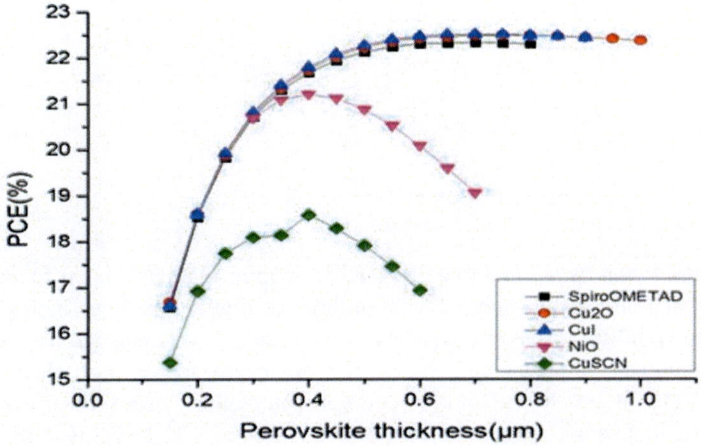

Fig. 5 Absorber layer thickness vs PCE

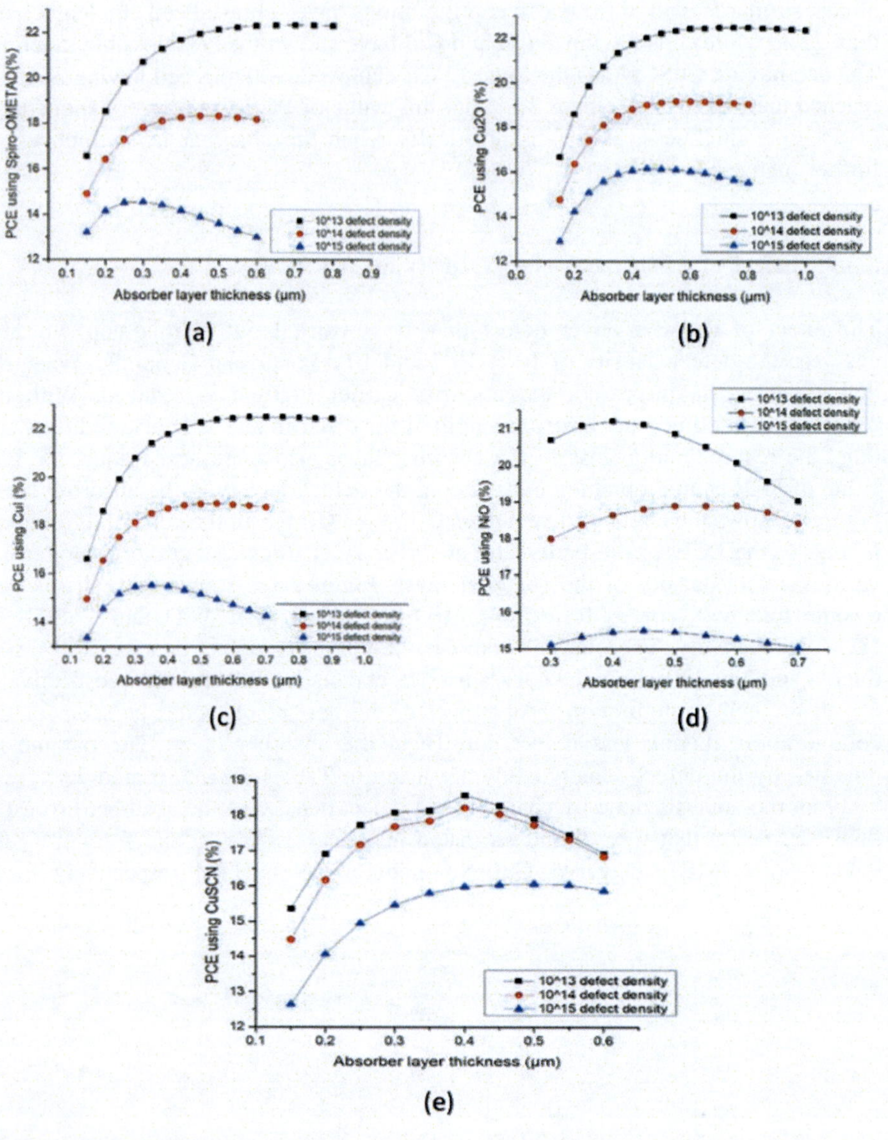

Fig. 6 Effect of defect density of absorber layer

these values, it can be noticed that the values for spiro-OMETAD, Cu_2O and CuI are very close. The fill factor values for these three configurations have increased as the defect density in the perovskite layer has been reduced. Whereas the fill factor value for cells having NiO and CuSCN as HTM has decreased as the defect density is increased. This is because of more recombination centers. From the above analysis, Cu_2O and CuI material can be chosen as HTM for the $CH_3NH_3SnI_3$ solar cell.

Table 3 Effect of defect density on spiro-OMETAD HTM-based cell

Findings	10^{15} defect density	10^{14} defect density	10^{13} defect density
Thickness of absorber	300 nm	450 nm	700 nm
Voc (V)	0.8587	0.9407	1.0357
Jsc (mA/cm^2)	28.340474	31.467798	33.489040
FF (%)	59.78	61.91	64.38
PCE (%)	14.55	18.33	22.33

Table 4 Effect of defect density on Cu_2O HTM-based cell

Findings	10^{15} defect density	10^{14} defect density	10^{13} defect density
Thickness of absorber	450 nm	550 nm	750 nm
Voc (V)	0.8325	0.9291	1.0318
Jsc (mA/cm^2)	31.403148	32.547903	33.695094
FF (%)	61.84	63.41	64.70
PCE (%)	16.17	19.18	22.50

Table 5 Effect of defect density on CuI HTM-based cell

Findings	10^{15} defect density	10^{14} defect density	10^{13} defect density
Thickness of absorber	350 nm	550 nm	650 nm
Voc (V)	0.8482	0.9315	1.0390
Jsc (mA/cm^2)	29.644683	32.535857	33.239298
FF (%)	61.85	62.48	65.19
PCE (%)	15.55	18.94	22.51

Table 6 Effect of defect density on NiO HTM-based cell

Findings	10^{15} defect density	10^{14} defect density	10^{13} defect density
Thickness of absorber	450 nm	600 nm	400 nm
Voc (V)	0.8179	1.4717	2.9791
Jsc (mA/cm^2)	32.370282	32.913131	30.717585
FF (%)	58.14	39.06	23.18
PCE (%)	15.39	18.92	21.21

3.3.3 Effect of Perovskite Thickness on Various Parameters

From the above analysis it could be said that, HTMs (CU_2O and CuI) can provide more power conversion efficiency (PCE) than traditionally used spiro-ometad HTM for solar cells using Perovskite material as $CH_3NH_3SnI_3$. The various parameters like Voc, Jsc and FF are found to be similar also for these three solar cell configurations. So, in this step, the analysis is carried out to find out the effect of

Table 7 Effect of defect density on CuSCN HTM-based cell

Findings	10^{15} defect density	10^{14} defect density	10^{13} defect density
Thickness of absorber	500 nm	400 nm	400 nm
Voc (V)	1.0742	2.6115	
Jsc (mA/cm^2)	31.954779	30.710304	30.717474
FF (%)	46.81	22.73	
PCE (%)	16.07	18.23	18.58

perovskite thickness on the open circuit voltage (Voc), short circuit current (Jsc), fill factor (FF) and PCE for the cells having Cu_2O and CuI as HTM. Figures 7 and 8 show the variation in the above-mentioned parameters with respect to change in absorber layer thickness for Cu_2O HTM cell and CuI HTM cell, respectively.

Figures 7 and 8 describe the effect of absorber perovskite layer thickness on the various parameters like: open circuit voltage, short circuit current, fill factor and photo conversion efficiency for the two types of cells, having Cu_2O and CuI as HTM, respectively. It is noticeable that, the both of the cells are having similar type of results for all the parameters. The Voc and FF curve tends to move downwards, as the thickness of the absorber layer is increased. This happens because of the creation of more recombination centers as the thickness of the absorber is increased.

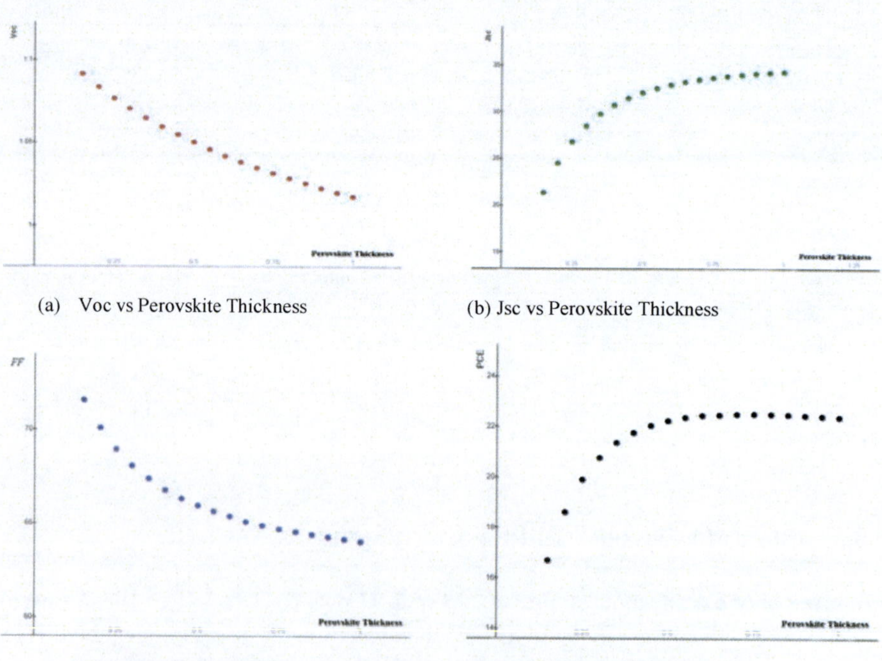

(a) Voc vs Perovskite Thickness

(b) Jsc vs Perovskite Thickness

(c) FF vs Perovskite Thickness

(d) PCE vs Perovskite Thickness

Fig. 7 Effect of perovskite thickness on various parameters for Cu_2O HTM cell

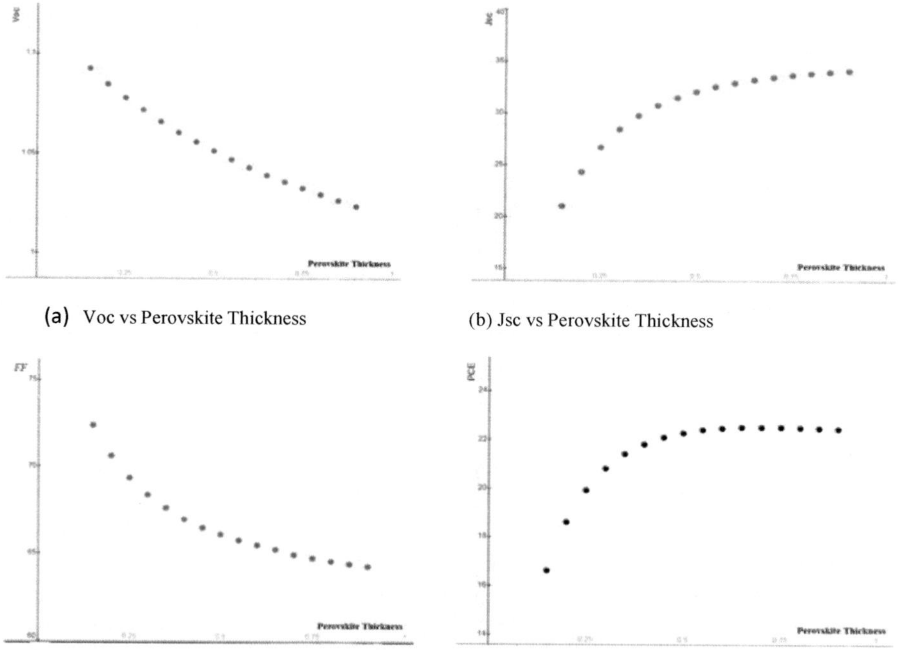

(a) Voc vs Perovskite Thickness

(b) Jsc vs Perovskite Thickness

(c) FF vs Perovskite Thickness

(d) PCE vs Perovskite Thickness

Fig. 8 Effect of perovskite thickness on various parameters for CuI HTM cell

The Jsc increases as the perovskite layer thickness is increased, because, as we increase the thickness of perovskite layer, more number of electron-hole couple is also generated. Same thing happens with the PCE curve, as the PCE value increases up to a certain point because of more number of e-h couple generations, and after that, because of creation of more recombination centers, the curve tends to move downwards.

4 Conclusion

Numerical modeling of non-toxic perovskite thin-film cell is carried out using solar cell capacitance simulator (SCAPS) software in this work. For prevention of degradation, different inorganic hole transport materials like Cu_2O, CuI, NiO and CuSCN are tried in the place of traditionally used organic hole transport material spiro-OMETAD. A comparison study of various parameters like photo conversion efficiency (PCE), open circuit voltage (Voc), short circuit current (Jsc) and fill factor (FF) is carried out for various cell configurations. From the J-V characteristics curve, generated by the SCAPS software, the data are collected, and maximum efficiency is found out for each configuration. It was found that, maximum efficiency for the cell configurations is 22.33, 22.50, 22.51, 21.21 and 18.58% for the cells having spiro-OMETAD, Cu_2O, CuI, NiO and CuSCN as HTM, respectively. It was also seen that, all the maximum efficiency was achieved when the absorber layer defect density was set to minimum (10^{13}). From this analysis, it can be concluded that, among the cell configurations, inorganic Cu_2O and CuI as HTM can outperform the organic spiro-OMETAD as HTM. But, only the J-V characteristic curve is not enough to understand any solar cell performance. To observe the performance of the cell, various other factors are also analyzed. A comparative study on the effect of the absorber layer on PCE of the cells has been carried out, and it was found that, the Cu_2O HTM cell and CuI HTM cell show similar behavior like spiro-OMETAD HTM cell, whereas the other inorganic HTM cells were lagging in case of PCE. From all these analysis, it could be concluded that, Cu_2O and CuI can be used as HTM as a replacement of spiro-OMETAD, to reduce the degradation along with $CH_3NH_3SnI_3$ perovskite absorber material to remove the toxicity from the solar cell. For better prediction of the cell performance, the effect of the absorber thickness is also analyzed on the various parameters like Voc, Jsc and FF for Cu_2O and CuI HTM cells.

Based on the different device structure modeling, solar cell with structure Cu_2O/$CH_3NH_3SnI_3$/TiO_2/FTO and CuI/$CH_3NH_3SnI_3$/TiO_2/FTO can exhibit an efficiency of 22.5% and the fill factor of more than 64% with the optimized physical parameters like HTM and ETM thickness of 100 nm for both configuration, and absorber thickness of 750 nm and 650 nm, respectively. The proposed result will give a valuable guideline for the feasible fabrication and designing of high power conversion efficiency non-toxic perovskite solar cell.

References

1. https://en.wikipedia.org/wiki/Svante_Arrhenius
2. https://www.energy.gov/eere/solar/perovskite-solar-cells#:~:text=Perovskite%20solar%20cells%20convert%20ultraviolet,efficiently%20convert%20lower%2Denergy%20light
3. Wang K, Liang Z, Wang X, Cui X (2015) Lead Replacement in $CH_3NH_3PbI_3$ Perovskites. Adv Electron Mater 1(10):1–8
4. Nogueira AF, Longo C, De Paoli MA (2004) Polymers in dye sensitized solar cells: Overview and perspectives. Coord Chem Rev 248(13–14):1455–1468
5. Pudzich R, Fuhrmann-Lieker T, Salbeck J (2006) Spiro compounds for organic electroluminescence and related applications. Adv Polym Sci 199(1):83–142
6. Malinauskas T et al (2015) Enhancing thermal stability and lifetime of solid-state dye-sensitized solar cells via molecular engineering of the hole-transporting material spiro-OMeTAD. ACS Appl Mater Interfaces 7(21):11107–11116
7. Badia L, Mas-Marzá E, Sánchez RS, Barea EM, Bisquert J, Mora-Seró I (2014) New iridium complex as additive to the spiro-OMeTAD in perovskite solar cells with enhanced stability. APL Mater 2(8)
8. Nguyen WH, Bailie CD, Unger EL, McGehee MD (2014) Enhancing the hole-conductivity of spiro-OMeTAD without oxygen or lithium salts by using spiro(TFSI)2 in perovskite and dye-sensitized solar cells. J Am Chem Soc 136(31):10996–11001

9. Bailie CD, Unger EL, Zakeeruddin SM, Grätzel M, McGehee MD (2014) Melt-infiltration of spiro-OMeTAD and thermal instability of solid-state dye-sensitized solar cells. Phys Chem Chem Phys 16(10):4864–4870
10. O'Brien DF, Burrows PE, Forrest SR, Koene BE, Loy DE, Thompson ME (1998) Hole Transporting Materials with High Glass Transition Temperatures for Use in Organic Light-Emitting Devices. Adv Mater 10(14):1108–1112
11. Subbiah AS, Halder A, Ghosh S, Mahuli N, Hodes G, Sarkar SK (2014) Inorganic hole conducting layers for perovskite-based solar cells. J Phys Chem Lett 5(10):1748–1753
12. Qin P et al (2014) Inorganic hole conductor-based lead halide perovskite solar cells with 12.4% conversion efficiency. Nat Commun 5(May):1–6
13. Christians JA, Fung RCM, Kamat PV (2014) An inorganic hole conductor for Organo-lead halide perovskite solar cells. improved hole conductivity with copper iodide. J Am Chem Soc 136(2):758–764
14. Ahmed S, et al (2019) Simulation studies of non-toxic tin-based perovskites: Critical insights into solar performance kinetics through comparison with standard lead-based devices. Superlattices Microstruct 130(December):20–27
15. Hossain MI, Alharbi FH, Tabet N (2015) Copper oxide as inorganic hole transport material for lead halide perovskite based solar cells. Sol Energy 120:370–380
16. Tan K, Lin P, Wang G, Liu Y, Xu Z, Lin Y (2016) Controllable design of solid-state perovskite solar cells by SCAPS device simulation. Solid State Electron 126:75–80
17. Snaith HJ (2013) Perovskites: The emergence of a new era for low-cost, high-efficiency solar cells. J Phys Chem Lett 4(21):3623–3630
18. Edri E, Kirmayer S, Mukhopadhyay S, Gartsman K, Hodes G, Cahen D (2014) Elucidating the charge carrier separation and working mechanism of CH 3 NH 3 PbI 3-x Cl x perovskite solar cells. Nat Commun 5:1–8

Optimization

Optimization of Process Variables of Twin-Screw Extruder Using Response Surface Methodology for the Production of Fish Added Extruded Snack Product

R. Pradeep, K. Rathnakumar, and P. Karthickumar

1 Introduction

The micro and small food processing industries gaining appreciation and acknowledgments from various governmental and non-governmental organizations; nowadays, it helps in achieving sustainable food production with nutritional stability for the nation [1]. Since it enables the opportunities for newcomers to start up a profitable business, Industry 4.0 will also help industries to automate the process and carrying out better management practices to such a trade [2]. The micro and small food industries, besides its small investment, always look further to develop new and innovative products to gain attention and appreciation from the consumers, which led to the competitiveness among the food processors to address the primary concern in any existing food products.

One such product called expanded crisp snack products, commercially called Kurkure, Cheetos, which is a famous snack among all age groups of people because of its lucrative and scrumptious nature [3]. Children are fonded of this snack than adults. The primary concern of this popular snack is lack of nutrition, even it has a high market value. Hence, fortification of such products is needed to improve its nutritional quality up to an acceptable level, which makes the product with a considerable market and nutritional value.

The primary fishery in India includes emperors, seer fish, groupers, tuna, snappers, goat fishes, and anchovies that fetch a reasonable price for fishers. Besides, these enormous quantities of low-value fishes like lesser sardines, oil sardine's sciaenid, skipjack tuna, leather jacket, silver bellies, lizard fishes, catfishes are also landed. However, they are not effectively utilized and predominantly shifted for fishmeal or the dry fish market as fish protein is rich in all essential amino acids irrespective of the size of the fish [4]. Small varieties of fishes are not relished

R. Pradeep (✉) · K. Rathnakumar · P. Karthickumar
College of Fisheries Engineering, Tamil Nadu Dr. J. Jayalalithaa Fisheries University, Nagapattinam, India

S. Bag et al. (eds.), *Next Generation Materials and Processing Technologies*, Springer Proceedings in Materials 9, https://doi.org/10.1007/978-981-16-0182-8_35

because of pin bones and are generally converted into meal for animal feed. Hence, the development of seafood-based snack products can enhance human consumption of these underutilized fishes rather than being utilized for other non-human utilizations [5]. This paves the way to the production of nutritious snacks, as mentioned above, by the addition of low-value fish to high market value snacks.

Several kinds of research were focused on the production of fish added extruded snacks, and its nutritional stability rendered the proper procedure for the production of good nutritional quality extruded snacks [6]. Extruded fish snack products were developed using extrusion technology with cooking formulation consisting of fish flour and cereals mixture were extruded at a moisture content of 15%, screw speed 480 rpm, sectional barrel temperature of 30, 60, 130, and 160 °C at four stages and 2 mm diameter of the die. The extruded product developed was fried using edible oil, and the resulting extruded product was analyzed for a physical characteristic, texture profile, proximate composition, microbiological analysis, and sensory acceptability [5]. However, still commercial fish added extruded snack products are not available in the market because of the less product acceptability and process variable optimization. The color, odor, puffiness, and other sensory properties of fish added extruded snacks are not as same as the higher market value typical extruded snacks. It happens because of the lack of optimization of machine parameters and raw material concentration [7] for the production of extruded fish snack product with high acceptability. The study investigated the effect of feed rate, feed moisture content, screw speed, and barrel temperature on the density, expansion, water absorption index (WAl) and water solubility index (WSI), and sensory characteristics such as hardness and crispness of an expanded rice snack [8] have concentrated machine parameter mainly for its product preparation.

Moreover, another primary concern is the types of machinery for the production of innovative products, for example, fish added extruded snack products. The micro and small food processing cannot afford to buy expensive machinery for every innovative product, and it is unrealistic too. Hence, the optimization of the existing conventional mechanism available already for the production of new and innovative products is necessary for the low capital food processing industry.

This paper aims to achieve such optimization of the very common twin-screw extruder available in every food processing industry for the production of fish added extruded snack products. The quality parameters are optimized by adjusting the process variable of the twin-screw extruder, for the production of good quality and highly acceptable extruded fish snack product.

2 Materials and Methods

2.1 *Raw Material Preparation*

Fresh lizardfish (Sauridatumbil) was purchased from Akkaraipettai fishing harbor, Nagapattinam. It cleaned, steam boiled (Fig. 1), and then bones removed. The cooked meat was allowed to dry under a temperature of about 50 °C. Then, the dried product was made into powder with the help of a mixer. The fish powder was divided equally and packed in aluminum foil laminated pouches and stored at room temperature. The cornflour (grounded maize kernels purchased on the local market, Nagapattinam) taken in a certain amount according to the trails, and 5% of lizardfish powder added along with it, and other ingredients like salt (2%) and water (10%) added to produce the experimental samples.

2.2 Twin-Screw Extruder

This study was carried out by the twin-screw extruder (M/s. Basic technology, Pvt Ltd, Kolkata, India). It consists of hopper, twin-screw (either co-rotating or counter-rotating in nature), barrel, heating element, die, and cutter. This twin-screw extruder helps us achieve the hot extrusion process very easy to produce the high acceptable puffed snack products. The pitch of the screw made decreasing along its length to apply gradual compression to the food material. Heaters (preferably two) help to achieve the desired temperature, thereby making the food material plastic enough to form the shape of our desire. The specification and image (Fig. 2) of the machine that has used throughout this research is given below.

2.3 Proximate Evaluation

The dried fish powder used for the extruded snack production has to be evaluated for its proximate composition, which has a severe impact on the final product and customer acceptance. Hence, the proximate analysis was done for both fresh fish and dried fish powder. Different laboratory tests conducted at the fish processing incubation center, Keechankuppam, Nagapattinam, to analyze the proximate

Fig. 1 Fresh, cooked, and powdered lizardfish

- Maximum screw speed - 400 rpm
- Capacity – 45 to 50 kg/hr
- No. of heaters – 2
- Power requirement – 440 v, 3 phases

Fig. 2 Twin-screw extruder used for the experiment

composition. Proximate composition viz., protein (AOAC 988.05), moisture content (AOAC 930.15), fat (948.22) and ash (AOAC 186) analyses were carried out according to the mentioned procedures.

2.4 *Design of the Experiment*

The experiment designed with the principal aim to optimize the process variables for the production of good quality extruded snack products. The values of process variables are fixed within the specified operating range decided by the capacity of the twin-screw extruder. The input machine variables like screw speed and heater temperature 1 and 2 are randomly selected for 28 trail sets within the operating ranges.

The quality and acceptability of the products rely on specific physical and sensory responses like expansion ratio, texture, color, bulk density, true density, and 50 product weights. Thus, an experiment is conducted with the 28 sets of different input machine parameter combinations, as discussed earlier. The responses of each trial are recorded and analyzed with design expert software version 11, which helps to optimize the machine parameters to get the desired product qualities. One additional parameter of maize concentration was also added as the input parameter since it also has its impact on the goodness of product developed.

Expansion ratio: Randomly selected 20 extruded samples subjected to this test. The diameter of the extruded samples measured with vernier caliper (M/s. Kannan Hydrol & tools, Chennai), from which the average diameter of the extruded snacks calculated. The expansion ratio calculated by the ratio of the average diameter of the product to the diameter of the die [9].

Sensory evaluation of extruded snack: Sensory evaluation of extruded snacks was carried out by a panel of 30 subject experts available in the College of Fisheries Engineering, Nagapattinam. They were asked to indicate their opinion on a 9-point hedonic scale (9 for extremely liked product to 1 for dislike extremely). The experts analyzed each product for its appearance, color, flavor, taste, and overall acceptability of the product.

Bulk density: Randomly selected 20 extruded samples considered for this test. The diameters of the extruded samples measured with Vernier caliper, from which the average radius of the extruded snacks calculated. Assuming the shape of the extruded snacks to be cylindrical, extruded snacks, the total volume was calculated by adding the individual lengths of 20 extruded samples. The bulk density calculated by dividing the average weight by the volume of extruded samples given in Eq. (10) [11].

$$\text{Bulk density} = \text{Weight}/\text{Volume of extruded samples} \quad (1)$$

True density: True density of the extruded product calculated by the mustard replacement method. In this method, extruded products filled up to a specific volume with the air voids between the products. After that, air voids filled with the mustard, and the volume occupied by the mustard calculated, and this volume subtracted from total volume occupied by the extruded products. Finally, the total weight of the product is divided by the reduced volume to find the true density of the extruded products given in Eq. (1) [12].

$$\text{True density} = \text{Weight}/\,(\text{Total volume} - \text{mustard volume}) \quad (2)$$

50 Product weight: 50 Product weight of the sample taken by measuring the weight of randomly selected 50 extruded products. It shows that the product will be more acceptable if this weight is minimal [11].

2.5 Response Surface Methodology

Response surface methodology is the technique to optimize the process variables systematically. Since most of the innovative foods products developed from modern machinery in food industries are lagging in its acceptability by the people, because of the inappropriate knowledge about the machine with its process variables for a particular product. This technique identifies the relationship between various machine parameters and their interaction. The specific technique called Box-Behnken mechanism applied to this problem, which helps us to find the right process variable in the particular range of operating conditions of twin-screw extruder

2.6 Box-Behnken Mechanism

It is the set of techniques applied to a controlled experiment to study the influence of different variables in the outcome [13]. Generally, the first step is identifying the independent variable or factors that affect the product or process and then studying their effects on dependent variables or the response.

It uses the corner selection, face, and central points by considering experimental space with fewer points. It is less expensive to run with the same number of factors. It can able to estimate efficiently first- and second-order coefficients.

3 Results and Discussion

3.1 Proximate Analysis

This study mainly concentrates on producing the acceptable quality in terms of sensory and physical characteristics of the product, which requires the actual nutritional composition of the raw material for better understanding to maintain the same nutritional quality after the extrusion process. The proximate analysis of both fresh and powdered fish conducted, and the result obtained as given below in Table 1. The results obtained are satisfactory enough to produce a nutritionally acceptable product. At the same time, it can accommodate the process of extrusion in all possible ways to create a product of acceptable quality. Previous studies reported the quality of fish [14] was satisfactory. The fish contains 1% fat used in this study, which helps to withstand the barrel's temperature without affecting the final product quality. For product quality, the responses recorded given in Table 2, which includes both physical and sensory responses of the product developed. The significance of developed model is given in Table 3, and coefficients of the process variable of the developed model are given in Table 4.

Table 1 Proximate composition of lizardfish analyzed at fresh and powdered state

S. no.	Parameters	% composition in fresh fish	% composition in powdered fish
1	Moisture	76.99	6.02
2	Protein	19.01	89.63
3	Fat	1.13	1.05
4	Ash	1.31	2.06

Expansion ratio: The expansion ratio is the decisive factor that always helps to improve the crunchy nature of the product. It is an important parameter to be considered for the better quality of the final product [15]. It is observed from Table 3 that the expansion ratio is significantly affected by heater temperature 2 linearly. It shows that the increase in temperature just before the die helps in achieving the increase in expansion ratio up to a certain level. The researchers obtained similar results in making puffed snacks from different ingredients [16–18].

The expansion ratio of the product is also quadratically affected by the screw speed. It gives us a clear view that the increase in screw speed, eventually increases the expansion ratio of the product (Fig. 3). A similar interpretation was found with carrot pomace as an ingredient for an extruded snack. The expansion ratio of the final product is controlled by varying the heater temperature 1 and 2.

Texture: The texture of the product is a sensitive indicator that meets the exceptional quality to be acceptable by consumers. From Table (11), it is affected quadratically affected by heater temperature 1 and 2. The same instance was recorded in [10]. Figure 4 confirms the set temperature for producing fish added extruded snack product and helps to obtain a good texture.

Color: The color is also a sensory attribute and an indirect measure of the quality of the product. The consumer choice of preference must play a vital role in judging this parameter. Since this experiment purely based on expert's survey on all 28 samples, the result could orient towards the high acceptable color of the product. The results obtained in Table 3 show that the color of the product is affected by linearly and quadratically on heater temperature 1. Figure 5 also supports the justification well with the effect of heater temperature against color.

Bulk density: It is the responsible parameter for the space occupancy of the product in the package. Since it has some more advantages in the packaging and logistic aspects, the product with considerable bulk density will help in the reduction of packaging cost. It is observed from Table 3 that it is affected linearly by heater temperature 1 and quadratically by screw speed. Bulk density of the product is induced by screw speed and heater temperature in a particular way (Fig. 6).

True Density: It plays a vital role in deciding the absolute matter in the product, which helps in determining the actual nutritional quality of the product. From Table 3, the true density is affected linearly by heater temperature 1 and quadratically by heater temperature 2 and screw speed. Figure 7 depicts the same above-said justification.

Table 2 Experimental trails with recorded responses

Sample	Screw speed (rpm)	Heater 1 temperature (°C)	Heater 2 temperature (°C)	Maize content (g)	Expansion ratio	Texture	Color	Bulk density (kg/m^3)	True density (kg/m^3)	50 product weight
1	250	60	100	110	3.3	3.2	4.4	150.88	251.46	13.48
2	350	60	100	80	3.36	5.5	5.5	84.3	126.45	13.85
3	350	80	100	110	3.16	3.4	3.5	166.23	249.35	14.29
4	300	70	105	95	3.49	5.8	7	140	200	28.05
5	250	80	100	80	3.59	7	5.3	74.79	123.94	13.55
6	300	70	105	95	3.69	5.5	6.9	113.4	195.52	22.5
7	250	80	110	110	3.2	3.3	4.6	210.56	310.5	12.68
8	350	80	110	80	3.38	3.3	5.3	220.4	305.8	12.25
9	350	60	110	110	3.61	7.8	7.8	75	150	13.25
10	250	60	110	80	4.16	6.6	6.9	105.03	152.3	13.91
11	250	60	110	110	4.14	7	8	91.16	136.75	12.34
12	350	80	100	80	3.73	7.25	5	78.23	134.11	14.04
13	250	80	100	110	2.86	3	3.75	177.4	266.1	12.72
14	350	80	110	110	3.55	3.5	4.5	178.46	237.95	17.83
15	350	60	100	110	2.89	4.8	5.3	106	181.71	13.38
16	300	70	105	95	3.52	6.3	6.3	98.83	156.05	21.44
17	350	60	110	80	3.64	6.8	8.8	70.03	105.05	24.54
18	250	60	100	80	3	6.5	3.3	103.5	147.85	8.27
19	300	70	105	95	3.19	6.8	6.8	128.33	220	19.37
20	250	80	110	80	3.26	5.8	3.5	158.4	264	14.6
21	300	70	115	95	3.08	5	4.5	221.65	354.64	13.18
22	300	70	105	125	3.16	5	3.8	200	300	10.09

(continued)

Table 2 (continued)

Sample	Screw speed (rpm)	Heater 1 temperature (°C)	Heater 2 temperature (°C)	Maize content (g)	Expansion ratio	Texture	Color	Bulk density (kg/m^3)	True density (kg/m^3)	50 product weight
23	300	70	105	95	3.3	8.25	8	124.6	213.6	22.18
24	300	70	105	65	2.99	6.5	5.5	176.85	353.7	9.97
25	300	70	95	95	2.98	4	2	254.46	386.52	14.54
26	300	50	105	95	3.05	3.8	2	178.5	214.83	15.37
27	300	90	105	95	3.07	4.3	3	209.9	419.8	13.47
28	300	70	105	95	3.5	7	8.8	105.66	181.14	22.02

Table 3 Significance values of the model developed for every response with significant terms

Source	Expansion ratio	Texture	Color	Bulk density	True density	50 product weight
Model	0.0360	0.0068	0.0044	0.0005	0.0002	0.0152
A-screw speed	0.8479	0.6439	0.3302	0.3822	0.3144	0.1067
B-heater temperature 1	0.2838	0.0609	0.0244	0.0009	<0.0001	0.7583
C-heater temperature 2	0.0171	0.3101	0.0015	0.4325	0.5469	0.3466
D-maize content	0.3847	0.0334	0.1347	0.0298	0.1195	0.7612
AB	0.0590	0.8447	0.1231	0.2046	0.5813	0.3531
AC	0.3362	0.9589	0.6943	0.6314	0.8351	0.5993
AD	0.6945	0.4444	0.4283	0.2817	0.4206	0.5982
BC	0.0095	0.0069	0.0132	0.0030	0.0061	0.3477
BD	0.3362	0.2895	0.1175	0.1948	0.7632	0.3911
CD	0.1392	0.0154	0.9458	0.0260	0.0208	0.3092
A^2	0.0162	0.1084	0.6924	0.0001	<0.0001	0.3826
B^2	0.0720	0.0006	0.0013	0.0032	0.0019	0.0074
C^2	0.0548	0.0025	0.0043	<0.0001	<0.0001	0.0048
D^2	0.0823	0.0398	0.1030	0.0054	0.0011	0.0003

50 product weight: This response is essential for the purpose when it comes to the packaging of the products. In the packaging area, the product weight plays a vital role in determining the package quantity, so by the economic view, this has to be optimally higher to achieve more benefits. From Table 3, it is understood that the quadratic factor of heater temperature 1 and 2 affects 50 product weight significantly. Figure 8 proves the dependency of 50 product weight on temperature 2 along with maize content. The temperature given to the raw material determines the weight of puffed snacks in addition to the maize concentration in it [17].

Among the process parameters, controlling the expansion ratio is a bit complicated than any other, since the expansion ratio is an essential parameter in deciding the product's ultimate quality. It is optimized with higher importance than any different response.

3.2 *Optimization*

The optimum condition was then determined using the Box-Behnken mechanism. The final acceptable product should have an increased expansion ratio, representing all other parameters with adjusted input values, and also, its model significance

Table 4 Process variable coefficients of the developed model for different responses

	Expansion ratio	Texture	Color	Bulk density	True density	50 product weight
	−63.78988	−476.63484	−472.90567	14767.74312	19803.50083	−1414.61606
A	−0.054925	−0.212750	0.082292	11.86261	24.58295	0.288742
B	0.480583	3.72354	2.82125	−84.14025	−110.52529	4.46746
C	1.11533	7.95042	6.47917	−266.39483	−383.64592	20.08975
D	0.019991	−0.670509	0.481435	7.47919	10.46917	3.65176
AB	0.000251	−0.000119	−0.000744	0.017176	0.010977	−0.001520
AC	−0.000243	−0.000063	−0.000363	0.012638	0.008245	0.001700
AD	−0.00032	0.000313	−0.000246	−0.009629	−0.010760	−0.000568
BC	−0.003687	−0.019062	−0.012937	0.468287	0.635150	−0.015375
BD	−0.000404	−0.002119	−0.002521	0.058604	0.019908	0.004667
CD	0.001275	0.011042	−0.000208	−0.215308	−0.340183	−0.011133
A^2	0.000109	0.000335	0.000060	−0.022656	−0.042330	−0.000466
B^2	−0.000971	−0.010917	−0.007479	0.18925	0.307325	−0.020433
C^2	−0.004183	−0.036167	−0.025417	1.19585	0.307325	−0.087333
D^2	−0.000415	−0.002463	−0.001435	0.077728	1.76195	−0.013959

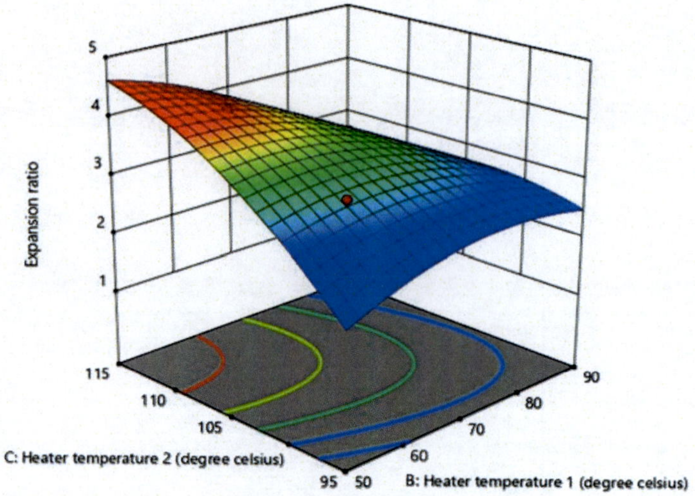

Fig. 3 Effect on expansion ratio—H2 versus H1

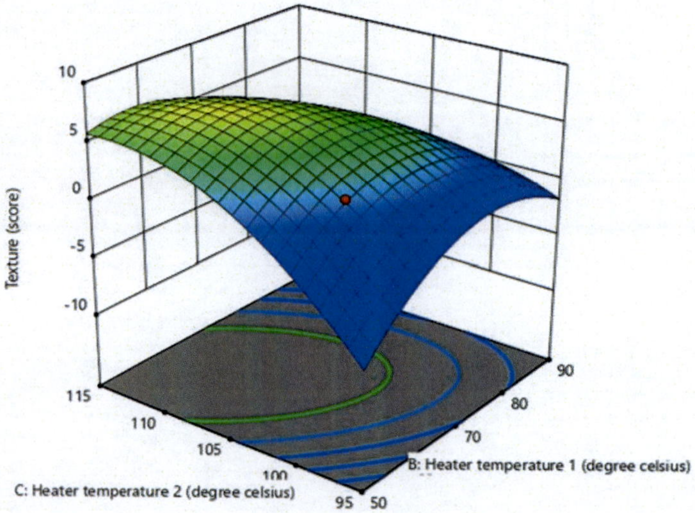

Fig. 4 Effect on texture—H2 versus H1

value is comparatively less than the different response model, as in Table 4. The emphasis placed on the expansion ratio, but the other parameters are maintained at the same level of importance to achieve the product of acceptable combination. The two best possible combinations derived from optimization for producing the product of high acceptability in twin-screw extruder within the operating range given

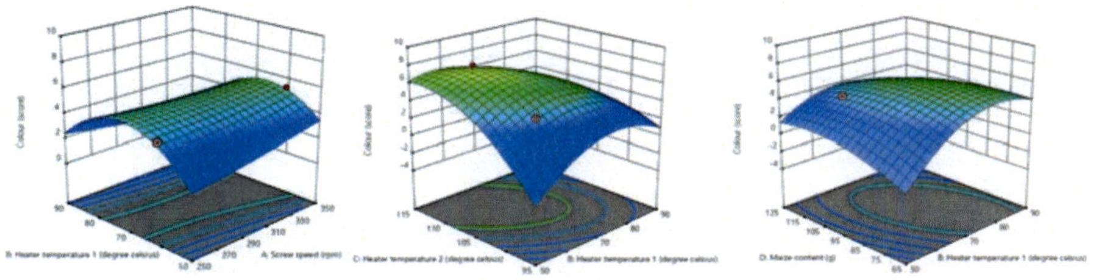

Fig. 5 Effects on color—H1 versus other variables

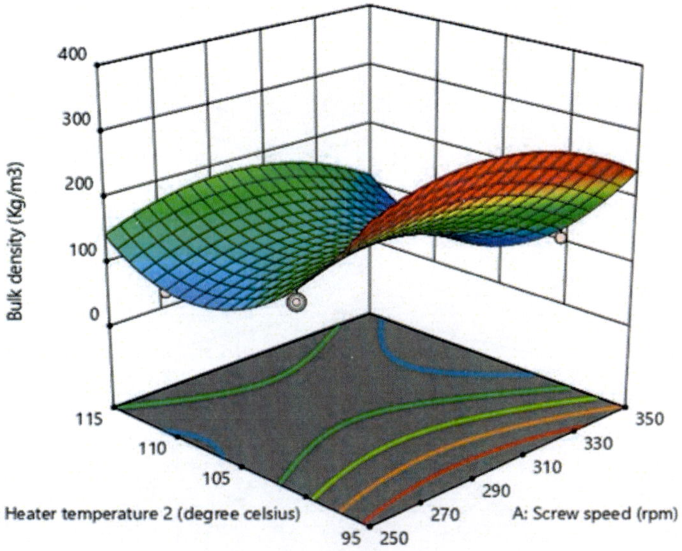

Fig. 6 Effects on bulk density—H2 versus screw speed

in Table 5. It was compared with the derived results in the optimization process. The accuracy of the optimized model is 95.31% for importance 4 and 93.68% for importance 5.

4 Conclusion

The nutrition-rich expanded crisp porous snack produced from the mixture of lizardfish, rice flour, and corn. This snack gaining recognition and appreciation due to their ready to eat nature, delicious taste, and appealing look. This paper is addressed and designed in a particular way to eliminate this issue by optimizing the most popular twin-screw extruder in its approach for producing the product of our

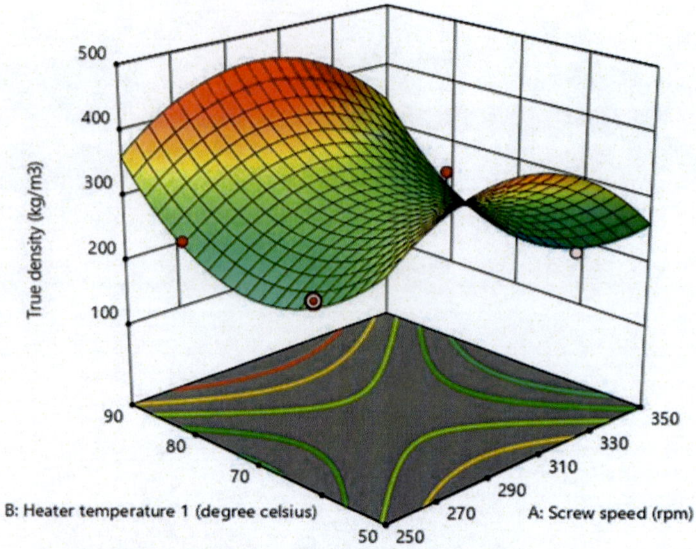

Fig. 7 Effects on true density—H1 versus screw speed

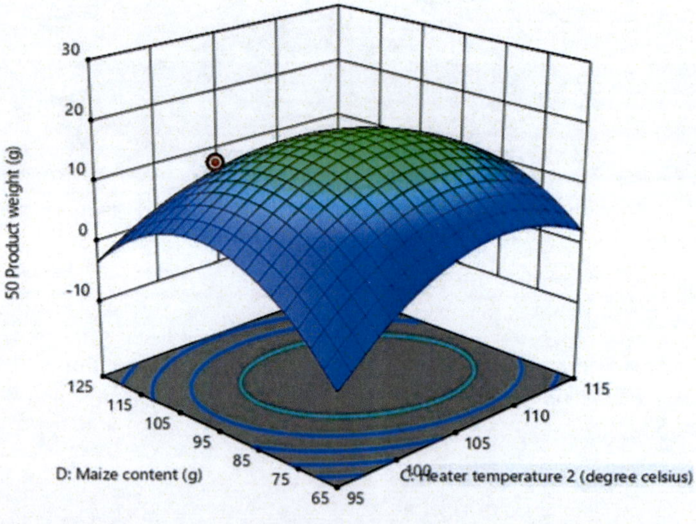

Fig. 8 Effect on 50 product weight—maize content versus H2

interest. This study determined the process condition for any food processing industries to formulate their combination for the production of nutritious and high market demand product called an extruded fish snack. In the future, this optimization may be included in the machine itself with the help of modern techniques

Table 5 Box-Behnken optimization and comparison with experimental results

	Importance 4		Importance 5	
Parameters	Obtained value	Experimental run	Obtained value	Experimental run
Screw speed	350.000		250.000	
Heater temperature 1	54.007		57.890	
Heater temperature 2	113.277		110.764	
Maize content	113.502		107.852	
Expansion ratio	3.763	3.98	4.227	4.32
Texture	7.459	7.8	6.991	7.5
Color	7.586	8	7.071	7.8
Bulk density	70.018	72.812	91.726	93.46
True density	125.143	124.8	129.160	132.4
50 product weight	8.270	9.09	11.134	13.15
Accuracy	0.9531		0.9368	

like programmable logic controller (PLC), which makes the machine intelligent enough to produce the high acceptable extruded fish snack product.

Acknowledgements This study is part of the thesis work to obtain a degree in Bachelor of Technology in Fisheries Engineering, which was developed in the Department of Fish Process Engineering of College of Fisheries Engineering, Nagapattinam, Tamil Nadu, IN. We acknowledge the Dean and faculties of the College of Fisheries Engineering for their kind support and motivation toward this project.

References

1. Avermaete T, Viaene J, Morgan EJ, Crawford N (2003) Determinants of innovation in small food firms. Eur J Innov Manage 6:8–17
2. Corallo A, Latino ME, Menegoli M (2018) From industry 4.0 to agriculture 4.0: a framework to manage product data in agri-food supply chain for voluntary traceability. Int J Nutr Food Eng 12:146–150
3. Method of making an expanded crisp snack food product (1973)
4. Kinsella JE, Melachouris N (1976) Functional properties of proteins in foods: a survey. Crit Rev Food Sci Nutr 7:219–280
5. Mulye VB, Zofair SM (2015) Utilization of croaker (Johnius dussumieri) to develop ready to eat puff snack product using extrusion technology. Food Sci Res J 6:184–191
6. Shaviklo GR, Olafsdottir A, Sveinsdottir K, Thorkelsson G, Rafipour F (2011) Quality characteristics and consumer acceptance of a high fish protein puffed corn-fish snack. J Food Sci Technol 48:668–676
7. Yağcı S, Göğüş F (2008) Response surface methodology for evaluation of physical and functional properties of extruded snack foods developed from food-by-products. J Food Eng 86:122–132

8. Ding Q-B, Ainsworth P, Tucker G, Marson H (2005) The effect of extrusion conditions on the physicochemical properties and sensory characteristics of rice-based expanded snacks. J Food Eng 66:283–289
9. Korkerd S, Wanlapa S, Puttanlek C, Uttapap D, Rungsardthong V (2016) Expansion and functional properties of extruded snacks enriched with nutrition sources from food processing by-products. J Food Sci Technol 53:561–570
10. Ajita T (2018) Extrusion cooking technology: an advance skill for manufacturing of extrudate food products. In: Extrusion of metals, polymers and food products
11. Case SE, Hamann DD, Schwartz SJ (1992) Effect of starch gelatinization on physical properties of extrused wheat-and corn-based products. Cereal Chem 69:401–404
12. Thymi S, Krokida MK, Pappa A, Maroulis ZB (2005) Structural properties of extruded corn starch. J Food Eng 68:519–526
13. Ferreira SC, Bruns RE, Ferreira HS, Matos GD, David JM, Brandao GC, da Silva EP, Portugal LA, Dos Reis PS, Souza AS (2007) Box-Behnken design: an alternative for the optimization of analytical methods. Anal Chim Acta 597:179–186
14. Rathnakumar K, Pancharaja N (2018) Development of health mix from lizard fish and its nutritional characteristics. Int J Curr Microbiol App Sci 7:3136–3144
15. Thachil MT, Subrato B, Chouksey MK, Gudipati V (2016) Modeling the inclusion of fish powder and fish oil into extruded snacks by response surface methodology. J Aquat Food Prod Technol 25:46–64
16. Giménez MA, González RJ, Wagner J, Torres R, Lobo MO, Samman NC (2013) Effect of extrusion conditions on physicochemical and sensorial properties of corn-broad beans (Vicia faba) spaghetti type pasta. Food Chem 136:538–545
17. Hagenimana A, Ding X, Fang T (2006) Evaluation of rice flour modified by extrusion cooking. J Cereal Sci 43:38–46
18. Yu L, Ramaswamy HS, Boye J (2012) Twin-screw extrusion of corn flour and soy protein isolate (SPI) blends: a response surface analysis. Food Bioprocess Technol 5:485–497

MCDM Optimization of Characteristics in Resistance Spot Welding for Dissimilar Materials Utilizing Advanced Hybrid Taguchi Method-Coupled CoCoSo, EDAS and WASPAS Method

Dilip Kumar Bagal, Antarjyami Giri, Ajit Kumar Pattanaik, Siddharth Jeet, Abhishek Barua, and Surya Narayan Panda

1 Introduction

Resistance spot welding (RSW) is a high-speed process, wherein the actual time of welding is a small fraction of second and it is one of the cleanest and most efficient welding process that has been widely used in sheet metal fabrication [1–6]. The high speed of process, the case of operation and its adaptability for automation in the production of sheet metal assemblies are its major advantages. Limitations of RSW are equipment cost and power requirements, difficulty of disassembly for maintenance or repair of RSW joints, and the nature of the design needed for the process (lap joints are required) [5–9]. Resistance spot welding has steadily gained importance over the years because of its ability to join the variety of materials and complicated shapes with high accuracy and great precision. Resistance spot welding (RSW) is a high-speed process, where the actual time of welding is a small fraction of second and it is one of the cleanest and most efficient welding processes that has been widely used in sheet metal fabrication [11–13]. The high speed of process, the case of operation and its adaptability for automation in the production of sheet metal assemblies are its major advantages. Over the last few years, the weight of

D. K. Bagal (✉) · A. Giri · A. K. Pattanaik
Department of Mechanical Engineering, Government College of Engineering, Kalahandi, Bhawanipatna, Odisha, India

S. Jeet · A. Barua
Department of Mechanical Engineering, Center for Advanced Post Graduate Studies, BPUT, Rourkela, Odisha, India

S. N. Panda
Department of Production Engineering, Birsa Institute of Technology Sindri, Dhanbad, Jharkhand, India

S. Bag et al. (eds.), *Next Generation Materials and Processing Technologies*, Springer Proceedings in Materials 9, https://doi.org/10.1007/978-981-16-0182-8_36

automobiles has increased considerably due to the addition of safety related items, such as impact resistance bumpers and door impact beams, emission control equipment and convenience items, such as air conditioning. At the same time, fuel consumption has increased significantly primarily due to emission control equipment [1–15].

In this study, the Taguchi parameter design phase is the most important design phase and served the objective of determining the optimal resistance spot welding parameters to achieve the lowest weld time and the highest tensile-shear strength and nugget diameter in dissimilar (steel + Al) materials under varying resistance spot welding parameter conditions. The following are the questions considered in this study the relationship between the control factors (squeeze time, welding time and current) and output response factors (tensile-shear strength and nugget diameter and weld time). In this investigation, three parameters such as squeeze time, welding time, current were chosen and also optimized to know about the change of mechanical properties around the welded nugget area. In this study, Taguchi's design of experiment was used for experimental design, and multi-response optimization techniques, i.e., combined compromised solution (CoCoSo), evaluation based on distance from average solution (EDAS) and weighted aggregated sum product assessment (WASPAS) method were used to find optimum results.

2 Experimental Analysis and Methodology

AA1200 aluminum alloy sheets with a thickness of 2.5 mm and 50HS stainless steel of 3.0 mm thickness were used as base alloys in this investigation. The sheets were cut to required size by shear-off machine, followed by surface grinding to remove oxides and scales. The dimensions of the AA1200 sheet and 50HS are 114.3 mm × 25.4 mm × 3 mm and 114.3 mm × 25.4 mm × 3 mm respectively. The sheets were resistance spot welded in a 25.4 mm overlap configuration. The chemical composition and mechanical properties of the base alloys are presented in Tables 1 and 2. Prior to welding, the surface of all specimens from both types of material were first ground by abrasive paper using acetone, then thoroughly cleaned, and finally spot welded to prepare the similar and dissimilar welded joints using a spot welding machine SIP type PPV50. A tensile test machine (Tinius Olsen) was used to carry out all the tensile-shear tests for the dissimilar spot-welded specimens. The procedure of experimental work was planned to be conducted in three groups according to the type of weld joint for dissimilar (steel + Al) materials. Nine specimens from each group were spot welded according to the experimental design employed in the current work. During welding the aluminum with steel, it was needed to insert a 0.3 thick sheet of copper (AISI C10200) as a filler metal between the dissimilar materials of the specimen to be welded [8] (Fig. 1).

Table 1 Chemical composition (wt%) of base metals

Element	50HS	AA1200
	Content (%)	Content (%)
Chromium, Cr	20.5–23.5	–
Nickel, Ni	11.5–13.5	–
Manganese, Mn	4–6	≤ 0.050
Molybdenum, Mo	1.5–3	–
Silicon, Si	1 max	≤ 1
Nitrogen, N	0.20–0.40	–
Niobium, Nb	0.10–0.30	–
Vanadium, Va	0.10–0.30	–
Phosphorous, P	0.04 max	–
Carbon, C	0.06 max	–
Sulfur, S	0.010 max	–
Zinc, Zn	–	≤ 0.10
Aluminum, Al	–	≥ 99
Iron, Fe	–	≤ 1
Copper, Cu	–	≤ 0.050
Titanium, Ti	–	≤ 0.050

Table 2 Parameters, codes and level values used for orthogonal array

Parameter	Unit	Code	Level 1	Level 2	Level 3
Squeeze time	(s)	A	13.75	15	16.25
Welding time	(s)	B	0.375	0.5	0.625
Current	(A)	C	60	65	70

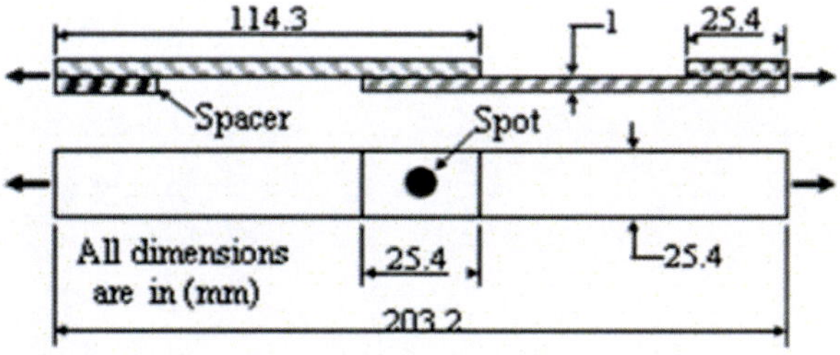

Fig. 1 Dimensions of RSW specimen

3 Optimization Methods

3.1 Combined Compromise Solution Method (CoCoSo)

The following steps are used to solve CoCoSo decision problem [16, 17]:

1. Determination of initial decision-making matrix using Eq. (1)

$$X_{ij} = \begin{bmatrix} x_{11} & x_{12} & \cdots & x_{1n} \\ x_{21} & x_{22} & \cdots & x_{2n} \\ \cdots & \cdots & \cdots & \cdots \\ x_{m1} & x_{m2} & \cdots & x_{mn} \end{bmatrix} \tag{1}$$

2. Using compromise normalization equation, normalization of criteria values is done:

$$r_{ij} = \frac{x_{ij} - \min x_{ij}}{\max x_{ij} - \min x_{ij}}; \text{ for benefit criterion;} \tag{2}$$

$$r_{ij} = \frac{\max x_{ij} - x_{ij}}{\max x_{ij} - \min x_{ij}}; \text{ for cost criterion :} \tag{3}$$

3. Determination of total weighted comparability sequence and whole of power of weight of comparability sequences for respective alternate as S_i and P_i, respectively:

$$S_i = \sum_{j=1}^{n} (w_j r_{ij}) \tag{4}$$

$$P_i = \sum_{j=1}^{n} (r_{ij})^{w_j} \tag{5}$$

4. Three appraisal score are used for generation of comparative weights of other options derived using Eqs. (6, 7, 8):

$$k_{ia} = \frac{P_i + S_i}{\sum_{i=1}^{m} (P_i + S_i)} \tag{6}$$

$$k_{ib} = \frac{S_i}{\min S_i} + \frac{P_i}{\min P_i} \tag{7}$$

$$k_{ic} = \frac{\lambda(S_i) + (1 - \lambda)(P_i)}{(\lambda \max S_i + (1 - \lambda)\max P_i)} \tag{8}$$

5. Ranking of all alternatives is determined from higher to lower based on k_i values:

$$k_i = (k_{ia}k_{ib}k_{ic})^{\frac{1}{3}} + (k_{ia} + k_{ib} + k_{ic}) \tag{9}$$

3.2 *Weighted Aggregated Sum Product Assessment Method (WASPAS)*

The chief technique of WASPAS method for solving MCDM problems is [18].

6. Initial decision matrix is set.
7. Decision matrix normalization using following Eqs. (10) and (11) for maximization and minimization criteria, respectively:

$$\overline{x}_{ij} = x_{ij}/\max_i x_{ij} \tag{10}$$

$$\overline{x}_{ij} = \min_i x_{ij}/x_{ij} \tag{11}$$

where x_{ij} is the assessment value of ith alternate with respect to jth measure.

8. Calculation of total comparative significance of ith alternate, based on weighted sum method (WSM) using Eq. (12):

$$Q_i^{(1)} = \sum_{j=1}^{n} \overline{x}_{ij} \cdot w_j \tag{12}$$

9. Calculation of total comparative significance of ith alternate, based on weighted product method (WPM) using Eq. (13):

$$Q_i^{(2)} = \prod_{j=1}^{n} \overline{x}_{ij}^{w_j} \tag{13}$$

10. Calculation of total relative significance of alternatives is done using Eq. (5) and ranked from higher value to lower value:

$$Q_i = \lambda \cdot Q_i^{(1)} + (1 - \lambda) \cdot Q_i^{(2)} \tag{14}$$

3.3 Evaluation Based on Distance from Average Solution Method (EDAS)

EDAS method was developed by M. Keshavarz Ghorabaee et al. [19] for multi-criteria inventory classification. The steps for using the EDAS method are presented as follows [20]:

Step 1: Select the most important criteria that describe alternatives.

Step 2: Construct the decision-making matrix (X), shown as follows:

$$X = \left[x_{ij}\right]_{nXm} = \begin{bmatrix} x_{11} & x_{12} & \dots & x_{1m} \\ x_{21} & x_{22} & \dots & x_{2m} \\ \dots & \dots & \dots & \dots \\ x_{n1} & x_{n2} & \dots & x_{nm} \end{bmatrix} \tag{15}$$

where X_{ij} denotes the performance value of ith alterative on jth criterion.

Step 3: Determine the average solution according to all criteria, shown as follows:

$$AV = \left[AV_j\right]_{1Xm} \tag{16}$$

where,

$$AV_j = \frac{\sum_{i=1}^{n} X_{ij}}{n} \tag{17}$$

Step 4: Calculate the positive distance from average (PDA) and the negative distance from average (NDA) matrixes according to the type of criteria (benefit and cost), shown as follows:

$$\text{PDA} = \left[\text{PDA}_{ij}\right]_{nXm} \tag{18}$$

$$\text{NDA} = \left[\text{NDA}_{ij}\right]_{nXm} \tag{19}$$

if jth criterion is beneficial,

$$\mathrm{PDA}_{ij} = \frac{\max\left(0, \left(X_{ij} - \mathrm{AV}_j\right)\right)}{\mathrm{AV}_j} \tag{20}$$

$$\mathrm{NDA}_{ij} = \frac{\max\left(0, \left(\mathrm{AV}_j - X_{ij}\right)\right)}{\mathrm{AV}_j} \tag{21}$$

and if jth criterion is non-beneficial,

$$\mathrm{PDA}_{ij} = \frac{\max\left(0, \left(\mathrm{AV}_j - X_{ij}\right)\right)}{\mathrm{AV}_j} \tag{22}$$

$$\mathrm{NDA}_{ij} = \frac{\max\left(0, \left(X_{ij} - \mathrm{AV}_j\right)\right)}{\mathrm{AV}_j} \tag{23}$$

where PDA_{ij} and NDA_{ij} denote the positive and negative distance of ith alternative from average solution in terms of jth criterion, respectively.

Step 5: Determine the weighted sum of PDA and NDA for all alternatives shown as follows:

$$\mathrm{SP}_i = \sum_{j=1}^{m} w_j \mathrm{PDA}_{ij} \tag{24}$$

$$\mathrm{SN}_i = \sum_{j=1}^{m} w_j \mathrm{NDA}_{ij} \tag{25}$$

where w_j is the weight of jth criterion.

Step 6: Normalize the values of SP and SN for all alterative, shown as follows:

$$\mathrm{NSP}_i = \frac{\mathrm{SP}_i}{\max_i(\mathrm{SP}_i)} \tag{26}$$

$$\mathrm{NSN}_i = 1 - \frac{\mathrm{SN}_i}{\max_i(\mathrm{SN}_i)} \tag{27}$$

Step 7: Calculate the appraisal score (AS) for all alterative, shown as follows:

$$\mathrm{AS}_i = \frac{1}{2}\left(\mathrm{NSP}_i + \mathrm{NSN}_i\right) \tag{28}$$

where $0 \leq \mathrm{AS}_i \leq 1$.

Step 8: Rank the alternatives according to the decreasing values of appraisal score (AS). The alternative with the highest AS is the best choice among the candidate alternatives [21].

4 Results and Considerations

Samples are prepared by using Taguchi's experimental design which is shown in Table 3 and as per design of experiment, nine experimental runs are carried out. The analysis of the results of the above-mentioned welding conditions is being done on basis of tensile-shear strength, nugget diameter and weld time of the work piece. Table 3 elucidates that the maximum resulted force of (Steel + Al) spot-welded specimens are 8.39 MPa.

4.1 *Optimization Using Combined Compromised Solution (CoCoSo)*

The first step demonstrates forming of the normalized decision-making matrix (using compromise equation (max–min)), which is shown in Table 4. The further step is to generate the comparability sequence matrix. In this process, the weights of decision-making criteria are involved in the algorithm. The S_i and P_i vectors must be generated, and the values of K_a, K_b, and K_c are calculated using equations of CoCoSo approach used to calculate the ranking score by k shown in Table 4.

From Table 4, for a values of input, parameter in experiment number 1 has the highest k_i value. Therefore, experiment number 1 is an optimal parameter combination for RSW operation according to CoCoSo technique optimization. Now the k_i

Table 3 Result table for tensile-shear strength and nugget diameter and weld time

Run no.	A	B	C	Tensile-shear strength (MPa)	Nugget diameter (mm)	Weld time (ms)
1	13.75	0.375	60	5.33	3.86	49
2	13.75	0.5	65	8.09	4.52	53
3	13.75	0.625	70	7.77	5.56	75
4	15	0.375	65	6.95	5.44	40
5	15	0.5	70	8.10	6.75	49
6	15	0.625	60	7.07	6.31	55
7	16.25	0.375	70	6.67	5.98	51
8	16.25	0.5	60	7.12	5.82	46
9	16.25	0.625	65	8.39	6.21	63

Table 4 Weighted comparability series (S_j), exponentially weighted comparability sequence (P_i), final aggregation and CoCoSo ranking of the alternatives

Run no.	S_i	P_i	k_{ia}	k_{ib}	k_{ic}	k_i	Rank
1	0.9317	2.9244	0.1663	9.8116	1.0000	4.8365	1
2	0.3253	2.0923	0.1042	4.1278	0.6270	2.2658	6
3	0.1775	1.5312	0.0737	2.5076	0.4431	1.4423	8
4	0.6137	2.5411	0.1360	6.8657	0.8181	3.5208	2
5	0.2585	1.5263	0.0770	3.1911	0.4629	2.0110	7
6	0.4294	2.2408	0.1151	5.1069	0.6925	2.7127	5
7	0.5535	2.4481	0.1294	6.2947	0.7784	3.2600	3
8	0.5170	2.4065	0.1261	5.9579	0.7581	3.1095	4
9	0.1180	1.5570	0.0722	2.0199	0.4344	1.2408	9

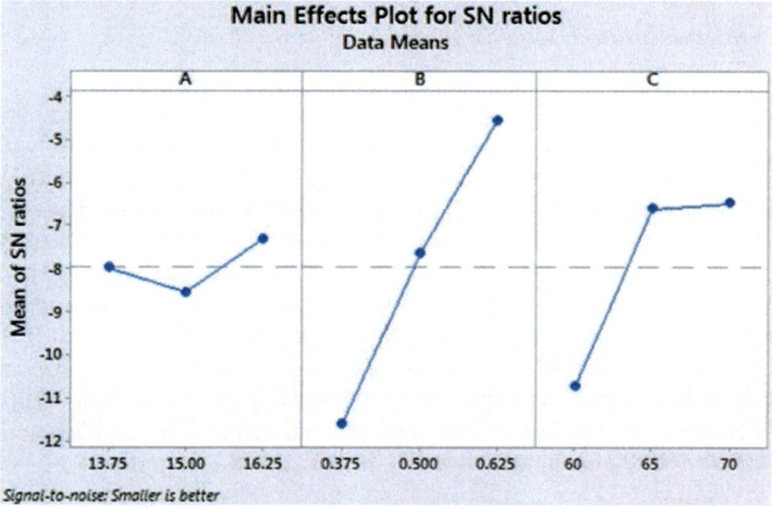

Fig. 2 S/N ratio by CoCoSo method

Table 5 ANOVA result for k_i

Source	DF	Seq SS	Adj MS	F	P	% influence
A	2	2.258	1.129	0.70	0.589	1.96
B	2	75.017	37.509	23.18	0.041	65.09
C	2	34.746	17.373	10.74	0.085	30.15
Residual error	2	3.236	1.618			2.81
Total	8	115.257				

values of alternatives were used to plot mean effect. In Fig. 2, A2 B1 C1 shows the smallest value combination in main effect plot for the three factors, i.e., A, B, C respectively which is optimum parameter arrangement for RSW operation.

Most influential factor

Table 5 gives the results of the ANOVA for the tensile-shear strength, nugget diameter and weld time using the calculated values from the k_i of alternatives of Table 4. According to Table 5, factor B, welding time with 65.09% is the most significant controlled parameters for RSW process followed by factor C, current with 30.15% of contribution and factor A, squeeze time with 1.96% of contribution if the minimization tensile-shear strength, nugget diameter and weld time are simultaneously considered.

$$S = 1.2720, R - \text{Sq} = 97.19\%\ R - \text{Sq(adj)} = 88.77\%$$

4.2 Optimization Using WASPAS

Since semantic terms, used to express the responses, have already been converted into crisp (real) values, the application of the WASPAS method starts with normalization of the decision matrix by applying WASPAS approach since the output has to be minimized. Subsequently, total relative importance of alternatives as per WSM and WPM is calculated by using equations of WASPAS approach. Finally, joint criterion of optimality of the WASPAS method is calculated by using WASPAS methodology. Table 6 provides the values of total relative importance (performance scores) for all the considered alternatives for a λ value of 0.5.

Based on the total relative importance values of alternatives, it is observed that trial 1 is determined as the best sample according to the ranking. Therefore, experiment no. 1 is an optimal parameter combination for RSW operation according to WASPAS technique optimization.

Table 6 Computational details of the WASPAS method

Run No.	$Q_i^{(1)}$	$Q_i^{(2)}$	Q_i	Rank
1	0.9499	0.9461	0.9480	1
2	0.7072	0.7042	0.7057	6
3	0.6457	0.6416	0.6437	8
4	0.8226	0.8159	0.8193	2
5	0.6892	0.6845	0.6869	7
6	0.7255	0.7239	0.7247	5
7	0.7721	0.7703	0.7712	3
8	0.7702	0.7674	0.7688	4
9	0.6329	0.6329	0.6329	9

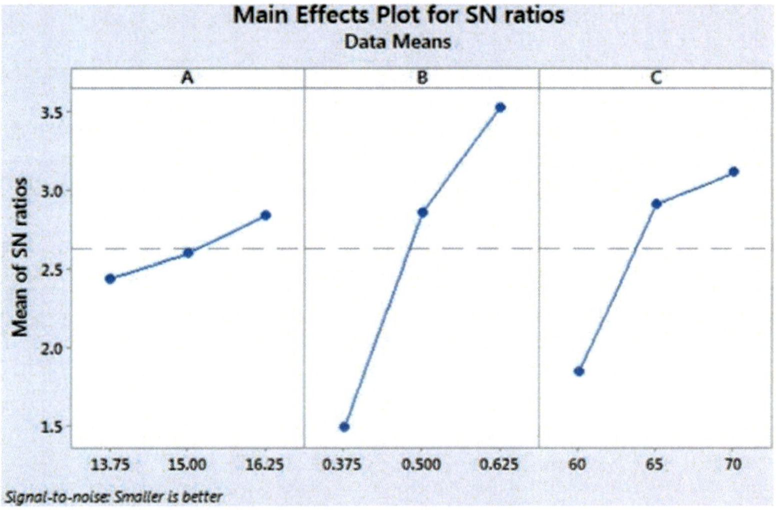

Fig. 3 S/N ratio by WASPAS method

Table 7 Analysis of variance for Q_i

Source	DF	Seq SS	Adj MS	F	P	% influence
A	2	0.24134	0.12067	9.79	0.093	2.52
B	2	6.53704	3.26852	265.30	0.004	68.24
C	2	2.77741	1.38870	112.72	0.009	28.99
Residual error	2	0.02464	0.01232			0.26
Total	8	9.58043				

Now the Q_i values of alternatives were used to plot mean effect. In Fig. 3, A1 B1 C1 shows the smallest value combination in main effect plot for the three factors, i.e., A, B, C respectively which is optimum parameter arrangement for RSW operation.

Most influential factor

Table 7 gives the results of the ANOVA for the tensile-shear strength, nugget diameter and weld time using the calculated values from the Qi of alternatives of Table 6. According to Table 7, factor B, welding time with 68.24%, is the most significant controlled parameters for RSW process followed by factor C, current with 28.99% of contribution and factor A, squeeze time with 2.52% of contribution if the minimization tensile-shear strength, nugget diameter and weld time are simultaneously considered.

$$S = 0.1110,\ R - \text{Sq} = 99.74\%,\ R - \text{Sq(adj)} = 98.97\%$$

Table 8 Positive distance from average and negative distance from average of all output responses

Expt. No.	Tensile-shear strength		Nugget diameter		Weld time	
	PDA_{ij}	NDA_{ij}	PDA_{ij}	NDA_{ij}	PDA_{ij}	NDA_{ij}
1.	0.2674	0.0000	0.3121	0.0000	0.0000	0.0000
2.	0.0000	0.1113	0.1939	0.0000	0.0000	0.0001
3.	0.0000	0.0668	0.0091	0.0000	0.0000	0.3993
4.	0.0458	0.0000	0.0305	0.0000	0.1344	0.0000
5.	0.0000	0.1131	0.0000	0.2030	0.0000	0.0000
6.	0.0288	0.0000	0.0000	0.1252	0.0000	0.0375
7.	0.0838	0.0000	0.0000	0.0663	0.0000	0.0000
8.	0.0224	0.0000	0.0000	0.0372	0.0221	0.0000
9.	0.0000	0.1529	0.0000	0.1068	0.0000	0.1747

Table 9 Weighted sum of PDA and NDA, normalized values of SP and SN, appraisal score and rank of all output responses

Expt. No.	SP	SN	NSP	NSN	AS	Rank
1.	0.2010	0.0000	1.0000	1.0000	1.0000	1
2.	0.0252	0.0668	0.1254	0.5630	0.3442	6
3.	0.0012	0.1479	0.0059	0.0987	0.0523	8
4.	0.0677	0.0000	0.3369	1.0000	0.6685	4
5.	0.0000	0.0942	0.0000	0.3832	0.1916	7
6.	0.0173	0.0264	0.7617	0.8272	0.7944	2
7.	0.0503	0.0086	0.2501	0.4496	0.3498	5
8.	0.0194	0.0048	0.5999	0.9683	0.7841	3
9.	0.0000	0.1528	0.0000	0.0000	0.0000	9

4.3 Optimization Using EDAS

The first step demonstrates forming of the normalized decision-making matrix and determines the average solution according to all criteria using EDAS method. The further step is to generate the positive distance from average (PDA) and the negative distance from average (NDA) matrixes according to the type of criteria, i.e., benefit criteria in this case using equation shown in table. Determination of the weighted sum of PDA and NDA for all alternatives was done in next step using equation shown in Table 8. After finding weighted sum of PDA and NDA, normalization is done using equation. Finally, the appraisal score (AS) was calculated for all alterative using equations of EDAS approach and ranking was done according to the decreasing values shown in Table 9.

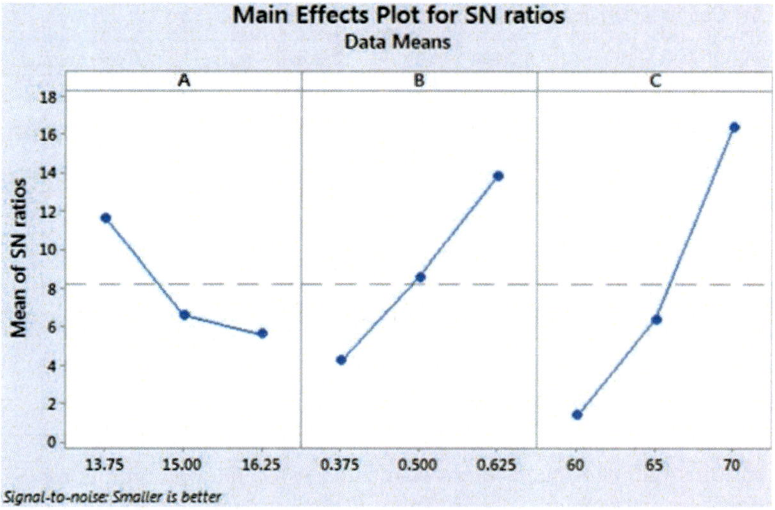

Fig. 4 S/N ratio by EDAS method

Table 10 Analysis of variance for AS

Source	DF	Seq SS	Adj MS	F	P	% influence
A	2	56.17	20.08	1.02	0.573	11.05
B	2	94.64	36.35	1.85	0.461	18.61
C	2	338.03	169.02	8.61	0.234	66.48
Residual error	2	19.64	19.64			3.86
Total	8	508.48				

Based on the total relative importance values of alternatives, it is observed that trial 1 is determined as the best sample according to the ranking. Therefore, experiment no. 1 is an optimal parameter combination for RSW operation according to EDAS technique optimization.

Now, the appraisal score (AS) calculated for all alterative was used to plot mean effect for SN ratios. Based on this study, one can select a mixture of the levels that provide the smaller average response. In Fig. 4, the combination of A3 B1 C1 shows the largest value of the SN ratio plot for the factors A, B and C respectively which is optimum parameter arrangement for RSW operation.

Most influential factor

Table 10 gives the results of the ANOVA for the tensile-shear strength, nugget diameter and weld time using the calculated values from the AS of alternatives of Table 9. According to Table 10, factor C, current with 66.48% is the most significant controlled parameters for RSW process followed by factor B, welding time of 18.61% contribution and factor A, squeeze time with 11.05% of contribution if

Table 11 Confirmatory test results

Optimization technique	Optimal setting	Predicted value	Experimental value
CoCoSo method	A2 B1 C1	4.75139	4.5208
WASPAS method	A1 B1 C1	0.956638	0.9480
EDAS method	A3 B1 C1	0.9807	0.9212

the minimization tensile-shear strength, nugget diameter and weld time are simultaneously considered.

4.4 Confirmation Experiment

The confirmation experiments were conducted using the optimum combination of the machining parameters obtain from Taguchi analysis. These confirmation experiments were used to predict and validate the improvement in the quality characteristics for RSW of AA1200 and 50HS. The final phase is to verify the predicted results by conducting the confirmation test [21–23]. The estimated total relative significance can be determined by using the optimum parameters as:

$$\mu_{\text{predicted}} = a_{2m} + b_{1m} - 3\mu_{\text{mean}} \tag{29}$$

where a_{2m} and b_{1m} are the individual mean values of total relative significance with optimum level values of each parameters and μ_{mean} is the overall total relative significance [21–23] where Table 11 shows the confirmatory test results.

5 Conclusions

This investigation clarifies the methodology for investigating the influence of the spot welding parameters on the tensile-shear force for dissimilar spot-welded joints of aluminum and steel materials. The "smaller is the better" approach was applied in Taguchi approach using Minitab 19 software to design the experiments and analyze the overall results. The Hybrid Taguchi methodologies, i.e., CoCoSo, WASPAS and EDAS were designed to predict which input variables give the optimum responses of resistance spot welding operation.

From this analysis, some important conclusions are drawn and listed below:

1. The optimum results can be achieved by a parametric optimization method, which provides a short period of time with a lower cost.
2. Analysis of the experimental results through the signal to noise ratio and means responses exhibited that the significant influence on the tensile-shear force for the similar material joint is the current. While, the squeeze time possesses a

major impact pursued by welding time and then current for the dissimilar material joint.
3. Tensile-shear force enhanced as the welding time was increased for the all welded joints. But the other parameters exhibited a different behavior, and the linear regression of the output results demonstrated this behavior. For the dissimilar joints, it is preferred to apply a lower squeezing time with a higher welding time and current.
4. The optimal setting of this investigation based on CoCoSo, WASPAS and EDAS is A_2 B_1 C_1, A_1 B_1 C_1 and A_3 B_1 C_1 respectively.
5. The results of confirmatory tests which were carried out at optimal setting are quite nearly come near the actual value with minimal error.
6. The CoCoSo and WASPAS methods have better result than EDAS because the P-value of input parameters comes less than 0.05 that means this experimental design fitted with 95% confidence interval.

It should be mentioned here that the current research can improve the spot welding process for similar and dissimilar welded joints through predicting the optimum input welding parameters for the optimal responses by applying Hybrid Taguchi approaches in order to avoid the encountered problems in the spot welding procedures of different structures as well as to reduce many expensive welding trials.

Acknowledgements This research work is jointly supported by National Institute of Technology, Rourkela, Odisha, India, and Central Tool Room and Training Center (CTTC), Bhubaneswar, Odisha, India.

References

1. Sahota DS, Singh R, Sharma R, Singh H (2013) Study of effect of parameters on resistance spot weld of ASS316 material. Mechanica Confab 2(2):67–78
2. Thakur AG, Rao TE, Mukhedkar MS, Nandedkar VM (2010) Application of Taguchi method for resistance spot welding of galvanized steel. ARPN J Eng Appl Sci 5(11):1298–1306
3. Sreeraj P (2016) Optimization of resistance spot welding process parameters using Moora approach. J Mech Eng Technol 8(2):81–94
4. Selvam G, Doss SA, Kannan MG, Thiruppathi R, Voppuru NS (2018) BIW resistance spot weld parameter standardization through parameter optimization across various sheet metal panel combinations. SAE Technical Paper 2018-28-0034, pp 1–7
5. Malik AR, Pani BB, Badjena SK (2018) Optimization in resistance spot welding of CR3 sheets by embedding WC powder at the lap joint. In: Proceedings of 9th international conference on mechatronics and manufacturing (ICMM 2018), IOP Conference Series: Materials Science and Engineering, vol 361, pp 1–7
6. Mookam N (2019) Optimization of resistance spot brazing process parameters in AHSS and AISI 304 stainless steel joint using filler metal. Defence Technol 15(3):450–456
7. Valera J, Miguel V, Martínez A, Naranjo J, Cañas M (2017) Optimization of electrical parameters in resistance spot welding of dissimilar joints of micro-alloyed steels TRIP sheets. Proc Manuf 13:291–298

8. Das AK, Nayak AR, Tiwari AK, Bagal DK (2019) Multi-objective optimization of resistance spot welding using MOORA technique. Int J Appl Eng Res 14(13):162–164
9. Mahmood NY (2020) Prediction of the optimum tensile–shear strength through the experimental results of similar and dissimilar spot welding joints. Arch Mech Eng 67(2): 197–210
10. Mahmood TR, Doos QM, Al-Mukhtar A (2018) Failure mechanisms and modeling of spot welded joints in low carbon mild sheets steel and high strength low alloy steel. Proc Struct Integrity 9:71–85
11. Hussein SK (2015) Analysis and optimization of resistance spot welding parameter of dissimilar metals mild steel and aluminum using design of experiment method. Eng Technol J 33(8 Part (A) Eng, 1999–1011
12. Lu Y, Mayton E, Song H, Kimchi M, Zhang W (2019) Dissimilar metal joining of aluminum to steel by ultrasonic plus resistance spot welding-microstructure and mechanical properties. Mater Des 165:107585
13. Sankar BV, Lawrence ID, Jayabal S (2016) Prediction of spot welding parameters for dissimilar weld joints. Bonfring Int J Ind Eng Manag Sci 6(4):123
14. Pradeep M, Mahesh N, Hussain R (2014) Process parameter optimization in resistance spot welding of dissimilar thickness materials. Int J Mech Mechatr Eng 8(1):80–83
15. Zedan MJ, Doos QM (2018) New method of resistance spot welding for dissimilar 1008 low carbon steel-5052 aluminum alloy. Proc Struct Integrity 9:37–46
16. Acharya KK, Murmu KK, Bagal DK, Pattanaik AK (2019) Optimization of the process parameters of dissimilar welded joints in FSSW welding process of aluminum alloy with copper alloy using Taguchi optimization technique. Int J Appl Eng Res 14(13):54–60
17. Barua A, Jeet S, Bagal DK, Satapathy P, Agrawal PK (2019) Evaluation of mechanical behavior of hybrid natural fiber reinforced nano sic particles composite using hybrid Taguchi-CoCoSo method. Int J Innov Technol Exploring Eng 8(10):3341–3345
18. Naik B, Paul S, Barua A, Jeet S, Bagal DK (2019) Fabrication and strength analysis of hybrid jute-glass-silk fiber polymer composites based on hybrid Taguchi-WASPAS method. Int J Manag, Technol Eng IX(IV):3472–3479
19. Ghorabaee MK, Zavadskas EK, Olfat L, Turskis Z (2015) Multi-criteria inventory classification using a new method of evaluation based on distance from average solution (EDAS). Informatica 26(3):435–451
20. Naik B, Paul S, Mishra SP, Rout SP, Barua A, Bagal DK (2019) Performance analysis of M40 Grade concrete by partial replacement of Portland Pozzolana Cement with Marble Powder and Fly Ash Using Taguchi-EDAS method. J Appl Sci Comput VI(VI):733–743
21. Jeet S, Barua A, Cherkia H, Bagal DK (2019) Comparative investigation based on MOORA, GRA and TOPSIS method of turning of Nickel-Chromium-Molybdenum Steel under the influence of low cost oil mist lubrication system. Int J Appl Eng Res 14(13):8–20
22. Bagal DK, Barua A, Pattanaik AK, Jeet S, Patnaik D (2020) Parametric optimization based on mechanical characterization of fused deposition modelling fabricated part using utility concept. In: Singh S, Prakash C, Ramakrishna S, Krolczyk G (eds) Advances in materials processing. Singapore, Lecture Notes in Mechanical Engineering. Springer, pp 313–325
23. Panda SN, Bagal DK, Pattanaik AK, Patnaik D, Barua A, Jeet S, Parida B, Naik B (2020) comparative evaluation for studying the parametric influences on quality of electrode using Taguchi method coupled with MOORA, DFA, and TOPSIS method for electrochemical machining. In: Parwani A, Ramkumar P (eds) Recent Advances in Mechanical Infrastructure. Singapore, Lecture Notes in Intelligent Transportation and Infrastructure. Springer, pp 115–129

MCDM Optimization of Karanja Biodiesel Powered CI Engine to Improve Performance Characteristics Using Super Hybrid Taguchi-Coupled WASPAS-GA, SA, PSO Method

Dilip Kumar Bagal, Anil Kumar Patra, Siddharth Jeet, Abhishek Barua, Ajit Kumar Pattanaik, and Dulu Patnaik

1 Introduction

Nowadays, biodiesel is produced from different derived from different organic substances to power compression Ignition engines. Biodiesel has been proved as a superior replacement for petroleum diesel fuel, but they need suitable engine modification or quality enhancement. Biodiesel has gained immense popularity due to its renewability and improved gas emission after combustion. Biodiesel such as Tamarind oil, Jatropha, Soyabeam, Mahua, Sunflower, Neem, Karanja, Rapeseed, etc. are popularly considered as a substitutes of petroleum diesel. Biodiesel made of these things has cleaner burning, non-toxic, renewable, environmentally friendly and biodegradable fuels which can be employed as a unadulterated form or in mingled form through petroleum diesel for different types of diesel engines [1–10].

The biodiesel development was done from Karanja tree seeds. In assessment of above-mentioned facts, the resolution of this investigation is to employ multi-objective optimization approach to optimize operative parameters of diesel engine to improve the performance powered with different blend of Karanja biodiesel and petroleum diesel. Different blends of Karanja biodiesel–diesel, engine load and compression ratio were considered as input operating parameters with different levels. Taguchi's method was used for experimental design using which six output

D. K. Bagal (✉) · A. K. Patra · A. K. Pattanaik
Department of Mechanical Engineering, Government College of Engineering, Kalahandi, Bhawanipatna, Odisha 766002, India

D. Patnaik
Department of Electrical Engineering, Government College of Engineering, Kalahandi, Bhawanipatna, Odisha 766002, India

S. Jeet · A. Barua
Department of Mechanical Engineering, Center for Advanced Post Graduate Studies, BPUT, Rourkela, Odisha 769004, India

S. Bag et al. (eds.), *Next Generation Materials and Processing Technologies*, Springer Proceedings in Materials 9, https://doi.org/10.1007/978-981-16-0182-8_37

responses, viz., brake thermal efficiency, exhaust gas temperature, oxides of Nitrogen (No_x), brake-specific fuel consumption, % of carbon monoxide (CO) and unburnt hydrocarbon (HC) were evaluated. A hybrid approach was used for optimizing engine operating parameters using weighted aggregated sum product assessment method (WASPAS) coupled with genetic algorithm, particle swarm optimization and simulate annealing. After prediction of the optimal operating parameters, confirmation experiments were implemented for validation of developed model.

2 Preparation of Karanja Biodiesel

Karanja tree seed has a prospective to be used for the creation of biodiesel which contains 27–39% of the oil. By transesterification practice at 65 °C with 1 wt% of KOH, 6:1 ratio of methanol to oil, Karanja oil methyl ester (KOME) of about 97% yield was acquired from Karanja oil in 2 h [1, 3]. After the completion of methyl ester formation, the heating was stopped, and the KOME was chilled and moved to a separating funnel where glycerol layer was separated. After washing oil with hot distilled water until the water, anhydrous sodium sulfate was added and left overnight moisture absorption. Hence, the sample of biodiesel was obtained after decantation which was further sent to laboratory for determination of its different properties [1, 3]. Table 1 displays the properties of Karanja Oil. The different properties of Karanja biodiesel and different blends of Karanja biodiesel + petroleum diesel have been stated in Table 2.

3 Experimental System

The experimentations were executed on four-stroke, one-cylinder compression ignition engine. Engine was coupled with rope brake dynamometer and eddy current dynamometer. A manometer was used to meter intake air mass flow rate and

Table 1 Properties of the Karanja oil

Properties	Values
Flash point	225 °C
Kinematic viscosity @ 40 °C	40.2 mm^2/s
Acid value	5.40 mg KOH/gm
Density	0.927 gm/cc
Cloud point	3.5 °C
Saponification value	184
Pour point	6 °C
Carbon residue	1.51 wt%
Calorific value	8742 MJ/kg
Specific gravity	0.936

Table 2 Different properties of Karanja biodiesel, petroleum diesel and their blends

Properties	Karanja biodiesel	Petroleum diesel	B10	B15	B20	B25
Density at 15 °C (kg/m^3)	912.43	824	841	845	849	855
Cetane number	57	51	48	47	46	46
Kinematic viscosity (m^2/s)	1.0×10^{-5}	2.3×10^{-6}	2.8×10^{-6}	3.3×10^{-6}	3.7×10^{-6}	3.7×10^{-6}
Calorific value (MJ/kg)	37.7	45	42.8	40.18	39.2	38.39
Flash point (°C)	97 °C	56 °C	60 °C	61 °C	65 °C	70 °C

fuel consumption flow meter was used for measuring rate of fuel consumption. Figure 1 depicts process of cleaning for Karanja biodiesel. Figure 2 shows the final washed sample of Karanja Bio-Diesel along with experimental setup and figure 3 shows the schematic layout of experimental setup. To gain variable compression ratio, addition of different number of gaskets between cylinder block and cylinder head was done. Exhaust gas temperature was monitored and measured by means of exhaust gas temperature gauge (EGT gauge) which is a thermocouple-type pyrometer fitted at the engine exhaust pipe. Table 3 shows the exhaust gas analyzer specification. Output response variables were recorded for different operating conditions planned according to response surface methodology. Table 4 displays the engine specification and Table 5 spectacles the different operating parameters with different levels.

4 Optimization Methods

4.1 WASPAS Method (Weighted Aggregated Sum Product Assessment)

The paramount practice of this this method is [11].

1. Preliminary decision matrix setting.
2. Decision matrix normalizing using Eq. (1) [maximization] and Eq. (2) [minimization]:

$$\bar{x}_{ij} = x_{ij}/\max_i x_{ij} \tag{1}$$

$$\bar{x}_{ij} = \min_i x_{ij}/x_{ij} \tag{2}$$

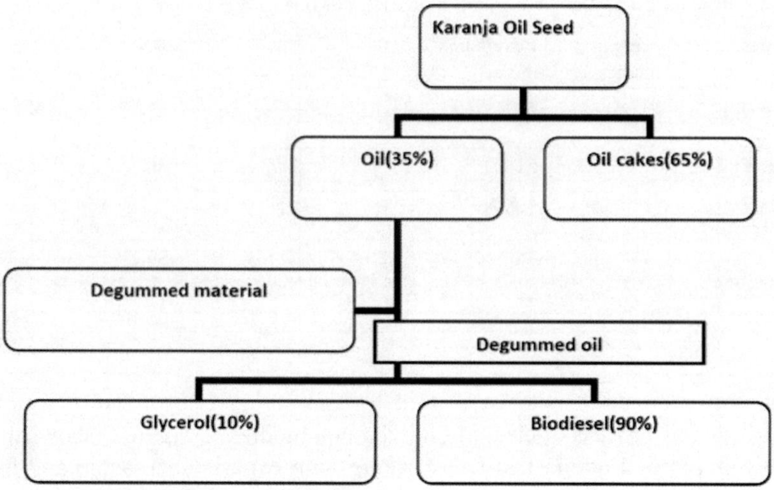

Fig. 1 Process of cleaning for Karanja biodiesel

Table 3 Exhaust gas measurement along with prescribed range

Item	Method	Measuring range
CO	Non dispersive infrared	0–9.99%
HC	Gas analyzer	0–5000 ppm
NO_x	Electrochemical cells	0–5000 ppm

Table 4 Compression ignition engine specification with the attachments

Component	Description
Make and model	KOEL DM10
Type of engine	Four-stroke, water-cooled, single-cylinder
Rated power	7 kW/10 hp @ 1500 rpm
Compression ratio	17.5:1
Bore × stroke	102 mm × 116 mm
Torque	0.048 kN m
Overall dimensions of the standard engine	617 mm × 532 mm × 850 mm
Device for engine loading	Eddy current dynamometer

Table 5 Different operating parameters with different levels

Parameters	Code	Level 1	Level 2	Level 3	Level 4
Compression ratio	A	17.5	17.7	17.9	18.1
Biodiesel blend	B	B10	B15	B20	B25
Engine load	C	20%	60%	80%	100%

Fig. 2 Final washed sample of Karanja Bio-Diesel along with experimental setup

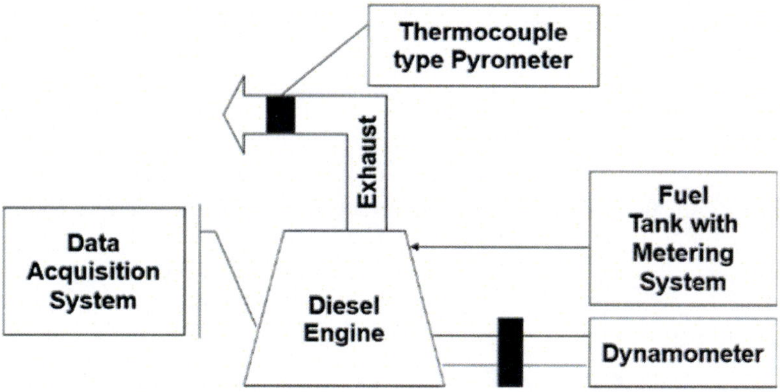

Fig. 3 Schematic layout of experimental setup

3. Total comparative significance calculation with weighted sum method by Eq. (3):

$$Q_i^{(1)} = \sum_{j=1}^{n} \bar{x}_{ij} \cdot w_j \tag{3}$$

4. Total comparative significance calculation with weighted product method by Eq. (4):

$$Q_i^{(2)} = \prod_{j=1}^{n} \bar{x}_{ij}^{w_j} \tag{4}$$

5. Total relative significance by Eq. (5) and ranking them from greater to lesser value:

$$Q_i = \lambda \cdot Q_i^{(1)} + (1 - \lambda) \cdot Q_i^{(2)} a \quad (5)$$

4.2 Genetic Algorithm

Genetic algorithm is based on the natural encroachment manner which is applied to advance responses for complex upgrade issues. Here, the wellbeing work reestablishes a single numerical health which is comparative with the utility or the limit of the individual which that chromosome addresses. Two guardians are picked and their chromosomes are recombined, usually using the instruments of half and half and change. Hybrid is progressively significant for quickly investigating a hunt space. Transformation gives just a modest quantity of arbitrary pursuit [12–17].

4.3 Simulated Annealing

Simulated annealing is a probabilistic method which reflects the path toward hardening (modest cooling of fluid metal) with a particular ultimate objective to achieve least salve regard in a minimization issue. The cooling wonder is surrendered out by administering a temperature like boundary gave the idea of the Boltzmann likelihood circulation [12–17].

4.4 Particle Swarm Optimization

PSO is an inhabitants-based stochastic augmentation scheme. The insight of multitude depends on the guideline of social and mental conduct of the multitude. The optimization method is instated with a populace of irregular arrangements and looks for optima by refreshing ages. There is just one bit of food in the region being looked. All the feathered creatures do not have the foggiest idea where the food is. In any case, they know how far the food is in their inquiry. So the best procedure to achieve the food is to just follow the fowl, which is closest to the food. In improvement issues, each feathered creature in the hunt space is referred to as molecule. All the particles are assessed by the wellness capacity to be optimized and have speeds for the particles. The particles fly through the issue space by following the current ideal particles [14].

5 Results and Considerations

The experimental analysis was conducted according to different operating parameter conditions prepared by means of Taguchi's experimental plan presented in Table 6 with results for brake thermal efficiency, oxides of nitrogen (NO_x), exhaust gas temperature, brake-specific fuel consumption, % of carbon monoxide (CO), and unburnt hydrocarbon (HC) were calculated andrecorded in Table 6. Table 7 shows the computational particulars of all alternatives using WASPAS.

From total relative significance values of alternatives, it was detected that investigational results obtained in experiment no. 2 are the best result according to the ranking.

Now, the total relative importance of responses was used to plot mean effect. Based on this study, one can select a mixture of the levels that provide the smaller average response. In Fig. 4, the grouping of A1, B1, C1 indicates the bottom value of main effect plot. Therefore, A1 B1 C1, i.e., fuel compression ratio of 17.5; B10 blend of Karanja biodiesel and petroleum diesel; and engine load of 20% is the optimum diesel engine operating parameter combination for Karanja biodiesel powered diesel engine.

5.1 *Optimization Using Nature-Based Algorithm*

Consequently, designing constraints are distinct in normal optimal format and deciphered using genetic algorithm, simulated annealing and particle swarm Optimization. The minimization problem expressed in mathematical form using the total relative significance of WASPAS method is as below:

Minimize

$$\frac{ab}{(-715.025 + 37.5482a + 1.27248ab + 0.0390438ac)} \tag{6}$$

all equations are subjected to constraints:

$17.5 \leq a \leq 18.1$

$10 \leq b \leq 25$

$20 \leq c \leq 100$

For GA, finest solution was achieved after 53 generations. For simulated annealing, finest solution was achieved after 2293 iterations. Optimum parameters obtained by GA, SA and PSO are shown in Table 8. Figures 5, 6 and 7 show best fitness value with generations using GA, SA and PSO, respectively.

Table 6 Taguchi's design for experimental runs and output responses

Run No.	A	B	C	BTE (%)	BSFC (kg/kWh)	EGT (°K)	CO (%)	NO_x (ppm)	HC (ppm)
1	17.5	10	40	32.34	0.25	457	0.12	1230	68
2	17.5	15	60	30.19	0.26	485	0.01	1169	71
3	17.5	20	80	30.51	0.29	496	0.20	1219	65
4	17.5	25	100	29.10	0.29	534	0.31	1152	85
5	17.7	10	60	32.34	0.26	473	0.33	1120	71
6	17.7	15	40	31.13	0.26	470	0.23	1112	66
7	17.7	20	100	30.77	0.30	524	0.50	1146	85
8	17.7	25	80	29.77	0.28	507	0.44	1058	63
9	17.9	10	80	30.03	0.30	475	0.35	1125	73
10	17.9	15	100	28.14	0.31	515	0.33	1101	93
11	17.9	20	40	30.06	0.31	473	0.19	1108	61
12	17.9	25	60	28.39	0.31	499	0.20	1004	64
13	18.1	10	100	32.93	0.33	510	0.54	1216	98
14	18.1	15	80	31.46	0.32	496	0.35	1171	76
15	18.1	20	60	32.71	0.34	497	0.28	1162	69
16	18.1	25	40	31.97	0.32	491	0.32	1111	64

Table 7 Computational details of the WASPAS method

Run No.	$Q_i^{(1)}$	$Q_i^{(2)}$	Q_i	Rank
1	0.7452	0.5058	0.6255	4
2	1.0000	1.0000	1.0000	1
3	0.6894	0.4238	0.5566	8
4	0.6558	0.3668	0.5113	11
5	0.7231	0.3897	0.5564	9
6	0.7244	0.4243	0.5744	6
7	0.6616	0.3265	0.4940	12
8	0.6887	0.3486	0.4314	14
9	0.6764	0.3634	0.4313	15
10	0.6355	0.3514	0.5999	5
11	0.6884	0.4268	0.5576	7
12	0.6720	0.4140	0.7031	2
13	0.6568	0.3170	0.3233	16
14	0.6659	0.3591	0.5439	10
15	0.6765	0.3820	0.6474	3

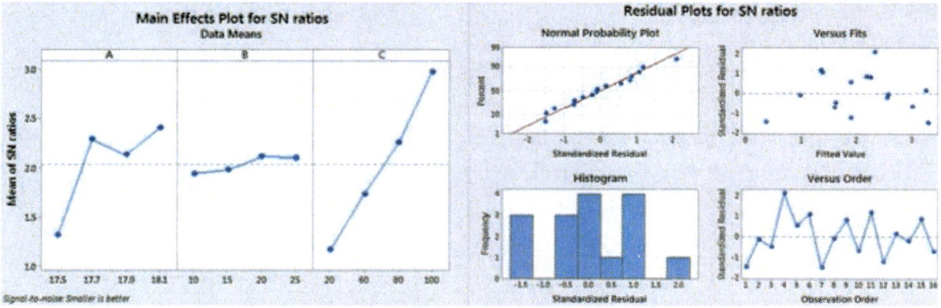

Fig. 4 Main effect plot and residual plots for total comparative significance

Table 8 Optimal parameter setting and fitness value obtained using GA, SA and PSO

Algorithm	Factor setting	Fitness value
Genetic algorithm		0.68151
Simulated annealing	A3 B1 C3	0.68294
Particle swarm optimization		0.68151

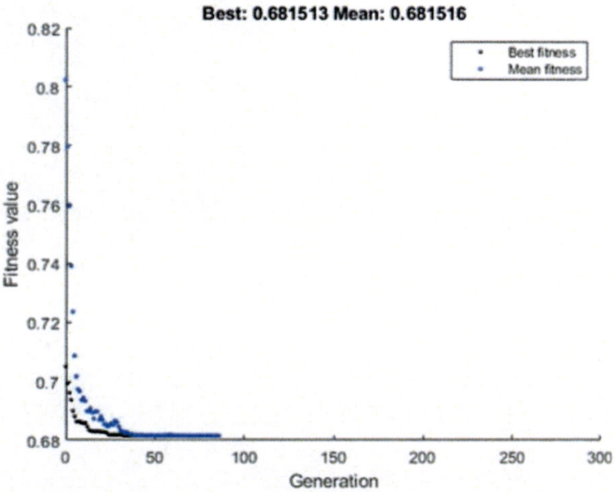

Fig. 5 Best fitness value with generations using GA

5.2 *Maximum Effective Factor*

Table 9 gives the ANOVA results for the calculated values of total relative importance of all six output responses. According to Table 9, factor B, blend of Karanja biodiesel and petroleum diesel, is the most significant controlled

Fig. 6 Best fitness value with generations SA

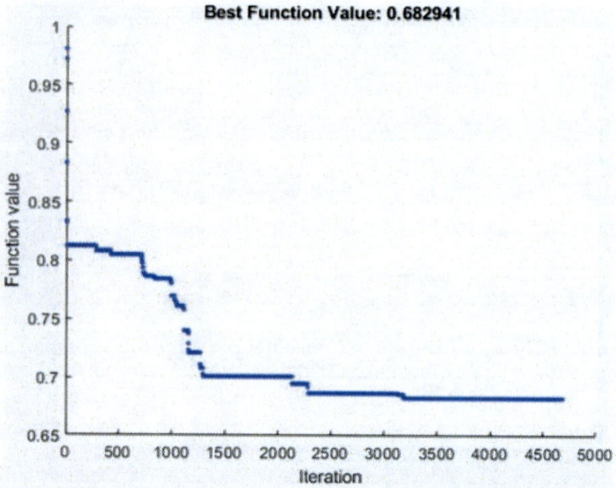

Fig. 7 Best fitness value with generations using PSO

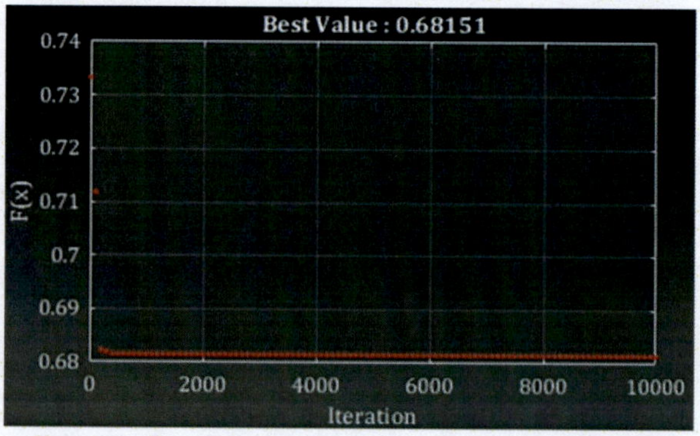

Table 9 ANOVA for Q_i

Source	DF	Seq SS	Adj MS	F	P	%
A	3	2.85410	0.95137	35.10	0.000	1.96
B	3	0.08626	0.02875	1.06	0.433	65.09
C	3	7.07034	2.35678	86.95	0.000	30.15
Residual error	6	0.16262	0.02710			2.81
Total	15	10.1733				

parameters for diesel engine operation followed by factor C, engine load and factor A, engine compression ratio, if the minimization of BTE, EGT, BSFC, NO_x, CO and HC is concurrently considered.

$S = 0.1646$, $R - \text{Sq} = 98.40\%$, $R - \text{Sq(adj)} = 96.00\%$

Table 10 Confirmatory test results

Technique	Best setting	Projected value	Experimental value
Taguchi-WASPAS	A1 B1 C1	0.95566	0.97531
Genetic algorithm		0.68151	0.68007
Simulated annealing	A3 B1 C3	0.68294	0.68173
Particle swarm optimization		0.68151	0.68007

Table 11 Optimal factor settings

Algorithm	Compression ratio	Biodiesel blend	Engine load
Taguchi-WASPAS	17.5	10% or B10	20%
Genetic algorithm	17.9	10% or B10	80%
Particle swarm optimization	17.9	10% or B10	80%
Simulated annealing	17.9	10% or B10	80%

5.3 Confirmation Experiment

The confirmation experiments were led for different operating parameters. Table 10 shows result of confirmatory test of different lubricating environment with different optimizing methods [17–27].

6 Conclusions

This study involves three different operating parameters of CI engine, namely different blend of Karanja biodiesel and petroleum diesel, and engine load and different compression ratio for investigation of performance characteristics. A hybrid approach was employed for optimizing engine operating parameters using a recent technique, i.e., weighted aggregated sum product assessment method (WASPAS) coupled with Taguchi's design of experiment. Genetic algorithm, simulated annealing algorithm and particle swarm optimization were also employed to predict optimal operating parameter setting by using the regression analysis. Table 11 shows the optimal operating parameter setting by using different techniques.

According to ANOVA, blend of Karanja biodiesel and petroleum diesel is most influencing parameter for controlling the enactment of the diesel engine than other two operating parameters if minimization of brake-specific fuel consumption, exhaust gas temperature, oxides of nitrogen (No_x), brake thermal efficiency, % of carbon monoxide (CO) and unburnt hydrocarbon (HC) was simultaneously considered.

References

1. Sharma A, Singh Y, Gupta SK, Singh NK (2019) Application of response surface methodology to optimize diesel engine parameters fueled with pongamia Biodiesel/diesel blends. Energy Sources, Part A: 1–12. https://doi.org/10.1080/15567036.2019.1623949
2. Sharma A, Singh Y, Singh NK, Singla A (2019) Sustainability of jojoba Biodiesel/diesel blends for DI diesel engine applications—Taguchi and response surface methodology concept. Ind Crops Prod 139:1–13
3. Pandian M, Sivapirakasam SP, Udayakumar M (2011) Investigation on the effect of injection system parameters on performance and emission characteristics of a twin cylinder compression ignition direct injection engine fuelled with pongamia Biodiesel–diesel blend using response surface methodology. Appl Energy 88:2663–2676
4. Sivaramakrishnan K, Ravikumar P (2014) Optimization of operational parameters on performance and emissions of a diesel engine using Biodiesel. Int J Environ Sci Technol 11(4):949–958
5. Vadlamudi S, Goud RNK, Krishnaiah T (2019) Performance of diesel engine using bio-fuel from sesame oil. Int J Innov Technol Exploring Eng 8(10):1554–1558
6. Bahar D, Kirti G, Mounika R, Rajesham S (2018) Study of performance and emission characteristics of a compression ignition engine using tamarind Biodiesel. Int J Adv Technol Eng Explor 5(43):134–139
7. Kumar MV, Babu AV, Kumar PR, Reddy SS (2018) Experimental investigation of the combustion characteristics of Mahua oil Biodiesel-diesel blend using a DI diesel engine modified with EGR and nozzle hole orifice diameter. Biofuel Res J 19:863–871
8. Singh VB, Yadav AK (2018) Optimization of performance and emission characteristics of CI engine fuelled with Mahua Oil methyl ester-diesel blend using response surface methodology. Int J Ambient Energy. https://doi.org/10.1080/01430750.2018.1484804
9. Ganapathy T, Murugesan K, Gakkhar RP (2009) Performance optimization of Jatropha Biodiesel engine model using Taguchi approach. Appl Energy 86:2476–2486
10. Karnwal A, Hasan MM, Kumar N, Siddiquee AN, Khan ZA (2011) Multi-response optimization of diesel engine performance parameters using thumba biodiesel-diesel blends by applying the Taguchi method and grey relational analysis. Int J Automot Technol 12(4): 599–610
11. Naik B, Paul S, Barua A, Jeet S, Bagal DK (2019) Fabrication and strength analysis of hybrid jute-glass-silk fiber polymer composites based on hybrid Taguchi-WASPAS method. Int J Manag, Technol Eng IX(IV):3472–3479
12. Jeet S, Barua A, Bagal DK, Pattanaik AK, Agrawal PK, Panda SN (2019) Multi-parametric optimization during drilling of aerospace alloy (UNS A97068) using hybrid RSM-GRA, GA and SA. Int J Manag, Technol Eng IX(II):2501–2509
13. Jeet S, Barua A, Parida B, Sahoo BB, Bagal DK (2018) Multi-objective optimization of welding parameters in GMAW for stainless steel and low carbon steel using hybrid RSM-TOPSIS-GA-SA approach. Int J Tech Innov Mod Eng Sci 4:683–692
14. Bagal, DK (2015) Experimental Investigation and Optimization of Cutting Parameters in Plasma Arc Cutting. NIT Rourkela Ethesis 1(2):1–452
15. Bagal DK, Barua A, Jeet S, Satapathy P, Patnaik D (2019) MCDM optimization of parameters for wire-EDM machined stainless steel using Hybrid RSM-TOPSIS, Genetic Algorithm and Simulated Annealing. Int J Eng Adv Technol 9(1):366–371
16. Barua A, Jeet S, Parida B, Sahoo BB, Bagal DK, Samantray A (2018) Virtual optimization of motorcycle sprocket material by using FEA and Taguchi Coupled TOPSIS-GA-SA. Int J Adv Sci Res Manag 3:54–63
17. Sahoo BB, Barua A, Jeet S, Bagal DK (2018) Multi objective optimization of WEDM process parameters using hybrid RSM-GRA-FIS, GA and SA approach. Int J Res Advent Technol 6(7):1752–1761

18. Barua A, Jeet S, Bagal DK, Satapathy P, Agrawal PK (2019) Evaluation of mechanical behavior of hybrid natural fiber reinforced nano sic particles composite using Hybrid Taguchi-Cocoso method. Int J Innov Technol Explor Eng 8(10):3341–3345
19. Jeet S, Barua A, Cherkia H, Bagal DK (2019) Comparative investigation based on MOORA, GRA and TOPSIS method of turning of nickel-chromium-molybdenum steel under the influence of low cost oil mist lubrication system. Int J Appl Eng Res 14(13):8–20
20. Panda SN, Bagal DK, Pattanaik AK, Patnaik D, Barua A, Jeet S, Parida B, Naik B (2020) Comparative evaluation for studying the parametric influences on quality of electrode using Taguchi method coupled with MOORA, DFA, and TOPSIS method for electrochemical machining. In: Parwani A, Ramkumar P (eds) Recent advances in mechanical infrastructure. Lecture Notes in Intelligent Transportation and Infrastructure, Springer, Singapore, pp 115–129
21. Naik B, Paul S, Mishra SP, Rout SP, Barua A, Bagal DK (2019) Performance analysis of M40 grade concrete by partial replacement of Portland Pozzolana Cement with Marble Powder and Fly Ash Using Taguchi-EDAS method. J Appl Sci Comput VI(VI):733–743
22. Bagal DK, Barua A, Pattanaik AK, Jeet S, Patnaik D (2020) Parametric optimization based on mechanical characterization of fused deposition modelling fabricated part using utility concept. In: Singh S, Prakash C, Ramakrishna S, Krolczyk G (eds) Advances in materials processing. Lecture Notes in Mechanical Engineering, Springer, Singapore, pp 313–325
23. Barua A, Jeet S, Parida B, Samantray A, Bagal DK (2018) Comparative evaluation and optimization of 4-cylinder CI engine camshaft material using finite element analysis: a hybrid MOORA technique and Taguchi based desirability function analysis approach. Int J Tech Innov Mod Eng Sci 4(11):105–114
24. Barua A, Jeet S, Bagal DK, Agrawal PK, Pattanaik AK (2019) Comparative analysis based on MCDM optimization of printing parameters affecting compressive and tensile strength of fused deposition modelling processed parts. Int J Tech Innov Mod Eng Sci 5(2):383–392
25. Bagal DK, Pattanaik AK, Patnaik D, Barua A, Jeet S, Panda SN (2020) Multi-objective optimization and experimental investigation of CNC oxy-fuel gas cutting parameters using Taguchi coupled data envelopment analysis. In: Parwani A, Ramkumar P (eds) Recent advances in mechanical infrastructure. Lecture Notes in Intelligent Transportation and Infrastructure. Springer, Singapore. https://doi.org/10.1007/978-981-32-9971-9_9
26. Bagal DK, Parida B, Barua A, Jeet S, Sahoo BB (2019) Multi-parametric optimization in CNC dry turning of chromoly steel using Taguchi coupled desirability function analysis and utility concept. Int J Appl Eng Res 14(13):21–26
27. Barua A, Jeet S, Cherkia H, Bagal DK, Sahoo BB (2019) Parametric optimization of FDM processed part for improving surface finish using MOORA Technique and Desirability Function analysis. Int J Appl Eng Res 14(13):1–7

Characterization of Electrically Hybridized Friction Stir Welding of Mild Steel and Optimization of Process Parameters

Dilip Kumar Singh, Kaushik Sengupta, Arpan Kumar Mondal, Debtanu Patra, and Arindam Dhar

1 Introduction

This manuscript presents the friction stir joining process to join the similar and dissimilar material of different grade and composition. It has been developed to overcome the fusing joining defects such as porosity, enhance the mechanical properties, reduce the heat affected zone (HAZ), reduce the weight of the components, and make economical by reducing the energy consumption and filler rod, etc. The weight reduction is important as the automobile and aerospace required to reduce the buy-to-fly ratio. The FSJ is the solid-state technique.

To take advantage of friction-stir-welded aluminum alloy, it has already been applied to rail cars and automobiles, in the aerospace and electronics industries and to a wide range of structural components. After that, the joining of high-temperature alloy is the challenge for the industrial development. In current scenario, many researches are going on for friction stir joining of high-temperature alloys such as stainless steel and titanium. There are major challenges found that the heat generation capability due to tool rotation in the form of friction to perform the solid-state joining. In such case, the high-temperature alloys are required to achieve higher temperature to develop the material flow required additional heat source, which may be induction, electrical resistance, plasma, laser, etc. Also, the tool geometry and material will play essential role for defects free joining at solid state. Some of the researchers specified the tool material such as WC and PCBN, and their pin profile for the joining of stainless steel. Therefore, materials with high melting points, such as steel, can be used as tungsten carbide (WC) tools. The mechanical properties of the joints were evaluated to change the rotation speed and the traveling speed, which are the main process parameters.

D. K. Singh (✉) · K. Sengupta · A. K. Mondal · D. Patra · A. Dhar
National Institute of Technical Teachers' Training and Research, Kolkata, West Bengal, India
e-mail: Dilipkumar.singh@springernature.com

S. Bag et al. (eds.), *Next Generation Materials and Processing Technologies*,
Springer Proceedings in Materials 9, https://doi.org/10.1007/978-981-16-0182-8_38

2 Principal of Friction Stir Joining

Friction stir welding is done by two steps, (I) initially, heat is generated by friction between the tool and the workpiece, and (II) the probe starts mixing the material. Transverse speed and the rotational speed of the tool is controlled to get the welding efficiency [1, 2]. Figure 1 shows the basic diagram of friction stir welding. At first, the base plate is fixed, after that the tool starts to rotate, and the tool pin is plunged into the joining line of the base plate. Then, downward force is applied to the tool, and when the temperature rise became up to the three-fourth of the melting temperature of the base metal, then the transverse speed is given.

It is a solid-state welding process, where the welding occurs by the stirring action of material beneath the tool called the nugget zone. So that the temperature rise always stays under the melting temperature of the base metal, which is to be welded [3]. When FSW was developed from that time, it was used on low temperature melting alloys like aluminum and magnesium [4]. In recent times, friction stir welding is applied on high-temperature melting point alloys. It is started to apply on advance material like steel, titanium alloy, etc. [5]. In the case of high-temperature melting alloys, high tool wear rate occurred very fast, where tooltip damages very excessively and shoulder is affected by friction [6]. So, FSW needs very high frictional force to generate enough heat. Also, tool tip faces heavy load on when it goes to the transverse direction. So, if we apply another heat source

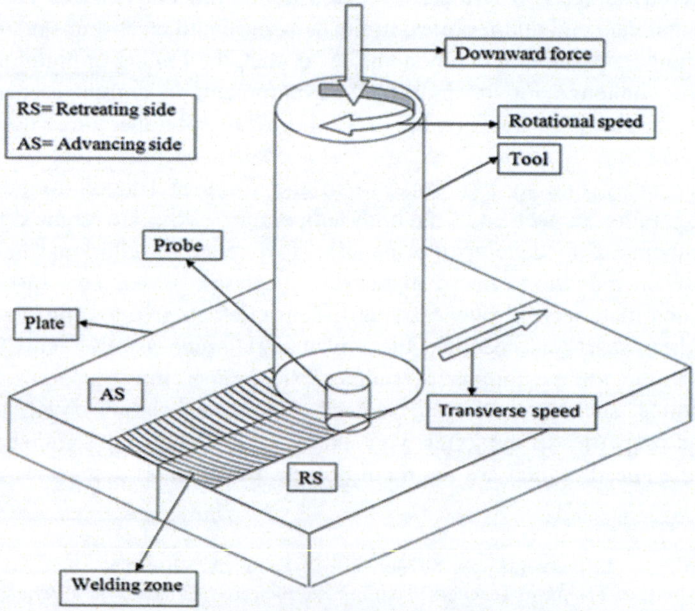

Fig. 1 Schematic diagram friction stir welding

to soften the metal, then the load on the tool will be decreased [6], and it has been found that preheating makes the weld very easy, and it reduces tool wear because of variation in flow stress value. The electrical resistance source of heat is found more successful and available source of energy in all industrial application. There are many other heat sources utilized to make successful high-temperature alloy material joining. The defect-free friction stir weld joint also depends on the process parameters of material joining.

The friction stir joining process parameters are rotational speed, welding speed, tilt angle, tool depth, and the force acting on the work piece as given in Table 1.

3 Experimental Setup

In this experimentation, all the operation is successfully carried out by the universal milling machine as shown in Fig. 2. Essential care has been taken during the welding operation. In Fig. 2, the milling machine has controlled its tool rotational speed by the levers 1 and 2. The levers 3 and 4 are given to control the speed variation of the bed. After setting the desired speed of the bed, the bed is then controlled by the electronics board and switch. The point 5 is the switch to control the speed of the bed. The experiment is carried out on 100 × 75 × 1.6 mm dimension sample with tungsten carbide tool, which has cylindrical 3 mm tool pin diameter with 1.4 mm pin length. The specimen material composition is given in Table 2. The electrical energy is utilized as additional source of energy to produce additional heat energy by using high-voltage transformer from a microwave oven.

As per the schematic shown in Fig. 3, the transformer has two ports which is connected to the milling machine setup as shown in Fig. 3. One port will connect to the welding plate, and another port is connected to the tool pin post. Then, the electric resistance heating exactly will occur at the touch point of the tool pin to the work piece. Then, localized heat will be occurred at the stir zone of the welding. The resistance connection point are in seven location (denoted R).

The welding performed successfully is shown in Fig. 4. The welded specimen has tested the mechanical properties such as tensile strength and hardness. Also, the base metal mechanical properties are tested before performing the welding. The welded sample's tested properties are given in Table 3.

Table 1 Variable parameters and its effects

Parameters	Effects
Rotational speed (rpm)	Frictional heat, stirring, mixing, and oxide layer removal
Welding speed (mm/min)	heat control, plastic flow
Tool tilting angle	The appearance of the weld
Tool plunge depth (mm)	Effective stir material, need moderate insertion depth
Down force (Newton)	Frictional heat, maintaining contact conditions

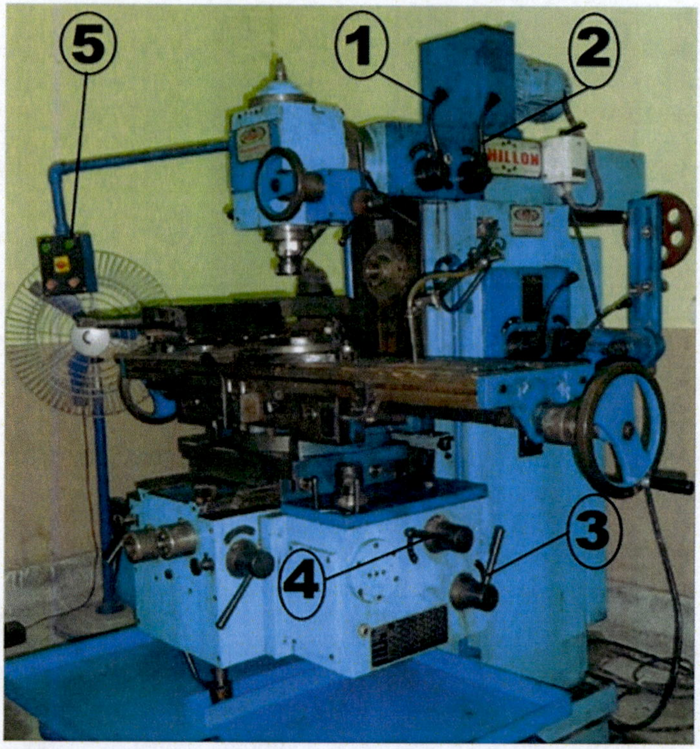

Fig. 2 Universal milling machine

Table 2 Mild steel composition

C	Mn	Si	S	P
0.14	0.92	0.20	0.021	0.028

All the welded plates of mild steel are cut according to the specimen ASTM E8. Figure 4 shows the design specimen of ASTM E8 for tensile testing.

4 Taguchi Optimization

Taguchi technique is a remarkable factual outline of trial tool used to assess the impact of process parameters on output parameters. In Taguchi method, optimization of process parameters is played a key role to getting the high-quality product without increasing cost. The optimization of process parameters in Taguchi method is improved the quality. The optimal process achieved from the Taguchi method is not sensitive to the variation of environment condition and it is not

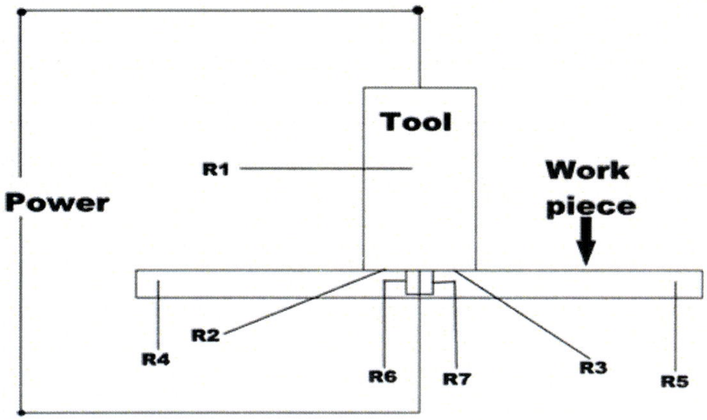

Fig. 3 Schematic diagram of the hybrid system

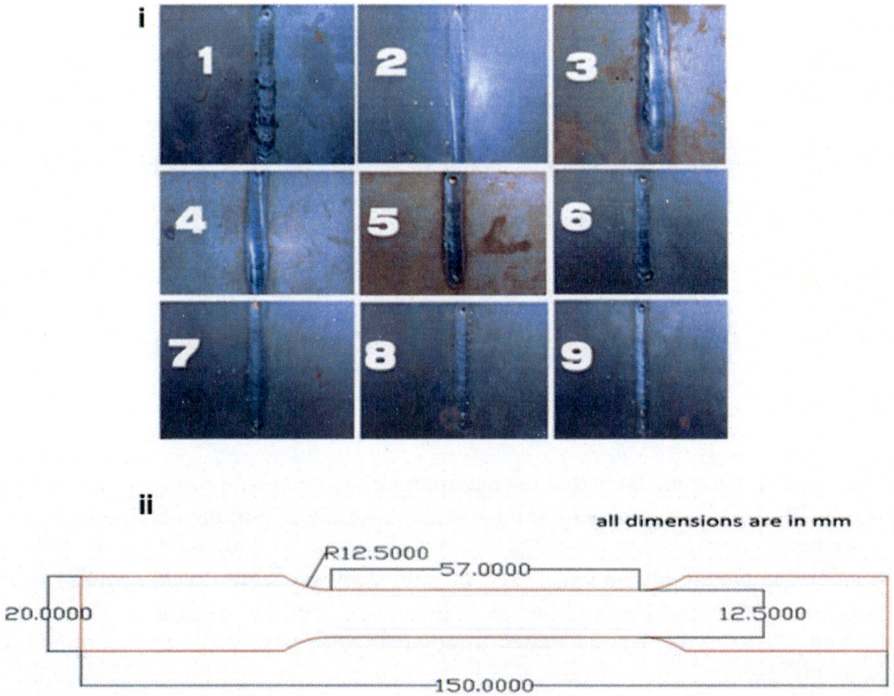

Fig. 4 **a** Schematic diagram of the hybrid system. **b** Tensile test specimen configuration according to the specimen ASTM E8

Table 3 Mechanical properties of welded sample

Sl. no.	Tool rotational speed (rpm)	Welding speed (mm/min)	UTS (MPa)	Hardness (Hv)	Welding efficiency
1	500	24	532	115	89.26
2	500	43	540	116	90.60
3	500	75	458	98	76.84
4	650	24	580	125	97.31
5	650	43	594	128	99.66
6	650	75	501	108	84.06
7	1050	24	573	124	96.14
8	1050	43	568	123	95.30
9	1050	75	500	108	83.89

Table 4 Taguchi L9 array

Taguchi array	L9(3^3)
Factors	3
Runs	9

Columns of L9(3^4) array: 1 2 3

sensitive to the other noise factors. Taguchi method has a great advantage that it gives emphasis to a final performance characteristics value is always nearest to the desired value rather than any value within some selected specified limits, so improvement in the product quality. The L9 orthogonal array selected based on the DOE calculated by using Minitab software (Table 4).

5 Result and Discussion

It can be seen from the table that the maximum ultimate tensile strength for the mild steel is 594 MPa which is very similar to the base metal. So, the obtained result is desirable.

After completing all the tensile test of mild steel, successful weld samples were tested. All the output result obtained for stainless steel are given in Table 2. The welding efficiency is also calculated in compare with base metal as mentioned in Eq. (1) (Fig. 5).

$$\text{Welding efficiency} = \frac{\text{UTS of weled sample}}{\text{UTS of base material}} \times 100 \quad (1)$$

In Table 3, we can see that the best welding efficiency for mild steel is at tool rotational speed 650 rpm and welding speed 43 mm/min. For the experiment, sample 5 has the welding efficiency near to 100%. It means that in 650 rpm, the rate

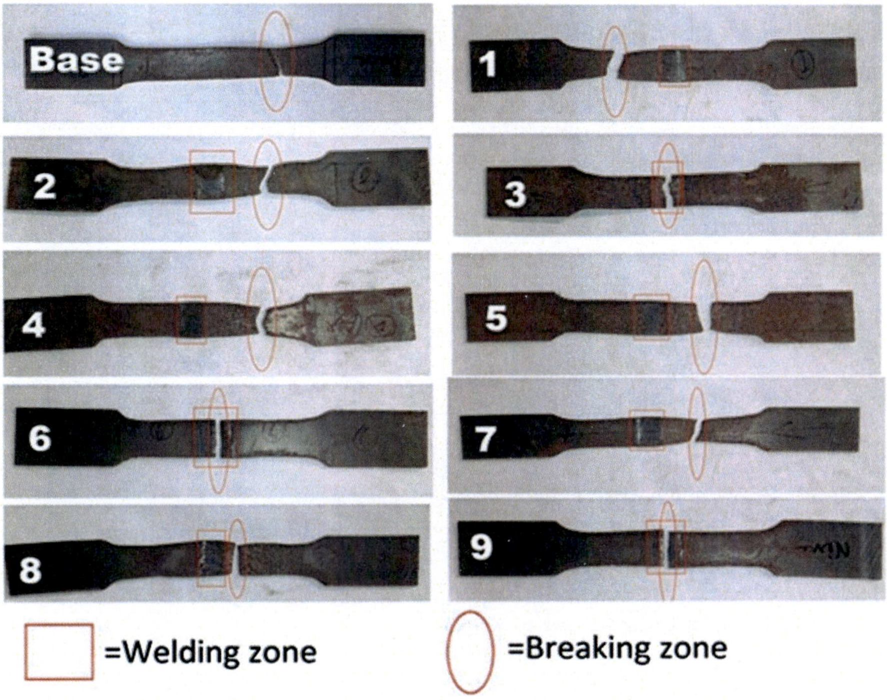

Fig. 5 Specimen tensile strength test

of heat input is perfect, so that the tensile strength of the welding is high. The poor tensile strength I got is at the 500 rpm. It is proven that the heat generation is lower at lower rotational speed.

In case of sample no 8, the failure occurred at the TMAZ in the advancing side because there is not much material flow occurred. The breaking zone for stainless steel is shown in Fig. 6. In the figure, it is clearly shown that nearly in every case, the tensile failure occurred at the welding zone (samples 1, 2, 3, 4, 5, 6, and 9). So, the results showed that for these cases, the heat input at the stir zone did not match with the required heat input. So due to this reason, the welding defects occurred at the welding zone. In case of welding samples 8 and 9, it can show in the above figure that the failure occurred at the advancing side of HAZ. It implies that the required heat input at the stir zone is fulfilled by the rotational speed 1050 rpm. Welding speed has effect on the welding strength. It has been observed that when the welding speed was 75 mm/min, then the welding strength is getting lesser. For minimum welding speed, both the steels have the strength just lesser than the best result. Similarly, the tool rotational speed has effects on the tensile strength of the joint. The optimization of tensile strength against the process parameters S/N ratio is given in Table 5.

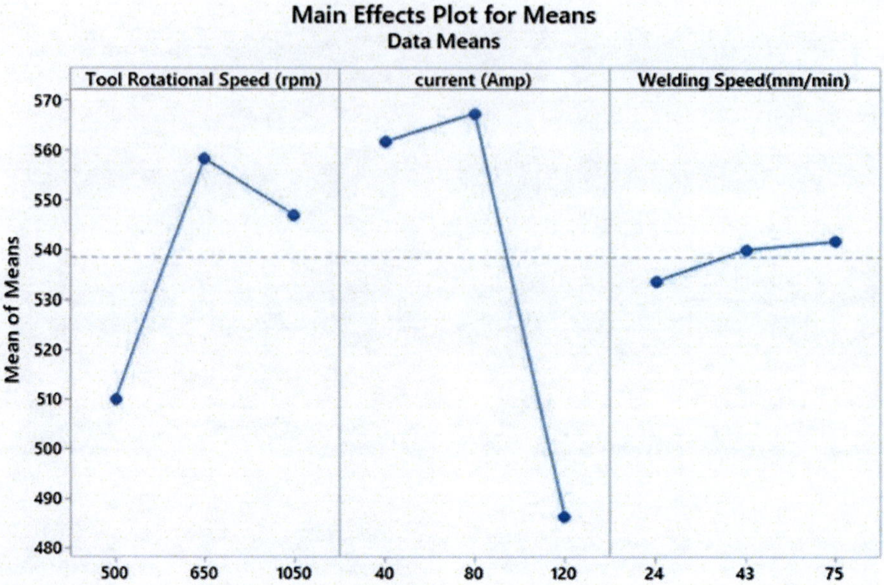

Fig. 6 S/N Ratio plot graph for tensile strength

Table 5 Tensile strength S/N ratio—larger is better

Tool rotation (rpm)	Current (A)	Welding speed (mm/min)	UTS (MPa)	SNRA1
500	40	24	532	54.51823265
500	80	43	540	54.6478752
500	120	75	458	53.21730956
650	40	43	580	55.26855987
650	80	75	594	55.4757289
650	120	24	501	53.99675452
1050	40	75	573	55.16309244
1050	80	24	568	55.08696671
1050	120	43	500	53.97940009

The process parameters optimization is performed to achieve maximum tensile strength by using Taguchi static tool.

Figure 6 represent the tensile strength optimization graph and found that maximum tensile strength is achieved at 650 rpm, current 80 A, and 75 mm/min welding speed. Similarly, the hardness optimization has been performed and is given in Table 6 (Fig. 7).

Similarly, the optimum hardness is achieved at 650 rpm, current 80 A, and 75 mm/min welding speed. Also, the optimization is performed for the welding

Table 6 Hardness S/N ratio—larger is better

Tool rotational speed (rpm)	Current (A)	Welding speed (mm/min)	Hardness (Hv)	SNRA1
500	40	24	115	41.21395681
500	80	43	116	41.28915978
500	120	75	102	40.17200344
650	40	43	125	41.93820026
650	80	75	128	42.14419939
650	120	24	108	40.66847511
1050	40	75	124	41.8684337
1050	80	24	123	41.79810223
1050	120	43	108	40.66847511

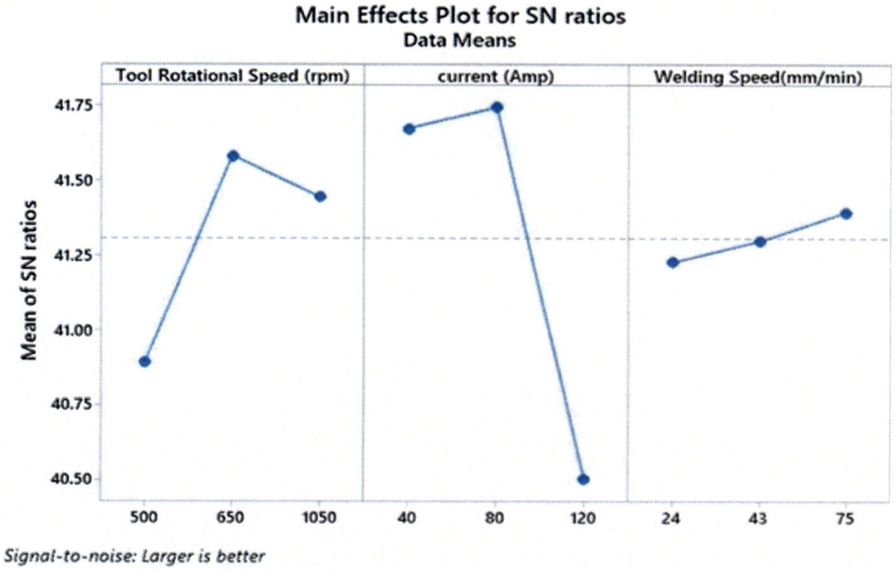

Fig. 7 S/N Ratio plot graph for hardness

efficiency and shows that the results higher is better in S/N ratio. The Taguchi static tool optimization confirmed that the larger is better signal-to-noise ratio found maximum at 650 rpm and 75 mm/min welding speed with 80 A current for additional heat generation (Fig. 8, Table 7).

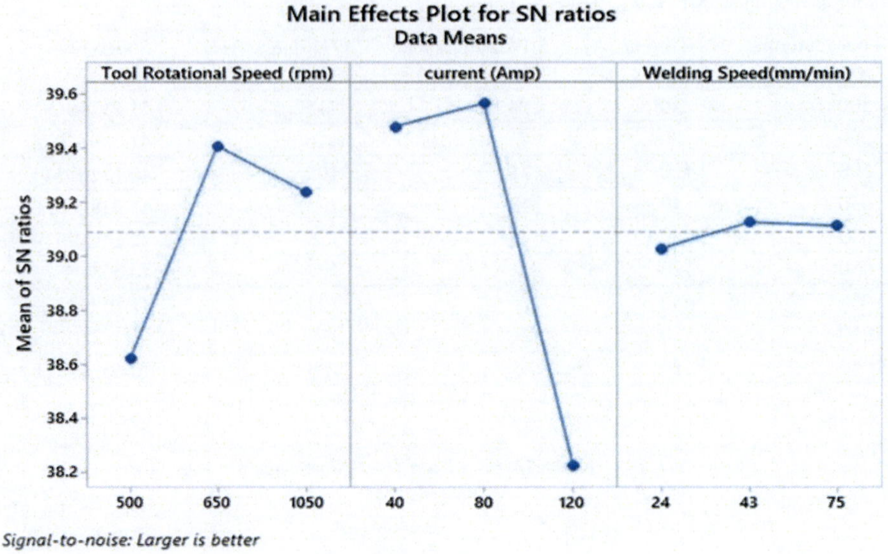

Fig. 8 S/N Ratio plot graph for welding efficiency

Table 7 Welding efficiency and S/N ratio—larger is better

Tool rotational speed (rpm)	Current (A)	Welding speed (mm/min)	Welding efficiency	SNRA1
500	40	24	89.26	39.01313765
500	80	43	90.6	39.14256395
500	120	75	76.84	37.71174712
650	40	43	97.31	39.76314945
650	80	75	99.66	39.97041766
650	120	24	84.06	38.49178771
1050	40	75	96.14	39.65808236
1050	80	24	95.3	39.58185801
1050	120	43	83.89	38.47420389

6 Microstructure Analysis

The material thus joined has been studied for the microstructure analysis. Figure 9 represents the microstructure of the joint. From the microstructural study of the base material, as shown in Fig. 9a, the base material contained disoriented ferrite grains with some fine pearlite grains. Figure 9b represents the microstructure of the HAZ, where it has been found that the ferrite-to-pearlite ratio has been found decreased. In HAZ, the upper grain size has been monitored to be 14.6 μm as equivalent circular diameter. However, in the stir zone, the upper size of the grains has been

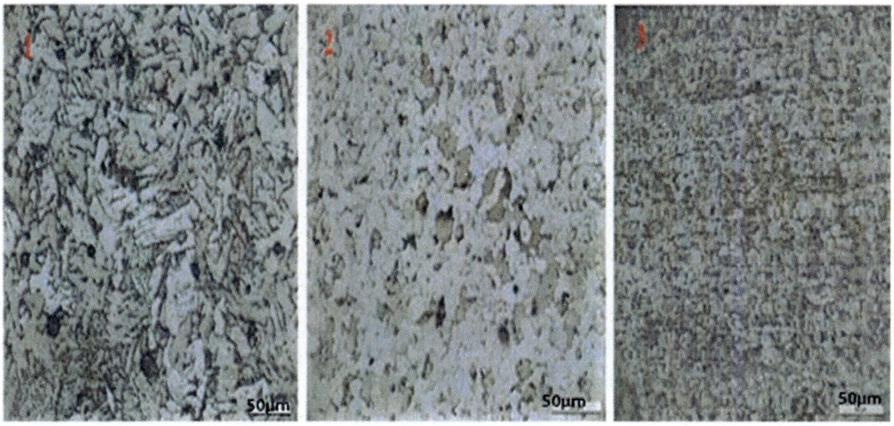

Fig. 9 **a** Base metal, 50 µ. **b** HAZ, 50 µ. **c** SZ, 50 µ

found to be 7.2 µm. The microstructure of the stir zone is shown in Fig. 9c, and structure analysis of this region has been found to be tempered martensitic. The martensitic structure has been formed from the austenitic structure during welding by stirring action with sufficient amount of cooling rate. The martensitic structure has been found to be the reason for the increase in hardness of the material in the stir zone compared to the other zones of the weld joint.

7 Conclusion

In this experimental analysis, the FSW weldability and mechanical properties of mild steel, which is an advanced high-strength steel, are investigated using the tungsten carbide tool. From the experiment results, following conclusions can be drawn.

(i) Defect-free joints were formed at 75 mm/min at 650 rpm using the WC tool. However, a groove-like defect is found along the joint line due to the insufficient heat input as shown in Fig. 4.
(ii) The welded joint tensile strength found maximum 594 MPa with welding efficiency 99% efficiency. The impact test characteristics were like the tensile test characteristics.
(iii) The hardness traverse measurement showed that the hardness at the stir zone found high 128 Hv which is near to the base metal. The base metal is deemed to have been reformed by the decrease in the ferrite matrix and the increase in the martensite due to the quenching effect after the FSW.
(iv) The process parameters optimization results are found at 650 rpm, 75 mm/min welding speed, and 80 A electric resistance heat source.

(v) The microstructure study of the joint represented that the size of the stir zone of the joint has been found to be 7.2 μm, least of the grain size among zones of weld.

(vi) The hardness of the stir zone increased due to conversion of the microstructure from austenitic to the martensitic due to the high cooling rate of the joint during joining process.

8 Future Scope of Study

(i) Forces acting on the tool can be computed for various welding and rotational speeds which will help to determine the radial and axial loads on the tool which may help in tool optimization and to increase in efficiency.

(ii) The mechanical properties such as residual stress, fracture, fatigue, and corrosion have good scope in future on ERFSW joint which may be needed to understand the failure and to calculate factor of safety (FOS) of the joint.

(iii) The effect of post-heat treatment of FSW joints with different aging treatments to improve the tensile strength can be studied.

(iv) Study of dissimilar friction stir welding between copper–brass and aluminum–magnesium can be attempted.

(v) More experiments with different tool materials and geometries should be attempted in order to improve the tensile strength and make the process acceptable to the welding industries.

(vi) Study of heat transfer analysis can be extended to lap friction stir welded joints.

(vii) The process of FSW can be extended to thermo-setting plastics.

(viii) ERFSW may be used for welding titanium alloys and with dissimilar high-strength metals with FSW by utilization of the additional heat input of the electric resistance. ERFSW may be proved to be one of the upcoming greens joining process for industries.

(ix) Automobile industries may use it with the robotic weld design for similar and dissimilar material which may be proved to be green technology joining process.

References

1. Reynolds AP, Tang W, Possada M, Deloch J (2003) Friction stir welding of DH36 steel. Sci Technol Weld Joining 8(6):455–460
2. Lienert TJ, Stellwag WL Jr, Grimmett BB, Warke RW (2003) Friction stir welding studies on mild steel. Weld J 82(1):1s–9s
3. Konkol PJ, Mathers JA, Johnson R, Pickens JR (2003) Friction stir welding of HSLA65 steel for shipbuilding. J Ship Prod 19(3):159–164

4. Reynolds AP, Tang W, Gnaupel-Herold T, Prask H (2003) Structure, properties and residual stress of 304L stainless steel friction stir welds. Scr Mater 48(9):1289–1294
5. Deb Roy T, De A, Bhadeshia HKDH, Manvatkar VD, Arora A (2012) Tool durability maps for friction stir welding of an aluminium alloy. Proc R Soc A 468:3552–3570
6. Nandan R, Roy GG, Lienert TJ, Deb Roy T (2007) Three-dimensional heat and material flow during friction stir welding of mild steel. Acta Mater 55:883–895

Optimized Design and Performance Testing of a 1.5 MW Wind Turbine Blade

Rajendra Roul and Awadhesh Kumar

1 Introduction

With the advent of technology, the demand for energy is increasing dramatically, and due to this, depletion of fossil fuel is increasing monotonically which further leads to pollution and becomes an environmental concern for engineers and scientist. To mitigate this issue, various renewable energies are being employed, for example, wind energy, solar energy, and hydro energy [1–4]. In the present scenario, wind energy is dominating worldwide for its innovative and fastest-growing clean energy approach [5, 6]. The effective production and usage of wind energy are highly important for improved energy infrastructure and sustainability. To harvest wind energy, typically two types of configuration are used: The one is horizontal axis wind turbine and other is vertical axis wind turbine. Due to various advantages, HAWT is dominating the wind industry [7]. To improve the competitiveness of wind energy with other sources of energy, researchers are focusing to develop wind turbines which show a reduction in cost of energy with increase in annual energy output and lower overall costs.

Wind turbine is composed of three important parts: The first is rotor which consists of two or three blades, the second is nacelle which is the house for gearbox, rotor shaft, generator, and control system, and the third is the tower. Among all the three parts highlighted above, blade plays a vital role in capturing wind energy because of its design, and because of this, a lot many researches are going on in designing and optimizing blades. The blade design process focuses on both aerodynamic and structural requirements of the blade. For aerodynamic point of view, wind turbine optimization is executed by using blade element momentum (BEM) theory without considering the three-dimensional changes nor modifying any airfoil shape [8]. The advantage of BEM theory is that it provides calculation very fast, and because of this, it is extensively used in blade optimization and

R. Roul (✉) · A. Kumar
National Institute of Technology Rourkela, Rourkela, Odisha, India

S. Bag et al. (eds.), *Next Generation Materials and Processing Technologies*, Springer Proceedings in Materials 9, https://doi.org/10.1007/978-981-16-0182-8_39

aero-elastic analysis [9–12]. However, such methods provide solution which is quite different from real-time solution [13], and this is due to the absence of viscous and 3D effect. To resolve this issue, researchers use the concept of parameterization of airfoils to modify the shape of the airfoil and optimize the airfoil [14, 15]. Parameterization of the airfoil is the method where the shape and coordinate of the airfoil are controlled by specific variable called control variables. These control variables then act as an input for an optimization of the airfoil. For parameterization of airfoils and utilizing it for optimization, various methods are available like parsec method [16], splines [17], B-splines [18, 19], and Bezier curve [20–24]. Few researchers also work on polynomial approach of parameterization technique and found that it is highly influencing the final optimized design [25]. More details of state of the art of parameterization technique can be found in the literature [26]. Although various parameterization techniques have been used for airfoil parameterization, to the author knowledge, parsec parameterization technique has not been applied for wind turbine application. For parameterization and optimization to carry out, it is necessary to choose the optimization algorithm, and this algorithm usually falls within two categories: One is gradient-based methods and another is heuristic algorithms. Former optimization method is a widely used optimization method due to its fast process and has been implemented to optimize wing section, airplanes and also to the turbo-machinery airfoil [27–29]. However, they never lead to global optimism. Therefore, testing various initial conditions becomes important to have some project accuracy. These methods are not reliable in this sense [30]. Although later optimization like genetic algorithms is considered to be slower, they are more efficient and powerful. A variety of individual cases, including airfoil [21, 22], wings [24], and entire airplanes [31], were applied. The combination of these two types of algorithms is often used by hybrid methods [32, 33].

In this paper, MATLAB code was written to implement parsec parameterization technique and coupling it with X-foil [34] and genetic algorithm to get the control variables, optimize the airfoil, and check the aerodynamic coefficients of the base and optimized airfoil. NREL S818 airfoil, NREL S825 airfoil, and NREL S826 airfoil are considered in this analysis for parameterization, optimization, and design of wind turbine blade.

2 Airfoil Parameterization

Parsec parameterization is used in this method. This method uses 12 variables to control the airfoil surface. These 12 variables act as a design variable which is further used in the optimization process to get an optimized airfoil. The name of the twelve variables is shown in Table 1. The upper and lower surfaces of airfoil are obtained by using six-order polynomial equation shown in Eqs. 1 and 2.

Table 1 Twelve design variables

(1)	Upper radius (leading edge-R_{leu})
(2)	Lower radius (leading edge-R_{lel})
(3)	Crest point (upper Y_{up})
(4)	Crest point (lower Y_{lo})
(5)	Position (upper crest X_{up})
(6)	Position (lower crest X_{lo})
(7)	Curvature (upper crest Yxx_{up})
(8)	Curvature (lower crest Yxx_{lo})
(9)	Edge offset (trailing-T_{off})
(10)	Edge thickness (trailing-TTE)
(11)	Edge direction angle (trailing-α_{TE})
(12)	Edge wedge angle (trailing-β_{TE})

$$Z_{up} = \sum_{i=1}^{n=6} a_{up}^{i} \cdot x^{i-\frac{1}{2}} \quad (1)$$

$$Z_{lo} = \sum_{i=1}^{n=6} a_{lo}^{i} \cdot x^{i-\frac{1}{2}} \quad (2)$$

where Z_{up} is the appropriate upper surface y-coordinate and Z_{lo} is the coordinate for the lower surface signifying y-direction, and a_{up}^{i} and a_{lo}^{i} are the coefficients representing 12 control variables that are essential to be determined. Apart from this, to solve the Eqs. 1 and 2, the following conditions are required which are given below

1. At $x(u, l)$ = maximum, $y(u, l)$ = maximum
2. At $x(u, l)$ = maximum, $\frac{dy_{(u,l)}}{dx} = 0$
3. At $x(u, l)$ = maximum, $\frac{d^2 y_{(u,l)}}{dx^2} = \text{maximum}$
4. At $x_u = 1,\ y_u = T_{off} + \frac{T_{TE}}{2}$
5. At $x_l = 1,\ y_l = T_{off} - \frac{T_{TE}}{2}$
6. At $x_u = 1,\ \frac{dy_u}{dx} = \tan\left(\alpha_{TE} - \frac{\beta_{TE}}{2}\right)$
7. At $x_u = 1,\ \frac{dy_l}{dx} = \tan\left(\alpha_{TE} + \frac{\beta_{TE}}{2}\right)$

3 Objectives and Geometry Constraints

The objectives are to maximize coefficient of lift and

$$F(Z_{up}, Z_{lo}, \mathrm{Re}) = \sum_{i=1}^{n} \frac{\mathrm{cl}_i}{\mathrm{cd}_i} \times \frac{1}{p}$$

To avoid overlap between two airfoil surfaces, the following condition was set

$$Z_{up} - Z_{lo} > 0$$

To provide structural strength to the wind turbine blade, the thickness is retained between 7.5 and 13.5% of the chord using the condition shown below.

$$0.135 > Z_{up} - Z_{lo} > 0.075$$

To avoid very highly cambered airfoil, few limits were set to the y-coordinates which are shown below. However, the *x*-coordinates were fixed to reduce the control variables,

$$Z_{up} \leq 0.18a$$

$$Z_{lo} \geq -0.1$$

4 Design of Wind Turbine Blade

The wind turbine model considered in this paper is shown in Fig. 1. The three-dimensional wind turbine blade is designed by considering the geometrical parameters given in the report [35]. This blade contains three types of the airfoil which are placed from root to tip, and those airfoils which are used in designing the wind turbine blade are given in Fig. 2. The specifications of the blade and rotor parameters are given in Table 2.

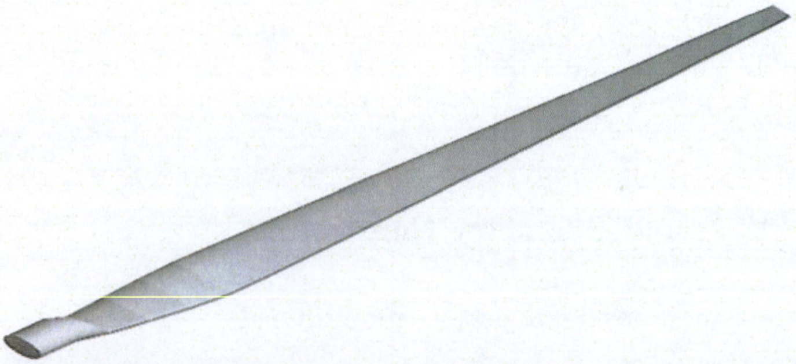

Fig. 1 Three-dimensional referenced blade

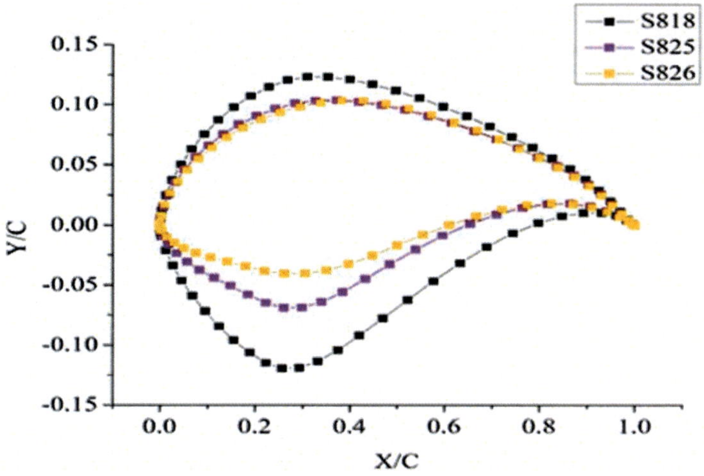

Fig. 2 Airfoils used in the blade

Table 2 Main parameters of referenced turbine

Parameters	Values	Units
Rated power, p_{rated}	1.5	MW
Number of blades, N_B	3	Not applicable
Rotor radius, R	41.25	m
Angular velocity, ω	2.22	rad/sec
Swept area	5346	m^2

5 Results and Discussions

5.1 Control Variables of Base Airfoils and Optimized Airfoil

Tables 3, 4, and 5 provide an information of parsec variables of original airfoil and optimized airfoil. By using Eqs. 1 and 2, the control variables are obtained. The original parsec variables were taken as input to carry out the optimization process.

5.2 Comparison of Base Airfoil and Optimized Airfoil

Figures 3, 4, and 5 are demonstrating the curve of original and optimized airfoil. Maximum thickness of S818 airfoil is 24.40% of chord occurred at the position 29.30% chord, whereas its optimized airfoil exerts maximum thickness as 21.40% of chord occurred at the position 28.40% of chord. Maximum thickness of S825 airfoil is 17.07% of chord occurred at the position 29.90% of chord, whereas its

Table 3 Parsec parameters for S818 airfoil

Control variables	Original outcome	Optimized outcome
(R_{leu})	0.0176	0.025418
(R_{lel})	0.0047	0.011372
(X_{up})	0.3606	0.359227
(YXX_{up})	−0.8238	−0.8241
(X_{lo})	0.2633	0.271728
(Y_{lo})	−0.0661	−0.06828
(YXX_{lo})	1.5987	1.59913
(TTE)	0	0.000319
(T_{off})	0.0004	−0.00358
(α_{TE})	16.5305	16.5305
(β_{TE})	11.5973	11.5973

Table 4 Parsec parameters for S825 airfoil

Control variables	Original outcome	Optimized outcome
(R_{leu})	0.0176	0.025418
(R_{leu})	0.0047	0.011372
(X_{up})	0.3606	0.359227
(Y_{up})	0.1034	0.115891
(YXX_{up})	−0.8238	−0.8241
(X_{lo})	0.2633	0.271728
(Y_{lo})	−0.0661	−0.06828
(YXX_{lo})	1.5987	1.59913
(TTE)	0	0.000319
(T_{off})	0.0004	−0.00358
(α_{TE})	16.5305	16.5305
(β_{TE})	11.5973	11.5973

Table 5 Parsec parameters for S826 airfoil

Control variables	Original outcome	Optimized outcome
(R_{leu})	0.0212	0.027152
(R_{lel})	0.0064	0.011853
(X_{up})	0.3904	0.391641
(Y_{up})	0.1226	0.130074
(YXX_{up})	−0.8927	−0.89293
(X_{lo})	0.207	0.208656
(Y_{lo})	−0.0338	−0.02531
(YXX_{lo})	0.5629	0.562999
(TTE)	0	0.012579
(T_{off})	0.0004	0.011381
(α_{TE})	14.4516	14.4516
(β_{TE})	15.2262	15.2262

optimized airfoil exerts maximum thickness as 14.04% of chord occurred at the position 33.70% of chord. Maximum thickness of S826 airfoil is 14.40% of thickness occurred at the position 33.70% of chord, whereas its optimized airfoil exerts thickness 15.09% of chord occurred at the position 33.90% of chord. Each of these airfoils has unique role in determining the efficiency of wind turbine, and because of this, they are placed at different spans of the blade according to their thickness. S818 airfoil is kept at the root position of the blade, S825 airfoil is kept at mid position of the blade, and S826 airfoil is placed at the tip position of the blade.

5.3 Comparison of Aerodynamic Coefficients

The ratio of cl/cd is very important parameter to consider while designing the wind turbine blade. The maximum cl/cd at a given angle of attack for a particular airfoil can be used as a twist angle for modifying the straight blade and also can be used for the twisted blade to get better aerodynamic efficiency. Figures 6, 7, and 8 show the cl/cd ratio curve of S818 airfoil, S825 airfoil, and S826 airfoil. The maximum cl/cd of S818 airfoil and its optimized airfoil occurred at 10° angle of attack. But its optimized airfoil shows 27.63% increment of cl/cd value as compared to the base one. Now, for S825 airfoil and its optimized airfoil, the best cl/cd ratio occurred at 6° angle of attack in which optimized airfoil shows an increment of 22.44% of cl/cd as compared to the base S825 airfoil. Finally, for S826 airfoil and its optimized airfoil, the best cl/cd ratio occurred at same 6° angle of attack, and it has been found that optimized airfoil of S826 airfoil shows a same increment as that of optimized S825 airfoil. Figures 9, 10, and 11 give an information about the stall point which plays a vital role in investigation of the performance parameter of airfoil and wind turbine blade. Stall angle is the angle at which coefficient of lift decreases after reaching maximum point. Stall angle for S818 airfoil occurred at 14° angle of attack, and for S825 airfoil and S826 airfoil, it occurred at 16° and 6° angle of attack.

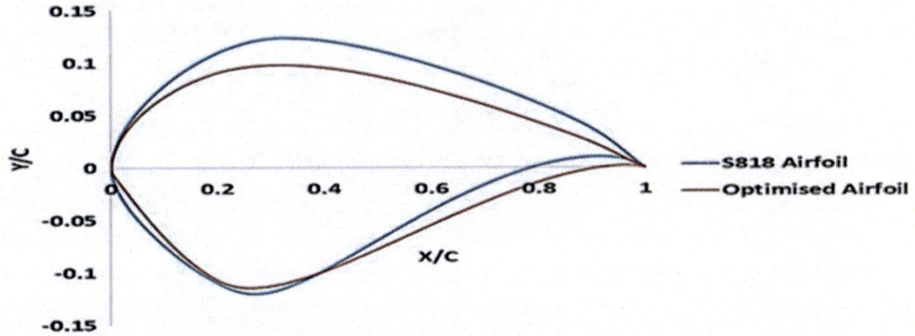

Fig. 3 Original NREL S818 airfoil and optimized airfoil

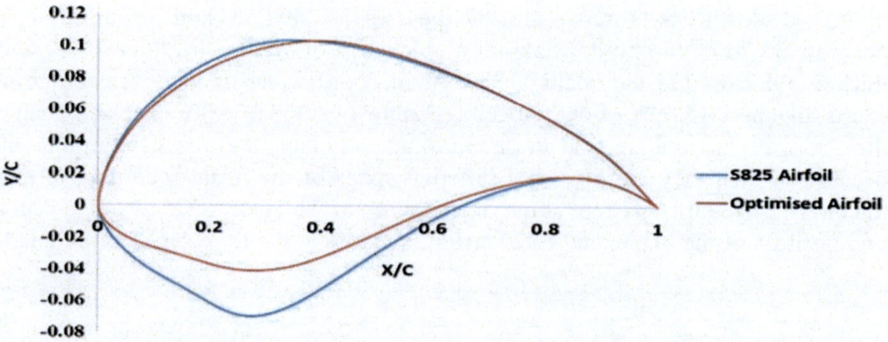

Fig. 4 Original NREL S825 airfoil and optimized airfoil

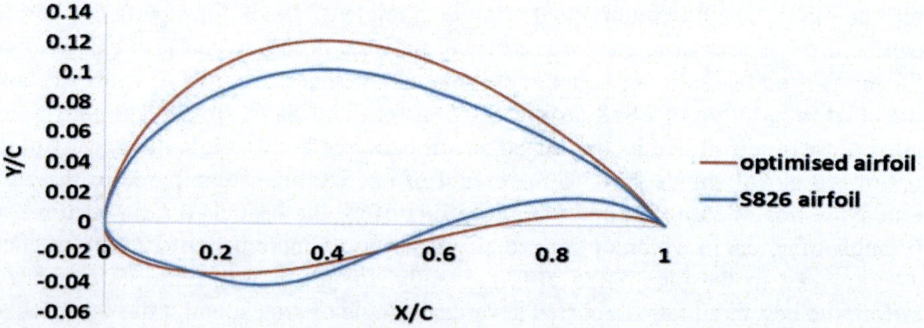

Fig. 5 Original NREL S825 airfoil and optimized airfoil

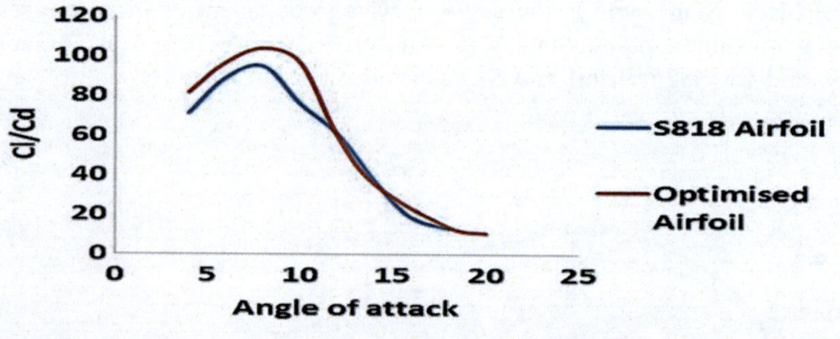

Fig. 6 Comparison of cl/cd ratio between S818 airfoil and optimized airfoil

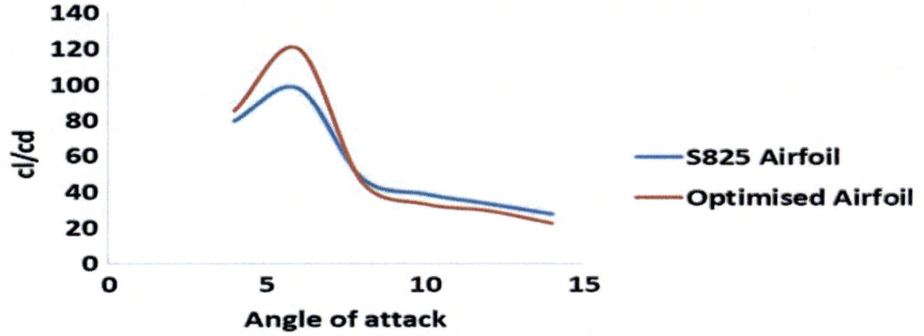

Fig. 7 Comparison of cl/cd ratio between S825 airfoil and optimized airfoil

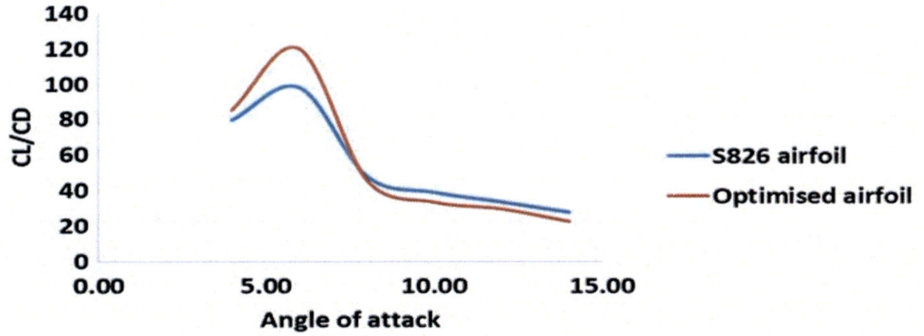

Fig. 8 Comparison of cl/cd ratio between S826 airfoil and optimized airfoil

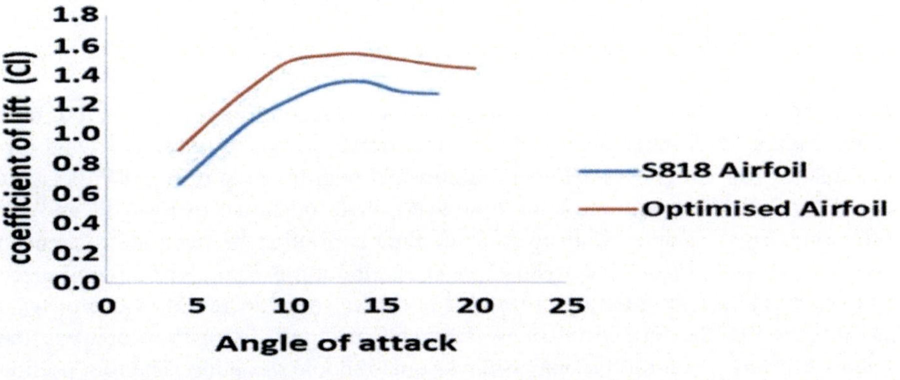

Fig. 9 Comparison of coefficient of lift between S818 airfoil and optimized airfoil

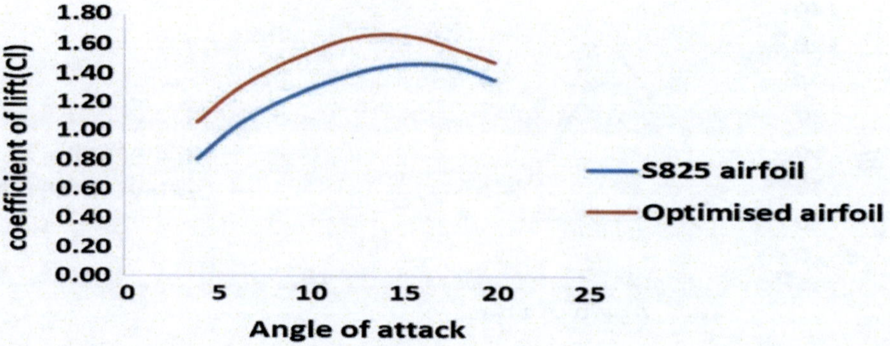

Fig. 10 Comparison of coefficient of lift between S825 airfoil and optimized airfoil

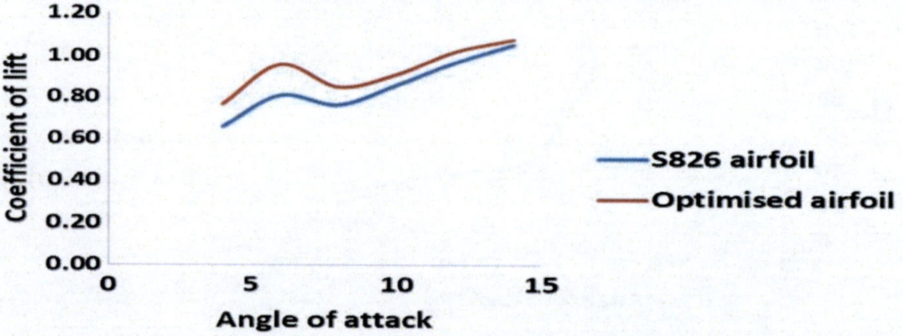

Fig. 11 Comparison of coefficient of lift between S826 airfoil and optimized airfoil

5.4 Aerodynamic Analysis

Aerodynamic analysis of the blade starts with the design of optimized wind turbine blade. Figure 12 is reflecting the three-dimensional design of optimized wind turbine blade. The design of blade is accomplished with the help of SolidWorks. The chord and twist distribution of the optimized blade is shown in Figs. 13 and 14. Moreover, Figs. 13 and 14 also demarcate the distribution between the referenced and the optimized blade. The trend of twist distribution is same for both optimized and original blades, but the optimized blade shows some decreasing value of twist as compared to the referenced one. The trend of chord distribution between the optimized blade and original blade shows some random variation. The root portion of the optimized blade shows more chord value as compared to the referenced one. This is because root portion of the wind turbine blade is more prone to get damage due to bending. Hence, it is necessary to keep the root portion thicker as compared to the other section of the blade.

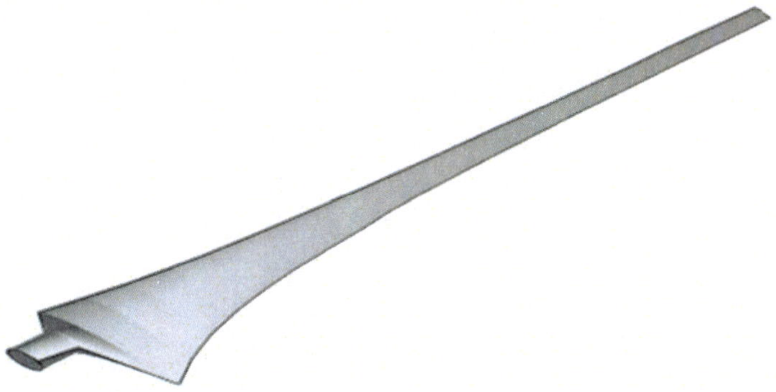

Fig. 12 Three-dimensional optimized blade

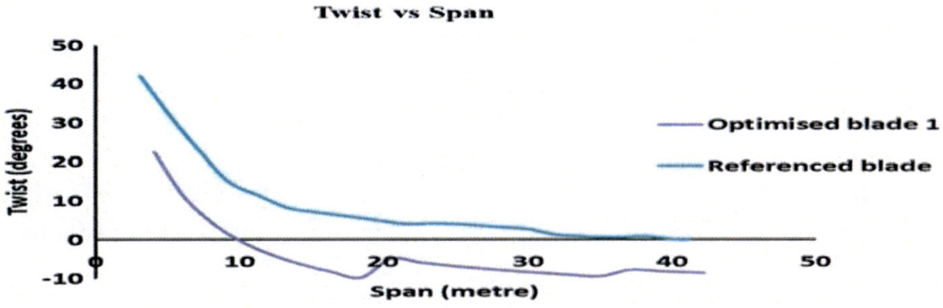

Fig. 13 Comparison of twist distribution between referenced blade and optimized blades

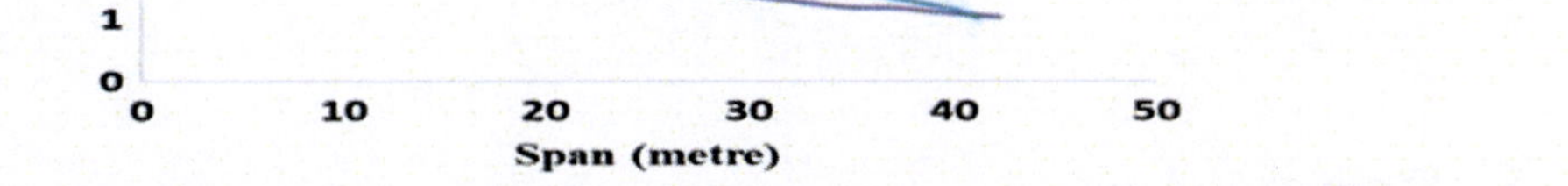

Fig. 14 Comparison of chord distribution between referenced blade and optimized blades

The performance parameter of wind turbine is determined according the solution of power coefficient and power output. Figures 15 and 16 show the comparison of power coefficient and power of original blade and referenced blade. From the figure, it has been found that at velocity 12 m/s, the optimized blade shows

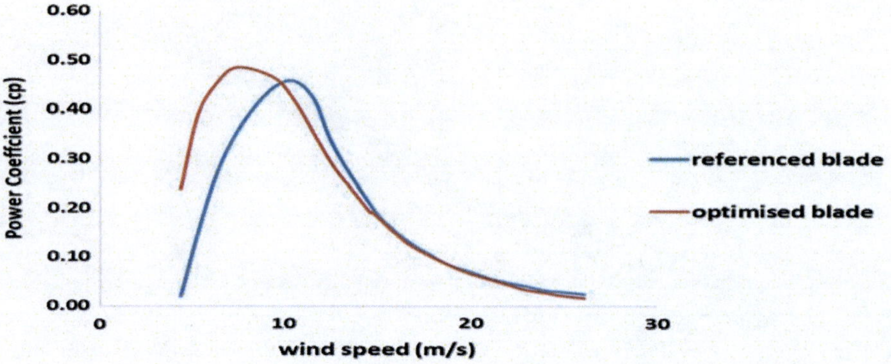

Fig. 15 Comparison of power coefficients between original blade and optimized blade

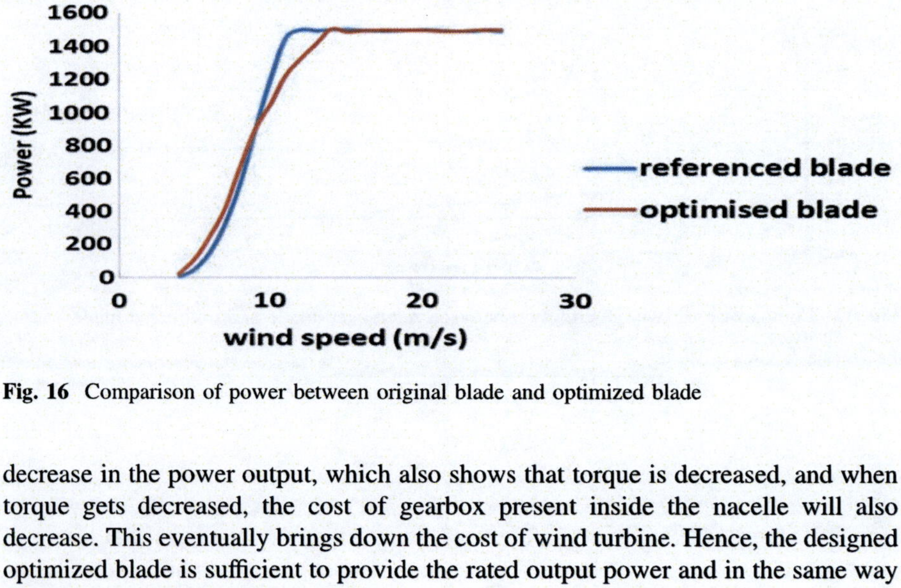

Fig. 16 Comparison of power between original blade and optimized blade

decrease in the power output, which also shows that torque is decreased, and when torque gets decreased, the cost of gearbox present inside the nacelle will also decrease. This eventually brings down the cost of wind turbine. Hence, the designed optimized blade is sufficient to provide the rated output power and in the same way also shows reduction in rotor cost.

5.5 Structural Analysis

In this section, the plot of flapwise deflection, edgewise deflection, and total thrust of the blade is shown. An open-source software called Q-Blade [36] has been used to perform the structural analysis. Wind speed of 12 m/s has been considered for the structural analysis. Figures 17 and 18 show that deflection in both directions of the optimized blade reflects less value as compared to the referenced blade. Less

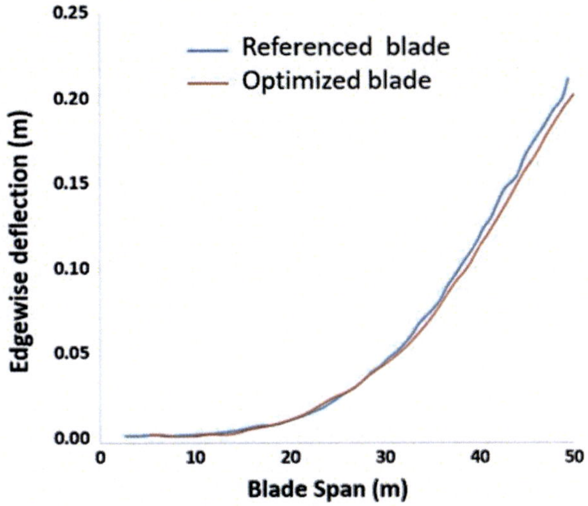

Fig. 17 Edgewise deflection of original and optimized blades

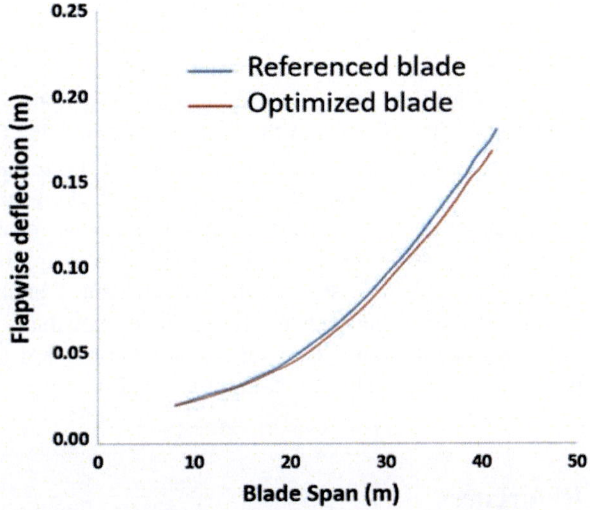

Fig. 18 Flapwise deflection of original and optimized blades

deflection is a good sign for the life span of the wind turbine blade. In addition to it, Fig. 19 demonstrates the curve of thrust against the wind speed, and it has been found that at 12 m/s, there is a decrement of 2.2% in the value of thrust in case of optimized one.

Decrease in thrust value can lead to reduction in the cost of the rotor.

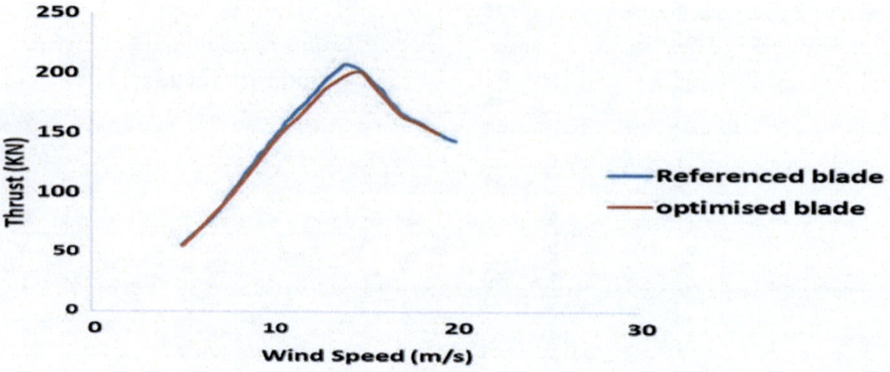

Fig. 19 Thrust of original and optimized blade systems

6 Conclusions

This work presents a multi-objective aerodynamic and structural optimization method to design HAWT blades. A procedure starts with the parameterization of airfoil, followed by optimization of airfoil and design of wind turbine blade. Genetic algorithm has been used to carry out optimization process. XFoil is integrated with the code to get the aerodynamic results. Comparison of optimized and original airfoil has been done and been observed that optimized airfoil shows an increment of more than 20% in the value of cl/cd ratio which shows great sign to get good aerodynamic output of optimized wind turbine. Moreover, the best angle of attack is also observed. A 1.5 MW blade is applied as the referenced configuration, and Q-blade is used for the structural analysis. Deflection in both the directions along the span of the blade is calculated and has been observed that optimized blade shows less deflection as compared to referenced one. Thrust and power output also reflected some meaningful output in the case of optimized blade.

References

1. Grujici M, Arakere G, Subramanian E, Sellappan V, Vallejo A, Ozen M (2010) Structural-response analysis, fatigue-life prediction, and material selection for 1 MW horizontal-axis wind-turbine blades. J Mater Eng Perform 19:790–801
2. Wang TG, Wang L, Zhong W, Xu BF, Chen L (2012) Large-scale wind turbine blade design and aerodynamic analysis. Chin Sci Bull 57:466–472
3. Ashuri T, Zaaijer MB, Martins JRRA, van Bussel GJW, van Kuik GAM (2014) Multidisciplinary design optimization of offshore wind turbines for minimum levelized cost of energy. Renew Energy 68:893–905
4. Tchakoua P, Wamkeue R, Ouhrouche M, Slaoui-Hasnaoui F, Tameghe TA, Ekemb G (2014) Wind turbine condition monitoring: state-of-the-art review, new trends, and future challenges. Energies 7:2595–2630

5. Yurdusev MA, Ata R, Çetin NS (2006) Assessment of optimum tip speed ratio in wind turbines using artificial neural networks. Energy 31:2153–2161
6. Maki K, Sbragio R, Vlahopoulos N (2012) System design of a wind turbine using a multi-level optimization approach. Renew Energy 43:101–110
7. Yang Z, Yin M, Xu Y, Zhang ZY, Zou Y, Dong ZY (2016) A multi-point method considering the maximum power point tracking dynamic process for aerodynamic optimization of variable-speed wind turbine blades. Energies 9:425
8. Fuglsang P, Madsen HA (1999) Optimization method for wind turbine rotors. J Wind Eng Ind Aerodin 80:191–206
9. Burton T, Sharpe D, Jenjins N, Bossanyi E (2001) Wind energy handbook. Wiley, West Sussex
10. Hansen MOL (2008) Aerodynamics of wind turbines, 2nd ed. Earthscan, London
11. Jureczko M, Pawlak M, Mezyk A (2005) Optimisation of wind turbine blades. J Mater Process Technol 167(2–3)–463–471
12. Wang X, Shen WZ, Wei JZ, Jens NS, Chen J (2009) Shape optimization of wind turbine blades. Wind Energy 12(8):781–803
13. Simms S, Schreck S, Hand M, Fingersh LJ (2001) NREL Unsteady aerodynamics experiment in the NASA-ames wind tunnel: a comparison of predictions to measurements. NREL/TP-500-29494, June 2001
14. Samareh JA (2001) Survey of shape parametrization techniques for high-fidelity multidisciplinary shape optimization. AIAA J 39(5):877–884
15. Wu HY, Yang SC, Liu F, Tsai HM (2003) Comparison of three geometric representations of airfoils for aerodynamic optimization. In: AIAA 2003-4095, 16th AIAA CFD Conference, 23–26 June, Orlando, FL
16. Shahrokhi A, Jahangirian A (2007) Airfoil shape parameterization for optimum Navier–Stokes design with genetic algorithm. Aerosp Sci Technol 11:443–450
17. Huyse L, Padula SL, Lewis RM, Li W (2002) Probabilistic approach to free-form airfoil shape optimization under uncertainty. AIAA J 40(9):1764–1772
18. Zhang F, Chen S, Khalid M (2003) Multi-point optimization of transonic wing by real-coded genetic algorithm. In: The eleventh annual conference of the CFD Society of Canada, Vancouver, May 2003
19. Duvigneau R, Visonneau M (2004) Hybrid genetic algorithms and artificial neural networks for complex design optimization in CFD. Int J Numer Methods Fluids 44:1257–1278
20. Karakasis MK, Giotis AP, Giannakoglou KC (2003) Inexact information aided, low-cost, distributed genetic algorithms for aerodynamic shape optimization. Int J Numer Methods Fluids 43:1149–1166
21. Cinnella P, Congedo PM (2008) Optimal airfoil shapes for viscous transonic flows of Bethe–Zel'dovich–Thompson fluids. Compu Fluids 37:250–264
22. López D, Angulo C, Macareno L (2008) An improved meshing method for shape optimization of aerodynamic profiles using genetic algorithms. Int J Numer Methods Fluids 56:1383–1389
23. Kampolis IC, Giannakoglou KC (2008) A multilevel approach to single-and multiobjective aerodynamic optimization. Comput Methods Appl Mech Eng 197:2963–2975
24. Vatandas E, Özkol I (2008) Coupling dynamic mesh technique and heuristic algorithms in 3-D-tapered wing design. Int J Numer Methods Eng 74:1771–1794
25. Balu R (1999) Natural evolution as a process of optimisation. In: Aerodynamics Research and Development Division, VSSC, ISRO, India
26. Samareh JA (1999) in CEA/AIAA/ICASE/NASA Langley international forum on aeroelasticity and structural dynamics: Proceedings CEA/AIAA/ICASE/NASA langley international forum on aeroelasticity and structural dynamics, Williamsburg, America, 22–25; NASA/CP: i1999–209136/PT1, pp. 333–343.
27. Nemec M, Zingg DW, Pulliam TH (2004) Multipoint and multi-objective aerodynamic shape optimization. AIAA J 39(6):1057–1065

28. Kroll N, Gauger NR, Brezillon J, Dwight R, Fazzolari D, Vollmer K, Becker H, Barnewitz V, Schulz S, Hazra S (2007) Flow smulation and shape optimization for aircraft design, J. Comput. Appl. Math. 203:397–411
29. Rai MM, Madavan NK (2000) Aerodynamic design using neural networks. AIAA J 38 (1):173–182
30. Obayashi S (1996) Aerodynamic optimization with evolutionary algorithms. In: Proceedings of ithe IEEE international conference on control, Dearborn, September
31. Chung H-S, Choi S, Alonso JJ Supersonic business jet design using a knowledge-based genetic algorithm with an adaptive, unstructured grid methodology. In: AIAA Paper 2003-3791
32. Foster NF Dulikravich GS (1997) Three dimensional aerodynamic shape optimization using genetic and gradient search algorithms, J. Spacecr. Rocket. 34(1):36–42
33. Vicini A, Quagliarella D, Airfoil and wing design through hybrid optimization istrategies, AIAA Paper 98–2729
34. Drela M, Youngren H (2001) XFOIL 6.94 user guide
35. Mezaal NA, Osintsev KV, Alyukov SV (2019) The computational fluid dynamics performance analysis of horizontal axis wind turbine. Int J Pow Elec Dri Syst ISSN, 2088 (8694), 1073
36. Marten D (2018) QBlade short manual

A Fuzzy Set-Based Energy Consumption Model of Selective Laser Sintering

Faladrum Sharma and Uday Shanker Dixit

1 Introduction

The ever-increasing environmental concerns have challenged the manufacturing industries to develop energy-efficient products and exploit sustainable manufacturing technologies. One such technology is additive manufacturing (AM). It is popularly known as 3D printing and is a foundation of the fourth industrial revolution, i.e., Industry 4.0. The ability of AM to manufacture a product by depositing material in layer-by-layer fashion makes it advantageous compared to other technologies especially for manufacturing complex-shaped geometries with minimal wastage of raw material.

During the process of converting a raw material to the final product, energy is consumed at every stage. The proper assessment of the energy requirement of a manufacturing process and reducing it is a significant strategy for achieving sustainability. Pertinent to AM, researchers have carried out numerous studies in the field of energy assessment. Mongol et al. [1] investigated the effect of different parameters on the energy consumption of three different AM processes, viz., thermojet, fused deposition modelling (FDM) and selective laser sintering (SLS). They considered a typical geometrical part in their analysis. The parameters in the study comprised orientation, position, height, layer thickness, quantity of support material and manufacturing time. It was observed that minimizing the height of the part minimizes energy consumption for thermojet and SLS processes. However, for an FDM process, reducing the volume of the support structure results in the minimum energy consumption. Sreenivasan et al. [2] studied the involvement of different components of an AM machine in energy consumption. They conducted experiments on a selective laser sintering (SLS) machine and recorded the power consumption by a current measuring instrument. They suggested to replace the

F. Sharma (✉) · U. S. Dixit
Indian Institute of Technology Guwahati, Guwahati, India
e-mail: fds80@iitg.ac.in

S. Bag et al. (eds.), *Next Generation Materials and Processing Technologies*,
Springer Proceedings in Materials 9, https://doi.org/10.1007/978-981-16-0182-8_40

heating system of the machine that consumes the maximum energy, with a better thermal control system. They also suggested to use an efficient laser system for reducing the energy consumption. Baumers et al. [3] classified the energy consumption of laser sintering AM machines based on the geometry and height of the part, time required to build the part and the energy consumed during the warm-up and cool-down of the machine chamber. Paul and Anand [4] discussed the dependence of energy consumption on layer thickness, orientation of the part and total area of sintering in an SLS process. However, their study was limited only to laser energy consumption based on the virtual manufacturing of a cube, a cylinder and a 3D shape comprising various geometrical surfaces. Meteyer et al. [5] developed a mathematical model of energy consumption in a binder jetting AM process. They validated the model by fabricating a small-sized part. Kellens et al. [6] presented parametric models to quantify the environmental footprint based on energy and resource consumption. The models were developed considering build height and volume of the parts to be produced. Peng et al. [7] analysed the energy consumption of an extrusion-based AM technology, viz., fused deposition modelling (FDM). They considered the energy consumed during the melting of raw material as a primary energy and the energy consumed by machinery components as a secondary energy. Yang et al. [8] proposed a mathematical model for the energy consumption of a stereolithography (SLA) process. They conducted experiments to validate the model and implemented design of experiments (DOE) to analyse the impact of different parameters on the overall energy consumption. They emphasized that proper setting of the machine parameters can reduce the energy consumption without compromising the quality of the product.

With respect to energy consumption, researchers have also carried out comparative studies with other manufacturing technologies to assess the sustainability of AM. Morrow et al. [9] carried out a comparative analysis of direct metal deposition (a laser-based AM process) and conventional manufacturing (end milling) in terms of energy consumption. They highlighted that the energy consumption of the two processes is influenced by the geometry of the part to be manufactured. Ullah et al. [10] compared SLA and a subtractive process, viz., wood-sawing technique with reference to emission of carbon dioxide gas (involved in model building material and model making process) and consumption of natural resources. They proposed a sustainability index as a function of energy consumption, emission of carbon dioxide and resource depletion. Telenko and Seepersad [11] made a comparative study of SLS and injection moulding (IM) based on energy requirements. The total energy consumption by SLS was considered as the energy required in the processing of feedstock and the energy consumption during the building process. It was found that energy consumption per part for IM reduces with an increase in the production volume, but for SLS, it remains almost constant. Also, the authors emphasized that full utilization of the machine chamber, reducing the powder loss, recycling the powder and reducing the scanning time of layers can make SLS process more energy-efficient.

The energy consumption models of AM technology developed over the years are deterministic. Often, the energy consumption has got an element of uncertainty.

A common approach to tackle uncertainty can be to develop a fuzzy set-based model. This work develops a fuzzy set-based energy consumption model for a specific AM technology, i.e., selective laser sintering (SLS).

After this brief introduction, Sect. 2 describes a deterministic energy estimation model. Section 3 fuzzifies the model developed in Sect. 2 and presents a fuzzy set-based energy consumption model. Two case studies are presented in Sect. 4. Finally, Sect. 5 concludes the paper.

2 Deterministic Model for Energy Estimation of Selective Laser Sintering

A deterministic model is a pre-requisite for developing a fuzzy set-based model. This section describes the deterministic energy consumption model of SLS that is later converted to a fuzzy set-based model in Sect. 3. For estimating the energy consumption of any AM process, it is necessary to have proper information of different energy consuming components of a 3D printing machine. Although every 3D printed parts are produced by the deposition of material in a layer-by-layer manner, the operating principles vary for different categories of AM. This work considers a popular AM process, viz., SLS. A schematic diagram of the process is depicted in Fig. 1.

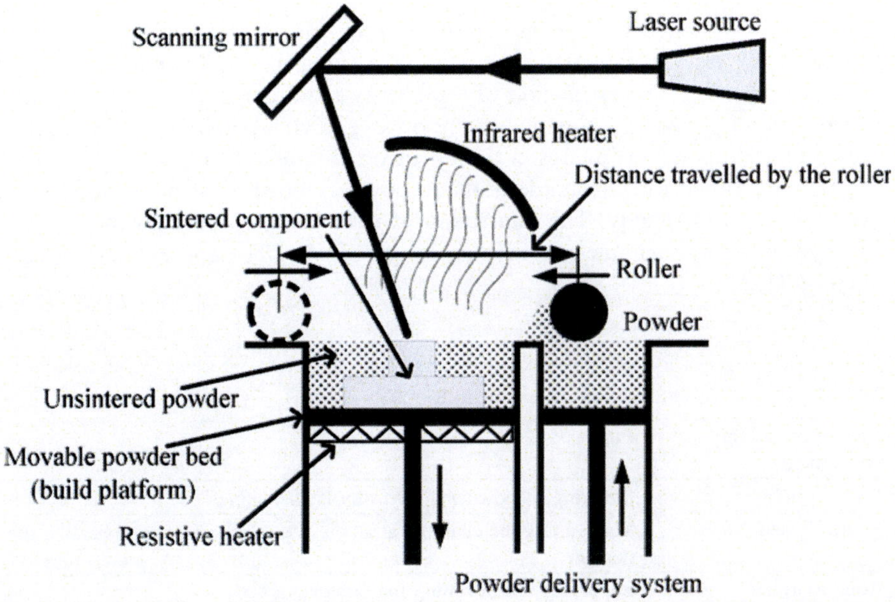

Fig. 1 Illustration of selective laser sintering (Modified with permission from Sharma and Dixit [12])

SLS falls under the category of powder bed fusion (PBF) process, one of the first commercialized AM processes. It uses raw material in the form of polymeric, metallic or ceramic powder. The entire building process takes place inside a closed chamber often referred as build chamber. There are several machinery components in the chamber. The rotating roller spreads the powder and forms a layer in the build area of the chamber. The heater system is composed of an infrared heater for heating the closed chamber and a resistive heater for maintaining the powder at an elevated temperature but below the melting point throughout the building process. The well-directed laser beam, generally of a CO_2 laser, sinters the powder particles according to the geometry of the part. After the laser sinters the particles of a layer, the movable piston lowers the powder bed vertically according to the layer thickness set by the operator. After this, the powder bed is again filled up with powder by the roller for sintering by the laser beam. The heap of powder is placed in front of the roller by the powder supply system that is controlled by another movable piston. This process repeats until the entire part is built. Unlike other AM processes, an advantage of SLS is that it does not require any support structures. The powder itself acts as a support for complicated geometries. After the building of the part is completed, it is removed from the build chamber and necessary post-processing activities are performed. Table 1 lists all the energy consuming elements of SLS.

2.1 *Energy Consumed by Laser System*

The energy consumed by the laser beam during the sintering of the powder is dependent on laser parameters, geometrical parameters of the part as well as material properties. The estimation of the energy requirement by the laser system considering all parameters simultaneously by an analytical model is a tedious task. Franco and Romoli [13] analysed the effect of different laser parameters on the energy requirement for polyamide polymer powder, which is also considered as a raw material in this study. The interaction of laser beam and the powder is characterized by a basic parameter, viz., the energy density. The energy density, E_d is given by [13]

Table 1 Energy consuming components of SLS

Energy consuming components	Purpose
Laser system	Sintering of powder in layer-by-layer manner
Heater system	Maintaining the chamber at an elevated temperature and heating the powder
Rotating roller	Supplying and levelling the powder in bed
Pistons in the movable platform	One piston for lowering of the powder bed and the other piston for supplying the powder in front of the roller

$$E_d = \frac{P_l}{V_s d_l}, \tag{1}$$

where P_l is the power rating of the laser, V_s is the scan velocity of the laser and d_l is the diameter of the laser beam. Franco and Romoli [13] conducted experiments on an SLS machine and provided data to estimate laser energy consumed per unit mass based on different values of E_d. Based on those data, the laser energy consumption can be estimated.

2.2 Energy Consumed by Heater System

The heater system is composed of an infrared heater and a resistive heater [14]. The closed chamber is maintained at a high temperature by the infrared heater. On the other hand, the resistive heater is used to maintain an elevated temperature of the build platform throughout the fabrication of the part. The energy consumed by the infrared heater, E_{ir} is given by [5]

$$E_{\text{ir}} = \frac{P_{\text{ir}} t_{\text{process}}}{\eta_{\text{ir}}}, \tag{2}$$

where P_{ir} is the power rating of the infrared heater, t_{process} is the time involved in warming up the machine chamber and building the part and η_{ir} is the efficiency of the infrared heater. On the other hand, a resistive heater is required to heat the powder in the build platform for proper sintering to occur. The energy is required in the form of heat. It depends on the powder material properties. The heat input Q_r to maintain the elevated temperature of the powder during building of the part is given by

$$Q_r = m_p c_p \left(T_f - T_i\right), \tag{3}$$

where m_p is the mass of the powder, c_p is the specific heat of the powder, T_f is the final attainable temperature of the powder and T_i is the initial temperature of the powder. The heat input Q_r is provided by the resistive heater and its energy consumption E_{rh} is given by

$$E_{\text{rh}} = \frac{Q_r}{\eta_r}, \tag{4}$$

where η_r is the efficiency of the resistive heater. The total energy consumed by the heater system, E_{heater} is given by

$$E_{\text{heater}} = E_{\text{ir}} + E_{\text{rh}}. \tag{5}$$

2.3 Energy Consumed by Roller

The roller spreads the powder and forms a layer in the machine bed. The roller moves according to the kinematic profile as illustrated in Fig. 2. The mechanical energy of the roller is estimated as per velocity profile shown in Fig. 2.

The energy spent in portion AB, E_1 is given by the change in kinetic energy and work done against powder:

$$\begin{aligned} E_1 &= \frac{1}{2}m_r v_r^2 + \frac{1}{2}I_r\omega_r^2 + F_r S_{\mathrm{AB}} \\ &= \frac{1}{2}m_r v_r^2 + \frac{1}{2}I_r\omega_r^2 + \frac{1}{2}F_r v_r t_1, \end{aligned} \tag{6}$$

where m_r is the mass of the roller, v_r is the maximum attainable translational velocity by the roller, I_r is the moment of inertia of the roller about its centre and ω_r is the maximum attainable angular velocity by the roller. The term F_r is the resistive force provided by the powder to the roller, S_{AB}, is the distance travelled by the roller from point A to B and t_1 is the time spent by the roller to reach the point B. The roller is considered to be a solid cylinder whose moment of inertia I_r is given by

$$I_r = \frac{1}{2}m_r R_r^2, \tag{7}$$

where R_r is the radius of the cylindrical roller. During travelling of the roller, it is assumed that rolling takes place without slipping that satisfies the relation

$$v_r = \omega_r R_r. \tag{8}$$

At point B, the roller attains its maximum velocity and continues to moves until point C. The energy spent in portion BC, E_2, is given by

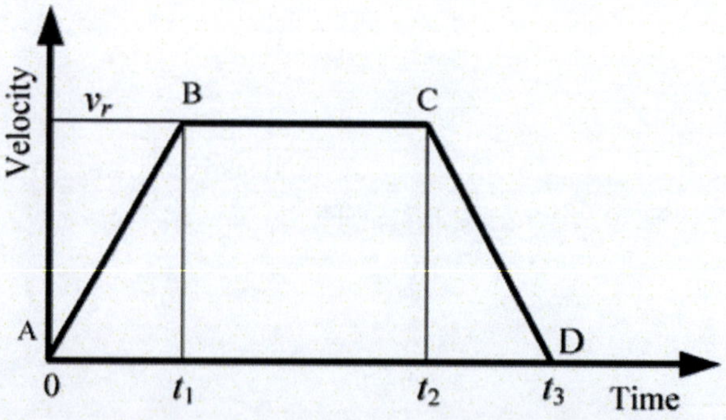

Fig. 2 Velocity versus time diagram of the roller

$$E_2 = F_r S_{\mathrm{BC}} = F_r v_r (t_2 - t_1), \tag{9}$$

where t_2 is the time required by the roller to reach the point C. From point C, the roller decelerates and comes to rest at point D. For bringing the roller to rest, it requires some amount of kinetic energy. However, during this period, the powder also offers resistance. Hence, the energy spent in portion CD, E_3 is less than the energy spent in portion AB. E_3 is given by

$$E_3 = \frac{1}{2} m_r v_r^2 + \frac{1}{2} I_r \omega_r^2 - F_r S_{\mathrm{CD}}, \tag{10}$$

where S_{CD} is the distance travelled by the roller from point C to D. The total energy spent, E_r is given by

$$E_r = E_1 + E_2 + E_3 = \frac{1}{2} m_r v_r^2 + \frac{1}{2} I_r \omega_r^2 + F_r S_{\mathrm{AB}} + F_r S_{\mathrm{BC}} + \frac{1}{2} m_r v_r^2 + \frac{1}{2} I_r \omega_r^2 - F_r S_{\mathrm{CD}}. \tag{11}$$

However, $S_{\mathrm{AB}} = S_{\mathrm{CD}}$, hence

$$E_r = m_r v_r^2 + I_r \omega_r^2 + F_r v_r (t_1 - t_2). \tag{12}$$

The study conducted by Nan et al. [15] revealed that the roller force, F_r, is approximately 20 times the total weight of the heap of powder generated in front of the roller. This heap of powder is generated approximately up to the half the diameter of the roller [16]. The diameter of the roller is evaluated based on the size of the polymer powder. Haeri et al. [16] stated that the size of the powder particle is related to the size of the roller as follows:

$$\frac{D_{\mathrm{roller}}}{D_{\mathrm{powder}}} = 1000, \tag{13}$$

where D_{roller} and D_{powder} are diameter of the roller and powder, respectively. Overall, the energy required by the roller is provided by a motor. Hence, the energy requirement by the motor to operate the roller, E_{roller} is given by

$$E_{\mathrm{roller}} = \frac{E_r}{\eta_m}, \tag{14}$$

where η_m is the efficiency of the motor.

2.4 Energy Consumed by the Movable Platform

The part fabricated in a build platform moves in the downward direction according to the prescribed layer thickness. Apart from this, a powder delivery platform is also present on the other side that supplies the powder in front of the roller. Both the platforms are controlled by piston. The piston moves and stops repeatedly according to the number of layers present in the part. Assuming that no energy is required in stopping, the mechanical energy of the piston, E_p is given by the summation of potential and total kinetic energy:

$$E_p = m_p gh + \frac{1}{2} m_p v_p^2 N_l, \tag{15}$$

where m_p is the mass of the piston, g is the acceleration due to gravity, h is the total height travelled by the piston, v_p is the velocity of the piston and N_l is the number of layers present in the part. Similar to the energy required by the motor, the energy required by the piston, E_{piston} is also provided by the motor:

$$E_{\text{piston}} = \frac{E_p}{\eta_m}, \tag{16}$$

where η_m is the efficiency of the motor.

Apart from the energy consumption mentioned in Sects. 2.1–2.4, some additional energy is also consumed by the computer, workstation, and some amount of energy is also lost during the process. This extra amount of miscellaneous energy, E_{misc}, is considered to be 5% of the total energy:

$$E_{\text{misc}} = 0.05 \left(E_{\text{laser}} + E_{\text{heater}} + E_{\text{roller}} + E_{\text{piston}} \right). \tag{17}$$

3 Implementation of Fuzzy Set Theory in the Energy Estimation

Based on the deterministic model developed in Sect. 2, this section describes the method to develop a fuzzy set-based model to consider uncertainties in the estimation. Fuzzy set theory is a popular technique for handling uncertainties. For developing a fuzzy set-based model, the uncertain variables are considered as fuzzy variables. A fuzzy variable is an imprecise value that is obtained based on the input from an expert's judgement. The uncertain or the fuzzy variable in a fuzzy set represents an interval. All the values of the interval are assigned membership grades that vary from 0 to 1, where the values 0 and 1 represent full non-membership and full membership, respectively. For simplicity, linear triangular fuzzy numbers are used in this work for considering uncertain parameters. A linear triangular fuzzy

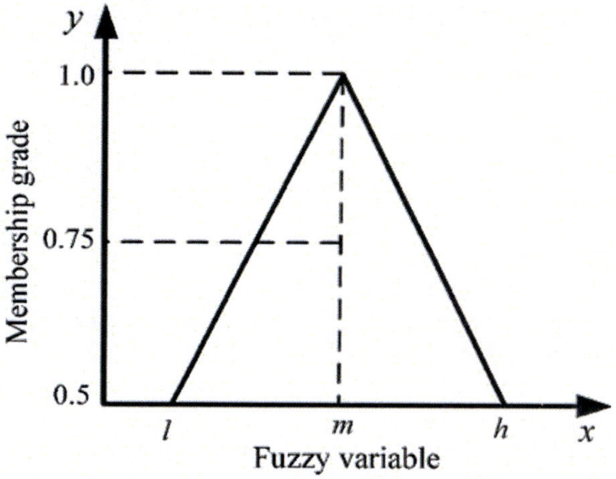

Fig. 3 Fuzzy variable as a function of membership grade

number assigns the membership grade of 1 to the most likely (m) estimate while the low (l) and the high (h) estimates of experts are assigned a membership grade of 0.5 as illustrated in Fig. 3.

A typical fuzzy variable (say A) at a particular value of α-cut is given by

$$A_\alpha = \left[a_1^\alpha, a_2^\alpha\right], \quad (18)$$

where A_α is the interval corresponding to membership grade α, a_1^α and a_2^α are the lower and upper estimates of the interval, respectively. Arithmetic operations of two fuzzy variables (say A and B) at a particular value of membership grade, i.e., at a particular value of α-cut are carried out as follows:

$$\text{Summation}: \quad \left(a_1^\alpha, a_2^\alpha\right) + \left(b_1^\alpha, b_2^\alpha\right) = \left(a_1^\alpha + b_1^\alpha, a_2^\alpha + b_2^\alpha\right) \quad (19)$$

$$\text{Subtraction}: \quad \left(a_1^\alpha, a_2^\alpha\right) - \left(b_1^\alpha, b_2^\alpha\right) = \left(a_1^\alpha - b_2^\alpha, a_2^\alpha - b_1^\alpha\right) \quad (20)$$

$$\text{Multiplication}: \quad \left(a_1^\alpha, a_2^\alpha\right) \times \left(b_1^\alpha, b_2^\alpha\right) = \left(a_1^\alpha \times b_1^\alpha, a_2^\alpha \times b_2^\alpha\right) \quad (21)$$

$$\text{Division}: \quad \left(a_1^\alpha, a_2^\alpha\right) \div \left(b_1^\alpha, b_2^\alpha\right) = \left(\frac{a_1^\alpha}{b_2^\alpha}, \frac{a_2^\alpha}{b_1^\alpha}\right) \quad (22)$$

It is assumed that the fuzzy variables considered in this study belong to a set of positive real numbers for which Eqs. (19)–(22) are valid. A fuzzy set-based energy consumption model is developed by considering the variables as fuzzy and performing fuzzy arithmetic operations. This gives the energy consumption as a fuzzy number. The addition and subtraction of A and B result in linear triangular fuzzy number. However, it is not the same for multiplication and division, but as an approximation, it can also be represented as linear triangular fuzzy numbers. More details are available in Ref. [12] where a fuzzy set-based cost model for SLS process is described. The same procedure can be applied for developing fuzzy set-based energy estimation model. The overall procedure is illustrated in Sect. 4 by taking two case studies.

4 Case Studies

The deterministic model will be implemented to estimate the energy consumption for manufacturing two parts, viz., part A and part B in an SLS machine having a build volume of 230 × 230 × 230 mm^3 [17]. Those two parts and the same SLS machine were also considered for cost estimation of SLS by Sharma and Dixit [12]. The raw material considered for making the parts is polyamide powder of density 840 kg/m^3 [13]. The details of the parts and other parameters used in energy estimation are listed in Table 2.

The first step is to estimate the energy consumption by the laser system. The laser parameters used in this study are listed in Table 3. Based on these parameters, the energy density, E_d is 0.07 J/mm^2. As per this value of E_d, the energy required for both the parts is obtained from Ref. [13]. Then, the energy consumed by the infrared and resistive heaters is estimated. The infrared heater consumes energy for heating the chamber during the machine warm-up and building the part. On the other hand, the resistive heater consumes energy for heating the powder in the machine bed. It depends on the mass of the powder and its material properties. The mass of the powder is estimated based on the size of the powder bed and height of the part. Table 3 lists the necessary parameters of the heater system used in this study.

The next step involves the energy estimation by the roller for spreading and levelling the powder. The kinematic behaviour of the roller is illustrated in Fig. 2. The complete procedure for estimating the time taken by the roller to travel from one end of the platform to the other considering necessary parameters is explained in the work of Sharma and Dixit [12]. The roller travels to and fro across the

Table 2 Details of parts and other necessary parameters [12]

Parameters	Value	
	Part A	Part B
Length of bounding box (mm)	120	74
Breadth of bounding box (mm)	120	34
Height of bounding box (mm)	28	34
Volume of bounding box (mm^3)	403,200	85,544
Volume of the part (mm^3)	184,893	9603
Mass of the part (gm)	155.31	8.07
Time for machine preparation (s)	1800	1800
Time to build the part (s)	4459.2	1865.04
Layer thickness (mm)	0.15	0.15

Table 3 Parameters for energy consumed by the laser and heater system

Parameters	Value	Basis
Power rating of the laser (W)	30	Ref. [12]
Scan velocity of the laser (mm/s)	700	Ref. [12]
Laser beam diameter (mm)	0.6	Ref. [13]
Power of infrared heater (W)	1000	Ref. [18]
Efficiency of infrared/resistive heater	0.9	Own judgement
Specific heat of powder (kJ/kgK)	2.5	Ref. [13]
T_{final} of powder (°C)	185	Ref. [13]
T_{initial} of powder (°C)	20	Ref. [13]

Table 4 Parameters for estimating the roller energy

Parameter	Value	Basis
Distance travelled by roller (mm)	506	Ref. [17], own judgement
Velocity of roller (mm/s)	700	Ref. [12]
Time, t_1–t_2 (s)	0.69	Ref. [12]
Mass of roller (kg)	5	Own judgement
Diameter of powder (μm)	100	Ref. [19]
Radius of the roller (mm)	50	Ref. [16]
Force required by roller (N)	435.92	Ref. [15]
Efficiency of the motor	0.75	Own judgement considering part load operation

platform based on the number of layers present in the part. For each travel, the roller consumes energy. The data used for estimating roller energy is given in Table 4.

Lastly, the energy consumed by the moving pistons is estimated. The total vertical distance travelled by the piston of the build platform is equal to the height of the part. For the build piston, an additional 20% increase is included in the mass of the piston to consider the mass of the powder deposited. The parameters for estimating the energy consumed by piston are given in Table 5. The piston lowers down according to the prescribed layer thickness and the number of layers present in the part. An additional 5% increase is included in the number of times the piston is lowered. However, the total distance travelled by piston of the powder delivery piston is much more than the piston of the build platform. The powder delivery piston feeds the powder to the height equal to the radius of roller, i.e., 50 mm. This process is repeated for every number of layers of the part. Also, the piston has to travel to the bottom frequently for complete filling of the feedstock with powder. This is done by an external delivery source. Considering these effects, an additional increase of 50% of the mass of the piston is also included. Hence, the energy of the piston for powder delivery is much more than the piston of the build platform. Finally, based on Eqs. (15) and (16), the energy consumed by the pistons is

Table 5 Parameters for estimating the energy consumed by the pistons

Parameters	Value	Basis
Mass of the platform (kg)	10	Own judgement
Velocity of the platform (mm/s)	700	Ref. [12], own judgement
Efficiency of the motor	0.75	Own judgement

Table 6 Energy consumption by different sources for part A and Part B

Energy components	Value	
	Part A	Part B
Energy consumed by laser system (kJ)	45.04	2.34
Energy consumed by infrared heater (kJ)	6954.67	4072.27
Energy consumed by resistive heater (kJ)	570.26	692.46
Energy consumed by the roller in spreading the powder (kJ)	158.55	192.53
Energy consumed by the piston by build platform (kJ)	0.78	0.94
Energy consumed by the piston for powder delivery (kJ)	4.58	4.89
Miscellaneous energy consumption (computer, losses)	386.69	248.27
Total energy consumption of the part (kJ)	8120.57	5213.7
Total energy consumption of the part (kWhr)	2.26	1.45
Total energy consumption per unit mass of the part (kJ/kg)	52286.2	646059.48
Total energy consumption per unit build time (kJ/hr)	4670.57	5121.18

estimated. The energy consumed by every component for manufacturing part A and part B considering a single quantity is listed in Table 6. As evident from the last two rows of the table, mass of the part and build time may give some indication about the energy consumption, but it also depends on the part complexity.

4.1 Estimation of Energy Considering Multiple Quantities of a Part in the Machine Chamber

The energy required by the laser and the infrared heater varies with the number of quantities manufactured in the machine chamber. However, the energy consumed by the resistive heater, roller and the pistons is independent of the number of quantities manufactured in the machine chamber. Hence, if more than one quantity is manufactured, the overall unit energy requirement is less as the energy is reduced among by the resistive heater, roller and pistons. As per the size of part A, only two quantities can be manufactured at a single setting of the machine chamber. In such a case, the total consumed energy is estimated as 12202.2 kJ, whereas unit energy consumption is 6101.1 kJ. Hence, in manufacturing two quantities of part A, energy consumption reduced by 25% for a part. On the other hand, considering the

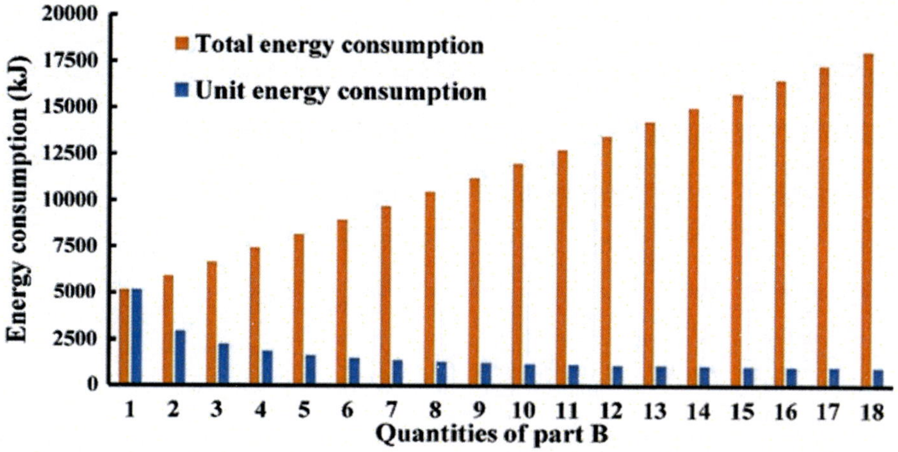

Fig. 4 Variation of energy consumption with quantities of part B

smaller sized part B, a maximum of eighteen quantities can be manufactured at a single setting. This is also referred as full utilization of the machine. The energy consumed for manufacturing a single quantity of part B is 5213.7 kJ, whereas the unit energy consumed for eighteen quantities is 1007.45 kJ reducing the energy consumption by 81%. The variation of energy with the quantities of part B is shown in Fig. 4.

4.2 Implementation of Fuzzy Arithmetic

This section considers all the uncertain parameters as fuzzy. All the fuzzy parameters are expressed as low (l), most likely (m) and high (h) estimates in Table 7. In the estimation of energy consumed by the roller, the lower limit of roller force is assumed to deviate by 20%, whereas the upper limit by 10% than the most likely estimate. The height of the heap of powder accumulated in front of the roller is assumed to vary by 10 and 20% of the radius of the roller for the lower and upper estimate, respectively. In the estimation of energy consumed by the pistons for build platform, the lower and upper limits of mass of the piston are additionally increased by 10 and 30% of the actual mass of the piston, respectively. However, for the piston of the powder delivery platform, the lower and upper limits are increased by 30 and 70%, respectively. This piston has to accumulate relatively more mass of the powder that is continuously fed by the feedstock.

Table 7 Values of different fuzzy parameters

Parameter	(*l*, *m*, *h*)
Power rating of the laser (W)	(20, 30, 50)
Scan velocity of the laser (mm/s)	(500, 700, 1000)
Power of infrared heater (W)	(800, 1000, 1200)
Efficiency of infrared/resistive heater	(0.8, 0.9. 0.95)
Mass of roller (kg)	(4.5, 5, 5.5)
Force required by roller (N)	(313.86, 435.92, 527.46)
Layer thickness (mm)	(0.12, 0.15, 0.21)

Table 8 Unit energy consumption as a fuzzy number considering a single part and full utilization of the machine

	Quantities in the machine chamber	Unit energy consumption (kJ)		
		Low estimate (*l*)	Most likely estimate (*m*)	High estimate (*h*)
Part A	1	4980.9	8120.57	15075.5
	2	3559.49	6101.1	11829.1
Part B	1	4202.84	5213.7	7666.88
	18	717.25	1007.45	1661.96

Table 8 shows the unit energy consumption of the parts considering a single quantity and full utilization of the machine chamber, i.e., two and eighteen quantities of part A and part B, respectively. The energy as a fuzzy number is obtained by applying the procedure mentioned in Sect. 3. The variation of energy as a fuzzy number with different membership grades is shown in Fig. 5.

5 Conclusion

In this work, a fuzzy set-based energy consumption model of additive manufacturing with a specific example of SLS is proposed. Initially, the deterministic energy estimation model based on the working principle of SLS is developed. The role of different energy consuming elements is briefly described. Subsequently, the fuzzy set-based model is proposed to tackle uncertainty in the energy consumption. The fuzzy set-based approach gives the energy consumption as an interval number with different membership grades, thus giving an approximate idea of the lower and the upper limits. The methodology proposed in this work is illustrated through two examples. The energy consumed for building a single part and multiple quantities of the part in the machine is estimated. It is observed that the maximum energy is

Fig. 5 Energy consumption as a fuzzy number considering: **a** single quantity of part A, **b** single quantity of part B, **c** full utilization of the machine chamber (two quantities of part A), **d** full utilization of the machine chamber (eighteen quantities of part B)

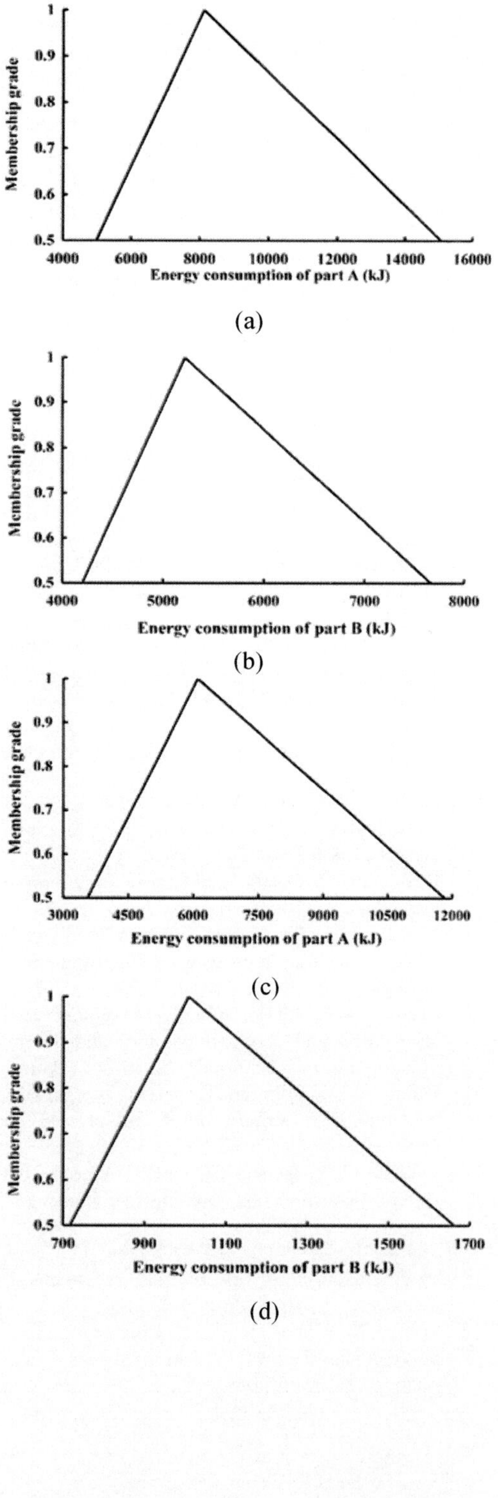

consumed by the heater system. Considering multiple quantities, unit energy consumption is the least when the full machine chamber is utilized. It is also seen that although the mass and build time of the product can give some indication of the energy consumption, but in only an approximate way, energy consumption is influenced by the part complexity.

The methodology presented in this paper will be validated by conducting a number of in-house experiments. Further, this work will also be extended to other types of manufacturing processes to compare their energy efficiencies. This will help in arriving at a sustainability index for each process.

References

1. Mognol P, Lepicart D, Perry N (2006) Rapid prototyping: energy and environment in the spotlight. Rapid Prototyp J 12(1):26–34. https://doi.org/10.1108/13552540610637246
2. Sreenivasan R, Bourell D (2010) Sustainability study in selective laser sintering—an energy perspective. Minerals, Metals and Materials Society/AIME, 420 Commonwealth Dr., P. O. Box 430 Warrendale PA 15086 USA. 14–18 Feb
3. Baumers M, Tuck C, Bourell DL, Sreenivasan R, Hague R (2011) Sustainability of additive manufacturing: measuring the energy consumption of the laser sintering process. Proc IMechE, Part B: J Eng Manuf 225(12):2228–2239. https://doi.org/10.1177/0954405411406044
4. Paul R, Anand S (2012) Process energy analysis and optimization in selective laser sintering. J Manuf Sys 31(4):429–437. https://doi.org/10.1016/j.jmsy.2012.07.004
5. Meteyer S, Xu X, Perry N, Zhao YF (2014) Energy and material flow analysis of binder-jetting additive manufacturing processes. Proc CIRP 15:19–25. https://doi.org/10.1016/j.procir.2014.06.030
6. Kellens K, Renaldi R, Dewulf W, Kruth JP, Duflou JR (2014) Environmental impact modeling of selective laser sintering processes. Rapid Prototyp J 20(6):459–470. https://doi.org/10.1108/RPJ-02-2013-0018
7. Peng T (2016) Analysis of energy utilization in 3D printing processes. Proc CIRP 40:62–67. https://doi.org/10.1016/j.procir.2016.01.055
8. Yang Y, Li L, Pan Y, Sun Z (2017) Energy consumption modeling of stereolithography-based additive manufacturing toward environmental sustainability. J Indus Eco 21(S1):168–178. https://doi.org/10.1111/jiec.12589
9. Morrow WR, Qi H, Kim I, Mazumder J, Skerlos SJ (2007) Environmental aspects of laser-based and conventional tool and die manufacturing. J Clean Prod 15(10):932–943. https://doi.org/10.1016/j.jclepro.2005.11.030
10. Ullah AS, Hashimoto H, Kubo A, Tamaki JI (2013) Sustainability analysis of rapid prototyping: material/resource and process perspectives. Int J Sustain Manuf 3(1):20–36. https://doi.org/10.1504/IJSM.2013.058640
11. Telenko C, Seepersad CC (2012) A comparison of the energy efficiency of selective laser sintering and injection molding of nylon parts. Rapid Prototyp J 18(6):472–481. https://doi.org/10.1108/13552541211272018
12. Sharma F, Dixit US (2019) Fuzzy set based cost model of additive manufacturing with specific example of selective laser sintering. J Mech Sci Technol 33(9):4439–4449. https://doi.org/10.1007/s12206-019-0840-x
13. Franco A, Romoli L (2012) Characterization of laser energy consumption in sintering of polymer based powders. J Mat Process Technol 212(4):917–926. https://doi.org/10.1016/j.jmatprotec.2011.12.003

14. Gibson I, Rosen DW, Stucker B (2014) Additive manufacturing technologies, vol 17. Springer, New York. https://doi.org/10.1007/978-1-4419-1120-9
15. Nan W, Pasha M, Ghadiri M (2020) Numerical simulation of particle flow and segregation during roller spreading process in additive manufacturing. Powder Technol 364:811–821. https://doi.org/10.1016/j.powtec.2019.12.023
16. Haeri S, Wang Y, Ghita O, Sun J (2016) Discrete element simulation and experimental study of powder spreading process in additive manufacturing. Powder Technol 306:45–54. https://doi.org/10.1016/j.powtec.2016.11.002
17. https://xyzwebsite.blob.core.windows.net/wwwportal/MfgPro230%20xS%20_SPEC_171006-2.pdf. Accessed on 5 July 2020
18. https://homeguides.sfgate.com/much-electricity-quartz-infrared-heater-use-87458.html#: ~ :text=Quartz%20infrared%20heaters%20use%20as,of%20750%20to%201500%20watts. Accessed on 5 July 2020
19. https://www.arptech.com.au/data/sls-duraform.pdf. Accessed on 5 July 2020

Improving the Performance Analysis of MPPT Controller Unit of a PV Generation System Using Optimization Technique Based on Spider Monkey Principle (SMO)

Sriparna Das and Kumari Namrata

1 Introduction

The importance of renewable energy (R.E) sources increases to a great extent. With the upcoming of concept known as microgrid, the use of R.E becomes an important source for production of power. The sources of R.E include energy production from PV unit which are combined to form PV arrays, wind turbines, tidal sources, waves, and various other types. On the evolvement of microgrid, DGs are basically placed at the load center for production of power when the utility grid is off due to occurrence of faults, so this DGs basically includes renewable sources of energy.

In India, the total power generation capacity from solar energy sources has increased to 37.627GW at the end of March 2020 [1]. The country successfully made 42 solar parks for creating solar plants [2]. There is implementation of solar panels on roof-tops, and it counts about 2.1GW, out of which 70% are utilized in industry for commercial use. And the other part includes the PV implementation, India having the plan to introduce off-grid generation for supplying the local demand of people. But the problem arises as the efficiency of the power production from PV units is less due to change in the atmospheric conditions and load demand. So, to fulfill this, the system needs some optimizing schemes for reduction of losses and running the system for MPP generation. The two important parameters on which the power production depends are the irradiance and the temperature. With the variation of these parameters, the change of the power production is shown subsequently as we move on through the paper. For tracking of MPP, there are many techniques, but here P&O is considered as it is independent on the environmental conditions. Further, there is only requirement of voltage and current parameters which is present for the considered system, and lastly, the performance

S. Das (✉) · K. Namrata
Department of Electrical Engineering, NIT Jamshedpur, Jamshedpur, Jharkhand 831014, India
e-mail: namrata.ee@nitjsr.ac.in

S. Bag et al. (eds.), *Next Generation Materials and Processing Technologies*, Springer Proceedings in Materials 9, https://doi.org/10.1007/978-981-16-0182-8_41

depends on the step size. If the step size is more, the time requirement for tracking of MPP is less. Here, new upcoming methods such as artificial neural network and fuzzy control can be used, but the disadvantage is for those techniques where there is requirement of large data set which has to be fed in the system, so it requires large memory. So, use of PI controller unit will make the system robust.

As for getting better output response, SMO technique is performed. It is one of the most simple and robust algorithms than all other algorithm like artificial bee colony, particle swamp, genetic, etc. [3]. The conditions of how this strategy is performed and is discussed in details. This technique is used to minimize the error of PI controller unit and generating the pulse for boost converter. So, while moving, first, the PV design is explained, followed by boost converter working, followed by description of P&O algorithm, and lastly explaining of SMO method [4], stating the reduction in error while obtaining the power.

2 Modeling of PV Cell

By applying Kirchhoff's current law from Fig. 1, we get:

$$I_1 = I_{\mathrm{ph1}} - I_{d1} \tag{1}$$

From the general equation of the diode, we can say:

$$I_{d1} = I_{s1}\left[\exp\left(\frac{q_1 V_{\mathrm{oc1}}}{N_{s1} K_1 A_1 T_0}\right) - 1\right] \tag{2}$$

Substituting the value of I_{d1} from Eq. (2) to Eq. (1), we get:

$$I_1 = I_{\mathrm{ph1}} - I_{s1}\left[\exp\left(\frac{q_1 V_{\mathrm{oc1}}}{N_{s1} K_1 A_1 T_0}\right) - 1\right] \tag{3}$$

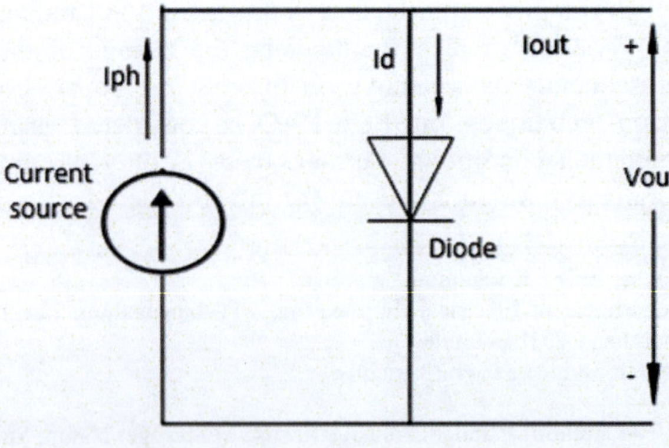

Fig. 1 Ideal structure of a PV cell [6]

But here, we cannot consider the actual ideal circuit as there is addition of series and parallel resistance in the circuit. The series resistance is due to losses for Joules' effect that is losses due to flowing of current in metal grid, semiconductor material, collecting bus, and terminal [5]. The parallel resistance is due to seepage of current due to cell thickness and surface effect [6]. So, the above equations are modified into following equations:

$$I_{d1} = I_{s1}\left[\exp\left(\frac{q_1(V_1 + I_1 R_{s1})}{N_{s1}K_1 A_1 T_0}\right) - 1\right] \tag{4}$$

$$I_1 = I_{\text{ph}1} - I_{s1}\left[\exp\left(\frac{q_1(V_1 + I_1 R_{s1})}{N_{s1}K_1 A_1 T_0}\right) - 1\right] \tag{5}$$

If we consider the solar cell which is acting in series parallel manner, then the above Eq. (5) is transformed to a new equation shown as:

$$I_1 = N_{p1} * I_{\text{ph}1} - N_{p1} * I_{s1}\left[\exp\left(\frac{q_1(V_1 + I_1 R_{s1})}{N_{s1}K_1 A_1 T_0}\right) - 1\right] \tag{6}$$

There are various parameters that we need to describe as they are not described in the above equation $I_{\text{ph}1}$ that is photocurrent of PV cell due to change in irradiance, the reverse saturation current $I_{\text{rs}1}$ and the saturation current I_{s1} which is given by the below equations:

$$I_{\text{ph}1} = [I_{\text{sc}1} + K_{i1}(T_{01} - T_{r1})] * \frac{G_1}{G_{\text{ref}1}} \tag{7}$$

$$I_{\text{rs}1} = \frac{I_{\text{sc}1}}{\left[\exp\left(\frac{q_1 V_{\text{oc}1}}{N_{s1}K_1 A_1 T_0}\right) - 1\right]} \tag{8}$$

$$I_{s1} = I_{\text{rs}1}\left[\frac{T_{01}}{T_{r1}}\right]^3 \exp\left[\left(\frac{q_1 E_{g1}}{A_1 K_1}\right)\left(\frac{1}{T_{r1}} - \frac{1}{T_{01}}\right)\right] \tag{9}$$

This are basically the equations of the model formulated for designing of a PV cell [7]. These formulas are placed in a MALTAB function, the inputs are taken from the datasheet of an ideal PV cell, and by varying the temperature and irradiation, the electrical parameters are noted and plotted for the prediction of the power generated from the cell.

By using the values given in Table 1, the PV cell is made in MATLAB, and by varying the temperature and irradiance, the effect of current and power that is produced is shown in Figs. 2 and 3, respectively.

From Figs. 2 and 3, there is a clear observation of how there is an effect on parameters with the change of irradiance and temperature, keeping one constant at a time. The graph clearly matches to the ideal behavior of a PV cell. On the first case,

Table 1 Given parameters of Solarex MSX 60 PV used module [8]

Name and units of parameters	Taken values
Peak power (W)	60
Maximum peak voltage (V)	17.1
Maximum peak current (A)	3.5
O.C voltage (V)	21.1
S.C current (A)	3.8
K_v (V/°C)	-80×10^{-3}
K_f (A/°C)	3×10^{-3}
N_p	1
N_s	36

as the irradiance is kept constant, the current remains almost constant, and the voltage gets decreased as the temperature is increased. Similarly, the power generation gets decreased as the temperature gets increased, while the case is different if the temperature is kept constant and irradiance is varied. With the increase in the amount of energy from the sun, more generation of power can be produced, so from Fig. 3, it is seen clearly that both the current as well as the power get increased with the increase in irradiance. This is the characteristic which is similar to any PV cell.

3 Boost Converter Unit

The output of the PV cell is limited, and there are many factors which affect the production, which has been discussed in the above section. So, the use of boost converter for boosting the voltage is necessary. It is a kind of mechanism which converts from lower level DC value to higher one [9]. Here it is better to consider IGBT as switch due to lower switching and conduction loss. It consists of an inductor, switch, diode, and capacitor for removing voltage ripples. By generating the duty, the MPP is basically tracked. The parameters are designed on the basis of output that is needed. The frequency remains fixed as that of grid's frequency. Basically, tracking of MPPT is performed on this converter section, and the optimization technique is done for error deduction [10]. The basic design of this converter is represented in Fig. 4.

4 MPPT Method

The importance of this mechanism is for extracting maximum power from the electrical system that can be utilized by the mankind. For running the PV system at higher efficiency, tracking of MPP is important followed by optimization technique

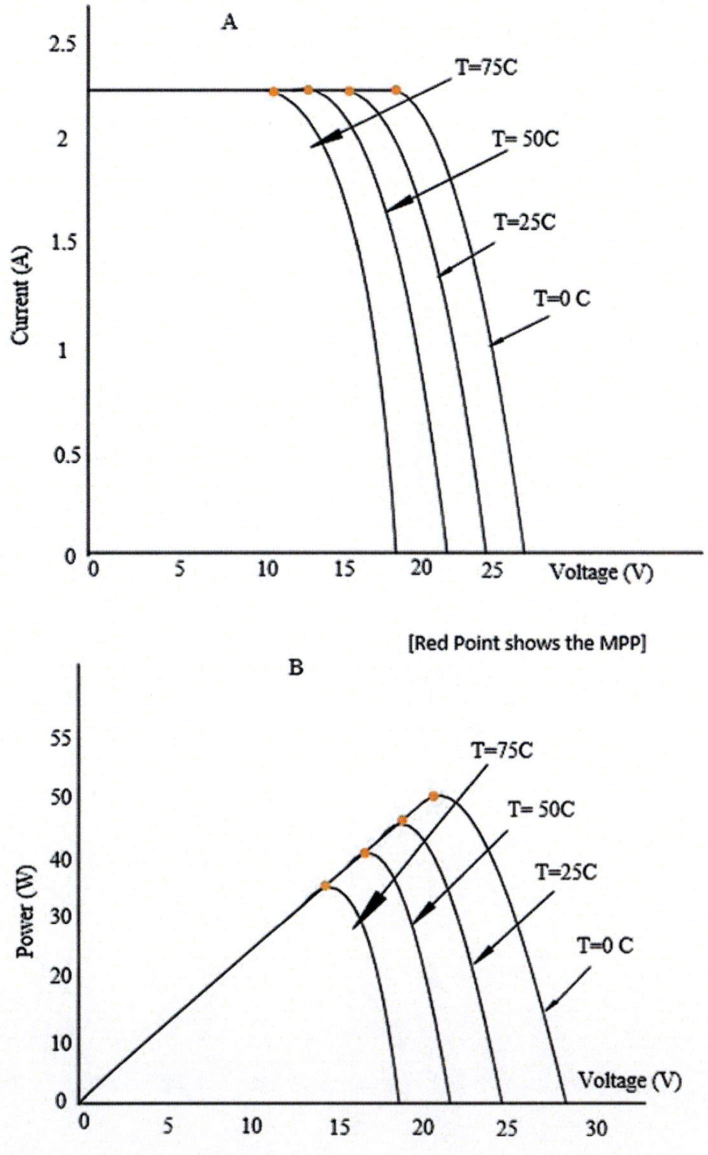

Fig. 2 Effective change of **a** current with change in voltage, **b** power with change in voltage keeping irradiance constant on both cases

to lessen the errors. In this paper, the focus has been given on P&O strategy [11], and the process is explained followed by a flowchart for easy understanding.

The main advantage is that the search for MPP has no dependency on temperature, climate, and the surroundings. However, only requirement is either current or

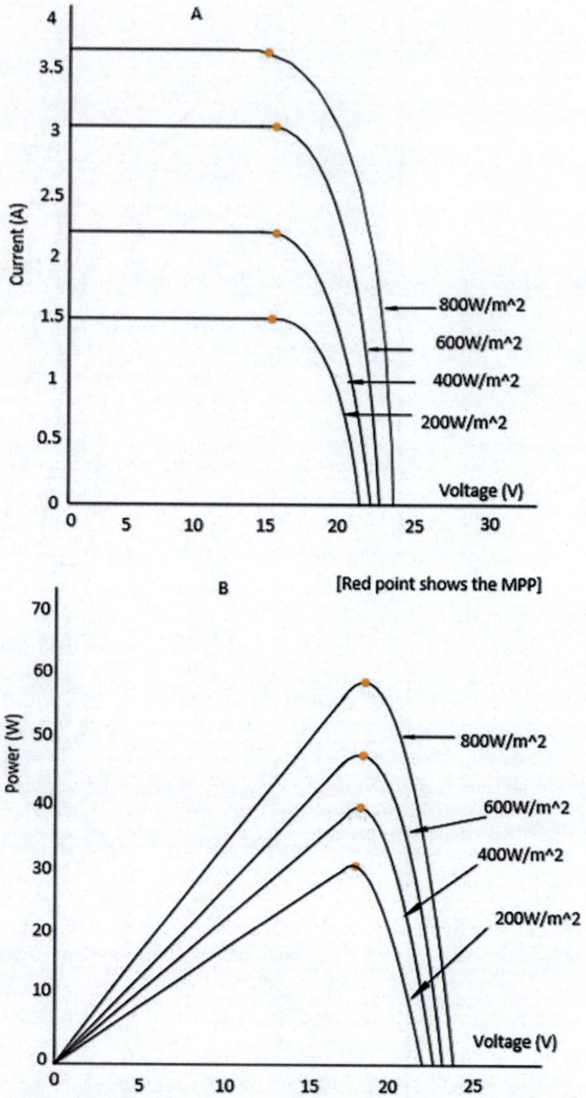

Fig. 3 Effective change of **a** current with change in voltage, **b** power with change in voltage keeping temperature constant on both cases

voltage sensor values. It helps in tracking the point which will indicate maximum power level. The algorithm is basically to find the final state power and previous to final state power. After the computation of these values, this method can easily be applicable. When the circuit is in operation, there will be occurrence of disturbance in the duty cycle. So, after introducing a perturbation, there will be change of duty cycle, the power that is P_n gets increased, P_n is greater than P_{n-1}, then the value is needed to be increased, and the line is to be moved in right direction to track that point.

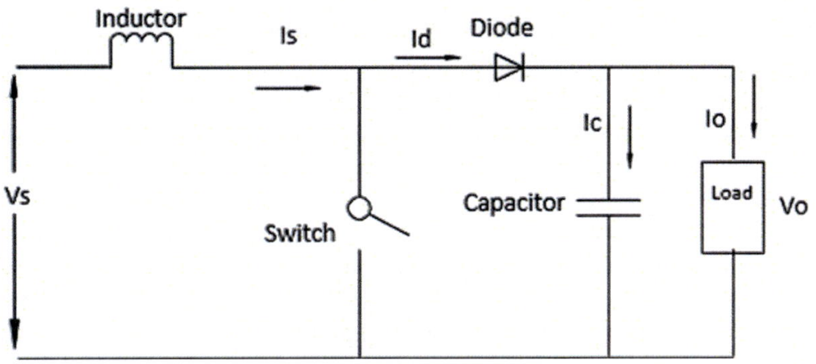

Fig. 4 Circuital view of boost converter section for stepping up the voltage profile

On other hand, if there is decrease in final state power than its previous state while change in the duty cycle, then the value or the perturbation is needed to be lessened, and movement should take place in the left-ward direction as while tracking the point, the maximum point where the power can be extracted is passed, so decrement of the value is necessary. The speed of this process depends up on the size of the increment or decrement. If the step size is assumed to be large value, then finding of the point will be first and vice versa [12]. For easy understanding of conditions, the visual diagram is shown in Fig. 5.

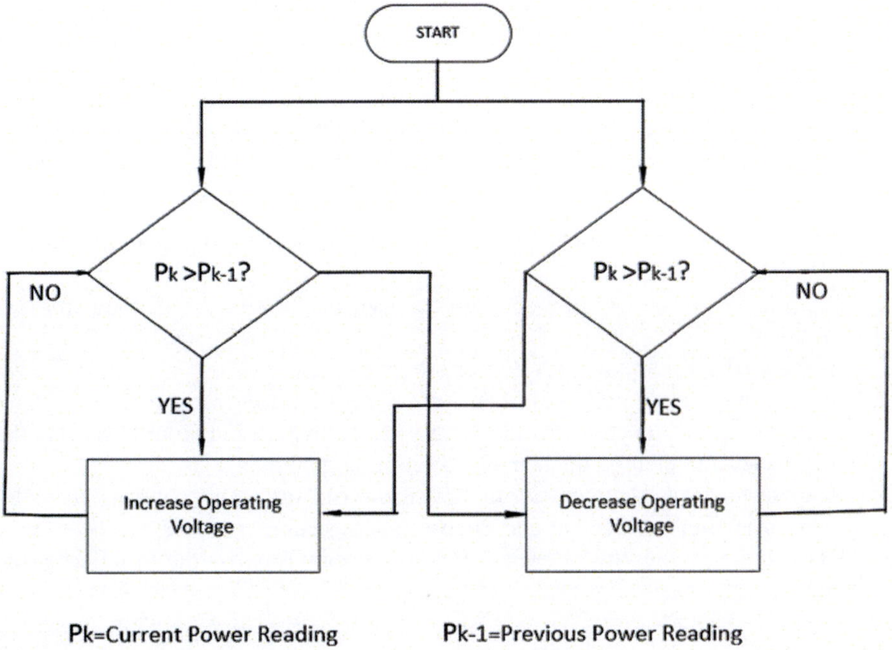

Fig. 5 Flowchart for MPPT algorithm

5 Spider Monkey Optimization Technique (SMO)

There are certain algorithms that originates from the imagination of survival of species in the nature, and from these ideas mathematical models are formulated, and these are known as nature-inspired algorithm. They originate from the idea to survive with the change in parameters like availability of food, change in weather condition, etc. These algorithms are classified as swamp intelligence (SI), evolutionary computation, and existence of artificial neural network. But SI becomes a subset of both computational intelligence as well as AI. SI is defined as the intelligence which occurs or comes up from a community or a group. Within Si, there are many such algorithms such as particle swamp, ant colony, artificial bee, and spider money [13]. In this paper, the stress has been given for the demonstration of spider monkey optimization due to its simple mechanism and less time complexity.

This algorithm is formulated from a simple strategy of movement of monkeys for finding of food in a forest. There is existence of two kinds of leaders local and global. In a group, a female monkey is considered as a leader. At the morning, the global leader will check the amount of food. If it is not sufficient, they listen to the ideas of local leader for the place where they can get enough food and will visit for obtaining food. If there is unavailability of food, they distribute among groups for finding and at the end of the day become united after collecting the food. Similarly, here this strategy has been used to optimize the error and to operate the PV cell at MPP for maximum power generation. The point at which it occurs is tracked, and the duty has been generated in such a way that the system should operate at that point only. The algorithm of SMO [14] which has been run in MATLAB is based on the algorithm as shown in Fig. 6.

The formulation and the searching mechanism of the algorithm is based on the above flowchart. But in real time, when it comes to the process of implementation, there are certain steps that are to be followed precisely [15], and those are illustrated below:

1. Take the input as the population set and define the local and global set which is to be defined along with pr.
2. The fitness is to be determined from the idea of the location of food from an individual.
3. Fix the leader in the local set and in the global set.
4. Until the final outcome is not obtained, perform the following steps:

a. For fulfilling the objective, the ideas are taken from each individual, and also the local leader and change of their old position is generated.
b. Compare the new changed position with the old once and compare them by greedy selection mechanism and choose that is better.
c. Generate the probability of each individual in the group by the formulas given below.

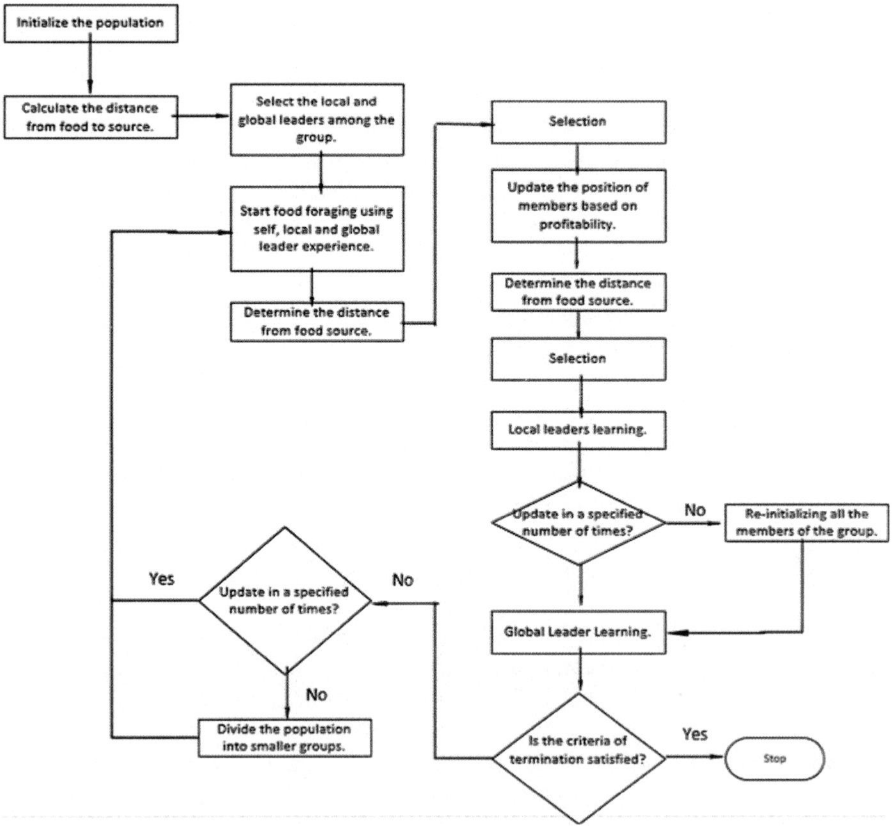

Fig. 6 SMO based error minimization algorithm

$$P_i = 0.9 * \frac{\text{Fitness}_i}{\text{Fitness}_{\text{maximum}}} + 0.1 \tag{10}$$

where, P_i denotes the probability of *i*th individual in a group.

a. The new positions are generated from this probability and the experience which the individual is having.
b. Similarly, change the position of the local and global leaders.
c. If after a number of iterations, the local leader place is not changed, then instruct the individuals of the group for foraging.
d. If the global leader position is not changed after a number of iterations, then a division in small groups is done for generating proper search mechanism.

6 Simulation Results

The system is to have higher efficiency, so a new technique is proposed which will minimize the errors of the controller unit obtaining sensors from P&O method that generates the duty for the boost converter to run at MPP. Further, SMO is implemented and simulated in MATLAB which is attached to the system for error deduction. The comparison has been made and also simulated, between a simple P&O with controller unit which is attached with the P&O with controller plus optimization mechanism that is attached. After attaching the SMO into the system, the loss has been minimized, and the system can track the MPP more accurately. The proposed model is shown in Fig. 7.

By creating the above model, the error of the system is minimized by tuning the controller efficiently and operating it at the line of maximum power point. The simulated result of the comparison between SMO based and normal based is shown in Fig. 8. In this, there is clear difference that can be seen which performance is better. The parameters like current, voltage, and power obtained with respect to time are shown. The duty is generated in such a way that clear the tracking of MPP which takes place. Though, after 1 s, there is abrupt change in parameters, it is well recovered at 2 s only.

In the new model that is proposed for optimization, there is an increase in the values of proportional and integral parameters of the controller. The proportional value gets increased from 0.5 to 2.5, and the integral value changes from 0.03 to 0.116. So, with the increase in those parameters, the steady-state error gets decreased, and the transient nature reduces, making the oscillation lesser than the previous system. There is also increase in the amount of power generation in this new technique.

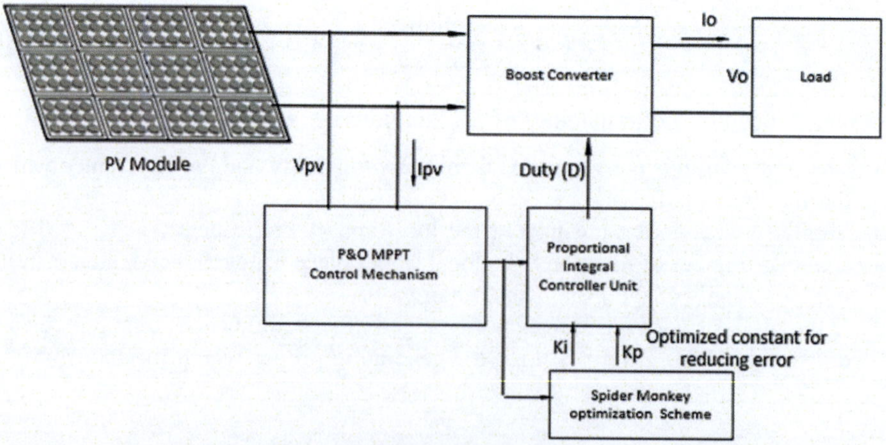

Fig. 7 Proposed technique for error detection

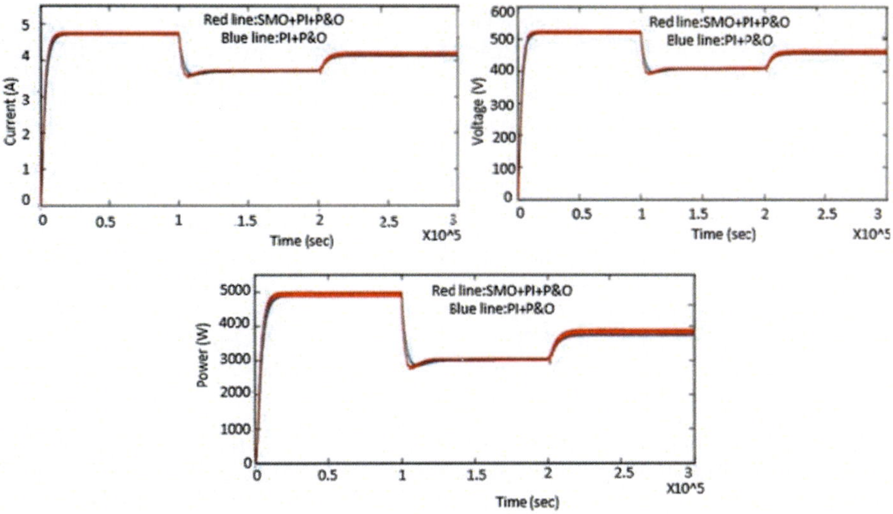

Fig. 8 Effect of change in current, voltage, and power parameters in case of SMO based as well as non-SMO based technique

7 Conclusion

In this paper, a reduction of error technique is explained. The new model with SMO is described [16] and is compared with the old set of models. After making the comparative analysis, it is clear that the tracking of MPP is more precise and accurate if the optimization technique SMO gets attached with the system. From the simulated results, there is a clear view of improvement of power generation capacity by reducing the errors. Due to the generation capacity of PV cell, it is not constant throughout the year, making the system run at higher efficiency which is much important. In this way, the generation capacity can be increased, and supplying of increasing demand can be possible.

References

1. Ministry of New & Renewable Energy. Physical Progress (Achievements). Retrieved 18 July 2019
2. List of Solar Parks in India. Retrieved 7 Sep 2019
3. Sachan RK, Gupta S, Kushwaha DS (2018) Evaluation and research directions in nature-inspired algorithm. In: 2018 5th IEEE Uttar Pradesh section international conference on electrical, electronics and computer engineering (UPCON), 2–4 Nov 2018, pp 1–5
4. Hazrati G, Sharma H, Sharma N, Bansal JC (2016) Modified spider monkey optimization. In: 2016 international workshop on computational intelligence (IWCI), 12–13 Dec 2016, pp 209–214

5. Perera PU, He L (2018) A simplified mathematical model of PV cell simulation. In: 2018 IEEE 7th world conference on photovoltaic energy conversion (WCPEC), 10–15 June 2018, pp 3204–3207
6. Park H, Kim H (2013) PV cell modelling on single-diode equivalent circuit. In: IECON 2013 39th annual conference of the IEEE industrial electronics society, 10–13 Nov 2013, pp 1845–1849
7. Dey BK, Khan I, Mandal N, Bhattacharjee A (2016) Mathematical modelling and characteristic analysis of solar PV cell. In: 2016 IEEE 7th annual information technology, electronics and mobile communication conference (IEMCON), 13–15 Oct 2016, pp 1–5
8. Bana S, Saini RP (2016) A mathematical modelling framework to evaluate the performance of a single diode and double diode based SPV systems. In: Energy reports, science direct, 19 July 2016, pp 171–187
9. Shulin L, Yibo M (2010) Optimization design of the inductance of the boost converters. In: 2010 international conference on electrical and control engineering, 25–27 June 2010, pp 3949–3952
10. Asif Iqbal SM, Mekhilef S, Soin N, Omar R (2011) Buck and boost converter design optimization parameters in modern VLSI technology. In: 2011 international conference and seminar on micro/nanotechnologies and electron devices proceedings, 30 June–4 July 2011, pp 125–128
11. Khaldi N, Mahmoudi H, Zazi M, Barradi Y (2014) The MPPT control of the PV system by using neural networks based on Newton Raphson Method. In: 2014 international renewable and sustainable energy conference (IRSEC), 17–19 Oct 2014, pp 1–6
12. Coelho RF, Concer FM, Martins DC (2010) A MPPT approach based on temperature measurements applied in PV systems. In: IEEE ICSET 2010, 6–9 Dec 2010, pp 1–6
13. Behera TK, Behera MK, Nayak N (2018) Spider monkey based improve P&O controller for photovoltaic generation system. In: IEEE international conference on technologies for smart-city energy security and power (ICSESP-2018), 28–30 Mar 2018, pp 1–6
14. Tomar AS, Dubey HM, Pandit M (2019) Spider monkey optimization for economic dispatch with diverse cost function. In: 2019 9th international conference on cloud computing, data science and engineering (Confluence), 10–11 Jan 2019, pp 349–354
15. Hazrati G, Sharma H, Sharma N (2016) Adaptive step-size based spider monkey optimization. In: 2016 IEEE 1st international conference on power electronics, intelligent control and energy systems (ICPEICES), 4–6 July 2016, pp 1–5
16. Kaur A, Sharma H, Sharma N (2017) Disruption operator-based spider monkey optimization algorithm. In: 2017 international conference on computing, communication and automation (ICCCA), 5–6 May 2017, pp 216–221

Industry 4.0

The Blueprint of Managing Manpower in the Framework of Industry 4.0

Shwati Sudha and Ankita Singh

1 Introduction

Industry, the mainstay of a country's economy, already witnessed three industrial revolutions. Friedrich Engels and Louis-Auguste Blanqui introduced the concept of "Industrial Revolution." The term industrial revolution comprises of two words industry and revolution. As cited from the literature of the Indonesian language dictionary, industry is viewed as an activity to execute the process of production, while the notion of revolution states that the changes are expeditious. Hence, industrial revolution is a rapid change in production process. These speedy changes in industrial processes stand for a pivotal point in the history of industrial revolutions. The industrial revolutions entrenched from different technological milestones over the past three centuries. Conventionally, the first stage of industrial revolution (Phase 1.0) emerged in Great Britain during the late seventeenth and early eighteenth century. It concentrated upon the introduction of machinery with a focus on mechanization. The second stage (Phase 2.0) emerged in Germany and America (1820–1870) which emphasized on mass production. The third stage (Phase 3.0) occurred during the multitude of industrialized countries in the late twentieth century which was intended toward computerized integration of mass customization. The advent of Industry 4.0 in the year 2011 belongs to the Germans, who were the pioneer to discuss it as a part of the development plan policy (High-Tech Strategy 2020) at Hannover Fair. This revolution introduced the concept of robotics, digitization, and automation. The study of West [1] states Industry 4.0 as an interdependent link between analog production and digital transformation. This confirms a pathway between manpower, materials, and machines. Industry 4.0 is usually mentioned as Smart Factory, Internet of Things, or Smart Industry. In 2013, the "Industry 4.0 Proclamation" was put into practicality and promulgates the

S. Sudha (✉) · A. Singh
DHSSM NIT Jamshedpur, Jamshedpur, Jharkhand 831014, India
e-mail: ssudha.hum@nitjsr.ac.in

S. Bag et al. (eds.), *Next Generation Materials and Processing Technologies*,
Springer Proceedings in Materials 9, https://doi.org/10.1007/978-981-16-0182-8_42

modernity of both strategic and intelligent industry in twenty-first century. However, the expansion of this revolution is illimitable. There is an urgent need to understand the significant demand of Industry 4.0 in context of innovative methods and techniques to accomplish organizational tasks effectively. Since, the organizations do no breathe without manpower, the role of manpower in gaining required skills and competencies is the key component for success of an organization in an aggressive competitive environment [2]. The new paradigm shift of Industry 4.0 is not solely a technological catastrophe but a conglomeration which results in different aftermath [3]. Thus, it is suggested to view the ultimatum of automation and digitalization as a door for enrichment and enhancement of manpower to comprise a whole new standard of skills and competencies. This new outlook enforces organizations to attract, recruit, and retain pool of non-imitable wisdom. This inculcates a culture of willingness and acceptance for application of new practices. An organization's effectiveness and excellence depend upon the strategy in managing manpower in the organization. To remain competitive and functional, it concentrates to fulfill the significant competency gap between existing capabilities of employees and the expected outcomes from them. This also conveys that the skills and competencies required in the subsequent years to accelerate are due to the high use of smart and intelligent techniques in the organizations. Manpower management is a tool to acquire the right person with the right skills in the right place at the right time. It focuses on the development of employees as a team leader and a player too. This way, manpower management plays a significant role to increase the boundary and foresee the future of smart businesses with a complete new set of competency standards. This inclination toward competent manpower proposes the concept of Smart Human Resources (SHR 4.0).

The proposed model (Fig. 1) focuses on the circumference where task overtakes the organization boundaries with an accomplished technology and competent manpower [4, 5].

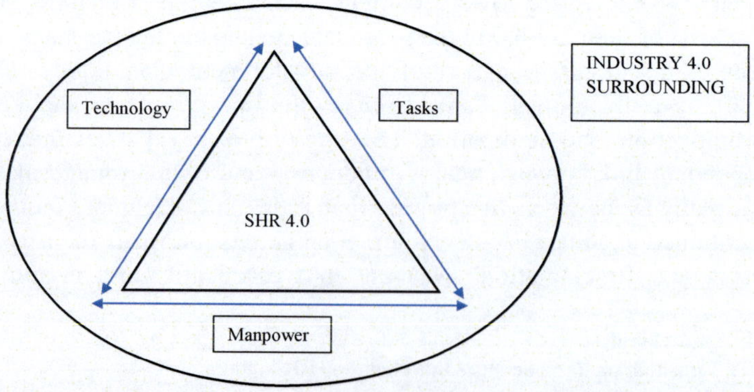

Fig. 1 Proposed structure of SHR 4.0 in Industry 4.0

2 Review of Literature

The study of Andrea Benesova and Jiri Tupa [2] found that the Industry 4.0 necessities the evolvement of skills and competencies due to the implementation of upcoming technologies and intelligent techniques.

Francesco Longo et al. [6] found that change and upgradation of technology provoke manpower management activities with high speed and intensity. It encourages the organization to move toward a modern concept of smart HRM for Industry 4.0.

The study of Katrin Fettig et al. [7] found that the influence of Industry 4.0 on thirty different German and French organizations focuses on the manpower management, activities of organization, organization structure, and work strategy.

The study of Guilherne Tortorella et al. [8] found that Industry 4.0 acts as a system of continuous improvement and reinforcement of manpower engagement and participation in the organization.

Geeta Rana and Ravindra Sharma [9] found that the effective management and utilization of manpower ensure success and survival of the organization in the period of technological advancement.

S. Fareri, G. Fantoni et al. [3] found that the new shift in the technology is not only a technological advancement but also a system of heterogeneous changes in the organization to efficiently manage the forthcoming trend.

3 Research Gap

Different studies focused on the impact of Industry 4.0 on the job profiles that require new standards of skills and competencies for forth coming changes and challenges of SHR 4.0 which requires further exploration and investigation. The study centers the strategies of managing manpower in the framework of Industry 4.0 with a prime focus on competencies, benefits, and challenges to re-address different roles of HRM. In addition to this, the study contributes to the forum of discussion with a paradigm shift to move simultaneously with the technological advancement.

4 Objectives

To build upon the objectives, the study specifies an inevitable amalgamation of new and present competencies to properly manage the rapid trend of digitization and automation. The velocity of technological advancement brought by Industry 4.0 generated a remarkable gap between the present competencies of manpower and the expeditious expansion of their roles. This ultimately stimulates to consider new, efficient, and effective approaches to manpower management and development. Hence, the study aims to find various strategies to manage manpower effectively in Industry 4.0, which requires the organizations to prepare for Smart Human Resources 4.0. So, the study further aims to specify the essential skills and competencies required for SHR 4.0 through identification of different changes, benefits, and challenges associated with its implementation.

5 Research Methodology

The exploratory study is based on qualitative analysis of secondary data. Systematic review technique is used for the collection of data from relevant research papers related to the objective of the study. Systematic review technique started with the collection of several peer-reviewed research papers through extensive research from different electronic database of journals. The next step involved the selection of relevant research papers. This was conducted to consider the current knowledge and forthcoming trends of manpower management to implement SHR 4.0 in the organizational setting of Industry 4.0.

6 Findings

Industry 4.0 instigates expeditious and marked changes in every sphere of organization. These modification leads toward the application of a new concept of SHR 4.0 which affects the organization in multifarious ways. The study pin points the pivotal changes in manpower management in this new time period with the application of principles and practices of SHR 4.0. The required modifications in organization structure, activities in managing manpower, changing role of human resources, requisite skills, and competencies are described with different challenges and benefits of SHR 4.0.

6.1 Smart Human Resources (SHR4.0)

SHR 4.0 revolves around few principles [10] based on the collaboration of manpower and technology to accomplish different tasks in and out the organization boundaries. The first principle is based on the concept of new work pattern in the organizations due to the existence of multigenerational manpower comprised of Generation Y (born between 1980 and 2000) and Generation Z (born after 2000) [9]. The second principle involves the concept of time and place of work. The traditional work format is removed and replaced by a flexible and dynamic format. It not only focuses on when and where work occurs but encloses the shift from "assignments focused on presence" to "assignments focused on results". The last principle comprises the nature of work. Though the traditional concept of manpower and supervisor is involved, the new trend of external workers creates a pathway in the organizations. It entails the enlarged job profiles of freelancers, traders, contractual employment, part time jobs, etc. The study of Bissola and Imperatori [10] states that the principles of digitalization and automation mentioned above affects the manpower and organization in a positive way. This facilitates a better work–life balance for the manpower. As everything comes with a price, similarly the implementation of SHR 4.0 also intimates certain set of changes, challenges, and benefits. Changes in organization and its process are often a response to change in the environment. It is either planned years in advance (proactive change) or exerts pressure on an organization because of a transformation in the environment (reactive change). The type of change and organization opts for deciding the advantage an organization gains in the long run. These advantages further provide competitive edge with a set of core competencies and skills to the organization in respect to the manpower. Few amendments which takes place in the organization to provide an aggressive competition in the era of Industry 4.0 requires changes in organization structure, change in manpower management, change in role of manpower management, and lastly, requisite change in skills and competencies.

6.1.1 Changes in Organizational Structure

The digital transformation in the fourth revolution causes the implementation of SHR 4.0 which leads to changes in the organization structure as well. These modification in the organization structure is established on the four design pillars of Industry 4.0 stated in the study of Hermann et al. [11]. The first pillar states the importance of connection in the organization structure which requires certain set of cooperation, communication, privacy, and standards. The second pillar states the pivotal role of lucidity through the organization structure. This helps to attain swiftness in the organization. The third pillar states the technical support at all levels of organization to promote flexibility and simplicity throughout the organization. Lastly, the pillar of decentralization is focused to embrace the motto of competitive advantage and innovativeness in the organization. The proposed pillars of organization structure due to different technological transformation contribute to the mobility of manpower. It promotes collaboration, transparency, flexibility, functionality, innovation, and decentralization to encourage a learning environment in the organization for effective manpower management. This change in the structure highlights the advantages of manpower involvement to cope up with the expeditious shift of Industry 4.0. Manpower deeply involved in their work profiles recommend strategies and competencies and also help in implementation of the same. Hence, Tortorella et al. [8] concludes the involvement of manpower in the organization as advantageous.

6.1.2 Changes in Manpower Management Activities

The study found that change in the organization structure promotes alterations in the manpower management activities. This involves the requirement of a more dynamic and interrelated combination of different activities. Changes in manpower activities lead to the up skill and re-skill of one's area of expertise. The significant changes in the field of recruitment and selection, training, development, appraisal, and remuneration are discussed below.

Traditional process of recruitment and selection is replaced by:

- Use of artificial intelligence to attain numerous information of the applicant.
- Enhancement in participation-based selection.

Present methods of training and development are modified by:

- Training focused on the emerging trends of digitalization and automation.
- On-demand competency attainment to deliver expected outcome.

Conventional appraisal technique is reframed by:

- Quick identification of the result.
- Importance to innovative and unique ideas.

Process of remuneration is redesigned to include:

- More of optional rewards than monetary rewards.
- Methods to build customized benefits.

6.1.3 Changes in the Role of Manpower Management

Due to tremendous change in technology, the roles of manpower management have changed simultaneously from 1996 to 2011. The significant change from the role of traditional workers to wisdom workers focuses now more on the administrative, motivational, and developmental areas to increase the efficiency and overall effectiveness of manpower. It also involves the strategic management of manpower to provide a competitive edge to the organizations through acquiring core competencies. The models of Ulrich clear the differences between the two time-periods objectively.

The initial model of Ulrich [12] states the roles as:

- Strategic partner: To deal with vision, mission, objectives, and strategy and inculcate it into manpower policies and practices.
- Administrative expert: To deal with the basic manpower management processes.
- Employee champion: To deal with the involvement and participation of manpower to accomplish success.
- Change agent: To deal with changes and challenges and act as catalyst or facilitator.

But, with the up gradation of technology, the role of manpower management changed automatically. The activities performed now are more centered toward value-added tasks through strategic selection and implementation of different programs to attain a competitive edge. This changed the concept of the initial model into a modified one.

In 2011, the new model presented six roles of manpower [13]:

- Strategic positioner: To understand the political, economic, social, technological, environmental, and legal concerns and act accordingly.
- Credible activist: To build and depend on the personal connection of trust and understanding.
- Capability builder: To inspect and participate in the formation of organizational capability to establish a competitive identity of the organization.
- Change champion: To support change that takes place at individual, team, and organizational level.
- Manpower pioneer and integrator: To implement the requisite manpower practices into consolidated outcomes over the long run.
- Technology proponent: To acquire, examine, and align new technology for efficiency and effectiveness in the organization.

6.1.4 Change in Skills and Competencies of Manpower

Manpower needs to acquire on demand skills and competencies in order to cope with drastic changes in roles and responsibilities. The adherence of these skills and competencies acts as an exclusive advantage to the individual. Use of these skills and competencies in their expertise area make them stand apart from the rest of the manpower in the organization. Hecklau et al. [14] discussed four major competencies with different focus mentioned as under:

- Social competency: Focused on communication and cooperation skills.
- Methodological: Focused on analytic competence, solution to complex problem, and enhancement in decision making.
- Personal competency: Focused on willingness to learn and excel skill.
- Domain competency: Focused on digital networking, digital security, coding competency, and interdisciplinary competence.

Changes in different realm of organizations due to dynamic and versatile characteristics of digitalization and automation enforce SHR 4.0 in the organization structure, manpower activities, role of manpower, and requisite skills and competencies. This also requires resilience techniques to overcome challenges in implementation of SHR 4.0. Though, few organizations survive without adaption to change, on the contrary, acceptance of these challenges with a new mindset results in the benefits of it.

6.2 Benefits of SHR 4.0

An organization acquaints different changes and challenges of SHR 4.0. Effective utilization of these acquaintances provides opportunity to grab maximum benefits in favor of the manpower. These benefits are summed up into the following points as discussed in different studies:

The benefits classified by Irianto [15] are:

- Creativity in ecosystem.
- Ruthless industrial base.
- Investment in technology.

Sivathanu and Pillai [16] mentioned the following different opportunities:

- Attract, arise, and appoint new age talent.
- Efficient and effective HR operations.
- Leaner manpower departments.

6.3 Challenges of SHR 4.0

To work with organizations in the forthcoming scenario requires the manpower to manage constant changes. This automatically leads to management of different challenges in order to attain the benefits of it. Identification of these challenges ahead of time, prepares the manpower to deal with it positively. SHR 4.0 possesses different challenges for manpower [17]. These challenges are described below on the basis of different studies.

The challenges stated by Irianto [15] are:

- Willingness for Industry 4.0.
- Believe in workforce.
- Availability of socio cultural setting.
- Heterogeneity and job creation.

Hecklau et al. [14] mentioned different challenges of manpower as under:

- Economic challenges labeled into globalization, innovation, increased service inclination, and requirement for cooperation and participation.
- Social challenges categorized into change in the values associated with multi-generational manpower, enhanced virtual work, and higher complexity.
- Technical challenges lead to implication of advanced technology and growth in partnership-based task.
- Environmental challenges represent climate and culture change and limitations of resources.
- Political challenges demonstrate standardization and data privacy.

Sivathanu and Pillai [16] mentioned the challenges as:

- Selection of the appropriate set of updated technological tools.
- To overcome the present organizational culture.
- To control the expectations of multigenerational manpower.

7 Conclusion

Advancement in technology through transformation of digitalization and automation provokes the concept of Smart Human Resources (SHR 4.0). Further, the implementation of SHR 4.0 requires the metamorphosis of manpower in aspects related to work pattern, time, and plan of work and nature of work. The study specifies that the above transformations lead to changes in the organization structure in form of effective communication, transparency, trust, cooperation, and decentralization. These changes promote innovation, participation, engagement, and commitment of manpower. The study also proposes changes in different manpower activities and roles within an organization for survival and sustenance. The contrasting roles of human resources are shaped as a role of strategic positioner, credible crusader, capability enhancer, change expert, innovator, and integrator and technocrat within the framework of Industry 4.0. The novel concept of SHR demands standard competencies in various areas of social, methodological, and personal domain. The study identified certain benefits and challenges in formulation and implementation of SHR. Appointment of new age talent, decentralization of authority, promotion of creativity and innovativeness, effective utilization of manpower, and leaner human resource departments are chief benefits of SHR. At the same time, selection of advanced appropriate technological tools, a control on the expectations of multigenerational manpower, and encouragement for a culture of willingness to adapt SHR 4.0 are identified as the key challenges of SHR. The study concludes that advancement in digitalization and automation penetrates the manpower management with an unexpected speed and intensity. This recommends for a dual role of SHR 4.0 which concentrates on different changes in the organization at one end and acts as a tool to successful accomplish these changes on the other.

8 Future Research Directions

The study presents the blueprint of manpower management within the framework of Industry 4.0 with a chief focus on SHR. The exploratory in-depth study based on different literatures chiefly centered on SHR acts as the major limitation of the study. This limitation if used as an opportunity leads to empirical and experimental investigation of the derived conclusion. As well, the practical applicability of

SHR.0 in Industry 4.0 is open for further exploration. The study further emphasizes to provide an open forum for the discussion on manpower management in this rapid evolving environment of the organizations to avoid and navigate the upcoming uncertainties.

References

1. West DM (2015) What happens if robots take the jobs? The impact of emerging technologies on employment and public policy. Centre for Technology Innovation at Brookings, Washington, DC
2. Benešová A, Tupa J (2017) Requirements for education and qualification of people in Industry 4.0. Procedia Manuf 11:2195–2202. Elsevier. https://doi.org/10.1016/j.promfg.2017.07.366
3. Fareri S, Fantoni G, Chiarello F, Coli E, Binda A (2020) Estimating Industry 4.0 impact on job profiles and skills using text mining. Comput Ind 118:103222. Elsevier. https://doi.org/10.1016/j.compind.2020.103222
4. Dregger J, Niehaus J, Ittermann P, Hirsch-Kreinsen H, ten Hompel M (2016). The digitization of manufacturing and its societal challenges: a framework for the future of industrial labor. In: 2016 IEEE international symposium on ethics in engineering, science and technology (ETHICS), pp 1–3. IEEE, May 2016. https://doi.org/10.1109/ETHICS.2016.7560045
5. Hirsch-Kreinsen H (2016) Digitization of industrial work: development paths and prospects. J Labour Market Res 49(1):1–14. Springer. https://doi.org/10.1007/s12651-016-0200-6
6. Longo F, Nicoletti L, Padovano A (2017) Smart operators in industry 4.0: a human-centered approach to enhance operators' capabilities and competencies within the new smart factory context. Comput Ind Eng 113:144–159. Elsevier. https://doi.org/10.1016/j.cie.2017.09.016
7. Fettig K, Gačić T, Köskal A, Kühn A, Stuber F (2018) Impact of industry 4.0 on organizational structures. In: 2018 IEEE international conference on engineering, technology and innovation (ICE/ITMC), pp 1–8. IEEE, June 2018. https://doi.org/10.1109/ICE.2018.8436284
8. Tortorella G, Miorando R, Caiado R, Nascimento D, Portioli Staudacher A (2018) The mediating effect of employees' involvement on the relationship between Industry 4.0 and operational performance improvement. In: Total quality management and business excellence, pp 1–15. Taylor and Francis. https://doi.org/10.1080/14783363.2018.1532789
9. Rana G, Sharma R (2019) Emerging human resource management practices in industry 4.0. Strateg HR Rev. Emerald Publishing. https://doi.org/10.1108/SHR-01-2019-0003
10. Bissola R, Imperatori B (eds) (2020) HRM 4.0 for Human-centered Organizations. Emerald Publishing. https://doi.org/10.1108/S1877-636120190000023001
11. Hermann M, Pentek T, Otto B (2016) Design principles for industry 4.0 scenarios. In: 2016 49th Hawaii international conference on system sciences (HICSS), pp 3928–3937. IEEE, Jan 2016. https://doi.org/10.1109/HICSS.2016.488
12. Ulrich D (1996) Human resource champions: the next agenda for adding value and delivering results. Harvard Business Press
13. Ulrich D, Younger J, Brockbank W, Ulrich M (2011) Competencies for HR professionals working outside-in. The RBL White Paper Series. http://rbl-net.s3.amazonaws.com/hrcs/2012/Competen-cies%20for%20HR%20Professionals%20Working, 20
14. Hecklau F, Orth R, Kidschun F, Kohl H (2017) Human resources management: Meta-study-analysis of future competences in Industry 4.0. In: Proceedings of the international conference on intellectual capital, knowledge management and organizational learning, pp 163–174, Dec 2017

15. Irianto D (2017) Industry 4.0: the challenges of tomorrow. In: Disampaikanpada Seminar Nasional TeknikIndustri, Batu-Malang
16. Sivathanu B, Pillai R (2018) Smart HR 4.0—how industry 4.0 is disrupting HR. Hum Resour Manage Int Digest. Emerald Publishing. https://doi.org/10.1108/HRMID-04-2018-0059
17. Holland P, Jeske D (2017) Changing role of social media at work: Implications for recruitment and selection. In: Electronic HRM in the smart era. Emerald Publishing. https://doi.org/10.1108/978-1-78714-315-920161011

3D and 4D Printing in Industry 4.0: Trends, Challenges, and Opportunities

S. Deepak Kumar, Shailesh Dewangan, S. K. Jha,
S. K. Parida, and Ajit Behera

1 Introduction

In the current era of twenty-first century, Industry 4.0 is a common term for the Internet of Things focused on cyber-physical systems (CPS) that combines digital and physical environments and has contributed to the idea of a 'Smart Factory'. Industry 4.0 is also the fourth wave of the industrial revolution, the latest trend of automation, and data sharing of production technology. The evolution of Industry 4.0 is as below [1].

The First Industrial Revolution: The first industrial revolution took place between the late 1700s and the early 1800s. Over this span of time, manufacturing has changed from relying on manual labor performed by humans and aided by working animals to a more optimized method of labor performed by people with the use of water and steam engines and other forms of machine tools [2].

The Second Industrial Revolution: The world started a second industrial revolution in the early part of the twentieth century, with the invention of steel and the use of electricity in factories. The advent of electricity has made it easier for factories to improve production and has helped to make factory equipment more mobile. It was during this period that mass manufacturing ideas such as assembly lines were adopted as a means of improving productivity.

S. Deepak Kumar (✉) · S. Dewangan · S. K. Jha
Department of Production Engineering, Birla Institute of Technology Mesra, Ranchi, Jharkhand 835215, India
e-mail: dks10@iitbbs.ac.in

S. K. Parida
Department of Manufacturing Engineering, National Institute of Foundry and Forge Technology, Ranchi, Jharkhand 834003, India

A. Behera
Department of Metallurgical and Materials Engineering, National Institute of Technology Rourkela, Rourkela, Odisha 769008, India

S. Bag et al. (eds.), *Next Generation Materials and Processing Technologies*, Springer Proceedings in Materials 9, https://doi.org/10.1007/978-981-16-0182-8_43

Third Industrial Revolution: Starting in the late 1950s, a new technological revolution steadily began to emerge as manufacturers continued to integrate more electronic and ultimately computer technology into their factories. During this time, factories started to undergo a change that centered less on analog and mechanical technologies and more on digital technology and automation tools.

The Fourth Industrial Revolution or Industry 4.0: A fourth industrial revolution, known as Industry 4.0, has arisen in the last few decades. Industry 4.0 focuses on emerging technologies from the last decades to a whole new stage with the aid of interconnectivity across the Internet of Things (IoT), access to real-time data, and the introduction of cyber-physical structures. Industry 4.0 presents a more detailed, interlinked, and holistic approach to production. It links physical and digital and facilitates greater communication and access through agencies, partners, suppliers, goods, and individuals. This covers cyber-physical networks, the Internet of Things, and cloud computing. Industry 4.0 creates the so-called "Smart Factory", and the evolution of Industry 4.0.is shown in Fig. 1 [3].

The main features of Industry 4.0 are described below [2]:

- *Interoperability*: Machines, computers, sensors, and people that interact and communicate with each other;
- *Knowledge transparency*: Systems build a simulated replica of the real world;
- *Technical assistance*: To citizens in decision-making and problem-solving;
- *Decentralized decision-making*: The capacity of cyber-physical systems to make basic decisions.

The goals and objectives of Industry 4.0:

The first strong aim for Industry 4.0 is to improve efficiency and enhance productivity by automation. In addition, Industry 4.0 can minimize pollution and

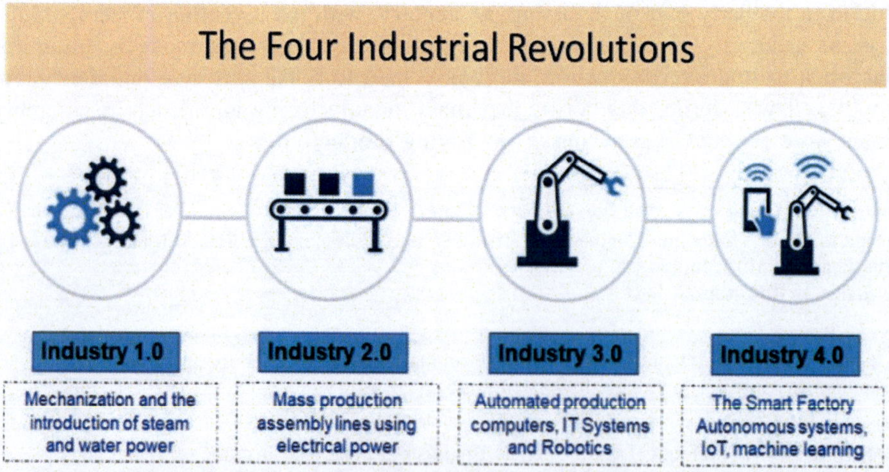

Fig. 1 Four generations of industrial revolutions [1]

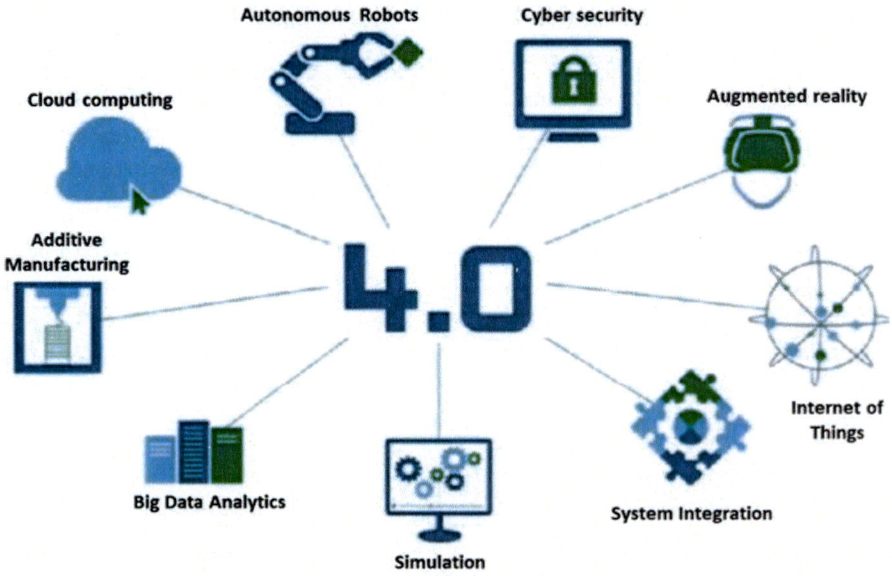

Fig. 2 Technology enablers of Industry 4.0 [2]

increase yields, all of which are very significant driving forces in all technologies. The advantages and drivers of this evolution are shown in Fig. 2.

Advantages/Benefits of Industry 4.0:

- Increased efficiency through optimization and automation.
- Real-time evidence for the supply chain in real-time economy.
- Improved quality products: real-time tracking, IoT-enabled quality management, and cobots.
- Useful in very unsafe working conditions.
- Supply chains should be more efficiently managed.
- Computer control could make efficiency and performance even more efficient and predictable.
- Increased sales, market shares, and earnings could result for several firms.
- Improved working practices and productivity.

Challenges of Industry 4.0 [3]:

- A high degree of reliability and stability.
- Data security.
- Proprietary production knowledge.
- Maintaining the quality of manufacturing process with less human control could become an obstacle.
- Loss of high paying human jobs.
- Avoiding technological issues that could cause costly production outages are still a worry.

- Systematic lack of expertise and manpower in the development and deployment of these programs.

General reluctance of partners and consumers is to spend heavily in emerging technology.

2 Additive Manufacturing Applications in Industry 4.0

Additive manufacturing (AM) (or 3D printing) is a method of creating three-dimensional solid objects from a digital image. The development of a 3D printed model is done by additive processes. In an additive process, an object is formed by laying down successive layers of material before the whole object is created. AM requires multiple phases, from a simulated CAD model to a final physical portion [4]. The general AM method, right from the CAD model to the final output of the component, is shown in Fig. 3 [5].

The virtual design of the object has been developed. Computer-assisted design (CAD) uses either a 3D modeling application or a 3D scanner for simulated design. The program splits the final model into hundreds or thousands of horizontal layers. The printer produces an object layer by layer, resulting in a single three-dimensional object.

3D Printing Techniques and Innovations [6].

The most popular 3D printing machine that uses polymer as a filament is fused deposition modeling. In addition, the main techniques of AM technologies are selective laser sintering (SLS) and selective laser melting (SLM).

Fused deposition modeling (FDM) uses a plastic filament or metal wire as an input medium for an extrusion nozzle. The nozzle is heated to melt the material and can be moved in both horizontal and vertical directions by means of CAM. The substance hardens instantly after being extruded from the nozzle.

Selective laser sintering (SLS) uses a high-power laser to fuse input materials such as plastic, metal, and glass and scans the powdered content layer by layer.

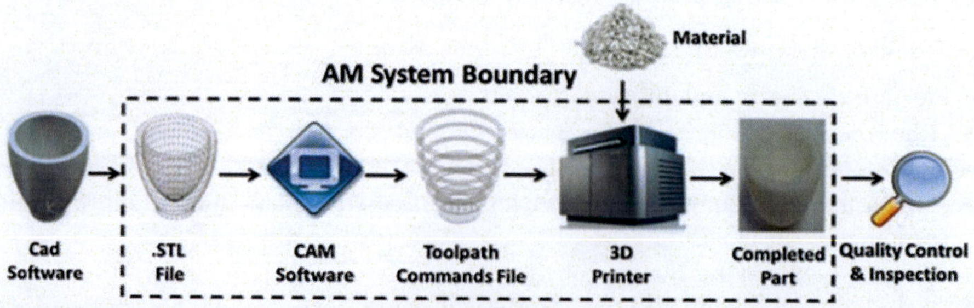

Fig. 3 Generic additive manufacturing process [5]

Stereolithography (SLA) photopolymerization is used to create a stable portion of a liquid. This technology uses a vat of liquid ultraviolet curable photopolymer resin and an ultraviolet laser to create the layers of the object one at a time. UV laser solidifying the pattern [7].

3 4D Printing

The fourth dimension of 3D printing technology is referred to as "4D printing." In this new dimension, 3D printed objects have the ability to change their form by themselves under the effect of external influences such as light, heat, electricity, and magnetic field. By adding the dimension of time, written objects change form dynamically depending on the needs and demands of the case, without any electromechanical components or moving parts. This form-changing phenomena of 3D printed objects is focused on the capacity of the material to transform in response to particular stimuli over time and does not require human intervention to assist the process [8].

3.1 Emergence of 4D Printing

3D printing, an additive manufacturing process, is known to be one of the most revolutionary developments in industrial manufacturing. It has totally changed the way parts/components and equipment are produced in the market, along with their design and development. 3D printing allows manufacturers and engineers to create intricate forms and systems that were historically thought difficult using conventional manufacturing methods. 3D printing technology has seen consistent developments over the past three decades and has grown significantly. Despite its potential to build dynamic, bio-inspired, multi-material structures, 3D printing is not yet able to be used in large-scale production. The increasing need for modular artifacts in a variety of applications, such as self-folding packaging and adaptive wind turbines has fueled the advent of 4D printing. Researchers are now looking forward to traditional 3D printing, which creates structures constructed from a single material, to build a meta-material structure. The meta-material structure is created by the combination of different materials that provide superimposed structural reactions when triggered by external stimuli. Congruent printing of various materials can form a material anisotropy that allows the object to alter its shape by folding, elongating, rotating, and corrugating along its axes. Researchers are now focusing on extending these technological improvements to build lockers, lifters, microtubes, soft robotics, dolls, etc. This ability of structures to transform their shape over time by using the behavior of various materials is referred to as 4D printing. The key distinctions in 3D printing and 4D printing are the use of the

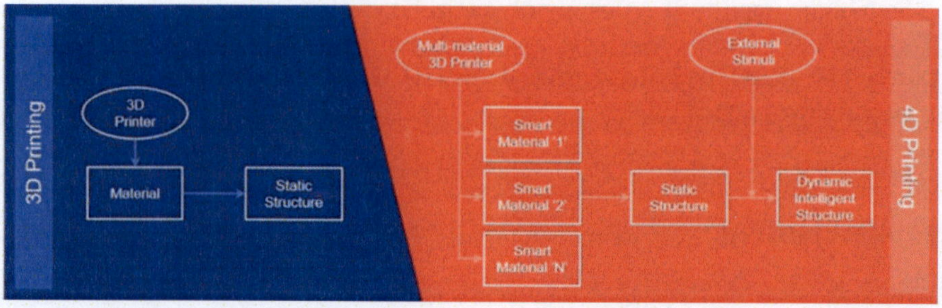

Fig. 4 3D versus 4D printing process [6]

Table 1 Differences between 3D and 4D printing [6]

Description	3D printing	4D printing
Dynamic shape change	No	Changes in color, shape, function, etc.
Materials used	• Thermoplastics (Acrylonitrile butadiene styrene (ABS), etc.) • Metals and alloys • Biomaterials and gels • Nanomaterials	• Smart materials—Shape memory alloys (SMA) and shape memory polymers (SMP) • Self-assembled materials • Hydrophilic polymers, biomaterials, and plant oil
Printing facility	• 3D Printer—Stereolithography (SLA) • Fused deposition modeling (FDM) • Selective laser sintering (SLS)	• 3D printer—Stereolithography (SLA) • Multi-material 3D printers

materials to be printed and the printing facility. Exhibits 1 and 2 below illustrate the main variations in 3D and 4D printing [9] (Fig. 4, Table 1).

3.2 4D Printing Materials and Technologies

Since 4D printing technology is still in its incipient period, the materials used for it are limited. However, research and progress in 3D printing are expected to create new possibilities for 4D printing. Main research areas currently in focus for 4D printing are shown in Fig. 5 [8].

Smart material is one of the highly focused research fields of 4D printing, where the deformation process of different materials is synthesized according to their reaction to different external stimuli. Researchers are currently using direct inkjet cure, fused deposition modeling, stereolithography, laser-assisted bioprinting, and selective 4D laser melting processes.

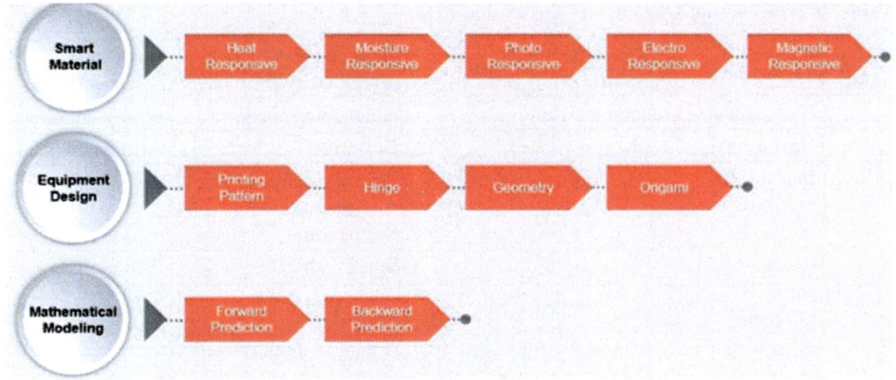

Fig. 5 Major research areas in 4D printing process [7]

Material selection: 4D printing materials are categorized on the basis of their environment or the external stimulus in which they respond. Current classes of smart materials are currently categorized in the sections below.

Thermo responsive materials: These materials work on the form memory effect (SME) process. They are known as shape memory alloys (SMA), shape memory polymers (SMP), shape memory hybrids (SMH), shape memory ceramics (SMC), and shape memory gels (SMG). Many researchers choose SMPs because it is easy to print on these samples. They shape and distort as heat or thermal energy is used as a stimulus.

Moisture responsive materials: These materials are commonly chosen by researchers, since water is available in abundance and can be used in a wide variety of applications. Hydrogel is one of the smart materials that fall into this group because it reacts strongly with water. For instance, hydrogels can increase its size by up to 200% of its original volume, when it comes in contact with water.

Photo/Electro/Magneto responsive Materials: These materials react to light, current, and magnetic fields. Magnetic nanoparticles are inserted in the printed object to obtain magnetic power over the object [10].

3.3 *Applications of 4D Printing*

Significant end-use applications of 4D printing technologies are projected to originate in the biomedical, automobile, aerospace, and consumer sectors. However, the ability of 4D printing is also predicted to have an effect on other markets, such as electronics, manufacturing, and industrial, in the immediate future [11, 12]. Table 2 illustrates the future effect of 4D printing on diverse applications through industries.

Table 2 Potential applications of 4D Printing [13]

Sl. no.	Industry	Near-term future products	Mid-term future products	Long-term future products
1	Electronics	Smart sensors	Adaptable sensors	Nanotechnology
2	Healthcare	Artificial tissues	Bio-prints	Artificial organs
3	Industrial machinery	Building and pipes	Machines and equipment, reverse engineering	
4	Automotive and aerospace		Space vehicles	Automotive body parts
5	Consumer appliances	Fashion and lifestyle (4D printed dress, shoes, etc.)	Home appliances	

4 Conclusions

Several research and development projects related to 4D printing are ongoing in sectors such as healthcare, electronics, automobile, aerospace and defense, and consumer goods. Despite being a novel technique, the possible possibilities offered by 4D printing are immense and recognized by a range of experts in the field. Due to various research and development activities, the 4D printing industry is starting to be developed.

Some of the major challenges facing the printing industry include the lack of capacity to provide service systems for complicated objects, the lack of multi-material printers, the lack of low-cost printers and smart materials, slow printing times, and the poor durability of printed objects over the long term. Though there are certain advancements in printing technology, such as five-axis printing equipment, which is expected to eliminate the problem of building support structures for complicated internal structures, other challenges still remain.

However, considering the interest shown by manufacturers and the high-intensity level of research and development activities with respect to 4D printing, the technology could make an exponential jump at a faster pace than the predicted pace. Finally, manufacturers who want to be at the forefront of technological changes and advancements should be abreast with the technological advancements and potential implications of 4D printing.

Acknowledgements The authors acknowledge the financial funding received from National Project Implementation Unit (NPIU), MHRD, India, under the Collaborative Research Scheme Project grant reference [CRS ID: 1-5732031971].

References

1. Dohale V, Kumar S (2018) A review of literature on Industry 4.0. https://www.researchgate.net/publication/328345685
2. Machado CG, Winroth MP, Ribeiro da Silva EHD (2020) Sustainable manufacturing in Industry 4.0: an emerging research agenda. Int J Prod Rese 58:1462–1484. https://doi.org/10.1080/00207543.2019.1652777
3. Bidnur CV (2020) Study on Industry 4.0 concept. IJERT 9(4):Paper ID IJERTV9IS040569. https://doi.org/10.17577/IJERTV9IS040569
4. Gibson I, Rosen DW, Stucker B (2010) Additive manufacturing technologies. Springer US, Boston, MA. https://doi.org/10.1007/978-1-4419-1120-9
5. Sahini DK, Ghose J, Jha SK, Behera A, Mandal A (2020) Optimization and simulation of additive manufacturing processes: challenges and opportunities—a review. In: Balasubramanian K, Senthil Kumar V (ed) Additive manufacturing applications for metals and composites, pp 187–209. IGI Global 2020. https://doi.org/10.4018/978-1-7998-4054-1.ch010
6. Mehrpouya M, Dehghanghadikolaei A, Fotovvati B, Vosooghnia A, Emamian SS, Gisario A (2019) The potential of additive manufacturing in the smart factory Industrial 4.0: a review. Appl Sci 9(3865):1–34. https://doi.org/10.3390/app9183865
7. Haleem A, Javaid M (2019) Additive manufacturing applications in Industry 4.0: a review. J Ind Intg Mgmt **04**:1930001. https://doi.org/10.1142/S2424862219300011
8. Zhang Z, Demir KG, Gu GX (2019) Developments in 4D-printing: a review on current smart materials, technologies, and applications. Int J Smart Nano Mater 10:205–224. https://doi.org/10.1080/19475411.2019.1591541
9. Zhang C, Chen F, Huang Z, Jia M, Chen G, Ye Y, Lin Y, Liu W, Chen B, Shen Q, Zhang L, Lavernia EJ (2019) Additive manufacturing of functionally graded materials: a review. Mater Sci Eng, A 764:138209. https://doi.org/10.1016/j.msea.2019.138209
10. Alcácer V, Cruz-Machado V (2019) Scanning the Industry 4.0: a literature review on technologies for manufacturing systems. Eng Sci Technol Int J 22:899–919. https://doi.org/10.1016/j.jestch.2019.01.006
11. Abdulhameed O, Al-Ahmari A, Ameen W, Mian SH (2019) Additive manufacturing: challenges, trends, and applications. Adv Mech Eng 11:1–27. https://doi.org/10.1177/1687814018822880
12. Kumar SD, Ghose J, Mandal A (2019) Thixoforming of light-weight alloys and composites: an approach toward sustainable manufacturing. In: Sustainable engineering products and manufacturing technologies, pp 25–43. Elsevier. https://doi.org/10.1016/B978-0-12-816564-5.00002-5
13. Chong S, Pan GT, Chin J, Show P, Yang T, Huang C-M (2018) Integration of 3D printing and Industry 4.0 into engineering teaching. Sustainability 10:3960. https://doi.org/10.3390/su10113960

Challenges in Implementation of Industry 4.0 in Manufacturing Sector

Nikita Sinha and Amaresh Kumar

1 Introduction

With the rapidly evolving technologies in the past few decades, remarkable transformations have taken place in the manufacturing sector that continues to play a critical role for both developing and developed nations. It is anticipated that the ongoing of Fourth Industrial Revolution will increase the share of manufacturing sector in Gross Domestic Product (GDP) wherein it currently accounts for nearly 16% of the global GDP. The Fourth Industrial Revolution, also termed as Industry 4.0 (I4) and first declared by the German government in 2011, heralds the new era of manufacturing that will unfold over the twenty-first century [1]. In essence, Industry 4.0 also known as "smart factory" is the convergence of physical and digital technologies which is expected to increase the flexibility, scalability, agility and productivity. The transformation is reinforced by key technological areas that include the Internet of things (IoT), artificial intelligence (AI), cognitive computing, the cloud, big data, simulation, additive manufacturing (3D printing), augmented reality, big data and analytics, advanced robotics, simulation and horizontal and vertical system integration [2].

The worldwide industrial environment has enormously varied as a result of successive technological development and innovation brought on by the inception of Industry 4.0. It has led to advanced globalization of industries, and in order to gain competitive advantage, the companies are propelled to adopt this new change. Industry 4.0 gave rise to the integration of production facilities, supply chains and service systems through cyber-physical system (CPS) technology that leads to the

N. Sinha (✉) · A. Kumar
Department of Production and Industrial Engineering, National Institute of Technology, Jamshedpur, Jamshedpur, India
e-mail: 2018ugpi040@nitjsr.ac.in

A. Kumar
e-mail: akumar.prod@nitjsr.ac.in

S. Bag et al. (eds.), *Next Generation Materials and Processing Technologies*, Springer Proceedings in Materials 9, https://doi.org/10.1007/978-981-16-0182-8_44

development of global value-creation network [3]. The application of CPS, that is, interaction between physical processes and computation networking, when applying it in manufacturing is characterized as cyber-physical production system (CPPS) [4]. The new approach to manufacturing has various benefits like smart and efficient production, reduced supply chain complexity and sustainable growth that can meet changing demands, without human intervention, which is the foremost objective of Industry 4.0. It has also been stated that technologies which initially have linear growth will follow the path of exponential growth trajectory instigating large-scale creative destruction [5].

However, opportunities come with their own set of challenges. Despite of I4 offering myriad of benefits that are crucial for an industry, there are roadblocks to its implementation which is the reason behind lack of manufacturing industries leveraging I4 technologies. The revolution is bound to cause disruption, affecting the ways of manufacturing and its business models as it was seen during the previous revolutionary phases. Today, with range of facilities available in context of accessibility, it might seem easy to adopt I4 technologies, but still there are significant barriers owing to which the SMEs as well as large enterprises are unable to make use of advanced technologies.

In order to identify the challenges, research papers have been analysed through extensive literature review. This paper is structured as follows. Section 2 throws light on the background and concept of I4 in manufacturing sector. Section 3 presents the methodology used to select research papers. Section 4 elaborates the challenges faced by manufacturing sector in implementation of I4. Lastly, Sect. 5 concludes the paper along with future research scopes of exploiting I4.

2 Industry 4.0 Concept

Industrial sector has witnessed four phases in the process of revolution. The use of steam-powered engines was marked as the First Industrial Revolution. The Second Industrial Revolution came with the introduction of concept of mass production along with the use of electricity. The Third Industrial Revolution is characterized by information technology (IT) system and the change from electronic to digital technology.

Today, the industries are in the midst of Fourth Industrial Revolution that was set off by the emergence of CPS and advanced interconnectivity. The Industry 4.0 concept first appeared in 2011 in high-tech strategy project by Germany's government whose final report was then presented at Hannover Fair on 8 April 2013.

The basic idea of Industry 4.0 is to exploit the potential of emerging technologies such as big data, cloud computing and manufacturing, additive manufacturing, IoT and robotics for the purpose of attaining transparency, decentralization, agility, flexibility, scalability and real-time functioning. These are the characteristics that enhance the competitiveness and ability to resist the fluctuations in global market [6]. Since the onset of I4, it has substantially changed the industrial outlook wherein companies adopting these technologies will be able to produce high quality and customized products with improved efficiency. The introduction of intelligent manufacturing system has optimized the production, logistics and supply chain management and reduced the need of human involvement which consequently lowers the production cost. It will also strengthen the customer relationship due to increase in transparency and faster response [7].

Many countries have taken initiative to promote Industry 4.0 in pursuit to digitalize manufacturing, such as Industrie 4.0 by Germany, Advanced Manufacturing Partnership (AMP) by USA, Industrial 4.1J by Japan, Make in India by India and Made in China 2025 by China. However, only few companies, mostly large enterprises have transitioned into smart factory. In further section, the literature explicates the causes for limited implementation and impact of new-age technology.

3 Methodology

The proposed extensive literature review is based on qualitative analysis of previous research work. The main aim is to assess the challenges in implementation of I4 in manufacturing sector and hence provide future scope of I4. This extensive literature review was carried out in the following order (Fig. 1).

The planning phase is the initial step to determine the database that will be utilized. To get access of wide range of authentic and peer-reviewed articles, ScienceDirect (www.sciencedirect.com) and Google Scholar (www.scholar.google.com) were selected as electronic database platform.

After planning, the objective was defined as to identify the underlying factors that have obstructed the companies, under manufacturing sector, from adopting I4 and analyse the social impact of industrial transformation. Eventually, the research question was formulated as challenges in implementation of I4.

The search was performed over the period of 2015 to June 2020 for the most recent research articles published only in journals and conference proceedings. The keywords were carefully selected in combination of words such as "Industry 4.0", "challenges", "manufacturing", "workforce", "automation", "robotics", "security", "augmented reality", "Internet of Things" and "proactive maintenance".

The methodology enabled us to find research articles that were close to the objective of this paper. As a result, 41 articles were comprehensively analysed

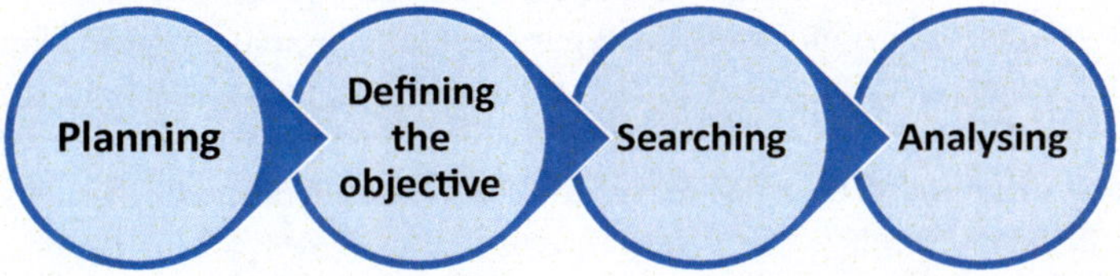

Fig. 1 Methodology of the literature review

under six major categories, that are, security, finance, human resources, operations management, education and training and government policies. The overview of selected literature is shown in Table 1.

4 Discussion

Industry 4.0 is on the verge of transforming traditional manufacturing process, proposing abundance of opportunities to progress. However, for successful implementation resulting into collaboration of human and intelligent machine to work in tandem, it is crucial to analyse the potential barriers. The primary reasons for some of the barriers are considered to be lack of awareness about I4 and lack of proper guidance for its implementation. In order to get deeper understanding of challenges, they have been categorized under six major segments influencing the manufacturing sector.

4.1 Security and Privacy

The increase in IoT and CPS-enabled manufacturing has improved the efficiency of production, but at the same time, a widely distributed network has made the industries extremely prone to cyber-attacks. A survey conducted in 2016 stated that 39% of the manufacturing industries faced security threats within twelve-month period [8]. Data privacy, data security, malware attack and identity theft are some of the major concerns in a data-driven manufacturing, and as it reaches the maturing stage, severity is increased, leading to disruption at larger scale [9–12]. Shift to the cloud storage poses security risk which can compromise the intellectual property rights of an organization and personal information of employees and customers [13, 14]. As the supply chain gets highly outsourced, communicating with two or more parties makes the server more vulnerable to hackers. In some cases, such as, additive manufacturing or robots employed in production, their systems can be attacked resulting into breakdown or data manipulation [15, 16]. If the operator is unable to identify these threats in the initial stage, it remains undetected until defective products have been manufactured. The carelessness of workers and lack of standard security software also create loopholes for cybercriminals [17, 18]. Cybersecurity is not only limited to technical domain; it has become a managerial problem too since the concerned managers remain passive in taking any serious measures towards security threat [19]. Manufacturing industries tend to give secondary preference to cybersecurity. These issues are not new, and it is probable that challenges will increase in order to secure the system from cyber-attacks.

Table 1 Overview of the reviewed literature

Sr. No.	Authors	Source of data	Qualitative analysis	Quantitative analysis
1	Abdur et al. [9]	Secondary data	✓	
2	Gonçalves et al. [10]	Case study	✓	
3	Khan and Turowski [11]	Survey/interviews	✓	
4	Raj et al. [12]	Case study		✓
5	Roman et al. [13]	Secondary data	✓	
6	Tupa et al. [14]	Secondary data	✓	
7	Rokka et al. [15]	Secondary data	✓	
8	Sturm et al. [16]	Case study	✓	
9	Zarreh et al. [17]	Secondary data	✓	
10	Ghadge et al. [18]	Secondary data	✓	
11	Culot et al. [19]	Secondary data/ interviews	✓	
12	Hamzeh et al. [21]	Case study		✓
13	Glass et al. [22]	Survey		✓
14	Kamble et al. [23]	Interview		✓
15	Kurt [24]	Secondary data	✓	
16	Centea et al. [25]	Secondary data	✓	
17	Fareri et al. [26]	Survey	✓	
18	Lee and Lee [27]	Secondary data	✓	
19	Luthra and Mangla [28]	Survey		✓
20	Ingaldi and Ulewicz [29]	Survey/interview		✓
21	Dora and Roland [30]	Secondary data/ interview	✓	
22	Sevinç et al. [31]	Survey		✓
23	Abu-bakr et al. [32]	Secondary data	✓	
24	Lynch et al. [33]	Secondary data	✓	
25	Baboli et al. [34]	Secondary data	✓	
26	Haddud et al. [36]	Survey	✓	
27	Kache and Seuring [37]	Case study	✓	
28	Fisher et al. [38]	Survey	✓	
29	Lee et al. [39]	Survey	✓	
30	Chen et al. [41]	Secondary data	✓	
31	Zhong et al. [42]	Secondary data	✓	
32	Andersen et al. [43]	Secondary data	✓	
33	Mourtzis et al. [44]	Case study	✓	
34	Suárez et al. [45]	Secondary data	✓	

(continued)

Table 1 (continued)

Sr. No.	Authors	Source of data	Qualitative analysis	Quantitative analysis
35	Seamus et al. [46]	Case study	✓	
36	Benešová and Tupa [47]	Secondary data	✓	
37	Motyl et al. [48]	Survey		✓
38	Manda and Ben [50]	Secondary data	✓	
39	Iyer [51]	Case study	✓	
40	Fatorachian and Kazemi [52]	Secondary data	✓	

4.2 Human Resources

The automation in industries will reduce the manual workload which can create alarming impact on the workforce, especially the unskilled laborers [9–11, 14]. McKinsey estimates that by 2030, 400–800 million workers globally can be displaced by automation and will need to find new jobs [20]. It is evident that as the number of smart factories will increase, there will be shortage of skilled labourers and the workers will be left unemployed, though for a shorter term, owing to the lack of training facility in technology-centred skills in addition to population boom in many countries [21–23]. Although there are a greater number of qualified people in large enterprises, training concept which is necessary to get familiar with new technologies lacks in both large enterprises and SMEs. Moreover, SMEs have significantly high number of unskilled labourers. Thus, retaining workforce and aligning the labour market in the right direction will be a challenge in near future [24]. It is found that lasers and hazardous materials used in additive manufacturing and other such technologies can endanger the user's safety, and therefore, Centea et al. [25] emphasizes on the importance of giving proper guidance to the workers which will require more time and effort. The future will also see a major shift in job profiles inclined towards IT skill set [26]. For example, big data will raise the demand for data scientists and self-driving vehicles and predictive maintenance will reduce the need for logistic personnel and traditional technicians, respectively. The challenge is to revise the prerequisites of the existing job profiles without impeding the development of workforce or giving rise to unemployment.

4.3 Finance

Financial aspect is one of the significant challenges hindering the adoption of Industry 4.0 which is greatly influential for the developing and under-developed countries [12, 27]. Initially, a large capital investment is required in appropriate equipment like sensors, actuators, control system and remodelling the existing infrastructure to incorporate better networking system [21]. Further, investment is also needed for security of the system from cyberthreats and research and development activities, but companies are not ready to bear the cost since they lack in technical knowledge and experts. Training and reskilling of employees to handle big data, cloud computing and other complex technologies which add on to the implementation cost of I4 [23, 28–30]. Moreover, the companies use advanced technologies with lack of clarity in goals; in such situation, the risk of failure is high with uncertainty of return as compared to companies using traditional methods [31]. Therefore, due to unstable global economic condition, investment in I4 is highly afflicting.

4.4 Operations Management

As the transition from centralized to decentralized manufacturing becomes prevalent, new outlook is necessary to improve the production and supply chain management [11, 32]. The use of IoT, cloud computing, robotics, etc., requires high Internet connectivity which has pushed the industries in adopting wireless systems. Therefore, sustainable restructuring of business operation is the need of the hour to accommodate new equipment by replacing the old infrastructure. The other most common problem that companies deal with is incompatibility between the existing system and I4 technologies which arises due to lack of infrastructure, planning and preparedness [10, 12, 30]. Consequently, transforming a company into a "smart factory" by using advanced technologies, for example, AGV and ASRS, will make the production efficient only when accurate and optimized navigation planning is done [33, 34]. A survey conducted by NewVantage Partners shows that data-driven companies have declined in the past three years from 37.1% in 2017 to 31% in 2019 [35]. The cause for decline in data-driven model, which is used for predictive analysis and other manufacturing processes, is that they require large volume of data where extraction and organizing the data can be time consuming [36, 37]. Also, non-availability and variability in data can pose challenges in meeting the required standard of product [38–41]. While looking from the managerial perspective, particularly SMEs, they are unable to assess the environmental turbulence due to lack of awareness and ignorance towards changing trends [29]. Also, the study has found that investing in research and development has had significant impact on the growth of organization which is important for sustainable growth, yet it is deemed less important by majority of the companies that can cease further progress [31].

4.5 Education and Training

During the implementation phase of Fourth Industrial Revolution, field of education will play an important role, as any other sector, in seamless transition into I4. The study suggests that future unemployment due to automation can be avoided by revising the course curriculum focusing on more practical knowledge rather than theoretical and inculcating necessary skills to make students capable of tackling new challenges [42, 43]. According to the current scenario, universities lack in interdisciplinary courses and the existing ones greatly differ from actual job requirements [22, 44]. Until engineering courses are upgraded, that are efficient in bridging the gap between academic and industries lack of awareness would persist among future industrialists. Accordingly, Raj et al. [12] has confirmed the fact that unawareness among majority of industrialists or having low knowledge regarding I4 is the key barrier. The rapid increase of artificial intelligence and machine learning in industries has raised the demand for experts who are equipped in mechanics as well as IT skill set, and therefore, Suarez et al. [45] conclude that it has now become necessary for students of mechanical engineering to master their core subjects along with IT. Not only universities but also vocational high schools should impart knowledge on coding, software, robotics, etc., so as to provide high-quality workforce [46, 47]. If proper measures are not taken, there can be shortage of workers skilled in IT required for emerging jobs in the near future. With rapidly advancing technologies, the challenge lies in keeping pace with the advancement, and it can be overcome by introducing the concept of lifelong learning [48].

4.6 Government Policies

The government plays an important role in the development of a business which consequently assists in economic growth of a country. Since I4 is the future of manufacturing sector, which is one of the highest contributors in GDP of a country, it is essential for the government to rethink its policies in favour of I4. A survey conducted by Glass et al. [22] found that SMEs as well as large enterprises require external support from the government, and in order to so, Kuo et al. [49] discuss how the government of some countries (China, Germany and USA) has come up with policies from the supply, demand and environmental sides to combat implementation challenges. Although it is observed that some political institutions have come to realize the importance of I4 and many developing countries have made progress in leveraging advanced technologies, nevertheless it still needs to travel a long distance to keep pace with other developed nations. For instance, in South Africa, National E Strategy focuses on expanding the reach of ICTs for developing digital industries; however, due to poor implementation of strategy and lack of coordination between industrialists and governments, policy reforms fail to take off as planned [50]. The "Make in India" and other similar initiatives will suffer a similar set back due to lack of research and development facility and lack of financial aid from the government, and above all, there is lack of clarity in providing standard procedure to integrate I4 with the existing infrastructure [51, 52]. The challenge lies not merely in initiating change in the existing policies and making long term plans to cultivate Industry 4.0 readiness, but also in implementing them while ensuring inclusive economic and social growth.

5 Conclusion and Future Research Prospects

This literature highlights the major challenges from technological, social and financial aspects that need to be focused on while advancing into new technological era. The discussed challenges seem to be interlinked, and each of them demands equal attention from managerial as well as employee's perspective. The lack of planning and preparation from the management side can defeat the purpose of implementing I4, whereas workers need to quickly adapt to the abruptly changing environment and grasp the skills in domains, wherein the technologies cannot replace the intelligence and ability of humans. Also, to bring extensive transformation and a long-term solution to the challenges, it is crucial to adopt change in education system to meet the requirements of a technology-driven society.

From the analysis, it can be determined that greater extent of challenges is mostly faced by SMEs, which constitutes 90% of the business population [53] and accounting for major part of global economy. Readiness to avoid these challenges can be consequential in shaping the future of manufacturing sector.

Most of the reviewed papers identified lack of knowledge as one of the major barriers which attributes to unwillingness of adopting I4 by the manufactures. Also, some surveys conducted by authors of reviewed papers are subjective in nature and limited to only few companies which tends to produce bias results. There are technological limitations too since many technologies are in their primary stage of development, and their compatibility and competence need to be tested before integration. For example, 3D printers require advancement before they can be used for large-scale production.

The future research scopes of Industry 4.0 are discussed below.

- More empirical study is needed in the context of integration of new technological paradigm to determine its practicality and feasibility.
- The study can be done to gain insight into the extent of sustainability benefits and drawbacks of I4 enablers since sustainable growth is an important factor that company strives to achieve.

- The inception of advanced technologies in manufacturing is expected to change the ways of working and managing. Examining the organizational structure in context of smart factory can help in seamless transformation.
- An in-depth analysis to identify the prerequisites of adopting I4 and model-based approach for reconfiguration of production system can greatly assist the industrialists.

References

1. Bahrin M, Othman F, Azli N, Talib M (2016) Industry 4.0: a review on industrial automation and robotic. Jurnal Teknologi. 78. https://doi.org/10.11113/jt.v78.9285
2. Rüßmann M, Lorenz M, Gerbert P, Waldner M (2015) Industry 4.0: the future of productivity and growth in manufacturing industries
3. Xu L, Xu E, Li L (2018) Industry 4.0: state of the art and future trends. Int J Prod Res 56:1–22. https://doi.org/10.1080/00207543.2018.1444806
4. Schlechtendahl J, Keinert M, Kretschmer F, Lechler A, Verl A (2014) Making existing production systems Industry 4.0-ready. Prod Eng Res Devel 9:1–6. https://doi.org/10.1007/s11740-014-0586-3
5. Industry 4.0 Challenges and Solution for Digital Transformation and use of Exponential Technologies. Deloitte. http://www2.deloitte.com
6. Zhong R, Xu X, Klotz E, Newman S (2017) Intelligent manufacturing in the context of Industry 4.0: a review. Engineering 3:616–630. https://doi.org/10.1016/J.ENG.2017.05.015
7. Qin J, Liu Y, Grosvenor R (2016) A categorical framework of manufacturing for Industry 4.0 and beyond. Procedia CIRP 52:173–178. https://doi.org/10.1016/j.procir.2016.08.005
8. Cyber risk in advanced manufacturing, Deloitte. https://www2.deloitte.com
9. Abdur RM, Habib S, Ali M, Ullah S (2017) Security issues in the Internet of Things (IoT): a comprehensive study. Int J Adv Comput Sci App 8. https://doi.org/10.14569/ijacsa.2017.080650
10. Gonçalves MC, Winroth M, Carlsson D, Almström P, Centerholt V, Hallin M (2019) Industry 4.0 readiness in manufacturing companies: challenges and enablers towards increased digitalization, 81:1113–1118. https://doi.org/10.1016/j.procir.2019.03.262
11. Khan A, Turowski K (2016) A perspective on Industry 4.0: from challenges to opportunities in production systems, 441–448. https://doi.org/10.5220/0005929704410448
12. Raj A, Dwivedi G, Sharma A, Jabbour AB, Rajak S (2019) Barriers to the adoption of Industry 4.0 technologies in the manufacturing sector: an inter-country comparative perspective. In J Prod Econ 224:107546. https://doi.org/10.1016/j.ijpe.2019.107546
13. Roman R, Lopez J, Mambo M (2016) Mobile edge computing, Fog et al.: a survey and analysis of security threats and challenges. Future Gener Comput Syst 78. https://doi.org/10.1016/j.future.2016.11.009
14. Tupa J, Šimota J, Steiner F (2017) Aspects of risk management implementation for Industry 4.0. Procedia Manuf 11:1223–1230. https://doi.org/10.1016/j.promfg.2017.07.248
15. Rokka CS, Rashid N, Faezi S, Al Faruque, Mohammad A (2017) Security trends and advances in manufacturing systems in the era of Industry 4.0, 1039–1046. https://doi.org/10.1109/iccad.2017.8203896
16. Sturm L, Williams C, Camelio J, White J, Parker R (2017) Cyber-physical vulnerabilities in additive manufacturing systems: a case study attack on the.STL file with human subjects. J Manuf Syst 44:154–164. https://doi.org/10.1016/j.jmsy.2017.05.007

17. Zarreh A, Wan H, Lee Y, Saygin C, Al Janahi R (2019) Cybersecurity concerns for total productive maintenance in smart manufacturing systems. https://doi.org/10.31224/osf.io/d6fh9
18. Ghadge A, Weib M, Caldwell N, Wilding R (2019) Managing cyber risk in supply chains: a review and research agenda. Supply Chain Manage 25(2):223–240. https://doi.org/10.1108/SCM-10-2018-0357
19. Culot G, Fattori F, Podrecca M, Sartor M (2019) Addressing Industry 4.0 cybersecurity challenges. IEEE Eng Manage Rev 47:79–86. https://doi.org/10.1109/EMR.2019.2927559
20. Jobs lost jobs gained: workforce transitions in the time of automation, Mckinsey and Company. https://www.mckinsey.com
21. Hamzeh SR, Zhong R, Xu X (2018) A survey study on Industry 4.0 for New Zealand manufacturing. Procedia Manuf 26:49–57. https://doi.org/10.1016/j.promfg.2018.07.007
22. Glass R, Meißner A, Gebauer C, Stürmer S, Metternich J (2018) Identifying the barriers to Industrie 4.0. Procedia CIRP 72:985–988. https://doi.org/10.1016/j.procir.2018.03.187
23. Kamble S, Gunasekaran A, Sharma R (2018) Analysis of the driving and dependence power of barriers to adopt Industry 4.0 in Indian manufacturing industry. Comput Ind 101:107–119. https://doi.org/10.1016/j.compind.2018.06.004
24. Kurt R (2019) Industry 4.0 in terms of industrial relations and its impacts on labour life. Procedia Comput Sci 158:590–601. https://doi.org/10.1016/j.procs.2019.09.093
25. Centea D, Singh I, Yakout M, Boer J, Elbestawi M (2020) Opportunities and challenges in integrating additive manufacturing in the SEPT learning factory. Procedia Manuf 45:108–113. https://doi.org/10.1016/j.promfg.2020.04.080
26. Fareri S, Fantoni G, Chiarello F, Coli E, Binda A (2020) Estimating Industry 4.0 impact on job profiles and skills using text mining. Comput Ind 118:103222. https://doi.org/10.1016/j.compind.2020.103222
27. Lee I, Lee K (2015) The Internet of Things (IoT): applications, investments, and challenges for enterprises. Bus Horiz 58:431–440. https://doi.org/10.1016/j.bushor.2015.03.008
28. Luthra S, Mangla S (2018) Evaluating challenges to Industry 4.0 initiatives for supply chain sustainability in emerging economies. Process Safety Environ Protect 117. https://doi.org/10.1016/j.psep.2018.04.018
29. Ingaldi M, Ulewicz R (2019) Problems with the Implementation of Industry 4.0 in Enterprises from the SME Sector. Sustainability 12:217. https://doi.org/10.3390/su12010217
30. Dóra H, Roland ZS (2019) Manufacturing of Industry 4.0: do multinational and small and medium-sized companies have equal opportunities? Technol Forecast Soc Chang 146:119–132. https://doi.org/10.1016/j.techfore.2019.05.021
31. Sevinç A, Gür Ş, Eren T (2018) Analysis of the difficulties of SMEs in Industry 4.0 applications by analytical hierarchy process and analytical network process. Processes 6:264. https://doi.org/10.3390/pr6120264
32. Abu-bakr M, Abbas A, Tomaz I, Soliman M, Mohammed M, Hegab H (2020) Sustainable and smart manufacturing: an integrated approach. Sustainability 12(6):2280. https://doi.org/10.3390/su12062280
33. Lynch L, McGuinness F, Clifford J, Rao M, Walsh J, Toal D, Newe T (2020) Integration of autonomous intelligent vehicles into manufacturing environments. Challenges 38:1683–1690. https://doi.org/10.1016/j.promfg.2020.01.115
34. Baboli A, Okamoto JJ, Tsuzuki M, Martins T, Miyagi P, Junqueira F (2015) Intelligent manufacturing system configuration and optimization considering mobile robots multi-functional machines and human operators: new facilities and challenge for industrial engineering. IFAC-PapersOnLine 48:1912–1917. https://doi.org/10.1016/j.ifacol.2015.06.366
35. Big data and AI executive survey 2019. https://www.newvantage.com
36. Haddud A, de Souza A, Khare A, Lee H (2017) Examining potential benefits and challenges associated with the Internet of Things integration in supply chains. J Manuf Technol Manage. https://doi.org/10.1108/JMTM-05-2017-0094

37. Kache F, Seuring S (2017) Challenges and opportunities of digital information at the intersection of Big Data Analytics and supply chain management. Int J Oper, Prod Manage 37:10–36. https://doi.org/10.1108/IJOPM-02-2015-0078
38. Fisher O, Watson N, Escrig EJ, Witt R, Porcu L, Bacon D, Rigley M, Gomes R (2020) Considerations challenges and opportunities when developing data-driven models for process manufacturing systems. https://doi.org/10.1016/j.compchemeng.2020.106881
39. Lee J, Davari H, Singh J, Pandhare V (2018) Industrial artificial intelligence for Industry 4.0-based Manufacturing Systems. https://doi.org/10.1016/j.mfglet.2018.09.002
40. World Trade Report 2016 Leveling the Trading Field for SMEs. http://www.wto.org
41. Chen B, Wan J, Shu L, Li P, Mukherjee M, Yin B (2017) Smart factory of Industry 4.0: key technologies application case and challenges. https://doi.org/10.1109/access.2017.2783682
42. Zhong R, Newman S, Huang G, Lan S (2016) Big Data for Supply Chain Management in the Service and Manufacturing Sectors: Challenges, Opportunities, and Future perspectives. Comput Ind Eng. https://doi.org/10.1016/j.cie.2016.07.013
43. Andersen AL, Brunoe T, Nielsen K (2019) Engineering education in changeable and reconfigurable manufacturing: using problem-based learning in a learning factory environment. Procedia CIRP 81:7–12. https://doi.org/10.1016/j.procir.2019.03.002
44. Mourtzis D, Vasilakopoulos A, Zervas E, Boli N (2019) Manufacturing system design using simulation in metal industry towards Education 4.0. Procedia Manufacturing 31:155–161. https://doi.org/10.1016/j.promfg.2019.03.024
45. Suárez FMS, Marcos-Bárcena M, Peralta ME, González F (2017) The challenge of integrating Industry 4.0 in the degree of Mechanical Engineering. Procedia Manuf 13:1229–1236. https://doi.org/10.1016/j.promfg.2017.09.039
46. Seamus G, Alan R, Shane L (2018) Meeting the needs of industry in smart manufacture—the definition of a new profession and a case study in providing the required skillset. Procedia Manuf 17:262–269. https://doi.org/10.1016/j.promfg.2018.10.045
47. Benešová A, Tupa J (2017) Requirements for education and qualification of people in Industry 4.0. Procedia Manuf 11:2195–2202. https://doi.org/10.1016/j.promfg.2017.07.366
48. Motyl B, Baronio G, Uberti S, Speranza D, Filippi S (2017) How will change the future engineers' skills in the Industry 4.0 framework? A questionnaire survey. Procedia Manuf 11:1501–1509. https://doi.org/10.1016/j.promfg.2017.07.282
49. Kuo C-C, Shyu J, Ding K (2019) Industrial revitalization via Industry 4.0—a comparative policy analysis among China, Germany and the USA. Global Transitions. https://doi.org/10.1016/j.glt.2018.12.001
50. Manda MI, Ben DS (2019) Responding to the challenges and opportunities in the 4th Industrial revolution in developing countries, 244–253. https://doi.org/10.1145/3326365.3326398
51. Iyer A (2018) Moving from Industry 2.0 to Industry 4.0: a case study from India on leapfrogging in smart manufacturing. Procedia Manuf 21:663–670. https://doi.org/10.1016/j.promfg.2018.02.169
52. Fatorachian H, Kazemi H (2020) Impact of Industry 4.0 on supply chain performance. Prod Plan Control 1–19. https://doi.org/10.1080/09537287.2020.1712487
53. Ayer Z, Akgül H (2020) Examining the impact of Industry 4.0 on education. J Awareness 5:159–168. https://doi.org/10.26809/joa.5.013